Studies in Computational Intelligence

Volume 1240

The series "Studies in Computational Intelligence" (SCI) publishes new developments and advances in the various areas of computational intelligence—quickly and with a high quality. The intent is to cover the theory, applications, and design methods of computational intelligence, as embedded in the fields of engineering, computer science, physics and life sciences, as well as the methodologies behind them. The series contains monographs, lecture notes and edited volumes in computational intelligence spanning the areas of neural networks, connectionist systems, genetic algorithms, evolutionary computation, artificial intelligence, cellular automata, self-organizing systems, soft computing, fuzzy systems, and hybrid intelligent systems. Of particular value to both the contributors and the readership are the short publication timeframe and the world-wide distribution, which enable both wide and rapid dissemination of research output.

Indexed by SCOPUS, DBLP, WTI AG (Switzerland), zbMATH, SCImago.

All books published in the series are submitted for consideration in Web of Science.

Mohamed Arezki Mellal · Yusuke Nojima ·
Naoki Masuyama
Editors

Computational Intelligence for a Greener Future: Innovations in Renewable Energy Systems

Integrating AI and Behavioral Insights to Drive Economic Efficiency and Sustainability

Springer

Editors
Mohamed Arezki Mellal
LMSS, Faculty of Technology
M'Hamed Bougara University
Boumerdes, Algeria

Yusuke Nojima
Osaka Metropolitan University
Osaka, Japan

Naoki Masuyama
Osaka Metropolitan University
Osaka, Japan

ISSN 1860-949X ISSN 1860-9503 (electronic)
Studies in Computational Intelligence
ISBN 978-3-032-06731-9 ISBN 978-3-032-06732-6 (eBook)
https://doi.org/10.1007/978-3-032-06732-6

This Springer imprint is published by the registered company Springer Nature Switzerland AG
The registered company address is: Gewerbestrasse 11, 6330 Cham, Switzerland

Contents

A Review of the Genetic Algorithm Approach in Predictive Maintenance and Energy Forecasting

Enis Muratović

Evolution is a perfect adaptation mechanism.

Abstract To explore the full potential of predictive maintenance and energy forecasting in renewable energy sectors, extensive data collection sets, obtained through various monitoring methods, are often used to detect the early signs of system outages and failures as well as to analyze and obtain parameters such as energy output, efficiency and overall system health. As high forecasting accuracy is mandatory in these systems, a large amount of data must be utilized and extracted in the form of unseen relations and features, which can be accomplished by using artificial intelligence (AI) frameworks, libraries, and algorithms that are beginning to penetrate all areas of industry and engineering. With the rapid development in AI and broad tools provided by machine learning (ML) algorithms, it is possible to obtain solutions to significant data set problems that are very difficult to solve by classical methods. One of the most powerful techniques to solve these problems is based on the genetic algorithm (GA) approach, which mimics the process of evolution and natural selection. This review chapter shows how this evolutionary-based approach can be successfully used in various renewable energy systems to optimize performance, forecast system failures and downtimes, and plan maintenance schedules.

Keywords Predictive maintenance · Energy forecasting · Renewable energy · Artificial intelligence · Genetic algorithm · Evolution · Evolutionary-based approach

---

E. Muratović (✉)
Department of Mechanical Design, Faculty of Mechanical Engineering, University of Sarajevo, Sarajevo, Bosnia and Herzegovina
e-mail: muratovic@mef.unsa.ba

M. A. Mellal et al. (eds.), *Computational Intelligence for a Greener Future: Innovations in Renewable Energy Systems*, Studies in Computational Intelligence 1240,
https://doi.org/10.1007/978-3-032-06732-6_1

1 Introduction

As contemporary industrial needs require the processing of large data sets, system engineering often encounters the challenges of extracting meaningful information into knowledge [1]. AI frameworks and libraries, which are used in almost every aspect of daily life, resulted from powerful ideas and a broad set of tools that have followed the development of scalable hardware and software solutions and represent the breakthrough in science as they can discover the hidden patterns in extensive collections of structured and unstructured data and can be used to automate decisions [2]. The self-learning algorithms developed within the field of ML can be used to efficiently derive rules from unseen relations in data to build predictive models and performance-driven decisions [3]. With these technologies, the solutions to the most complex system engineering problems that cannot be solved by human experience, by any stretch of the imagination, can be obtained [4].

One of the most powerful algorithms to solve such complex problems is GA, which is based on the principles of evolution and natural selection. This algorithm directs the search for the solution based on the principle of survival of the fittest, as it mimics the mechanisms of evolution and natural genetics [5]. The GA evolutionary modelling approach consists of several building blocks, represented with naturally inspired terms, that construct the basic steps and stages of the GA flow. The potential solution to the problem is represented by an individual that is extracted from the population, or, more specifically, many generations of individuals based on the fitness function [6]. The fitness function, as a criterion that describes how well an individual is adapted to the environment, labels the best individuals according to their maximum or minimum properties for the next population of individuals. Therefore, according to the principle of natural selection, the most adapted individuals participate in the evolution process, that is, the creation of the offspring population epoch, which is then adapted to the problem being solved by applying the same principle over many epochs. In each generation, the weak individuals, that is, the least possible solutions to the problem, are removed from the population pool, allowing the GA to converge toward the optimal solution [7, 8]. This convergence is not only achieved by selection based on the fitness function evaluation but also by the crossover and mutation methods, which are used to exchange the experience from the intermediate to the offspring population to enhance the speed of the algorithm and to add diversity in the future epochs of individuals by adding a random change for additional improvements in adaptation to environment, respectively [9].

With industrial architecture being shifted to sustainable development, the global need for ecologically friendly solutions places renewable energy systems at the center of the transition toward less carbon-intensive sources [10]. The complexity of renewable energy systems, which include power sources such as wind and solar power, bioenergy, hydroelectric, and tidal energy, etc., is reflected in the predictive maintenance, energy optimization, and anomaly detection based on large data sets obtained from real-time monitoring [11, 12]. These complex industrial systems undergo various dynamic changes and reconfigurations as they must address adaptability and scalability issues and confront data variability, and minimize decision uncertainty to enhance forecasting accuracy, which can be fully integrated with the transformative capabilities of AI [13, 14]. By using state-of-the-art methods and technologies, which include AI-based models, future events on energy output optimization, resource allocation, and fault detection can be quickly challenged by using data-driven algorithms based on feature selection to provide valuable insights and automate decisions and maintenance strategies in real time [15–17]. The easy implementation and versatility of GA make it an ideal match for sectors of renewable energy systems, as they are less demanding about the nature of the problem being solved and are very convenient to parallelize multiple operations on large data sets [18]. Additionally, GA is invariant to dynamic changes in system data, and with self-adaptation ability, it is easily applicable to the conditions of the addressed problem even in the absence of previous solution experience [19, 20]. The cutting-edge methods used in contemporary renewable energy systems that rely on AI-based configurations that involve GA can be thematically grouped into several topics, which include predictive maintenance and energy forecasting, as will be described and showcased in more detail within the following text of this chapter.

2 Predictive Maintenance

2.1 System Optimization and Planning

As the renewable energy sectors require thorough construction planning in dynamic conditions that support the needs for green development, the development of future power plants should be based on optimizing specific components to enhance efficiency and reduce losses [21]. With wind power being recognized as one of the major renewable energy sources of the future, a lot of attention is directed toward wind power plants (WPPs), and, more specifically, to wind turbines (WT) that represent only a tiny fraction of the complex WPPs problems which involve a significant number of variables [22]. Optimization problems of the WTs are mainly focused on the placement location, performance, and maintenance strategies. Based on the latest literature, a certain amount of investigation is dedicated to the WT design that includes the adjustment of WT parameters, such as hub height, rotor diameter, and rated power in correspondence to varied wind conditions [23]. These design-based problems have been solved by several optimization algorithms, including GA [24]. Furthermore, recent studies have been focused on the simultaneous investigation of multiple models and algorithms in terms of a comparative analysis of optimization results [25]. These studies are mostly based on reducing energy costs and improving annual energy production (AEP), that is, a multi-objective fitness function, by optimizing WT design following wind statistic data and various placement constraints [26, 27]. The typical model for the optimal WT design can be determined based on the variable wind conditions, primarily reflected in wind speed defined by the Weibull distribution, as expressed by Eq. (1):

$$f(V) = \frac{k}{c}\left(\frac{V}{c}\right)^{k-1} \cdot e^{-\left(\frac{V}{c}\right)^k} \quad (1)$$

where k is the shape parameter and c is the scale parameter of the Weibull probability function $f(V)$ which defines the wind speed probability. The proper WT design also requires the proper description of the wind profile, that is, the wind speeds at different altitudes in relation to a reference point, as described by the wind profile showcased in Eq. (2):

$$V = V_{ref}\left(\frac{h}{h_{ref}}\right)^{\alpha} \quad (2)$$

where V is the wind speed at the hub height h, V_{ref} is the wind speed at the reference point, h_{ref} is the altitude of the reference point, and α is the wind shear exponent with different values for specific terrain configurations. Equation (2) is of significant importance for the description of the WT power output at various wind speeds, or, more specifically, the relation between the WT power output and hub height h, as expressed in Eq. (3):

$$P_p(V_i) = \begin{cases} 0;\ V_i < V_{ci} \vee V_i > V_{co} \\ q(V_i);\ V_{ci} \le V_i < V_n \\ P_n;\ V_n \le V_i < V_{co} \end{cases} \quad (3)$$

where $P_p(V_i)$ is the produced electric power at the specific wind speed V_i, V_{ci} is the cut-in speed, V_{co} is the cut-out speed, $q(V_i)$ is a function between the WT output power and wind speed, P_n is the nominal power, and the V_n is the rated wind speed. The values of the cut-in and the cut-out wind speeds are defined according to the terrain conditions. Based on the described parameters, the AEP can be obtained by multiplying the WT power at a certain operating wind speed by the number of operating hours, T_i, at the same speed, as expressed by Eq. (4):

$$W_t = \int_0^{T_{year}} P(t)dt \approx \sum V_i \cdot T_i \cdot P_i(V_i) = T_{year} \cdot \sum V_i \cdot F(V_i) \cdot P_p(V_i) \cdot \Delta V \quad (4)$$

where W_t is the energy produced at the annual level t, T_{year} is the number of hours in a year, and ΔV is the integration step. Accordingly, the AEP can be described as a function of wind parameters k, c, and α, as well as the WT design parameters h, D, and P_n, as expressed by Eq. (5):

$$W_t(h, D, P_n) = T_{year} \cdot \sum F(V_i) \cdot q(V_i) \cdot \Delta V + T_{year} \cdot P_n \cdot \sum F(V_i) \cdot \Delta V$$

$$= T_{year} \cdot \sum V_n \cdot \left[e^{-\left(\frac{V_i - \frac{\Delta V}{2}}{c}\right)^k} - e^{-\left(\frac{V_i + \frac{\Delta V}{2}}{c}\right)^k} \right]$$
$$\cdot \left(C_1 + C_2 \cdot V_i + C_3 \cdot V_i^2\right) \cdot \Delta V$$
$$+ T_{year} \cdot P_n \cdot \sum V_n \cdot \left[e^{-\left(\frac{V_i - \frac{\Delta V}{2}}{c}\right)^k} - e^{-\left(\frac{V_i + \frac{\Delta V}{2}}{c}\right)^k} \right] \cdot \Delta V \quad (5)$$

where C_1, C_2, and C_3 are the evaluated coefficients for the $q(V_i)$ function. The rated wind speed, V_n, can be expressed by Eq. (6):

$$V_n = \sqrt[3]{\frac{8 \cdot P_n}{\pi \cdot \rho \cdot D^2 \cdot C_p}} \quad (6)$$

where ρ is the density of the air at the hub height h, D is the diameter of the rotor, and C_p is the coefficient of power. To fully address the optimization problem of the WT design, the WT costs have to be implemented into the model. Correspondingly, investment costs, I_{tot}, can be described by Eq. (7):

$$I_{tot} = I_0^{(D,h,P_n)} + I_{transport}^{(P_n)} + I_{install}^{(D,h)} \quad (7)$$

where $I_0^{(D,h,P_n)}$ are the costs of the rotary turbine system, $I_{transport}^{(P_n)}$ are the costs of the transportation, and $I_{install}^{(D,h)}$ are the installation costs, that is, assembly costs. Additional costs that need to be taken into account are the maintenance costs, M_t, which are usually considered constant based on the annual amount of produced energy, as presented in Eq. (8):

$$M_t = k_t \cdot W_t \quad (8)$$

where k_t is the constant that is estimated based on the type and the size of the rotary turbine system, and the terrain conditions. Finally, the main objective of the WT optimization problems is to minimize the levelized cost of energy, LCE, based on the rate of interest, p, which can be described by Eq. (9):

$$LCE = \frac{I_{tot} + \sum_{t=1}^{n} \frac{M_t}{(1+p)^t}}{\sum_{t=1}^{n} \frac{M_t}{(1+p)^t}} \quad (9)$$

The minimization of the LCE, that is, the formulation of the WT optimization problem, is defined with the objective function denoted by Eq. (10):

$$f_{LCE}^{(min)} = min(LCE(h, D, P_n)) = min\left(\frac{\alpha I_{tot} + M_t}{W_t}\right)$$

$$= min \frac{\alpha\left(I_O^{(D,h,P_n)} + I_{transport}^{(P_n)} + I_{install}^{(D,h)}\right)}{T_{year} \cdot \sum F(V_i) \cdot q(V_i) \cdot \Delta V + T_{year} \cdot P_n \cdot \sum F(V_i) \cdot \Delta V} + k_t \quad (10)$$

The WT design constraints, that is, the geometric boundaries of hub height, h, and rotor diameter, D, as well as the limitations of the rated power, are described by Eq. (11). Optimization results of the WT design problem are shown in Fig. 1.

$$\begin{cases} h_{min} \leq h \leq h_{max} \\ D_{min} \leq D \leq D_{max} \\ P_{nmin} \leq P_n \leq P_{nmax} \end{cases} \quad (11)$$

A similar approach is used for other renewable energy systems, such as solar power plants, where the GA is used to optimize the efficiency of solar panels by adapting the optimal parameter configuration to solar conditions [28]. This way, by

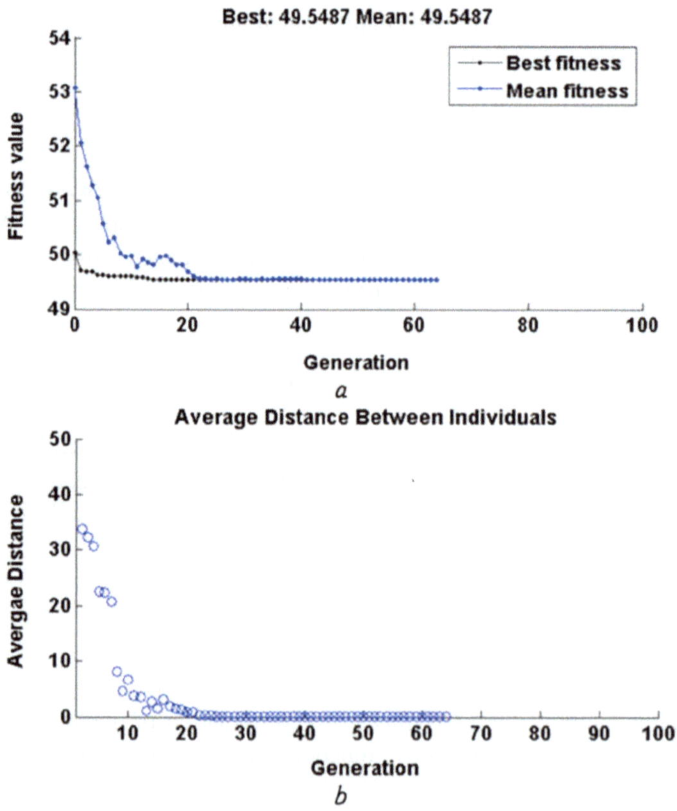

Fig. 1 GA flow: **a** fitness value change, **b** average solution spread [22]

affecting parameters such as tilt and azimuth angle, temperature, and solar irradiance, panel performance can be increased to a notable percentage [29].

Other studies have been focused on investigating various infrastructures, that is, the potential of assessing solar energy locations in correspondence to data on geo-layers, site slopes, solar radiation, and elevation [30]. The optimal location for solar energy power plants has also been investigated by integrating the GA into the Analytical Hierarchy Process (AHP) [31]. This GA-based model outperformed the classic AHP as it can identify a considerable area of land eligible for solar power plants. Furthermore, by using data on topographic complexity, the identified area can be categorized into several regions based on the suitability factor. The mapping of the land surface temperature, *LST*, for the considered area with the specified spatial resolution is described by Eq. (12):

$$LST = \frac{T_B}{\left(1 + \left(0.00115 \cdot x \cdot \frac{T_B}{1.4388}\right) \cdot x \cdot ln(pv)\right)} - 273.15 \tag{12}$$

where T_B is the brightness temperature and pv is the vegetation proportion. The structure of the GA algorithm, which aims to detect areas suitable for solar power plants, is based on minimizing specific parameters while maximizing others, therefore, resulting in two objective functions. The GA approach tends to minimize the distance to urban areas, roads, and transit lines to reduce the investments and improve power plant efficiency. Likewise, the slope of the area needs to be considered, as a greater slope angle will result in decreased efficiency. The minimization function can be formulated by Eq. (13):

$$f_n(x, y) = min \sum_{i=1}^{n} D_{(x_j, y_j)}, C_i \tag{13}$$

where$f_n(x, y)$ is the minimization function, $D_{(x_j, y_j)}$ and C_i are the image, that is, pixel parameters, which define distances to areas of interest. The maximization function is based on improving the performance and viability of a power plant by exploiting solar radiation, as described by Eq. (14):

$$f_m(x, y) = max \sum_{i=1}^{n} V_{(x_j, y_j)} \tag{14}$$

where $V_{(x_j, y_j)}$ are the values of the image, that is, pixel parameters. The GA-driven optimization used in solar systems can be further refined by extracting the parameters of solar photovoltaic (PV) cells, as the data on voltage conditions is essential for the simulation and control of PV systems [32]. The newer studies use an improved GA, that is, a genetic algorithm based on non-uniform mutation (GAMNU), to effectively approximate the parameters of PV cells and solar modules [33]. Based on the comparison of experimental data and results of recent studies, GAMNU has proven to be an invaluable and efficient tool in the renewable energy sector. The error function, $h(V_k, I_k, \Psi)$, used for the optimization of GA parameters, of the single diode model (SMD) can be expressed by Eq. (15):

$$\begin{aligned} h(V_k, I_k, \Psi) = I_{ph} - I_{01}\left[exp\left(\frac{V_k + R_s \cdot I_k}{a_1 \cdot V_t}\right) - 1\right] \\ - I_{02}\left[exp\left(\frac{V_k + R_s \cdot I_k}{a_2 \cdot V_t}\right) - 1\right] - \frac{V_k + R_s \cdot I_k}{R_{sh}} - I_k \end{aligned} \quad (15)$$

where: V_k is the measured voltage, I_k is the measured current, Ψ is the vector parameter, I_{ph} is the photo-current, I_{01} and I_{02} are the saturation currents of the diodes, a_1 and a_2 are the ideal factors of diodes, V_t is the thermal voltage, and R_{sh} is the shunt resistance.

2.2 GA-driven Maintenance Strategies

The movement to sustainable development, which required the transition from reactive to predictive maintenance strategies, has led to remarkable progress in cost-effective methods of monitoring system anomalies and predicting system failures, with the most powerful technologies resulting from the areas of hydroelectric power plants (HPPs). The latest approach in this area combines multi-criteria and meta-heuristic methods to optimize the maintenance problem of the generator, as the most vital element in HPPs [34]. In this sense, GA, previously integrated with the mutation-based weight factors, is defined based on the minimum delay in possible disruption occurrences. The described novelty has a great impact on the decision-making system in terms of reliability, efficiency, and sustainability. As all the maintenance actions need to be carried out in optimal time frame intervals to minimize the process cost, it is necessary to fully reflect possible schedule delays and maintenance deadlines [35]. The complexity of the HHPs maintenance strategies is reflected in a great number of variables and scheduling tasks that need to be optimized to ensure a constant power supply. These variables, considering just the generator as the most vital component of the system, need to address insulation conditions, cooling performance, vibration and balance issues, rotor windings monitoring, automatic part replacement, and operating efficiency. By exploiting data on these parameters, the GA-driven models prioritize the maintenance tasks by employing the weighting functions, that is, the ranking system based on the importance. Even though such a complex system requires rigorous data handling, the maintenance tasks are based on a relatively simple approach. The objective function is based on the weighting functions, w_i, and maintenance time, T_i, that is, the prioritization of the task with the highest ranks and minimization of the maintenance completion, as described by Eq. (16):

$$min = \sum_{i=1}^{n} \omega_i \cdot T_i \tag{16}$$

The simultaneous evaluation of timing, severity of intervention, and prompt maintenance action, which can be easily employed by the parallel-processing nature of GA, shows that the less severe interventions, indicated by the predictive model, are beneficial in the long-term cost-efficiency. This behavior can be explained by the framework of the GA, which treats the severity of intervention as the dependent variable of maintenance actions, where, even at lower severity levels, a reduction of system failures is expected [36].

Continuous growth of previously highlighted WT makes the maintenance tasks more difficult, as the general problems of maintenance actions, resources, prediction, and resource allocation can result in conflicting and inconsistent patterns in decision-making. With the ever-growing need for industrial competitiveness, the dynamic scheduling of wind farm maintenance tasks, which is becoming increasingly unpredictable, can be solved by combinatorial optimization with a non-dominated sorting genetic algorithm-II (NSGA-II) [37]. The maintenance tasks and resource allocation indicators can be obtained in correspondence with the monitored data on WT. Based on operation-cycle data, which is usually collected monthly, NSGA-II can conduct the combinatorial optimization process over a large data set of monitored resources by guiding the GA search to minimize required resources for maintenance actions and reduce penalty costs. The multi-objective optimization function, for these types of problems, can be expressed by Eq. (17a):

$$minF = \omega_1 \cdot F_1(\Delta P_{loss}) + \omega_2 \cdot F_2(\Delta U_{ave}) + \omega_3 \cdot N_{TAP} + \omega_4 \cdot SATHD \tag{17a}$$

where $\omega_1, \omega_2, \omega_3$, and ω_4 are the weighting factors, $F_1(\Delta P_{loss})$ and $F_2(\Delta U_{ave})$ are the satisfactory functions related to system power losses and voltage deviations, N_{TAP} is the total tap step ranging from the initial to on-load tap changer positions (OLTCs), which is one of the most important control variables, and *SATHD* is the average of the total harmonic distortion (THD). The power losses and voltage deviations, as the most influential functions on an overall performance, can be expressed by Eqs. (17b) and (18), respectively:

$$\Delta P_{loss} = \sum P^{grid} + \sum P^{WT} - \sum P^{loads} \tag{17b}$$

where $\sum P^{grid}$, $\sum P^{WT}$, and $\sum P^{loads}$ are powers of generators, WTs, and loads, respectively.

$$\Delta U_{ave} = \frac{\sum |U_i - U_{inom}|}{M} \tag{18}$$

where U_{inom} is the nominal voltage of the M system. The $F_1(\Delta P_{loss})$ and $F_2(\Delta U_{ave})$ functions can be described by Eqs. (19) and (20), respectively:

$$F_1(\Delta P_{loss}) = \begin{cases} 0; & \Delta P_{loss} \le a \\ 2\left(\frac{\Delta P_{loss}-a}{b-a}\right)^2; & a \le \Delta P_{loss} \le \frac{a+b}{2} \\ 1-2\left(\frac{\Delta P_{loss}-b}{b-a}\right)^2; & \frac{a+b}{2} \le \Delta P_{loss} \le b \\ 1; & \Delta P_{loss} \ge b \end{cases} \tag{19}$$

where a and b are the minimum and maximum power losses, respectively.

$$F_2(\Delta U_{ave}) = e^{\frac{(\Delta U_{ave}-c)^2}{\sigma^2}} \tag{20}$$

where c and σ are the mean and standard values of deviations in voltage. The constraints of the GA driven optimization problem for the power loss and voltage deviation functions, which must satisfy the technical feasibility, are described in Eqs. (21) and (22), respectively:

$$\Delta P_{lossmin} \le \Delta P_{loss} \le \Delta P_{lossmax} \tag{21}$$

$$U_{imin} \le U_i \le U_{imax} \tag{22}$$

where $\Delta P_{lossmin}$ and $\Delta P_{lossmax}$ are the minimum and maximum power losses, and U_{imin} and U_{imax} are the minimum and maximum voltage deviations, respectively. The required maintenance time for operation cycles over the monthly data set is presented in Fig. 2.

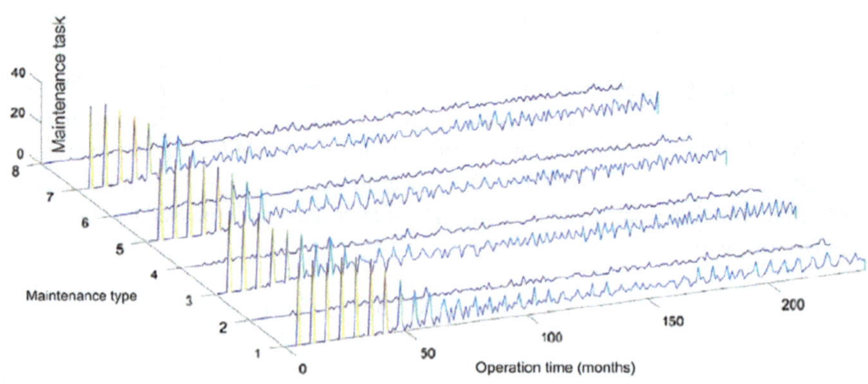

Fig. 2 Maintenance task of the wind farm over operation time [37]

2.3 Fault Detection

One of the major issues that needs to be addressed in modern solar systems is the fault detection and diagnosis (FDD) of the PV modules, which, due to numerous working regimes and system faults, require rigorous monitoring attention for enhancing their availability, efficiency, and safety [38]. Faults such as partial shading, cell aging, cell damage, short and open-circuit faults, and temperature faults need to be addressed by computational intelligence and data-driven techniques following historical knowledge on PV operational data [39]. The development of the modern solar system, following the movement of a green future, requires the FDD approach based on the classification methods provided by artificial neural networks (ANN), with the operational metrics being provided by the GA. This hybrid approach enables the extraction, selection, and classification of FDD features with great efficiency and low computational resources [40]. The fitness function for the previously described hybrid approach can be expressed by Eq. (23):

$$fitness = W_{CA} \cdot accuracy + W_f \cdot \left(\sum_{i=1}^{n_f} F_i \right)^{-1} \tag{23}$$

where W_{CA} represents the classification accuracy weight factor, W_f is the subset weight factor, and F_i is the feature of the subset. The ANN-based equation of the fitness function considers two attributes, which are combined by weight factors of subset features with classification accuracy.

Additionally, the GA-based approach can be used as a supplementary aid to conventional FDD algorithms, such as maximum power point tracking (MPPT), which extracts the necessary data on the operating current and voltage of PV cells [41]. This data is then additionally refined to subset features by the Lasso penalty. The fault classification stage, which mostly relies on logistic regression (LR), support vector machine (SVM), and k-nearest neighbors (KNN), can be effectively improved in terms of accuracy by employing the GA to optimize the algorithm-assigned weight factors for the PV status prediction. According to the SVM method, the minimization problem can be described with the Eq. (24):

$$min : \frac{1}{2}\|w\|^2 + C\sum_{i=1}^{n}(\xi_i + \xi_i^*) \tag{24}$$

where $\frac{1}{2}w^2$ is the empirical risk, $C\sum_{i=1}^{n}(\xi_i + \xi_i^*)$ is the confidence risk, ξ_i and ξ_i^* are the relaxation variables. Similarly, the same approach can be used in wind farms, where the early detection of WT faults can positively influence the maintenance cost and prevent catastrophic events. Supervisory control and data acquisition system (SCADA), as the most common WT condition monitoring approach, can implement GA for the selection of optimal features [42]. This way, in correspondence to threshold values on system data, WT anomalies are discovered and recorded.

3 Energy Forecasting

3.1 Optimizing Energy Forecasting Models in Solar Systems

The second most crucial topic in the renewable energy sector is energy forecasting. The data on the output power of any power plant and corresponding components has a great impact on future energy demands, operational behavior of energy grids, backup schedules, and energy consumption. Based on historical data, weather conditions, economic factors, and system constraints, the modern AI-based approach can be used to develop cost-effective, environmentally friendly, and effective renewable energy systems. At present, suitable energy forecasting can greatly reduce the need for the usage of fossil fuels, therefore contributing to sustainable development by strategically implementing less carbon-intensive sources [43]. The state-of-the-art approach in this field is based on the evolutionary mating algorithm (EMA), a derived version of classic GA adjusted to optimize the weight and bias parameters of a feed-forward neural network (FFNN), which can be used in solar systems to forecast output power over a longer period. The meta-heuristic approach, implemented by the EMA, can establish a relation between solar irradiance, ambient conditions, and PV cell temperature with astonishing accuracy [44]. Additionally, the comparison of EMA with some well-established algorithms showcased an exceptional performance in solar system energy forecasting, as well as the potential of integrating meta-heuristic algorithms with contemporary neural network frameworks [45]. The operating mechanism of the EMA is based on the selection method, I_{mates}, which can be described by Eq. (25):

$$I_{mates} = 1 + \left[var\left(X_{m,*}^{T}\right) - var\left(X_{f,*}^{T}\right)\right] \tag{25}$$

where $var\left(X_{m,*}^{T}\right)$ and $var\left(X_{f,*}^{T}\right)$ represent the variance of the selected mating variables at the specific iteration T. The offspring can be defined with Eq. (26):

$$X_{child}^{T} = \begin{cases} p \cdot X_{m,*}^{T} + q \cdot x_{f,*}^{T};\ I_{mates} \geq 0 \\ p \cdot X_{f,*}^{T} + q \cdot x_{m,*}^{T};\ I_{mates} < 0 \end{cases} \tag{26}$$

where p represents the random variable and q represents the complementary value to p. At each iteration, that is, a new epoch of offspring, the EMA search is directed by the crossover probability, as expressed by Eq. (27):

$$X_{child}^{T+1} = rand(1, d) \cdot X_{j}^{best} \tag{27}$$

where d represents the problem dimension and X_{j}^{best} is the best solution for the current generation. Exemplary effectiveness of the PV panel location, after the computation process based on the efficiency score, is presented in Fig. 3.

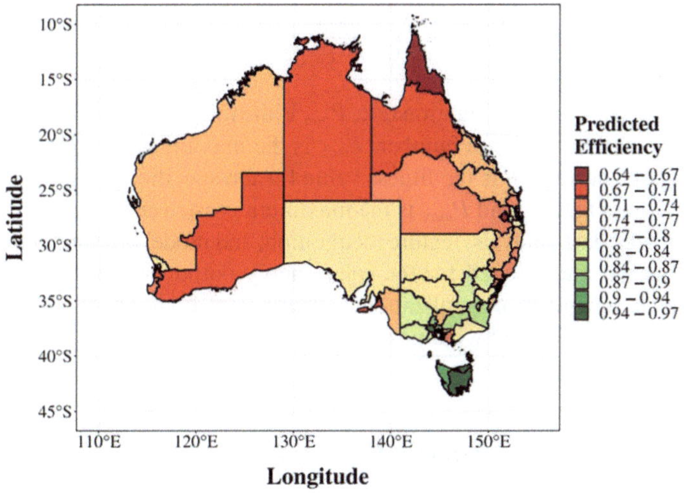

Fig. 3 Predicted efficiencies of the PV panels installed in the Australian region in 2020 [45]

Short-term energy forecasting in solar power plants can also be accomplished by the predictive performance of the GA, with the employment of recurrent neural Network (RNN) and the KNN models as the benchmark models, due to their proven forecasting accuracy. Results from many studies suggest that the GA-driven model is by far the best for conditional short-term energy forecasting in solar energy systems [46]. Multiple uses of GA, which include their involvement in the solar systems, that is, PV state evaluation metrics, are also reflected in the forecasting processes of energy control and the optimization of the power flow.

As contemporary solar power plants struggle to integrate and adapt into existing power networks, an accurate energy forecasting model is mandatory to improve power grid reliability and stability. Alongside the optimal parameter configuration, these issues are usually addressed by optimization algorithms. With the classic algorithms being limited due to the narrow field of utility and fixed search patterns, an auto-evolution hyper-heuristic algorithm, AE-GAPB, which integrates a high-complexity level GA, particle swarm optimization (PSO), and the bat algorithm (BA) as the low-complexity level features, is suitable for continuous search of the best hyperparameters for PSO and BA by following prediction accuracy [47]. The specific adaptability of crossover and mutation features characteristic of the AE-GAPB, which evolve over algorithm epochs, iteration frames, and fitness function fields, results in excellent forecasting datasets on renewable energy generation and power consumption. The adaptation of the crossover and mutation rates is guided by the fitness function values, that is, feedback information, which significantly reduces the possibility of the search algorithm getting stuck in the local optima, which can be expressed by Eq. (28):

$$P = P_{min} + 2\left(1 - \frac{T}{T_{max}}\right) \cdot \frac{f_i}{f_{total}} \cdot P_{max} \tag{28}$$

where P is the crossover or mutation rate, P_{min} is the minimum crossover or mutation rate, T is the current iteration number, T_{max} is the maximum iteration number, that is, the stoppage criteria, f_i is the fitness value for the specific individual, f_{total} is the sum of all fitness values, and P_{max} is the maximum crossover or mutation rate.

Additionally, data analysis, feature recognition, and model training of the PV cells can be enhanced and utilized for the solar energy output by the GA-SVM-driven model, which integrates normalization and principal component analysis (PCA) techniques with the PV and wind power (WP) factors.

3.2 GA-driven Wind Energy Forecasting

Renewable energy management in the sustainable development of wind power plants demands a strong computational framework to approach and describe the complex wind fluctuations, as accurate forecasting is mandatory for the maximum energy output. By addressing this vital task of managing the nature of wind patterns, based on the historical data on wind conditions, a firm and reliable integration into energy grids can be accomplished. These difficult predictive tasks can be solved by diverse AI-driven models that contribute to the theoretical grasp and development of enhanced energy-management strategies, therefore contributing to sustainable development transition by gradually phasing out fossil-based power plants and improving grid stability. Acceptable energy forecasting requires the selection of feasible features from the available data pool, which can be accomplished by integrating the GA into multiple models to optimize the stage of feature extraction. The optimal features, commonly based on historical datasets on energy consumption, weather properties, and temporal conditions, can be derived with the GA and then integrated into various prediction models, such as long short-term memory (LSTM), bi-directional long short-term memory (BiLSTM), and gated recurrent unit (GRU) [48]. All these GA-integrated deep learning (DL) models showcase exceptional results that outperform any standard single-model framework for energy forecasting. When it comes to uncertainty issues due to the data on non-linear wind properties, the short-term wind power can be effectively predicted with the GA-driven LSTM model due to its capability of extracting features from physical and statistical sequential data, with the GA performing global optimization for the adjustments of size and neuron number in LSTM layers [49]. The multi-objective optimisation procedure, implemented within these models, considers the coupled nature of the objectives, that is, levelized cost of electricity, LCH, and levelized cost of heat, LCH, with the main system components represented as the design variables, as expressed by Eq. (29):

$$\vec{x} = \left\{P_{\omega t,nom}, A_{pv}, N_{bat}, P_{chp,n}, M_{st}, P_b\right\} \tag{29}$$

where $P_{\omega t,nom}$ is the nominal power of the WT, A_{pv} is the surface area of panels, N_{bat} is the number of batteries, $P_{chp,n}$ is the power output of the combined heat and power (CHP) unit, M_{st} is the mass of the water in the tank, and P_b is the capacity of the battery. The ineffective components are eliminated from the variable field, as the lower bound, $\overrightarrow{x}^L$, of the search space is set to zero, while the upper bound, $\overrightarrow{x}^U$, of each component can be calculated according to Eq. (30):

$$\overrightarrow{x}^U = \left\{ \begin{array}{c} \frac{\alpha_{\omega t} \cdot max(L_{el})}{1000}, \frac{8 \cdot \alpha_{pv} \cdot max(L_{el})}{1000}, \\ \alpha_{bat} \sum_{i=1}^{24} \frac{L_{el,i}}{P_{bat,max}}, \frac{\alpha_{chp} \cdot max(L_{el})}{1000}, \alpha_{st} \cdot \sum_{i=1}^{24} \frac{L_{th,i}}{H_{st,max}}, \frac{\alpha_b \cdot max(L_{th})}{1000} \end{array} \right\} \tag{30}$$

where the α parameters are the adjusted search factors, L_{el} and L_{th} are the electrical and thermal loads, respectively, $P_{bat,max}$ is the maximum battery capacity and $H_{st,max}$ is the maximum storage tank thermal energy. The 1000 division factor in Eq. (30) indicates the 1 kW step size. The GA-driven optimisation problem can effectively be described by Eq. (31):

$$min \left\{ LCE(\overrightarrow{x}), LCE(\overrightarrow{x}) \quad s.t. \sum_{i}^{N} BO_i \leq BO_{all} \right\} \tag{31}$$

where $\sum_i^N BO_i$ is the total number of hours of thermal and electrical outages on the annual level. Further optimization of GA parameters is also possible and can be implemented in numerous time series forecasting problems for high-performance renewable energy systems.

The accurate representation of wind properties for the needs of medium and long-term energy forecasting must also be coordinated through computational modelling. As unpredictable occurrences, which arise from the complex nature of wind patterns, can greatly affect power systems in terms of planning and scheduling power flow in the long run, an approach that integrates numerical weather prediction (NWP) models can be used [50]. An open-source NWP model, namely, the weather research and forecasting (WRF) model, can be implemented to forecast weather for the needs of wind farms in an arbitrary area. With the millions of unique ways to set up such a model, a GA-driven optimizer for the forecasting of wind density can make a significant contribution in the adjustment of individual control parameters to improve energy forecasting and scalability of any NPW framework. By employing the data on terrain constraints and analysing various WT layouts with corresponding spacings, the GA-driven optimizer successfully proposed the layout of the wind farm, with the capacity factor (CF), as the main objective closely related to the LCOE, being estimated by Eq. (32):

$$CF = \frac{E_{gen}}{C \cdot T} \tag{32}$$

where E_{gen} is the generated energy, C is the capacity, and T is the period.

3.3 GA for Enhancing Energy Forecasting in HPPs

With the increased development of wind power systems and the demand for accurate WP prediction, the usage of smart electric systems (ES) is rapidly increasing. Even though the nonlinear nature and time-varying properties of WP can be accurately predicted with previously described models, the modern ES frameworks need to satisfy multiple criteria, which, on one hand, need to correspond to forecasting accuracy, and on the other, follow the cheaper computational resources, that is, the usage and development of the frameworks which do not require staggering computational power. In that sense, due to economic convenience, GA can be integrated into a conventional backpropagation neural network (BPNN) to optimize the weights and threshold values, by modifying the crossover and mutation processes, to perform an ES scheduling strategy within the reasonable accuracy limits. As most HPPs require accurate tank inflow for effective forecasting on energy planning and reservoir management, contemporary methods can be used to predict the energy share of the estimated tank inflow in the short-term time frame. By integrating the ANN framework with the GA-driven optimization model, the historical data on meteorological events can be used for effective energy prediction. Compared to the similar models in the HPPs energy forecasting field, such as feed-forward backpropagation network (FFBPNN), and nonlinear autoregressive network with exogenous inputs (NARX), GA-driven ANN has proven to be more accurate, as evaluation metrics data demonstrated reduced error rates [51]. By utilizing meteorological data, conventional ML models can be upgraded with the hybrid GA approach, namely, genetic grey wolf optimizer (GGWO), which can accurately predict hydroelectric power output. Compared to the other popular AI-driven techniques in this area, such as support vector regression (SVR), extreme learning machine (ELM), random forest regression (RFR), and weighted k-nearest neighbor (WKNNR), the GGWO exhibited dominant performance in feature selection and data extraction, that is, decision-making management, which significantly contributes to the field of energy forecasting in HPPs [52]. The typical single-objective problems in this area are based on cost minimization, with detailed consideration of power, flow, and feasibility constraints. The power constraint, P, is defined by Eq. (33):

$$P \geq P_{min} \tag{33}$$

where P_{min} is the minimum power supply necessary to satisfy the energy needs. The flow constraint, limited by the design of the HPPs, is conditioned by the extraction parameter, that is, the flow of the turbine, Q, as defined by Eq. (34):

$$Q \leq \kappa \cdot Q_{supp} \tag{34}$$

where κ is a factor of the water supply flow rate Q_{supp}. The feasibility constraints are defined based on the terrain properties, and they usually cover the height of the penstock, $z_{P,i}$, at specific coordinates, s_i, as expressed by Eq. (35):

$$\begin{aligned} z_{P,i} - z_i &\leq \epsilon_{sup};\ i = 1 \ldots N \\ z_j - z_{P,j} &\leq \epsilon_{exc};\ i = 1 \ldots N \end{aligned} \tag{35}$$

where ϵ_{sup} is the maximum support length and ϵ_{exc} is the maximum excavation depth, where both ϵ_{sup} and ϵ_{exc} are determined based on the terrain properties. The fitness function, which extracts the best solutions based on the cost level, implements the death penalty function, which is applied to the invalid solutions, that is, individuals of the population, therefore restricting their participation in the further evolution process. Finally, the fitness function can be described by Eq. (36):

$$\begin{cases} if\ solution\ is\ valid\ find\ F \\ else\ F = \infty \end{cases} \tag{36}$$

where F is the optimal obtainable power P extracted based on the minimum costs of plant parameters, which include costs of piping systems, penstocks, distribution lines, wiring, etc. Accordingly, the obtainable power is defined by Eq. (37):

$$P = \frac{\eta \cdot \rho}{2 \cdot S_{noz}^2} \sqrt{\left(\frac{H_g}{\frac{1}{2 \cdot g \cdot S_{noz}^2} + k_{fric} \cdot \frac{L}{D_P^5}} \right)^3} \tag{37}$$

where η is the equipment efficiency, ρ is the density, S_{noz}^2 is the section area of the nozzle, H_g is the gross height, k_{fric} is the material constant of frictional losses, L is the penstock length, and D_P is the pipe diameter.

Small hydroelectric power plants (SHPPs), which were developed as one of the ways to confront energy issues in isolated areas, also require precise short-term forecasting of energy production based on the data variability concerning the inflow parameters, weather conditions, as well as the retention tank status. The most efficient use and flow of resources, as in the large-scale hydroelectric power systems, can be derived with the use of GA, which can be structured as a single or multi-optimization model to minimize the cost, or to simultaneously reduce the cost expenses while maximizing the energy output, respectively [53].

4 Conclusion

As presented in this review chapter, the global need for a green future requires extensive data on numerous parameters, which must be ordered and structured to provide a high level of accuracy in the automated decision-making processes of renewable energy systems. The GA-driven models that are extensively used in these systems in conventional or hybrid frameworks represent a cutting-edge approach for data optimization, feature selection, and pattern extraction, in a manner convenient for prompt and efficient failure predictions and energy forecasting. Regardless of the framework being used, as shown in this chapter, GA has proven to be an effective and irreplaceable tool in various renewable energy systems. With the ever-increasing demand for renewable energy systems, AI will most certainly have a significant role in system maintenance and development, with an undeniable integration of GA models at the forefront of various techniques and methods being used.

References

1. Raschka, S., & Mirjalili, V. (2019). *Python machine learning: Machine learning and deep learning with Python, scikit-learn, and TensorFlow 2.*
2. Gridin, I. (2022). *Automated deep learning using neural network intelligence: Develop and design pytorch and tensorflow models using Python.*
3. Liu, Y. (Hayden) (2020). *Python machine learning by example—Build intelligent systems using Python, TensorFlow 2, PyTorch, and scikit-learn.*
4. Raschka, S., Patterson, J., & Nolet, C. (2020). Machine learning in python: Main developments and technology trends in data science, machine learning, and artificial intelligence. *Information, 11.*
5. Carloman, A. N., Bermudo, U. V. W., Estilloso, E. M., & Llantos, O. E. (2024). Bundle AI: An application of multiple constraint knapsack problem (MCKP) through genetic algorithm (GA). *Procedia Computer Science.*
6. Gridin, I. (2022). *Machine learning methods, systems, challenges.*
7. Zhong, X., Song, X., Zhang, X., & Zhang, Z. (2025). A hybrid approach of genetic algorithm and truncated branch-and-bound for seru scheduling problem with sequence-dependent setup time. *European Journal of Industrial Engineering, 19*(2). https://doi.org/10.1504/ejie.2025.10060760.
8. Domashova, J. V., Emtseva, S. S., Fail, V.S., & Gridin, A. S. (2021). Selecting an optimal architecture of neural network using genetic algorithm. *Procedia Computer Science.*
9. Kashyap, P. (2024). *Machine learning for decision makers.*
10. Lai, J. P., Chang, Y. M., Chen, C. H., & Pai, P. F. (2020). A survey of machine learning models in renewable energy predictions. *Applied Sciences, 10.*
11. Benti, N. E., Chaka, M. D., & Semie, A. G. (2023). Forecasting renewable energy generation with machine learning and deep learning: Current advances and future prospects. *Sustainability, 15.*
12. Al-Dahoud, A., Fezari, M., & Aldahoud, A. (2023). Machine learning in renewable energy application: Intelligence system for solar panel cleaning. *WSEAS Transactions on Environment and Development, 19.* https://doi.org/10.37394/232015.2023.19.45.
13. Gogas, P., & Papadimitriou, T. (2023). Machine learning in renewable energy. *Energies, 16.*
14. Tiwari, S. (2023). Implications of machine learning in renewable energy. *Engineering Proceedings, 37*(1). https://doi.org/10.3390/ECP2023-14610.
15. Talwariya, A., Singh, P., Jobanputra, J. H., & Kolhe, M. L. (2023). Machine learning based renewable energy generation and energy consumption forecasting. *Energy Sources, Part A: Recovery, Utilization and Environmental Effects, 45*(2). https://doi.org/10.1080/15567036.2023.2194257.
16. Ajibade, S. S. M., Bekun, F. V., Adedoyin, F. F., Gyamfi, B. A., & Adediran, A. O. (2023). Machine learning applications in renewable energy (MLARE) research: A publication trend and bibliometric analysis study (2012–2021). *Clean Technologies, 5*
17. Bin Abu Sofian, A. D. A., Lim, H. R., Siti Halimatul Munawaroh, H., Ma, Z., Chew, K. W., & Show, P. L. (2024). Machine learning and the renewable energy revolution: Exploring solar and wind energy solutions for a sustainable future including innovations in energy storage. *Sustainable Development, 32.*

18. Sharma, P., Said, Z., Kumar, A., Nižetić, S., Pandey, A., Hoang, A. T., Huang, Z., Afzal, A., Li, C., Le, A. T., Nguyen, X. P., & Tran, V. D. (2022). Recent advances in machine learning research for nanofluid-based heat transfer in renewable energy system. *Energy and Fuels, 36*.
19. Khanna, A., Lamba, B. Y., Jain, S., Bolshev, V., Budnikov, D., Panchenko, V., & Smirnov, A. (2023). Biodiesel production from Jatropha: A computational approach by means of artificial intelligence and genetic algorithm. *Sustainability, 15*(12). https://doi.org/10.3390/su15129785.
20. Mansoury, I., Bourakadi, D. El., Yahyaouy, A., & Boumhidi, J. (2024). Optimized extreme learning machine using genetic algorithm for short-term wind power prediction. *Bulletin of Electrical Engineering and Informatics, 13*(2). https://doi.org/10.11591/eei.v13i2.6476.
21. Zhang, C., & Yang, T. (2021). Optimal maintenance planning and resource allocation for wind farms based on non-dominated sorting genetic algorithm-II. *Renewable Energy, 164*, 1540–1549. https://doi.org/10.1016/J.RENENE.2020.10.125
22. Petrović, A., & Đurišić. (2021) Genetic algorithm based optimized model for the selection of wind turbine for any site-specific wind conditions. *Energy, 236*, 121476. https://doi.org/10.1016/J.ENERGY.2021.121476.
23. Yu, Q., Bangalore, P., Fogelström, S., & Sagitov, S. (2024). Optimal preventive maintenance scheduling for wind turbines under condition monitoring. *Energies, 17*(2). https://doi.org/10.3390/en17020280.
24. do Couto, T. G., Farias, B., Diniz, A. C. G. C., & de Morais, M. V. G. (2013). Optimization of wind farm layout using genetic algorithm. In *10th World Congress on Structural and Multidisciplinary Optimization.*
25. Bin Ali, M., Ahmad, Z., Alshahrani, S., Younis, M. R., Talib, I., & Imran, M. (2022). A case study: Layout optimization of three gorges wind farm Pakistan, using genetic algorithm. *Sustainability, 14*(24). https://doi.org/10.3390/su142416960.
26. Qureshi, T. A., & Warudkar, V. (2023). Wind farm layout optimization through optimal wind turbine placement using a hybrid particle swarm optimization and genetic algorithm. *Environmental Science and Pollution Research, 30*(31). https://doi.org/10.1007/s11356-023-27849-7.
27. Kirchner-Bossi, N., & Porté-Agel, F. (2024). Wind farm power density optimization according to the area size using a novel self-adaptive genetic algorithm. *Renewable Energy, 220*. https://doi.org/10.1016/j.renene.2023.119524.
28. Turja, A. I., Hasan, M. M., Ehsan, M. M., & Khan, Y. (2024). Multi-objective performance optimization & thermodynamic analysis of solar powered supercritical CO2 power cycles using machine learning methods & genetic algorithm. *Energy and AI, 15*. https://doi.org/10.1016/j.egyai.2023.100327.
29. Shoaei, M., Hajinezhad, A., & Moosavian, S. F. (2023). Design, energy, exergy, economy, and environment (4E) analysis, and multi-objective optimization of a novel integrated energy system based on solar and geothermal resources. *Energy, 280*. https://doi.org/10.1016/j.energy.2023.128162.
30. Al Neyadi, S., Al Kuwaiti, M., Al Daheri, S., Al Marzroei, M., Al Ketbi, A., & Alnajjar, F. (2023). Optimization of large-scale solar panels distribution using genetic algorithm. In *2023 Middle East and North Africa Solar Conference, MENA-SC 2023—Proceedings*.
31. Qasimi, A. B., Toomanian, A., Nasri, F., & Samany, N. N. (2023). Genetic algorithms-based optimal site selection of solar PV in the north of Afghanistan. *International Journal of Sustainable Energy, 42*(1). https://doi.org/10.1080/14786451.2023.2246081.
32. Saadaoui, D., Elyaqouti, M., Assalaou, K., Ben hmamou, D., & Lidaighbi, S. (2021). Parameters optimization of solar PV cell/module using genetic algorithm based on non-uniform mutation. *Energy Conversion and Management X, 12*, 100129. https://doi.org/10.1016/J.ECMX.2021.100129.
33. Durez, A., Ali, M., Waqas, A., Nazir, K., & Kumarasamy, S. (2023). Modelling and optimization of phase change materials (PCM)-based passive cooling of solar PV panels in multi climate conditions. *Frontiers in Energy Research, 11*. https://doi.org/10.3389/fenrg.2023.1121138.
34. Bulut, M., & Özcan, E. (2024). Planning of prescriptive maintenance types for generator with fuzzy logic-based genetic algorithm in a hydroelectric power plant. *Expert Systems with Applications, 240*, Article 122480. https://doi.org/10.1016/J.ESWA.2023.122480

35. Pereira, F. H., de Melani, A. H., Kashiwagi, F. N., da Rosa, T. G., dos Santos, U. S., & de Souza, G. F. M. (2023). Imperfect preventive maintenance optimization with variable age reduction factor and independent intervention level. *Applied Sciences, 13*(18). https://doi.org/10.3390/app131810210.
36. Pîraianu, V. F., Dragoi, C., Vasiliu, N., & Pavel, G. L. (2017). Optimizing maintenance works for hydroelectric power plants by genetic algorithms. In *15th International Industrial Simulation Conference 2017, ISC 2017.*
37. Liu, Y., Ćetenović, D., Li, H., Gryazina, E., & Terzija, V. (2022). An optimized multi-objective reactive power dispatch strategy based on improved genetic algorithm for wind power integrated systems. *International Journal of Electrical Power & Energy Systems, 136*. https://doi.org/10.1016/j.ijepes.2021.107764.
38. Eskandari, A., Milimonfared, J., Aghaei, M., & Reinders, A. H. M. E. (2020). Autonomous monitoring of line-to-line faults in photovoltaic systems by feature selection and parameter optimization of support vector machine using genetic algorithms. *Applied Sciences, 10*(16). https://doi.org/10.3390/app10165527.
39. Hichri, A., Hajji, M., Mansouri, M., Abodayeh, K., Bouzrara, K., Nounou, H., & Nounou, M. (2022). Genetic-algorithm-based neural network for fault detection and diagnosis: Application to grid-connected photovoltaic systems. *Sustainability, 14*(17). https://doi.org/10.3390/su141710518.
40. Boubakr, G., Smith, A., Gu, F., Ball, A., & Alrweg, M. (2023). Advanced neural network with optimized training parameters based on the cuckoo search for detecting and classifying faults in PV power systems. In *Mechanisms and Machine Science.*
41. Zhang, L., Wang, K., Lin, W., Geng, T., Lei, Z., & Wang, Z. (2019). Wind power prediction based on improved genetic algorithm and support vector machine. In *IOP Conference Series: Earth and Environmental Science.*
42. Qu, Z., Hou, X., Hu, W., Yang, R., & Ju, C. (2023). Wind power forecasting based on improved variational mode decomposition and permutation entropy. *Clean Energy, 7*(5). https://doi.org/10.1093/ce/zkad043.
43. Zheng, J., & Zeng, B. (2023). Unleashing the influencing factors of solar energy adoption to combat climate change: A roadmap toward sustainable energy technologies. *Sustainable Energy Technologies and Assessments, 57*. https://doi.org/10.1016/j.seta.2023.103303.
44. Sulaiman, M. H., & Mustaffa, Z. (2024). Forecasting solar power generation using evolutionary mating algorithm-deep neural networks. *Energy and AI, 16*, Article 100371. https://doi.org/10.1016/J.EGYAI.2024.100371
45. Cattani, G. (2023). Combining data envelopment analysis and Random Forest for selecting optimal locations of solar PV plants. *Energy and AI, 11*. https://doi.org/10.1016/j.egyai.2022.100222.
46. Ratshilengo, M., Sigauke, C., & Bere, A. (2021). Short-term solar power forecasting using genetic algorithms: An application using South African data. *Applied Science, 11*(9). https://doi.org/10.3390/app11094214.
47. Cao, Y., Yu, J., Zhong, R., & Munetomo, M. (2025). Forecasting renewable energy and electricity consumption using evolutionary hyperheuristic algorithm. *Science and Reports, 15*(1), 2565. https://doi.org/10.1038/s41598-025-87013-8
48. Kahwash, F., Barakat, B., & Maheri, A. (2023). Coupled thermo-electrical dispatch strategy with AI forecasting for optimal sizing of grid-connected hybrid renewable energy systems. *Energy Conversion and Management, 293*. https://doi.org/10.1016/j.enconman.2023.117460.
49. Sward, J. A., Ault, T. R., & Zhang, K. M. (2022). Genetic algorithm selection of the weather research and forecasting model physics to support wind and solar energy integration. *Energy, 254*, Article 124367. https://doi.org/10.1016/J.ENERGY.2022.124367
50. Dhunny, A. Z., Timmons, D. S., Allam, Z., Lollchund, M. R., & Cunden, T. S. M. (2020). An economic assessment of near-shore wind farm development using a weather research forecast-based genetic algorithm model. *Energy, 201*. https://doi.org/10.1016/j.energy.2020.117541.
51. Kongpaseuth, V., & Kaewarsa, S. (2023). Nam Theun 2 hydropower plant energy prediction using artificial neural network and genetic algorithm. In *15th Asia-Pacific Power and Energy Engineering Conference.*

52. Sahin, M. E., & Ozbaykarakus, M. (2024). Smart hydropower management: Utilizing machine learning and deep learning method to enhance dam's energy generation efficiency. *Neural Computing and Applications, 36*(19), 11195–11211. https://doi.org/10.1007/s00521-024-09613-1
53. Tapia, A., Reina, D. G., del Nozal, A. R., & Millán, P. (2020). Application of genetic algorithms for designing micro-hydro power plants in rural isolated areas—A case study in San Miguelito, Honduras. In *Studies in Computational Intelligence.*

Deep Reinforcement Learning in Energy Management System for Fuel Cell Hybrid Vehicles: A Review on Reward Design and Testing Framework

Ali Sayah, Marwa Ben Saïd-Romdhane, and Sondes Skander-Mustapha

Abstract Hybrid Electric Vehicles (HEVs), Fuel Cell Hybrid Electric Vehicles (FCHEVs) and Battery Electric Vehicles (BEVs) are key solutions for sustainable transportation, requiring advanced Energy Management Systems (EMS) to enhance efficiency and system longevity. Deep Reinforcement Learning (DRL)-based EMS offers an adaptive, data-driven approach for real-time energy optimization, yet challenges remain in designing robust reward functions that balance energy efficiency, powertrain durability, and operational stability. This chapter reviews reward function design and normalization strategies in DRL-based EMS, highlighting their impact on learning performance and system efficiency. Additionally, it examines state-of-the-art hardware testing methodologies, including simulation frameworks, Model-in-the-Loop (MIL), Hardware-in-the-Loop (HIL), and real-world implementations, enabling the transition from conceptual developments to real-world implementation. Furthermore, this chapter discusses key challenges in RL design and testing methodologies, explores potential solutions, and outlines future research directions, contributing to the advancement of intelligent energy management framworks for HEVs.

Keywords Fuel cell hybrid electric vehicles · Energy management system · Deep reinforcement learning · Reward function design · Normalization · Testing frameworks

---

A. Sayah · M. B. Saïd-Romdhane · S. Skander-Mustapha (✉)
LR11ES15 Laboratoire de Systèmes Électriques, Université de Tunis El Manar, Ecole Nationale d'Ingénieurs de Tunis, 1002 Tunis, Tunisie
e-mail: sondes.skander@enit.utm.tn

S. Skander-Mustapha
Université de Carthage, Ecole Nationale d'Architecture et d'Urbanisme, 2026 Sidi Bou Said, Tunisie

M. B. Saïd-Romdhane
Institut Supérieur des Sciences Appliquées et de Technologie de Gabès, Université de Gabès, 6029 Gabès, Tunisie

M. A. Mellal et al. (eds.), *Computational Intelligence for a Greener Future: Innovations in Renewable Energy Systems*, Studies in Computational Intelligence 1240,
https://doi.org/10.1007/978-3-032-06732-6_2

1 Introduction

Fossil fuels are a primary driver of environmental pollution and climate change, underscoring the urgent need for cleaner transportation solutions. The challenges associated with conventional vehicles have accelerated the development of hybrid systems that leverage multiple energy sources [1–4]. As a result, the automotive industry has witnessed a growing diversification in electrified propulsion technologies, incorporating various energy sources and advanced power management strategies [5]. Battery Electric Vehicles depend entirely on electrical energy for propulsion, while Plug-in Hybrid Electric Vehicles (PHEVs) pair a fuel-based engine with an electric motor for flexible energy use. Hybrid Electric Vehicles utilize a combination of batteries, supercapacitors, and another energy source for improved energy distribution. While Fuel Cell Vehicles (FCVs) rely solely on hydrogen fuel cells for propulsion, Fuel Cell Hybrid Electric Vehicles generate electricity through a hydrogen fuel cell and integrate it with electrical energy storage for enhanced performance.

With the increasing complexity of hybrid and electrified powertrains, effective energy management becomes essential to ensure optimal performance and longevity [6, 7]. An Energy Management System plays a pivotal role in efficiently regulating power flow between energy sources in hybrid vehicles, ensuring efficient energy utilization and system longevity.

By intelligently balancing energy flow between multiple sources, EMS enhances performance, reduces fuel consumption, and minimizes component degradation. Its role is particularly vital in real-time operation, where dynamic adjustments are needed to maintain stability and efficiency under varying driving conditions.

Rule-based energy management strategies are computationally efficient but heavily depend on expert knowledge, making it challenging to achieve globally optimal solutions. These methods often lack adaptability, limiting their effectiveness in dynamic and time-varying scenarios due to their reliance on predefined rules [8–10]. To address this issue, optimization-based methods have been developed [11–13], which can be classified into two groups: global optimization techniques and local (real-time) optimization techniques [14]. Techniques such as Dynamic Programming (DP) [15] and Heuristic Dynamic Programming (HDP) [16], can identify globally optimal solutions but are computationally intensive. In contrast, the latter methods, including Equivalent Consumption Minimization Strategy (ECMS) and Model Predictive Control (MPC) [6, 17, 18], require less computational effort but are limited to achieving only locally optimal results.

To achieve a better balance between precision and computational efficiency, various advanced methodologies have been introduced, such as game theory [19] and Artificial Intelligence (AI) algorithms [20–22]. The emergence of advanced Machine Learning (ML) techniques has positioned Reinforcement Learning as a prominent solution for energy management in hybrid vehicles. Unlike traditional methods, RL leverages a trial-and-error framework to learn optimal power split policies through environmental interactions, effectively bridging global and local optimization.

RL frameworks automatically extract control policies while continuously refining their decision-making, enabling them to realize optimal control performance through agent and environment interaction, as well as learning through trial and error [23–25]. RL-based EMSs are model-free, meaning they do not need prior knowledge of the dynamics of the system, and they utilize the Bellman equation to optimize decision-making over time. These systems do not depend on conventional rule-based or optimal control approaches, allowing them to address the limitations of both rule-based and optimization-based EMSs. Specifically, RL-based EMSs overcome the lack of control optimality in rule-based systems and the challenges of real-time implementation in optimization-based systems [26, 27]. Moreover, RL-based EMSs demonstrate strong adaptability to dynamic and time-varying environments, making them highly effective under varying driving conditions. As a result, considering real-time performance, control optimality and adaptability, RL-based EMSs can be viewed as a balanced and effective solution for energy management. RL strategies vary in their learning mechanisms and applications within EMS. Value-based methods like Q-Learning and Deep Q-Networks (DQN) estimate optimal action values, while policy-based approaches such as Proximal Policy Optimization (PPO) and Deep Deterministic Policy Gradient (DDPG) directly optimize action policies. Temporal Difference (TD) learning bridges dynamic programming and Monte Carlo methods, enabling faster and more efficient learning.

The reward setting is critical as it provides the agent with feedback on the effectiveness of its actions, significantly influencing the success of training convergence [28]. In RL-based EMS, reward functions are typically designed to minimize hydrogen consumption while ensuring battery SOC remains within safe limits. Additionally, some studies integrate objectives such as component longevity, drivability, and emission reduction to enhance overall system performance.

Normalization strategies in RL-based EMS are essential for stabilizing the learning process and improving convergence. By scaling reward values and input states within predefined ranges, these methods ensure consistent gradients during training, preventing excessive penalties or rewards. Common strategies include min–max normalization, z-score standardization, and adaptive scaling based on real-time driving conditions.

The development of energy management schemes for HEVs necessitates the use of advanced simulation tools and robust testing frameworks [29, 30]. Researchers have extensively utilized a model-based design approach, incorporating Software-in-the-Loop, Model-in-the-Loop, Hardware-in-the-Loop and digital twin simulations, followed by component and system bench testing, to assess the energy efficiency achieved through powertrain electrification. These methodologies enable the automotive industry to validate designs more comprehensively, paving the way for advancements in energy-efficient and complex alternative powertrains [31–33].

This chapter conducts a comprehensive critical review and analysis of reinforcement learning-powered energy allocation frameworks for fuel cell hybrid electric vehicles powertrains, with a particular focus on reward function design. Through an in-depth literature study, various reward design approaches are analyzed, highlighting their primary objectives, such as fuel efficiency, battery SOC management,

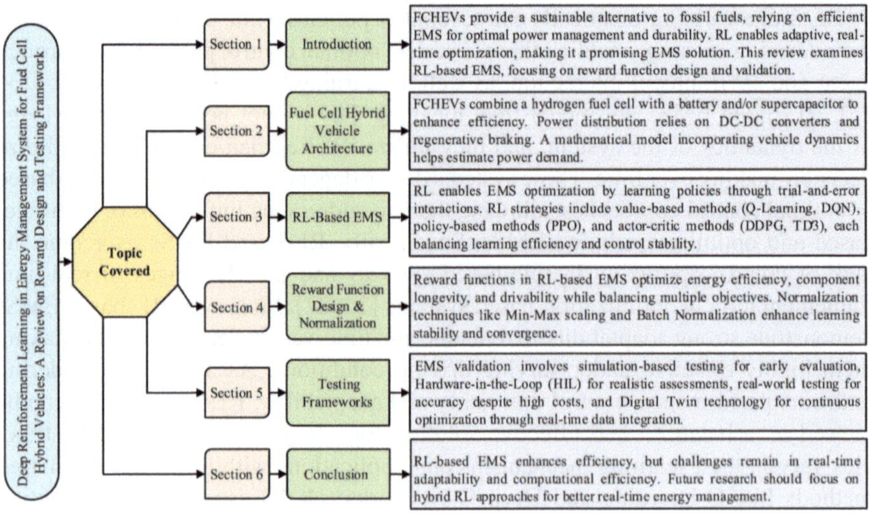

Fig. 1 Overview of the chapter structure

component longevity, and emissions reduction. Additionally, different RL strategies applied to EMS are examined, along with testing frameworks used for validation.

Figure 1 presents the structure of this chapter, outlining the key topics covered. The chapter begins in Sect. 1 with an introduction to FCHEVs and their energy management challenges. Section 2 explores the architecture of FCHEVs, detailing various powertrain configurations and components. Section 3 delves into reinforcement learning methodologies for EMS, covering strategies such as Deep Deterministic Policy Gradient, Proximal Policy Optimization, Temporal Difference Learning, Q-Learning and Deep Q-learning. Section 4 focuses on reward function design, classifying different approaches based on optimization objectives and reviewing normalization strategies for reward shaping in RL-EMS. Section 5 discusses various testing frameworks, including simulation-based testing, hardware-in-the-loop experiments, and real-world validation. Finally, Sect. 6 synthesizes the chapter's core findings and proposes promising avenues for further investigation in hybrid electric vehicle energy management.

2 Fuel Cell Hybrid Vehicle Architecture

Fuel Cell Hybrid Electric Vehicles integrate a hydrogen fuel cell with additional energy storage components to enhance efficiency and performance. FCHEVs offer distinct advantages over both pure FCVs and BEVs by synergistically combining their strengths while mitigating individual limitations. FCHEVs achieve superior range and faster refueling compared to BEVs through hydrogen fuel cell integration, while the hybrid battery system enhances power delivery and recaptures braking energy addressing the slow transient response of standalone FCVs.

The primary architectures include FC + Battery, where the battery supplements the fuel cell during peak power demands; FC + Supercapacitor (SC), which leverages SCs for rapid energy bursts and regenerative braking; and FC+ Battery + SC, combining both storage elements for optimal power management [3]. These configurations require DC–DC converters to regulate the flow of power between the energy sources and the electrical motor, ensuring stable operation and efficient energy distribution. The electric motor serves as the primary propulsion unit, drawing energy from the hybrid system while regenerative braking helps recover lost energy [7]. The choice of architecture depends on application-specific requirements, balancing efficiency, cost, and performance.

An examination of the longitudinal dynamics of the vehicle was conducted to evaluate the power requirements as outlined in Eq. (1) [34].

$$P_d = mgfv\cos(\theta) + mgv\sin(\theta) + \frac{1}{2}C_d A\rho v^3 + \delta mv\frac{dv}{dt} \tag{1}$$

where the instantaneous power demand P_d of a vehicle is determined by three primary factors: (1) inertial and gravitational forces from the vehicle mass (m), gravitational acceleration (g), and road inclination angle (θ); (2) motion resistance including the rolling resistance coefficient (f) and equivalent mass factor (δ) that accounts for rotational inertia; and (3) aerodynamic losses influenced by air density (ρ), frontal area (A), drag coefficient (C_d), and the vehicle's velocity (v). These interdependent factors collectively shape the energy needs under dynamic conditions, including variations in vehicle speed and external environmental influences.

Figure 2 shows the hybrid energy storage system design for FCHEV. The system uses a FC stack as the main power source, generating electricity from hydrogen and producing only water as a clean byproduct, while additional energy storage components (batteries, supercapacitors, or both) handle sudden power needs and temporary energy demands. The fuel cell provides steady-state energy, while the battery and supercapacitor dynamically support high-power acceleration, absorb regenerative braking energy, and buffer load fluctuations. Power from the DC bus is routed to the vehicle's drive system, where a DC motor transform electricity into mechanical torque for propulsion.

3 Reinforcement Learning-Based Energy Management System

Reinforcement Learning is a machine learning approach in which an agent improves its decision-making by interacting with an environment. It learns by receiving rewards for beneficial actions and penalties for unfavorable ones, gradually optimizing its

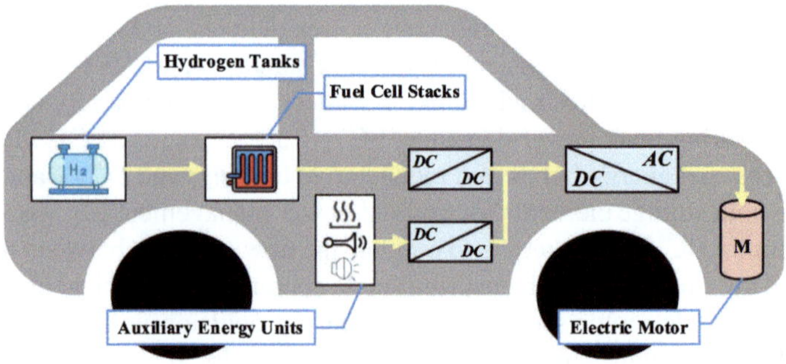

Fig. 2 Fuel cell hybrid electric vehicle architecture

strategy over time, the agent continuously improves its policy to maximize long-term performance. RL is particularly effective in dynamic and complex systems, where predefined rules may be insufficient. Its adaptability makes it well-suited for applications like energy management strategies. RL-based algorithms designed for electrified powertrains address the limitations of conventional control strategies. Like Dynamic Programming, RL integrates the equation of Bellman to attain global optimization by combining the immediate reward from the current state transition with the accumulated rewards from future steps, ensuring long-term performance improvement throughout the episode. RL methods are grounded in the Markov Decision Process, which assumes that future states depend solely on the current state and not on any previous states. Like human learning, an RL agent learns by actively engaging with its environment through interactions, exchanging information about states, rewards, and actions. Using this information, the RL agent learns decision-making rules to maximize the accumulated reward over an extended period, making it highly effective for dynamic and complex systems.

Figure 3 illustrates the workflow of an RL-based decision-making framework for vehicle energy management. It starts with vehicle state observation, followed by RL-based decision-making using value-based, policy-based, or hybrid methods, and concludes with action selection and reward calculation. The process iterates until convergence or the end of the drive cycle, updating the policy and value function to optimize energy management.

3.1 Q-Learning

Q-Learning is a reinforcement learning algorithm centered on values, utilizing a Q-table to predict the anticipated rewards for every possible action in each state. It follows an off-policy approach, meaning it learns from actions outside the current policy, allowing efficient exploration. In [35], a Q-learning based EMS was designed for a FC/battery electric vehicle. The propulsion load was modeled based on a new European drive cycle (NEDC). The results demonstrated that the novel algorithm

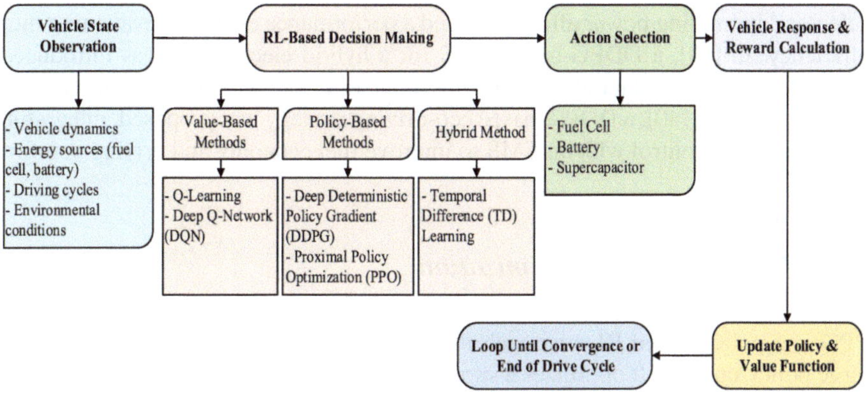

Fig. 3 Reinforcement learning scheme for Fuel cell hybrid electric vehicle

achieved an extended battery life and optimized power delivery efficiency. A dynamic rule-based EMS [36] was also proposed for a FC vehicle. This approach demonstrated superior performance in minimizing hydrogen consumption while maintaining batterie SOC with acceptable range compared to traditional fuzzy logic-based methods.

3.2 *Deep Q-Network*

DQN is a developed version of Q-Learning that integrates deep neural networks to predict Q-values, allowing it to manage large and continuous state spaces. It stabilizes training through methods such as experience replay and target networks. A RL-based DQN was developed for splitting power in a hybrid electric bus [37]. The proposed method achieved 89% of the fuel economy compared to dynamic programming-based approach while effectively dealing with state variable dimensionality challenges. A DRL-based EMS was designed for a PHEV in [38] based on DQN algorithm. This research implements a Q-learning-based energy management system, demonstrating a 16.3% improvement in energy efficiency over conventional control approaches in real-time operation.

3.3 *Deep Deterministic Policy Gradient*

DDPG is an advancement of the DQN approach, retaining its model-free and off-policy characteristics. This approach employs an actor-critic architecture specifically optimized for continuous control domains, making it ideal for EMS optimization. It combines policy-based and value-based methods, using two networks: an action selector determining power allocations and a performance estimator evaluating their efficiency. In [39], a DDPG-based EMS for a hybrid electric bus was introduced, incorporating traffic data and passenger count to create a more realistic environmental model. In [40], a DDPG-based eco-driving strategy was proposed, integrating adaptive cruise control with an EMS to improve fuel economy and driving comfort.

3.4 Proximal Policy Optimization

PPO is a policy-based RL approach that improves training stability by restricting large policy updates. It optimizes a surrogate objective function and balances exploration and exploitation effectively. Nowadays, limited research has been conducted applying the PPO in EMS for FCHEVs. In [41], A PPO-based RL EMS was developed for a hybrid electric bus. The proposed EMS demonstrated significant improvements across three key metrics: training efficiency achieved convergence faster than baseline methods, fuel economy reached 94.65% of theoretical maximum efficiency, and fuel cell degradation was reduced by 80.96% compared to conventional PPO-based approaches. In [42], a PPO-based EMS for a FCHEV was investigated, achieving superior fuel economy, lower fuel cell degradation, and faster convergence (93.55% and 97.17%) compared to DQN and TD3-based EMSs. The results also demonstrate its feasibility for embedded system deployment.

3.5 Temporal Difference Learning

Temporal Difference Learning (TD) is a fundamental RL approach where value estimates are updates based on observed rewards and future predictions. Unlike Monte Carlo methods, TD learning does not require waiting until the end of an episode, making it more suitable for real-time EMS adjustments. In [43], a TD3-based EMS for HEVs was employed, demonstrating superior fuel efficiency, the fastest convergence, and the highest robustness compared to other DRL-based EMSs. The integration of a local controller and hybrid experience replay further enhances convergence speed and resilience to environmental disturbances. A similar analysis of EMSs for HEVs using TD3 and DDPG was conducted in [44], where it was found that TD3 provided enhanced efficiency than DDPG regarding learning speed and fuel efficiency. The study also focused on tackling the challenges posed by discrete–continuous hybrid action spaces in complex hybrid systems.

Figure 4 illustrates the classification of the most common RL strategies used in literature into three main categories: Policy-Based Methods, Value-Based Methods and Actor-Critic Methods. The first category strategies include Q-Learning and Deep Q-Networks, which rely on estimating value functions. The second category approaches, such as PPO, optimize policies without using value functions. And finally, the third category methods combine both approaches, represented by Deep

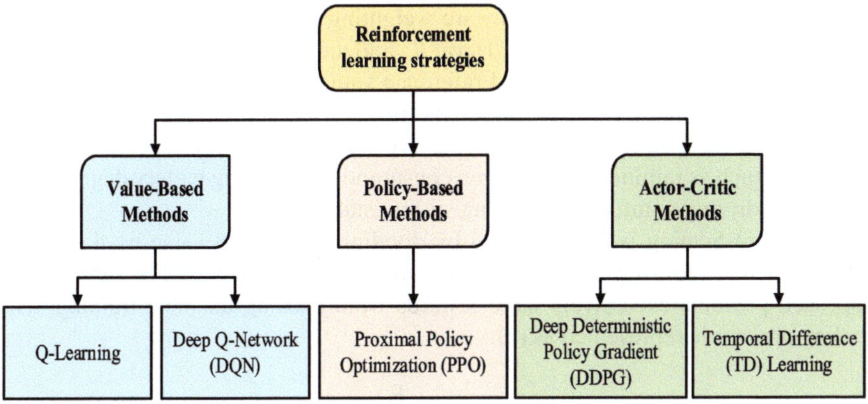

Fig. 4 Classification of most common reinforcement learning methods for energy management in FCHEVs

Deterministic Policy Gradient and Temporal Difference Learning, leveraging both value and policy updates for improved learning stability [7, 31–39].

4 Reward Function Design and Normalization Strategies

4.1 Reward Function Design

The reward function is a fundamental component in RL-based EMS, as it directly shapes the learning behavior of the agent. It defines the optimization objectives and ensures that the RL agent makes energy-efficient and stable decisions in real time. Poorly designed reward functions can lead to unstable training, suboptimal energy allocation, or excessive wear on vehicle components. A well-structured reward function must consider multiple factors, including energy efficiency, component longevity, driving comfort, and operational safety [45]. In RL-based energy management systems for FCHEVs, reward functions are primarily designed to optimize fuel usage and keep the SOC of the battery within an acceptable limit. A literature review shows that these two objectives are consistently present across different studies (Table 1), forming the foundation of reward design. Additional objectives, such as extending component lifespan, improving drivability, reducing emissions or enhancing system stability, are often integrated depending on the study's focus. This multi-objective nature ensures a well-balanced energy management strategy that meets both efficiency and operational requirements. To balance multiple objectives, the reward function integrates the previous objectives using a weighted sum approach [46]. The equation of the total reward is expressed as (2):

$$R_{total} = w_1R_1 + w_2R_2 + \cdots + w_n(SOC - SOC_{ref})^b \quad (2)$$

where R_{total} is the overall reward, w_i are weighting coefficients that determine the importance of each individual reward term R_i, and the final term penalizes deviations of the battery's SOC from a desired reference value SOC_{ref}. The exponent b is an even integer (usually 2) that amplifies the penalty as the deviation grows larger, ensuring that the SOC remains close to the target. Each R_i can correspond to specific objectives such as minimizing hydrogen consumption, reducing battery degradation, improving driving comfort, or ensuring system stability.

A Reward Scaling is often applied by dividing the reward value by a constant scaling factor, ensuring that rewards remain within a manageable range. This approach prevents excessively large rewards from causing unstable training. The scaling process is represented as (3):

$$\mathrm{R_{scaled}} = \frac{\mathrm{R_i}}{\mathrm{C_i}} \tag{3}$$

where $\mathrm{R_{scaled}}$ is the scaled reward, R_i represents the individual reward term and C_i is the corresponding constant scaling factor.

Table 1 Comparison of reward function designs for DRL-based EMS in fuel cell hybrid vehicles

Reward function design	Objective(s)	Advantages	Limitations	References
Energy efficiency-oriented	Minimize energy consumption while ensuring power balance	Reduces hydrogen consumption, keeps battery SOC within acceptable range	May neglect drivability and component degradation	Sayah et al. [60], Dezhou et al. [61], Mauro et al. [62], Qi et al. [63], Shuai et al. [64], Sun et al. [65], Lian et al. [66], Xu et al. [67], Qi et al. [68], Fu et al. [69]
Component longevity-based	Reduce stress on components	Extends components lifespan, improves reliability	Can compromise short-term efficiency for long-term durability	Rostami et al. [28], Reddy et al. [35], Guo et al. [70], Tang et al. [71], Huang et al. [72], Huo et al. [73]
Drivability and stability	Ensure smooth and stable power distribution	Enhances driving comfort, avoids abrupt power fluctuations	May not prioritize energy efficiency optimally	Hsu et al. [47], Xu et al. [74], Zhuo et al. [75], Yazar et al. [76], Liu et al. [77], Jia et al. [78]
Multi-objective	Balance energy, longevity, and drivability	Provides a balanced trade-off between efficiency, durability, and comfort	Complexity in weight selection requires fine-tuning	Yang et al. [79], Wu et al. [80], Zhang et al. [81], Zhou et al. [82], Hu et al. [83], Li et al. [84]

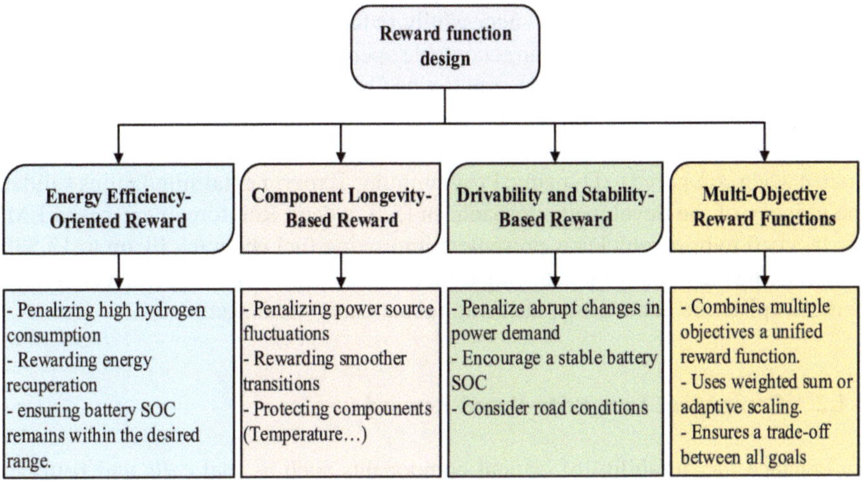

Fig. 5 Classification of reward function design methods for RL-based energy management in FCHEVs

Figure 5 illustrates the classification of reward function design methods for RL-based energy management in FCHEVs. It categorizes rewards into four main types: Energy Efficiency-Oriented, Component Longevity-Based, Drivability and Stability-Based, and Multi-Objective Reward Functions. Each category highlights key reward strategies, such as penalizing hydrogen consumption, ensuring battery longevity, maintaining stable power demand, and balancing multiple objectives.

4.1.1 Energy Efficiency-Oriented Reward

This type of reward function Focuses solely on minimizing hydrogen consumption while ensuring battery SOC remains within the desired range. This is the fundamental objective in most studies, ensuring optimal fuel economy without compromising battery performance, the reward function equation can be presented as (4):

$$\mathrm{R_{total}} = \mathrm{w_1 R_{fuel}} + \mathrm{w_2(SOC - SOC_{ref})^b} \tag{4}$$

where R_{total} is the overall reward, w_1 and w_2 are the weighing factors of the H_2 consumption minimization and battery SOC stability. R_{fuel} represents the penalty for fuel consumption and can be defined based on either total hydrogen consumption, as seen in [47], or instantaneous hydrogen consumption, as used in [48] or hydrogen consumption value cost [49]. SOC is the battery's state of charge and SOC_{ref} represents the control reference of SOC and b is an even number. In [50], an adaptive DRL-based energy management strategy for an FCHEV, which takes driving behavior into account, is proposed. Simulation results show that the approach employs a dual-objective reward function, successfully reducing hydrogen consumption while keeping the battery SOC within an acceptable operational bounds. In [51], a RL-based EMS for FCHEVs was developed, employing Q-Learning to manage the power allocation between fuel cells and lithium batteries. The strategy focuses on minimizing fuel consumption and ensuring battery sustainability, achieving near-optimal performance when compared to Dynamic Programming. Experimental simulations validate the efficacy of the developed approach. In [52], a deep Reinforcement-based EMS for fuel cell hybrid vehicles is presented, improving fuel economy by up to 13.53% over existing methods while reducing training time by 38%. The approach effectively handles epistemic uncertainty, ensuring robust energy management.

4.1.2 Component Longevity-Based Reward

To enhance the durability of critical components such as fuel cells and batteries, the reward function should extend the energy efficiency reward by incorporating component longevity considerations, penalize excessive power fluctuations to reduce fuel cell degradation and battery wear and encourage smoother power transitions to prevent rapid charging and discharging phases [53]. In [54], an energy management strategy utilizing the Soft Actor-Critic (SAC) algorithm, a form of deep reinforcement learning, is proposed to optimize fuel cell vehicle performance. The strategy considers both thermal characteristics and component durability, aiming to enhance fuel efficiency while extending the lifespan of powertrain components. Similarly, [55] present a DRL-based EMS that accounts for fuel cell aging. The reward function is designed to balance fuel consumption, SOC stability, and the remaining life of the fuel cell, thereby promoting component longevity.

4.1.3 Drivability and Stability-Based Reward

A smooth and responsive driving experience is essential for user satisfaction and vehicle performance. This category of reward function integrates energy efficiency with driving comfort and vehicle stability, penalizes abrupt power changes to prevent drivability issues and improve user experience and considers road conditions to optimize power management for rough terrains. In [56], a PPO-based energy management approach for FC hybrid electric buses operating in mountainous regions is proposed. A terrain-aware model, trained using real-world driving data, optimizes power distribution while considering road fluctuations. Experimental results show a 26.16% enhancement in fuel efficiency and a 44.98% convergence speed increase relative to conventional approaches, with further validation confirming the strategy's robustness. A Scalable Learning in Novel Environment (SLNE)-based energy management strategy (EMS) for fuel cell electric vehicles is presented in [57], aiming to optimize hydrogen consumption, fuel cell lifespan, and battery durability. By integrating a Dirichlet based memory library clustering approach with DQN, the proposed EMS adapts to novel and dynamic environments more effectively. The simulated outcomes reveal three key improvements: fuel consumption efficiency gains (5%), mitigated fuel cell degradation (4.5%), and superior battery durability compared to conventional DQN-based EMS.

4.1.4 Multi-Objective Reward Functions

Since EMS must balance multiple competing objectives, multi-objective reward functions combine various sub-goals into a single function. Approaches include weighted Sum Method consisting of assigning different weights to each objective Ensures an optimized trade-off between efficiency, longevity, and real-world driving performance. A RL-based EMS for a fuel cell hybrid bus is investigated in [58] improving energy sources durability along with increasing fuel economy. The Q-learning algorithm optimizes fuel consumption, maintains battery SOC, and reduces FC system power fluctuations through a 3-objectives reward function, validated through HIL testing. Results show a 7.8% improvement in energy economy. In [59], authors propose a deep reinforcement learning-based ecological driving scheme for FCHEV to optimize fuel consumption, car-following performance, and fuel cell power output stability. The reward function is formulated to achieve equilibrium among these competing objectives, and different weight coefficients are analyzed. Simulation results demonstrate the adaptability and effectiveness of the proposed method under various driving cycles. Table 1 provides a recap of the literature discussed in this section on reward function design methods, objectives, advantages and limitations.

4.2 Normalization Strategies

Normalization is essential in DRL-based EMS to ensure stable learning, improve convergence speed, and prevent dominance of certain reward components over others. Without proper normalization, large reward values can destabilize training, leading to inefficient energy management decisions. Various normalization techniques are used depending on the specific application and algorithm.

A review of the literature reveals that most studies on RL-based EMS for FCHEVs do not explicitly mention the normalization strategies applied in their frameworks. Normalization is a crucial step in stabilizing RL training, yet its role is often overlooked or implicitly integrated into preprocessing steps. Among the reported works that do specify a normalization method, min–max normalization is the most commonly used due to its simplicity and effectiveness in scaling input features within a fixed interval, usually between 0 and 1 or -1 and 1, ensuring stable learning dynamics.

Figure 6 presents mainly normalization strategies used in RL-based EMS for FCHEVs. Techniques such as min–max scaling, tanh scaling, mean–variance and batch normalization are commonly applied to enhance the learning process, ensuring

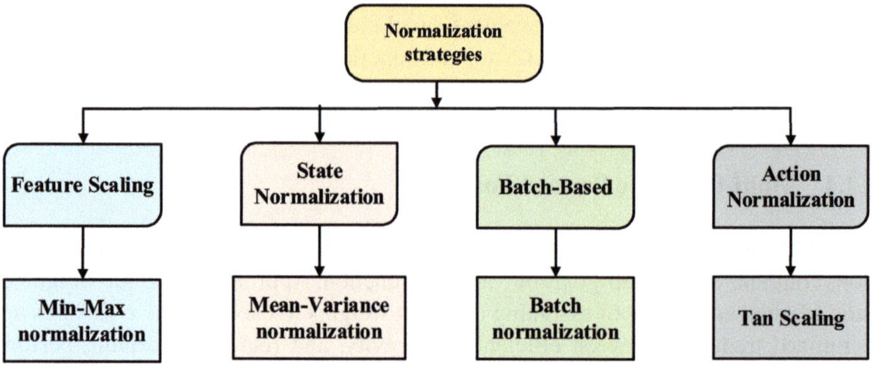

Fig. 6 Classification of normalization strategies for RL-based energy management in FCHEVs

that the RL model can efficiently adapt to the dynamic and complex environment of FCHEVs, ultimately leading to better performance and energy optimization.

4.2.1 Min–Max Normalization

This strategy scales input features within a specific limits, commonly [0, 1] or [−1, 1]. This method is effective for normalizing variables such as SOC, vehicle speed, and power demand, ensuring that all state variables contribute proportionally to the decision-making process. The normalization is defined as (5):

$$X_{\mathrm{norm}} = \frac{(\mathrm{x} - \mathrm{x}_{\mathrm{min}})}{(\mathrm{x}_{\mathrm{max}} - \mathrm{x}_{\mathrm{min}})} \tag{5}$$

where X_{norm} represents the normalized value of the variable x, scaled to a specific range while x_{min} and x_{max} are the minimum and maximum values of x.

4.2.2 Mean–Variance Normalization (MVN)

This method standardizes input data by removing the Mean and scaling it to unit Variance. This technique is especially valuable when input variables have non-uniform measurement scales and magnitudes, as it prevents any single variable from dominating the learning process. The equation is expressed as (6):

$$X_{\mathrm{norm}} = \frac{x - \mu}{\sigma} \tag{6}$$

where X_{norm} represents the normalized value of the variable x, μ is its mean and σ is its standard deviation.

4.2.3 Batch Normalization

This normalization method is integrated into neural networks within DRL algorithms such as DDPG and PPO. By normalizing each mini batch of data, this technique reduces internal covariate shifts and accelerates convergence. The normalization is performed as follows (7):

$$X_{\text{norm}} = \frac{x - \text{E}[x]}{\sqrt{Var[x] + \varepsilon}} \quad (7)$$

where $E[x]$ and $Var[x]$ denote the input batch mean and variance, and ε is a negligible constant added to ensure denominator stability in division operations.

4.2.4 Tanh Normalization

Tanh Scaling is a normalization technique used in RL to constrain values within a bounded range, typically between −1 and 1. It is commonly applied to action outputs in continuous control problems to ensure smooth transitions and prevent abrupt changes. The tanh function is defined as (8):

$$X_{\text{norm}} = \tanh(\text{x}) \quad (8)$$

where x is the original value and X_{norm} is the scaled output.

Table 2 provides a summary of the literature discussed in this section on normalization methods.

Table 2 Overview of normalization methods for DRL-based EMS in fuel cell hybrid vehicles

Category	Normalization method	Description	References
Feature scaling	Min–Max normalization	Normalizes data to a predetermined interval (e.g., [0, 1] or [−1, 1])	Zhang et al. [54], Fu et al. [69], Huang et al. [85], Bilgin et al. [86], Zhang et al. [87]
State normalization	Mean–Variance normalization	Normalizes state inputs using their mean and variance	Hou et al. [88], Peng et al. [89], Li et al. [90], Du et al. [91]
Batch-based	Batch normalization	Normalizes the input data within a batch to stabilize training	Hou et al. [73], Fayyazi et al. [92], Sun et al. [93], Lü et al. [94]
Action normalization	Tan scaling	Ensures action values remain within valid bounds	Li et al. [45], Guo et al. [51], Liu et al. [95], Yang et al. [96]

5 Testing Frameworks

Validating the performance of the EMS requires a structured testing framework that ensures reliability, robustness, and real-world applicability [97]. Testing frameworks could be categorized into various levels, ranging from purely simulation-based environments to real-world implementations. These frameworks play a key role in connecting theoretical innovations with practical implementation. On the other hand, real-world testing involves implementing EMS in actual operational settings, offering insights into system performance under practical conditions, which is crucial for assessing the EMS's adaptability and effectiveness in dynamic environments [98]. A comprehensive validation approach often integrates both simulation and real-world testing facilitating the transition from theoretical progress to real-world application, ensuring that EMS are both effective in controlled settings and resilient in real-world applications [99–101].

Figure 7 presents a classification of testing frameworks for energy management strategies in FCHEVs. It categorizes testing into four main types: Hardware-in-the-Loop, Simulation-Based Testing, Real-World Testing, and Digital Twin. Each category describes its testing level and purpose, such as validating EMS with hardware interaction (HIL), evaluating algorithms in a simulated environment (SIL and MIL), conducting physical performance assessments (bench and on-road testing), and integrating virtual-physical synchronization for real-time monitoring (Digital Twin).

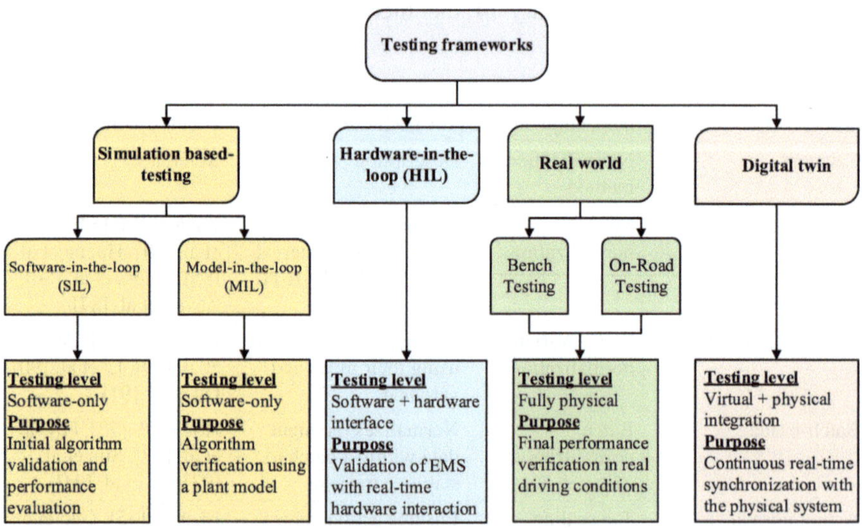

Fig. 7 Overview of testing frameworks for FCHEVs

5.1 Simulation-Based Testing

Simulation environments provide a controlled and cost-effective platform to evaluate DRL-based EMS before real-world implementation.

5.1.1 Software-in-the-Loop (SIL)

Software-in-the-Loop testing is crucial during the development stage of the EMS for hybrid and electric vehicles, offering numerous advantages such as cost reduction, enhanced safety, and accelerated development cycles. By utilizing simulation tools, researchers can model hybrid powertrain components and optimize EMS parameters before real-world implementation, leading to significant improvements in fuel economy across various driving cycles [102]. Additionally, simulation environments simplify adoption of cutting-edge algorithms, particularly reinforcement learning, by enabling efficient training and validation of adaptive control strategies without relying on physical prototypes [103]. Furthermore, virtual test platforms allow for the optimization of EMS using artificial neural networks, incorporating real-world data to refine energy distribution strategies in a controlled setting [104]. Together, these investigations emphasize the significance of simulation testing. In ensuring the cost-effective development, safety, and efficiency of EMS for hybrid and electric vehicles.

5.1.2 Model-in-the-Loop

In MIL, the control strategy interacts with a high-fidelity vehicle model that simulates real-world behavior. This method allows early validation of energy management algorithms before integrating them into hardware. It provides a more realistic validation than SIL, considering vehicle dynamics and energy flows but still limited to virtual conditions, lacking real sensor-actuator interactions. Additionally, MIL offers an efficient approach for validating mathematical models and functional logic, particularly during the initial phases of the development process [105]. In [106], an extended-range electric vehicle is evaluated using model-in-the-loop simulations to validate a power loss model. A proposed adaptive real-time dynamic programming (ARTDP) algorithm enhances energy consumption efficiency by up to 9.8% compared to conventional methods, providing adaptability and near real-time optimization for energy management. In [107], researchers developed energy management controllers (EMCs) for EVs, including intelligent, hybrid, PID and supervisory strategies, using MATLAB environment and MIL simulations. The supervisory controller outperforms others, achieving 68.4% SOC retention and 33.4 Wh/km energy consumption, significantly improving battery efficiency and energy savings.

5.2 Hardware-in-the-Loop

HIL testing introduces real hardware components into the loop, allowing evaluation of the energy management strategies under real-time conditions. The vehicle's powertrain, including the fuel cell, battery, and supercapacitor, can be partially or fully integrated into the test setup. High-performance real-time simulators (e.g., dSPACE, OPAL-RT) are used to replicate vehicle behavior while interacting with the EMS controller. This method allows testing under realistic conditions without requiring a full prototype vehicle but requires specialized hardware and is more expensive than pure simulation. By employing a Model-Based Design approach [108, 109], the software and hardware configurations of the Vehicle Control Unit can be effectively tested through HIL simulation. As outlined in [110], HIL architectures vary in complexity, depending on the extent to which vehicle subsystems are represented by real hardware versus real-time numerical simulations. A hardware testing platform of a vehicle model was developed in [111], and the impacts of various parameters on efficiency and driving range were investigated. In [112], a rule-based EMS for a plug-in hybrid electric bus is developed and tested using a hardware-in-the-loop setup with CAN bus monitoring, evaluating fuel consumption, battery SOC, and overall bus performance.

5.3 Real-World

Final validation requires on-road testing under real driving conditions. The DRL-based EMS is deployed in a prototype or commercial FCHEV to assess its real-time adaptability, efficiency, and robustness. Data collected from real-world tests helps refine the model for better generalization. This strategy provides the most accurate validation of system performance, but it is expensive, time-consuming, and requires compliance with safety regulations. Real-world EMS testing for hybrid vehicles involves a combination of laboratory setups and on-road testing to validate the performance, efficiency, and reliability of the control strategies. Laboratory testing provides a controlled environment to evaluate EMS performance under various conditions without the risks associated with on-road testing. It includes Bench testing where individual components (battery, motor, inverter) are tested on dedicated test benches to evaluate their performance and integration with the EMS [113]. In [114], a combination of wavelet transforms, neural networks, and fuzzy logic is used to optimize power distribution between a battery and a supercapacitor. The strategy is validated on a real-time hardware platform with a 72 V battery and a 96 V supercapacitor, demonstrating improved system efficiency and battery lifetime. In [115], stochastic dynamic programming energy management controller is implemented along with the shortest path approach in a prototype vehicle, focusing on real-time hardware execution while addressing challenges such as algorithmic computability, signal synchronization and operator perception.

5.4 *Digital Twin*

EMS investigations now incorporate digital twin methodologies. Essentially, a digital twin is a computer-based replica designed to mirror the functionality and attributes of an actual object or process [116]. By integrating data extraction from sensors and Internet of Things-based (IoT) devices in real-time, digital twins facilitate the emulation, evaluation, and enhancement of their corresponding physical systems [117]. The digital twin concept gained prominence through a NASA program, where it was utilized as a comprehensive digital mapping of a physical entity in the virtual domain [118]. As evidenced in [119], digital twin methodologies enable accurate performance simulation of FCHEVs, generating essential findings about energy efficiency correlations with control approaches to inform superior electric vehicle developments. In [118], the advantages of electrified powertrains were emphasized, such as higher energy efficiency, lower emissions and driving automotive electrification. The study proposes a digital twin-enhanced RL-based EMS for battery-ultracapacitor vehicles, improving real-time control by linking virtual models with physical systems. Results show a 7.08% boost in energy efficiency and a 25.28% reduction in battery degradation, highlighting its effectiveness in optimizing performance and battery lifespan.

Table 3 provides a summary of the literature discussed in this section on testing frameworks for FCHEVs.

6 Conclusion

This chapter provided a comprehensive analysis of reinforcement learning-based energy management approaches for fuel cell hybrid electric vehicles, with a particular focus on reward function design, RL strategies, normalization techniques, and testing frameworks. Through an extensive literature analysis, it was observed that while fuel efficiency remains the primary objective in reward function design, additional factors such as battery SOC management, component longevity, drivability, and emissions reduction play crucial roles in optimizing vehicle performance. The classification of reward designs showed that most studies employ multi-objective functions, integrating trade-offs between different operational goals.

Different RL strategies, including Q-Learning, Deep Q-learning, Deep Deterministic Policy Gradient, Proximal Policy Optimization, and Temporal Difference Learning, have been widely explored in EMS development. While model-free approaches demonstrate strong adaptability, challenges such as sample inefficiency, convergence speed, and real-time applicability remain key limitations. To enhance the learning process, various normalization strategies have been proposed to stabilize rewards, improve convergence, and ensure more efficient policy learning.

Testing frameworks play a crucial role in evaluating EMS performance. Simulation-based methods provide a fast and cost-effective way to validate control

Table 3 Comparison of testing frameworks for fuel cell hybrid electric vehicles

Framework	Advantages	Limitations	Software	References
Software-in-the-loop	Low cost, fast iterations, flexible environment	Limited real-world application	MATLAB, Simulink, CarSim, ADVISOR	Reddy et al. [35], Li et al. [45], Du et al. [120], Liu et al. [121], Liu et al. [122], Li et al. [123]
Model-in-the-loop	Fast, reproducible, easy debugging	Less realistic than hardware testing	MATLAB, Simulink, OPAL-RT	Huang et al. [105], Kalia and Fabien [106], Saiteja et al. [107], Zhou et al. [124], Abdelaal et al. [125], Sadekar et al. [126]
Hardware-in-the-loop	High realism, safe testing, real-time capability	Complex setup, higher cost	dSPACE, NI PXI, OPAL-RT, Labview	Yi et al. [112], He et al. [127], Mocera [128], Pereira et al. [129], Mayyas et al. [130], Zhang et al. [131, 132]
Digital twin	High accuracy, predictive capabilities, real-time monitoring	Requires accurate models and infrastructure	ANSYS Twin Builder, Siemens Simcenter	Ye et al. [116], Bartolucci et al. [119], Li et al. [133], Bartolucci et al. [134], Zhang et al. [135], Hadraoui et al. [136]
Real-world testing	Most accurate results, captures unexpected factors	Expensive, time-consuming, safety risks	Real vehicle	Sun et al. [65], Tao et al. [113], Zhang et al. [114], Zhang et al. [137], Kermani et al. [138], Khalatbarisoltani et al. [139], Wang et al. [140]

policies, while hardware-in-the-loop and real-world testing are essential for bridging the gap between theoretical developments and practical implementation. Digital twin technology is emerging as a promising approach for real-time system validation and adaptive learning in dynamic environments.

Despite the significant advancements in RL-EMS for FCHEVs, several research gaps remain open for further exploration, future studies should focus on dynamic reward function tuning that adapts to different driving conditions and vehicle aging effects. Additionally, the computational complexity of deep RL-based EMSs needs to be further optimized for real-time applications in embedded vehicle controllers.

In summary, RL-based EMS for FCHEVs holds great potential for achieving intelligent, adaptive, and highly efficient energy management. However, addressing the existing challenges in reward design, real-time implementation, and validation will be crucial for the successful deployment of these strategies in next-generation sustainable transportation systems.

References

1. El Harouri, K., El Hani, S., Naseri, N., Elbouchikhi, E., Benbouzid, M., & Skander-Mustapha, S. (2023). Hybrid control and energy management of a residential system integrating vehicle-to-home technology. *Designs, 7*(2), 52.
2. Said-Romdhane, M. B., Skander-Mustapha, S., & Belhassen, R. (2024). Streamlined tuning of H∞ control for electric vehicle motors. In *2024 14th International Conference on Electrical Engineering (ICEENG)* (pp. 367–372). IEEE.
3. Mahjoubi, C., Olivier, J. C., Skander-Mustapha, S., Machmoum, M., & Slama-Belkhodja, I. (2019). An improved thermal control of open cathode proton exchange membrane fuel cell. *International Journal of Hydrogen Energy, 44*(22), 11332–11345.
4. Ben Saïd-Romdhane, M., & Skander-Mustapha, S. (2025). A critical study on the environmental impacts of various types of green vehicles. *Euro-Mediterranean Journal for Environmental Integration (EMJE).*
5. Saïd-Romdhane, M. B., Skander-Mustapha, S., & Slama-Belkhodja, I. (2023). Optimizing solar-powered electric vehicle management with distributed decision making under shading conditions. In *2023 International Conference on Engineering and Emerging Technologies (ICEET)* (pp. 1–6). IEEE.
6. Said-Romdhane, M. B., Skander-Mustapha, S., & Belhassen, R. (2023). Adaptive deadbeat predictive control for PMSM-based solar-powered electric vehicles with enhanced stator resistance compensation. *Science and Technology for Energy Transition, 78*, 35.
7. Said-Romdhane, M. B., Skander-Mustapha, S., & Belhassen, R. (2022). Adaptive deadbeat predictive control for PMSM based electric vehicle. In *2022 IEEE International Conference on Electrical Sciences and Technologies in Maghreb (CISTEM)* (vol. 4, pp. 1–6). IEEE.
8. Zandi, M., Payman, A., Martin, J. P., et al. (2011). Energy management of a fuel cell/supercapacitor/battery power source for electric vehicular applications. *IEEE Transactions on Vehicular Technology, 60*(2), 433–443.
9. Song, Z., Hofmann, H., Li, J., et al. (2014). Energy management strategies comparison for electric vehicles with hybrid energy storage system. *Applied Energy, 134*, 321–331.
10. Mahjoubi, C., Olivier, J. C., Skander-Mustapha, S., Slama-Belkhodja, I., & Machmoum, M. (2018). An improved open cathode proton exchange membrane fuel cell control ensuring a simultaneous regulation of temperature and air flow. In *2018 9th International Renewable Energy Congress (IREC)* (pp. 1–6). IEEE.
11. Xu, L., Ouyang, M., Li, J., et al. (2013). Application of pontryagin's minimal principle to the energy management strategy of plugin fuel cell electric vehicles. *International Journal of Hydrogen Energy, 38*(24), 10104–10115.
12. Xu, L., Yang, F., Li, J., et al. (2012). Real time optimal energy management strategy targeting at minimizing daily operation cost for a plug-in fuel cell city bus. *International Journal of Hydrogen Energy, 37*(20), 15380–15392.
13. Sayah, A., Ben Saïd-Romdhane, M., & Skander-Mustapha, S. (2024). Advanced energy management system with road gradient consideration for fuel cell hybrid electric vehicles. *Results in Engineering, 23.*
14. Dellaly, M., Skander-Mustapha, S., & Slama-Belkhodja, I. (2023). Optimization of a residential community's curtailed PV power to meet distribution grid load profile requirements. *Renewable Energy, 218*, Article 119342.
15. Li, L., Yang, C., Zhang, Y., et al. (2015). Correctional DP-based energy management strategy of plug-in hybrid electric bus for city-bus route. *IEEE Transactions on Vehicular Technology, 64*(7), 2792–2803.

16. Li, G., & Gorges, D. (2018). Ecological adaptive cruise control and energy management strategy for hybrid electric vehicles based on heuristic dynamic programming. *IEEE Transactions on Intelligent Transportation Systems, 20*(9), 3526–3535.
17. Sun, C., Sun, F., & He, H. (2017). Investigating adaptive-ECMS with velocity forecast ability for hybrid electric vehicles. *Applied Energy, 185*, 1644–1653.
18. Huang, Y., Wang, H., Khajepour, A., et al. (2017). Model predictive control powermanagement strategies for HEVs: A review. *Journal of Power Sources, 341*, 91–106.
19. Gielniak, M. J., & Shen, Z. J. (2004). Power management strategy based on game theory for fuel cell hybrid electric vehicles. In *IEEE 60th Vehicular Technology Conference* (Vol. 6, no. 6, pp. 4422–4426).
20. Song, K., Li, F., Hu, X., et al. (2018). Multi-mode energy management strategy for fuel cell electric vehicles based on driving pattern identification using learning vector quantization neural network algorithm. *Journal of Power Sources, 389*, 230–239.
21. Sayah, A., Saïd-Romdhane, M. B., & Skander-Mustapha, S. (2024). Performance evaluation of neural network models for fuel cell power prediction in hybrid electric vehicles. In *2024 IEEE International Conference on Artificial Intelligence & Green Energy (ICAIGE)*
22. Saïd-Romdhane, M. B., Boussaada, Z., Curea, O., & Skander-Mustapha, S. (2023). Enhanced NARX neural network model for specific time-series prediction. Case study: battery state of charge in electric vehicles. In *2023 14th International Renewable Energy Congress (IREC)* (pp. 1–6). IEEE.
23. Zheng, C., Zhang, D., Xiao, Y., & Li, W. (2022). Reinforcement learning-based energy management strategies of fuel cell hybrid vehicles with multi-objective control. *Journal of Power Sources*. https://doi.org/10.1016/j.jpowsour.2022.231841.
24. Dellaly, M., Skander-Mustapha, S., & Slama-Belkhodja, I. (2022). Data-driven emulation of a peer-to-peer energy exchanged for interconnected residential nanogrids. In *2022 IEEE International Conference on Electrical Sciences and Technologies in Maghreb (CISTEM)* (Vol. 4, pp. 1–7). IEEE.
25. Takriti, M., Skander-Mustapha, S., Boussaada, Z., & Mrabet Bellaj, N. (2025). Data-driven optimization for efficient integration of photovoltaic agents in residential microgrid systems. *Euro-Mediterranean Journal for Environmental Integration*. https://doi.org/10.1007/s41207-025-00808-z.
26. Said-Romdhane, M. B., Skander-Mustapha, S., & Slama-Belkhodja, I. (2021). Analysis study of city obstacles shading impact on solar PV vehicle. In *2021 4th International Symposium on Advanced Electrical and Communication Technologies (ISAECT)* (pp. 01–06). IEEE.
27. Ben Saïd-Romdhane, M., Hadded, M., & Slama-Belkhodja, I. (2025). Innovative adaptive virtual impedance for resonance frequency mitigation in grid-connected converters. *Electric Power SystemsResearch, 239*.
28. Rostami, S. M. R., Al-Shibaany, Z., Kay, P., et al. (2024). Deep reinforcement learning and fuzzy logic controller codesign for energy management of hydrogen fuel cell powered electric vehicles. *Science and Reports, 14*, 30917. https://doi.org/10.1038/s41598-024-81769-1
29. Petersen, L., Iov, F., & Tarnowski, G. C. (2019). A model-based design approach for stability assessment, control tuning and verification in off-grid hybrid power plants. *Energies*.
30. Said-Romdhane, M. B., & Skander-Mustapha, S. (2024). Optimizing solar vehicle performance in urban shading conditions with enhanced control strategies. *Ain Shams Engineering Journal, 15*(10), Article 102985.
31. Alexander, V., Georg, L., & Lehfuss, F. (2011). Stabilization of power hardware-in-the-loop simulations of electric energy systems. *Simulation Modelling Practice and Theory, 19*(7), 1699–1708.
32. Ehsan, T. (2012). *Modeling, optimization and hardware-in-loop simulation of hybrid electric vehicles*. University of Manitoba.
33. Wang, L., Zhang, Y., Yin, C., Zhang, H., & Wang, C. (2012). Hardware-in-the-loop simulation for the design and verification of the control system of a series–parallel hybrid electric city-bus. *Simulation Modelling Practice and Theory, 25*, 148–162.

34. Liu, Y., Liu, J., Zhang, Y., Wu, Y., Chen, Z., & Ye, M. (2020). Rule learning based 630 energy management strategy of fuel cell hybrid vehicles considering multi-objective optimization. *Energy, 207*, Art. no. 118212.
35. Reddy, N. P., Pasdeloup, D., Zadeh, M. K., & Skjetne, R. (2019). An intelligent power and energy management system for fuel cell/battery hybrid electric vehicle using reinforcement learning. In *2019 IEEE Transportation Electrification Conference and Expo (ITEC)* (pp. 1–6). https://doi.org/10.1109/ITEC.2019.8790451.
36. Hsu, R. C., Chen, S., Chen, W., & Liu, C. (2016). A reinforcement learning based dynamic power management for fuel cell hybrid electric vehicle. In *2016 Joint 8th International Conference on Soft Computing and Intelligent Systems (SCIS) and 17th International Symposium on Advanced Intelligent Systems (ISIS)* (pp. 460–464). https://doi.org/10.1109/SCIS-ISIS.2016.0104.
37. Wu, J., He, H., Peng, J., Li, Y., & Li, Z. (2018). Continuous reinforcement learning of energy management with deep Q network for a power split hybrid electric bus. *Applied Energy, 222*, 799–811. https://doi.org/10.1016/j.apenergy.2018.03.104
38. Qi, X., Luo, Y., Wu, G., Boriboonsomsin, K., & Barth, M. (2019). Deep reinforcement learning enabled self-learning control for energy efficient driving. *Transportation Research Part C: Emerging Technologies, 99*, 67–81. https://doi.org/10.1016/j.trc.2018.12.018
39. Wu, Y., Tan, H., Peng, J., Zhang, H., & He, H. (2019). Deep reinforcement learning of energy management with continuous control strategy and traffic information for a series-parallel plug-in hybrid electric bus. *Applied Energy, 247*, 454–466.
40. Peng, J., Fan, Y., Yin, G., & Jiang, R. (2022). Collaborative optimization of energy management strategy and adaptive cruise control based on deep reinforcement learning. *IEEE Transactions on Transportation Electrification, 9*(1), 34–44. https://doi.org/10.1109/TTE.2022.3177572
41. Huang, R., He, H., Zhao, X., et al. (2023). Longevity-aware energy management for fuel cell hybrid electric bus based on a novel proximal policy optimization deep reinforcement learning framework. *Journal of Power Sources, 561*, Article 232717.
42. Li, B., Cui, Y., Xiao, Y., Fu, S., Choi, J., & Zheng, C. (2025). An improved energy management strategy of fuel cell hybrid vehicles based on proximal policy optimization algorithm. *Energy, 317*. https://doi.org/10.1016/j.energy.2025.134585.
43. Zhou, J., Xue, S., Xue, Y., Liao, Y., Liu, J., & Zhao, W. (2021). A novel energy management strategy of hybrid electric vehicle via an improved TD3 deep reinforcement learning. *Energy, 224*. https://doi.org/10.1016/j.energy.2021.120118.
44. Wang, H., Ye, Y., Zhang, J., & Xu, B. (2022). A comparative study of 13 deep reinforcement learning based energy management methods for a hybrid electric vehicle. *Energy, 266*(July), 2023. https://doi.org/10.1016/j.energy.2022.126497
45. Li, Q., Meng, X., Gao, F., Zhang, G., Chen, W., & Rajashekara, K. (2023). Reinforcement learning energy management for fuel cell hybrid systems: A review. *IEEE Industrial Electronics Magazine, 17*(4), 45–54. https://doi.org/10.1109/MIE.2022.3148568
46. Li, Q. et al. (2021). Online extremum seeking-based optimized energy management strategy for hybrid electric tram considering fuel cell degradation. *Applied Energy, 285*(2), 116, 505. https://doi.org/10.1016/j.apenergy.2021.116505.
47. Hsu, R. C., Chen, S.-M., Chen, W.-Y., & Liu, C.-T. (2016). A reinforcement learning based dynamic power management for fuel cell hybrid electric vehicle. In *2016 Joint 8th International Conference on Soft Computing and Intelligent Systems (SCIS) and 17th International Symposium on Advanced Intelligent Systems (ISIS)*, Sapporo, Japan (pp. 460–464). https://doi.org/10.1109/SCIS-ISIS.2016.0104.
48. Zheng, C., Li, W., Xiao, Y., Zhang, D., & Cha, S. W. (2021). A deep deterministic policy gradient-based energy management strategy for fuel cell hybrid vehicles. In *2021 IEEE Vehicle Power and Propulsion Conference (VPPC)*, Gijon, Spain (pp. 1–6). https://doi.org/10.1109/VPPC53923.2021.9699156.
49. Lu, Y. (2024). Research on vehicle energy management based on deep reinforcement learning algorithm under typical working conditions. In *2024 IEEE 7th Information Technology,*

Networking, Electronic and Automation Control Conference (ITNEC), Chongqing, China (pp. 915–920). https://doi.org/10.1109/ITNEC60942.2024.10733109.
50. Sun, H., Tao, F., Fu, Z., Gao, A., & Jiao, L. (2023). Driving-behavior-aware optimal energy management strategy for multi-source fuel cell hybrid electric vehicles based on adaptive soft deep-reinforcement learning. *IEEE Transactions on Intelligent Transportation Systems, 24*(4), 4127–4146. https://doi.org/10.1109/TITS.2022.3233564
51. Guo, L., Li, Z., & Outbib, R. (2021). Reinforcement learning based energy management for fuel cell hybrid electric vehicles. In *IECON 2021—47th Annual Conference of the IEEE Industrial Electronics Society*, Toronto, ON, Canada (pp. 1–6). https://doi.org/10.1109/IECON48115.2021.9589725.
52. Jouda, B., Al-Mahasneh, A. J., & Mallouh, M. A. (2024). Deep stochastic reinforcement learning-based energy management strategy for fuel cell hybrid electric vehicles. *Energy Conversion and Management, 301*. https://doi.org/10.1016/j.enconman.2023.117973.
53. Skander-Mustapha, S., & Slama-Belkhodja, I. (2020). Energy management of rooftop PV system including battery storage: Case study of ENIT building. In *2020 International Conference on Electrical and Information Technologies (ICEIT)* (pp. 1–6). IEEE.
54. Zhang, Y., Zhang, C., Fan, R., Deng, C., Wan, S., & Chaoui, H. (2023). Energy management strategy for fuel cell vehicles via soft actor-critic-based deep reinforcement learning considering powertrain thermal and durability characteristics. *Energy Conversion and Management*. https://doi.org/10.1016/j.enconman.2023.116921.
55. Shuai, Q., Hua, Q., Ma, J., Wang, Y., & Jiang, Z. (2025). Deep reinforcement learning of energy management strategy for aging fuel cell power systems. In *Proceedings of SPIE 13513, The International Conference Optoelectronic Information and Optical Engineering (OIOE2024)*, 1351323. https://doi.org/10.1117/12.3045626.
56. Tang, T., Peng, Q., Shi, Q., Peng, Q., Zhao, J., Chen, C., & Wang, G. (2024). Energy management of fuel cell hybrid electric bus in mountainous regions: A deep reinforcement learning approach considering terrain characteristics. *Energy, 311*. https://doi.org/10.1016/j.energy.2024.133313.
57. Wang, D., Mei, L., Xiao, F., Song, C., Qi, C., & Song, S. (2024). Energy management strategy for fuel cell electric vehicles based on scalable reinforcement learning in novel environment. *International Journal of Hydrogen Energy, 59*. https://doi.org/10.1016/j.ijhydene.2024.01.335.
58. Deng, P., Wu, X., Yang, J., Yang, G., Jiang, P., Yang, J., & Bian, X. (2024). Optimal online energy management strategy of a fuel cell hybrid bus via reinforcement learning. *Energy Conversion and Management*. https://doi.org/10.1016/j.enconman.2023.117921.
59. Chen, W., Yin, G., Fan, Y., Zhuang, W., Zhang, H., & Peng, J. (2022). Ecological driving strategy for fuel cell hybrid electric vehicle based on continuous deep reinforcement learning. In *2022 6th CAA International Conference on Vehicular Control and Intelligence (CVCI)*, Nanjing, China (pp. 1–6). https://doi.org/10.1109/CVCI56766.2022.9964786.
60. Sayah, A., Saïd-Romdhane, M. B., & Skander-Mustapha, S. (2025). Deep reinforcement learning-based energy management system for fuel cell hybrid electric vehicles under rough pavement constraints. In *2025 15th International Renewable Energy Congress (IREC)*, Hammamet, Tunisia (pp. 1–6). https://doi.org/10.1109/IREC64614.2025.10926823.
61. Xu, D., Cui, Y., Ye, J., Cha, S. W., Li, A., & Zheng, C. (2022). A soft actor-critic-based energy management strategy for electric vehicles with hybrid energy storage systems. *Journal of Power Sources*. https://doi.org/10.1016/j.jpowsour.2022.231099.
62. Carignano, M. G., Costa-Castelló, R., Roda, V., Nigro, N. M., Junco, S., & Feroldi, D. (2017). Energy management strategy for fuel cell-supercapacitor hybrid vehicles based on prediction of energy demand. *Journal of Power Sources, 360*. https://doi.org/10.1016/j.jpowsour.2017.06.016.
63. Qi, X., Luo, Y., Wu, G., Boriboonsomsin, K., & Barth, M. J. (2017). Deep reinforcement learning-based vehicle energy efficiency autonomous learning system. In *2017 IEEE Intelligent Vehicles Symposium (IV)*, Los Angeles, CA, USA (pp. 1228–1233). https://doi.org/10.1109/IVS.2017.7995880.

64. Shuai, B., Hua, M., Li, Y., Shuai, S., Xu, H., & Zhou, Q. Optimal energy management of plug-in hybrid electric vehicles through ensemble reinforcement learning with exploration-to-exploitation ratio control. *IEEE Transactions on Intelligent Vehicles*. https://doi.org/10.1109/TIV.2024.3377215.
65. Sun, H., Fu, Z., Tao, F., Zhu, L., & Si, P. (2020). Data-driven reinforcement-learning-based hierarchical energy management strategy for fuel cell/battery/ultracapacitor hybrid electric vehicles. *Journal of Power Sources, 455*, Article 227964. https://doi.org/10.1016/j.jpowsour.2020.227964
66. Lian, R., Peng, J., Wu, Y., Tan, H., & Zhang, H. (2020). Rule-interposing deep reinforcement learning based energy management strategy for power-split hybrid electric vehicle. *Energy*. https://doi.org/10.1016/j.energy.2020.117297.
67. Xu, B., Rathod, D., Zhang, D., Yebi, A., Zhang, X., Li, X., & Filipi, Z. (2020). Parametric study on reinforcement learning optimized energy management strategy for a hybrid electric vehicle. *Applied Energy, 259*. https://doi.org/10.1016/j.apenergy.2019.114200.
68. Qi, C., Zhu, Y., Song, C., Yan, G., Xiao, F., Wang, D., Zhang, X., Cao, J., & Song, S. (2022). Hierarchical reinforcement learning based energy management strategy for hybrid electric vehicle. *Energy, 238*(Part A). https://doi.org/10.1016/j.energy.2021.121703.
69. Fu, Z., Wang, H., Tao, F., Ji, B., Dong, Y., & Song, S. (2022). Energy management strategy for fuel cell/battery/ultracapacitor hybrid electric vehicles using deep reinforcement learning with action trimming. *IEEE Transactions on Vehicular Technology, 71*(7), 7171–7185. https://doi.org/10.1109/TVT.2022.3168870
70. Guo, L., Li, Z., & Outbib, R. (2022). A lifetime extended energy management strategy for fuel cell hybrid electric vehicles via self-learning fuzzy reinforcement learning. In *2022 10th International Conference on Systems and Control (ICSC)*, Marseille, France (pp. 161–167). https://doi.org/10.1109/ICSC57768.2022.9993916.
71. Tang, X., Zhou, H., Wang, F., Wang, W., & Lin, X. (2022). Longevity-conscious energy management strategy of fuel cell hybrid electric vehicle based on deep reinforcement learning. *Energy*. https://doi.org/10.1016/j.energy.2021.121593.
72. Huang, R., He, H., Zhao, X., & Gao, M. (2023). Longevity-aware energy management for fuel cell hybrid electric bus based on a novel proximal policy optimization deep reinforcement learning framework. *Journal of Power Sources, 561*. https://doi.org/10.1016/j.jpowsour.2023.232717.
73. Huo, W., Chen, D., Tian, S., Li, J., Zhao, T., & Liu, B. (2022). Lifespan-consciousness and minimum-consumption coupled energy management strategy for fuel cell hybrid vehicles via deep reinforcement learning. *International Journal of Hydrogen Energy*. https://doi.org/10.1016/j.ijhydene.2022.05.194.
74. Kang, X., Wang, Y., & Chen, Z. (2024). A reinforcement learning energy management strategy for fuel cell hybrid electric vehicles considering driving condition classification. *Sustainable Energy Grids and Networks*. https://doi.org/10.1016/j.segan.2024.101298.
75. Zhou, J., Guo, A., Wang, J., Wang, C., & Zhao, W. (2024). Energy management strategy of fuel cell vehicles with hybrid energy sources: A novel framework via deep reinforcement learning and transfer learning. *Proceedings of the Institution of Mechanical Engineers, Part D: Journal of Automobile Engineering, 238*(14), 4659–4675. https://doi.org/10.1177/09544070231195402
76. Yazar, O., Coskun, S., & Zhang, F. (2024). Adaptive learning-based energy management for HEVs using soft actor-critic DRL algorithm. *Mechatronics Technology*. https://doi.org/10.55092/mt20240005.
77. Liu, T., Tan, W., Tang, X., Chen, J., & Cao, D. (2020). Adaptive energy management for real driving conditions via transfer reinforcement learning. https://doi.org/10.48550/arXiv.2007.12560.
78. Jia, C., Liu, W., He, H., & Chau, K. T. (2024). Deep reinforcement learning-based energy management strategy for fuel cell buses integrating future road information and cabin comfort control. *Energy Conversion and Management, 321*. https://doi.org/10.1016/j.enconman.2024.119032.

79. Yang, D., Wang, L., Yu, K., & Liang, J. (2022). A reinforcement learning-based energy management strategy for fuel cell hybrid vehicle considering real-time velocity prediction. *Energy Conversion and Management, 274*. https://doi.org/10.1016/j.enconman.2022.116453.
80. Wu, J. J., Song, D. F., Zhang, X. M., Duan, C. S., & Yang, D. P. (2023). Multi-objective reinforcement learning-based energy management for fuel cell vehicles considering lifecycle costs. *International Journal of Hydrogen Energy*. https://doi.org/10.1016/j.ijhydene.2023.06.145.
81. Zhang, B., Chang, L., Teng, T., Chen, Q., Li, Q., Cao, Y., Yang, S., & Zhang, X. (2024). Multi-objective optimization with Q-learning for cruise and power allocation control parameters of connected fuel cell hybrid vehicles. *Applied Energy, 373*. https://doi.org/10.1016/j.apenergy.2024.123910.
82. Zhou, Y., Ravey, A., & Péra, M.-C. (2020). Multi-objective energy management for fuel cell electric vehicles using online-learning enhanced Markov speed predictor. *Energy Conversion and Management*. https://doi.org/10.1016/j.enconman.2020.112821.
83. Hu, H., Yuan, W.-W., Su, M., & Ou, K. (2023). Optimizing fuel economy and durability of hybrid fuel cell electric vehicles using deep reinforcement learning-based energy management systems. *Energy Conversion and Management, 291*. https://doi.org/10.1016/j.enconman.2023.117288.
84. Li, X., He, H., & Wu, J. (2025). Knowledge-guided deep reinforcement learning for multiobjective energy management of fuel cell electric vehicles. *IEEE Transactions on Transportation Electrification, 11*(1), 2344–2355. https://doi.org/10.1109/TTE.2024.3421342
85. Huang, Y., Hu, H., Tan, J., Lu, C., & Xuan, D. (2023). Deep reinforcement learning based energy management strategy for range extend fuel cell hybrid electric vehicle. *Energy Conversion and Management, 277*. https://doi.org/10.1016/j.enconman.2023.116678.
86. Bilgin, T., İzgü, S., Yalçin, A. K., & Yilmaz, S. (2024). Energy management strategy of fuel cell hybrid electric vehicle with deep reinforcement learning. In *2024 11th International Conference on Electrical and Electronics Engineering (ICEEE)*, Marmaris, Turkiye (pp. 270–274). https://doi.org/10.1109/ICEEE62185.2024.10779320.
87. Zhang, D., Cui, Y., & Xiao, Y., et al. (2024). An improved soft actor-critic-based energy management strategy of fuel cell hybrid vehicles with a nonlinear fuel cell degradation model. *International Journal of Precision Engineering and Manufacturing-Green Technology, 11*, 183–202. https://doi.org/10.1007/s40684-023-00547-y.
88. Hou, S., Yin, H., Pla, B., Gao, J., & Chen, H. (2023). Real-time energy management strategy of a fuel cell electric vehicle with global optimal learning. *IEEE Transactions on Transportation Electrification, 9*(4), 5085–5097. https://doi.org/10.1109/TTE.2023.3238101
89. Peng, H., Li, J., Deng, K., & Hameyer, K. (2022). Machine learning-based control for fuel cell hybrid buses: From average load power prediction to energy management. *Vehicles, 4*, 1365–1390. https://doi.org/10.3390/vehicles4040072
90. Li, M., Wang, Y., Yu, P., Sun, Z., & Chen, Z. (2023). Online adaptive energy management strategy for fuel cell hybrid vehicles based on improved cluster and regression learner. *Energy Conversion and Management, 292*. https://doi.org/10.1016/j.enconman.2023.117388.
91. Du, G., Zou, Y., Zhang, X., Guo, L., & Guo, N. (2022). Energy management for a hybrid electric vehicle based on prioritized deep reinforcement learning framework. *Energy*. https://doi.org/10.1016/j.energy.2021.122523.
92. Fayyazi, M., Sardar, P., Thomas, S. I., Daghigh, R., Jamali, A., Esch, T., Kemper, H., Langari, R., & Khayyam, H. (2023). Artificial intelligence/machine learning in energy management systems, control, and optimization of hydrogen fuel cell vehicles. *Sustainability, 15*, 5249. https://doi.org/10.3390/su15065249
93. Sun, M., Zhao, P., & Lin, X. (2022). Power management in hybrid electric vehicles using deep recurrent reinforcement learning. *Electrical Engineering, 104*, 1459–1471. https://doi.org/10.1007/s00202-021-01401-7
94. Lü, X., Qian, S., Zhai, X. R., Wang, P., & Wu, T. (2024). Adaptive energy management strategy for FCHEV based on improved proximal policy optimization in deep reinforcement learning algorithm. *Energy Conversion and Management, 321*. https://doi.org/10.1016/j.enconman.2024.118977.

95. Liu, Z. E., Li, Y., Zhou, Q., Shuai, B., Hua, M., Xu, H., Xu, L., Tan, G., & Li, Y. (2025). Real-time energy management for HEV combining naturalistic driving data and deep reinforcement learning with high generalization. *Applied Energy, 377*(Part A). https://doi.org/10.1016/j.apenergy.2024.124350.
96. Yang, N., Ruan, S., Han, L., Liu, H., Guo, L., & Xiang, C. (2023). Reinforcement learning-based real-time intelligent energy management for hybrid electric vehicles in a model predictive control framework. *Energy, 270*. https://doi.org/10.1016/j.energy.2023.126971.
97. Peng, H., et al. (2021). Validation of robustness and fuel efficiency of a universal model-based energy management strategy for fuel cell hybrid trains: From analytical derivation via simulation to measurement on test bench. *Energy Conversion and Management*. https://doi.org/10.1016/j.enconman.2020.113734.
98. Guo, Q., Zhao, Z., Shen, P., et al. (2019). Optimization management of hybrid energy source of fuel cell truck based on model predictive control using traffic light information. *Control Theory and Technology, 17*(4), 309–324.
99. Ravey, A., Blunier, B., & Miraoui, A. (2012). Control strategies for fuel-cell-based hybrid electric vehicles: From offline to online and experimental results. *IEEE Transactions on Vehicular Technology, 61*(6), 2452–2457.
100. Yun, H., Liu, S., Zhao, Y., et al. (2015). Energy management for fuel cell hybrid vehicles based on a stiffness coefficient model. *International Journal of Hydrogen Energy, 40*(1), 633–641.
101. Zhang, W., Li, J., Xu, L., et al. (2017). Optimization for a fuel cell/battery/capacity tram with equivalent consumption minimization strategy. *Energy Conversion and Management, 134*, 59–69.
102. Wu, J., & Nelson, D. (2023). Development and testing of a hybrid vehicle energy management strategy. SAE Technical Paper 2023-01-0552. https://doi.org/10.4271/2023-01-0552.
103. Yoon, H.-S. (2022). Review on reinforcement learning-based energy management strategies for hybrid electric vehicles. *Evolutions in Mechanical Engineering*. https://doi.org/10.31031/EME.2022.04.000579.
104. Altun, Y. E., & Kutlar, O. A. (2024). Energy management systems' modeling and optimization in hybrid electric vehicles. *Energies, 17*, 1696. https://doi.org/10.3390/en17071696
105. Huang, D., Zhu, Z., Tang, H., & Li, C., et al. (2018). Model-in-loop automated test based on function requirements of BMS. SAE Technical Paper 2018-01-0434. https://doi.org/10.4271/2018-01-0434.
106. Kalia, A. V., & Fabien, B. C. (2020). On Implementing optimal energy management for EREV using distance constrained adaptive real-time dynamic programming. *Electronics, 9*, 228. https://doi.org/10.3390/electronics9020228
107. Saiteja, P., Ashok, B., Mason, B., & Kumar, P. S. (2024). Assessment of adaptive self-learning-based BLDC motor energy management controller in electric vehicles under real-world driving conditions for performance characteristics. *IEEE Access, 12*, 40325–40349. https://doi.org/10.1109/ACCESS.2024.3375753
108. Gao, H., Zhang, T., Chen, H., Zhao, Z., & Song, K. (2018). Application of the X-in-the-loop testing method in the FCV hybrid degree test. *Energies, 11*, 433.
109. Mayet, C., Delarue, P., Bouscayrol, A., & Chattot, E. (2017). Hardware-In-the-Loop simulation of traction power supply for power flows analysis of multitrain subway lines. *IEEE Transactions on Vehicular Technology, 66*, 5564–5571.
110. Bouscayrol, A., Guillaud, X., Delarue, P., & Lemaire-Semail, B. (2009). Energetic macroscopic representation and inversion-based control illustrated on a wind-energy-conversion system using hardware-in-the-loop simulation. *IEEE Transactions on Industrial Electronics, 56*, 4826–4835.
111. Xie, Y., Li, Y., Zhao, Z., Dong, H., Wang, S., & Liu, J., et al. (2020). Microsimulation of electric vehicle energy consumption and driving range. *Applied Energy, 267*, 115081.
112. Yi, L., He, H., & Peng, J. (2016). Hardware-in-loop simulation for the energy management system development of a plug-in hybrid electric bus. *Energy Procedia*. https://doi.org/10.1016/j.egypro.2016.06.118.

2.2 Enhancing Seasonal Exponential Smoothing Through Particle Swarm Optimization

Deng et al. [16] explores the integration of Particle Swarm Optimization (PSO) with Seasonal Exponential Smoothing (SES) models [17] to enhance electricity consumption forecasting. Traditional SES models rely on predefined smoothing parameters, which may not always be optimal for varying seasonal trends and fluctuations in electricity demand. Equations (2) represents the mathematical formulation of the SES model:

$$\begin{aligned} L_t &= \alpha\left(\frac{X_t}{S_{t-s}}\right) + (1-\alpha)(L_{t-1} + T_{t-1}) \\ T_t &= \beta(L_t - L_{t-1}) + (1-\beta)T_{t-1} \\ S_t &= \gamma\left(\frac{X_t}{L_t}\right) + (1-\gamma)S_{t-s} \\ \hat{X}_{t+h} &= (L_t + hT_t)S_{t+h-s(k+1)} \end{aligned} \tag{2}$$

where:
L_t is the level component at time t.
T_t is the trend component.
S_t is the seasonal component.
X_t is the observed value.
$\hat{X}_{t+h}$ is the forecasted value for h steps ahead.
s is the seasonal period.
α, β, γ are smoothing parameters.

To address this issue, the research leverages PSO to dynamically optimize these parameters, thereby improving forecasting accuracy. The proposed methodology involves utilizing PSO to search for optimal smoothing parameters in SES models. The PSO algorithm, inspired by the social behavior of bird flocks, iteratively adjusts parameter values based on a fitness function that minimizes forecasting errors. This approach allows the model to adapt more effectively to different seasonal electricity consumption patterns, reducing prediction errors compared to traditional SES models. In the particle swarm optimization (PSO) procedure, each particle i is described by two D-dimensional vectors: a position vector $P_i = (P_i^1, P_i^2, \ldots, P_i^D)$ and a velocity vector $V_i = (V_i^1, V_i^2, \ldots, V_i^D)$. Here, D denotes the total number of parameters employed by the SES model. We denote the best position discovered so far by the i-th particle as $pbest_i$, while $gbest$ stands for the optimal position identified by the entire swarm.

Specifically, for the d-th dimension of particle i, the velocity V_i^d and the position P_i^d are iteratively updated using Eqs. (3) and (4), respectively:

$$V_i^d = wV_i^d + c_1\text{rand}_1^d(pbest_i^d - P_i^d) + c_2\text{rand}_2^d(gbest^d - P_i^d) \tag{3}$$

$$P_i^d = P_i^d + V_i^d \tag{4}$$

where w is the inertia factor, c_1 and c_2 are acceleration constants, and $rand_1^d$, $rand_2^d$ are independent random scalars uniformly distributed in [0, 1]. As shown in Eq. (5), the inertia weight w typically decreases from a higher value w_{max} to a lower value w_{min} over the course of T iterations, where t is the current iteration index:

$$w = (w_{max} - w_{min})\frac{(T - t)}{T} + w_{min} \tag{5}$$

By tuning w in this manner, PSO strikes a balance between broad exploration at earlier stages and more refined local searching when nearing convergence.

Boundary Constraints

In addition, each dimension of the particles' velocity and position is typically restricted to the intervals $[V_{min}, V_{max}]$ and $[P_{min}, P_{max}]$, respectively. Thus, after each update, we enforce:

$$V_i^d = min(V_{max}, max(-V_{max}, V_i^d)) \tag{6}$$

$$P_i^d = min(P_{max}, max(-P_{max}, P_i^d)) \tag{7}$$

This ensures that particles remain within a reasonable search region for all dimensions, and Fig. 1 is the schematic diagram of the PSO exploration process.

The study conducts extensive empirical evaluations using real-world electricity consumption datasets. Comparative experiments are performed between the PSO-enhanced SES model and conventional forecasting approaches. The results indicate that the optimized SES model consistently outperforms traditional SES models and other baseline methods in terms of accuracy, as measured by error metrics such as Mean Absolute Error (MAE) and Root Mean Square Error (RMSE). Additionally, the research discusses the computational efficiency of the PSO-based optimization

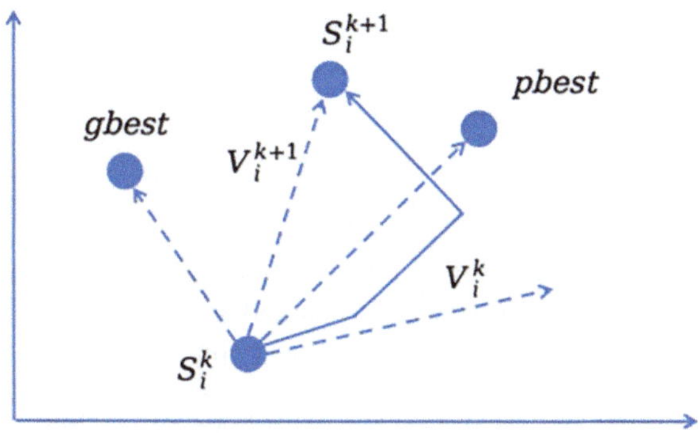

Fig. 1 Schematic diagram of the PSO exploration process

process, highlighting its feasibility for real-world applications. The findings suggest that integrating PSO with SES models provides a practical and effective method for improving electricity demand forecasting, which can benefit power system planning and management. Overall, this study demonstrates that equipping SES models with PSO enhances their adaptability and accuracy, offering a valuable tool for forecasting electricity consumption in dynamic environments.

2.3 Short Term Electricity Load Forecasting Using a Hybrid Model

Zhang et al. [18] focuses on the Short-Term Electricity Load Forecasting (STLF) problem. The authors emphasize that short-term load forecasting often exhibits substantial nonstationary and nonlinear characteristics, making single-model approaches inadequate. To address these challenges, they propose a hybrid methodology that integrates several techniques to capture both the overall trend and high-frequency fluctuations of the electricity load. The main idea is:

- Use Improved Empirical Mode Decomposition (IEMD) to decompose the original load series into a "near-linear or low-frequency" component and a "high-frequency, nonlinear" component.
- Fit the low-frequency part with an ARIMA (AutoRegressive Integrated Moving Average) model, and apply a Wavelet Neural Network (WNN) to model the high-frequency component.
- Employ the Fruit Fly Optimization Algorithm (FOA) to optimize WNN parameters, in order to avoid overfitting and local minima.
- Finally, combine the forecasts of these two components to obtain a more accurate and robust STLF result.

Traditional EMD (Empirical Mode Decomposition) successively identifies local maxima and minima, applies envelope interpolation, and extracts multiple Intrinsic Mode Functions (IMFs) plus one residual term. This work adopts the mirror method to mitigate the end-effect problem in standard EMD.

If we denote the original load series as L(t), then EMD (or IEMD) can be formulated as Eq. (8):

$$L(t) = \sum_{i=1}^{n} c_i(t) + r_n(t) \tag{8}$$

where $c_i(t)$ are IMFs, and $r_n(t)$ is the residual. Ultimately, the authors extract a main low-frequency component, denoted R(t), and treat the remainder N(t) = L(t) − R(t) as the high-frequency or nonlinear component.

For R(t), which exhibits near-linear or slowly varying trends, the authors choose an ARIMA model. In general, ARIMA can be expressed as:

$$\nabla^d y_t = a_0 + \sum_{i=1}^{p} b_i \nabla^d y_{t-i} + \sum_{j=1}^{q} a_j \varepsilon_{t-j} + \varepsilon_t \tag{9}$$

where $\nabla^d y_t$ is the d-th difference of the series, and ε_t is a white noise error term. The ARIMA orders (p, d, q) are determined via ACF/PACF analysis and information criteria such as AIC/BIC.

Wavelet Neural Network (WNN)

WNN integrates wavelet transform with a neural network, leveraging wavelet basis functions to capture nonstationary behaviors. For an input vector x_t, the network output can be:

$$\hat{y}(t) = \sum_{i=1}^{n} w_i \phi\left(\frac{x(t) - b_i}{a_i}\right) \tag{10}$$

where a_i, b_i are scale/translation parameters of the wavelet function, and w_i are connection weights. Figure 2 is the structure of WNN.

Fruit Fly Optimization Algorithm (FOA)

FOA is a swarm intelligence algorithm that imitates how fruit flies seek out food sources. It initializes a population of fruit flies within the parameter space, iteratively adjusts their positions based on "smell concentration" (inversely tied to the model error), and converges to a global optimum. In this study, FOA optimizes the WNN parameters $\{a_i, b_i, w_i\}$, thus reducing overfitting and local minima issues.

The Hybrid Model

Bringing all components together, the overall hybrid framework can be summarized as:

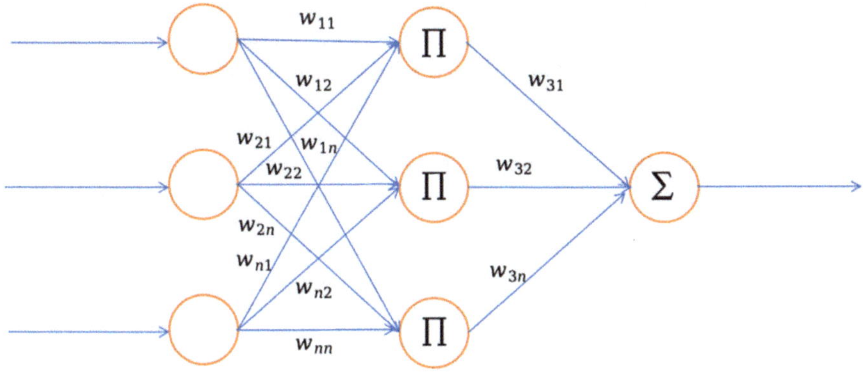

Fig. 2 The structure of WNN

- Decomposition via IEMD: separate the original load L(t) into R_t (low-frequency part) and Eqs. (11).

$$N(t) = L(t) - R(t) \tag{11}$$

- ARIMA for R(t): model and predict $\widehat{R}$(t).
- WNN–FOA for N(t): input the nonlinear portion N(t) (plus external factors such as temperature) to WNN; use FOA to optimize the wavelet network parameters, yielding $\widehat{N}$(t).
- Reconstruction: sum up the two forecasts:

$$\hat{L}(\mathrm{t}) = \hat{R}(\mathrm{t}) + \hat{N}(\mathrm{t}) \tag{12}$$

In the experiments, the authors validate their hybrid model with real load and temperature data from both the Australian and New York power markets, encompassing multiple seasons and load fluctuations. They employ Mean Absolute Percentage Error (MAPE), Mean Absolute Error (MAE), and Root Mean Square Error (RMSE) as primary metrics, while also examining the standard deviations of errors to assess model robustness. Compared with single-model approaches (e.g., standalone ARIMA or neural networks) and other wavelet- or EMD-based combination methods, the proposed IEMD + ARIMA + WNN (with FOA) framework consistently yields lower forecasting errors and greater stability. In certain tests, the MAPE remains at approximately 1% or below, thus demonstrating effectiveness across various day types, including weekdays and holidays.

The conclusions highlight several key points. First, the decomposition-and-coupling strategy clearly benefits from IEMD, which effectively separates linear from nonlinear characteristics. ARIMA and WNN–FOA are each well suited to modeling these respective components, leading to more precise fits. Second, the Fruit Fly Optimization Algorithm (FOA) enhances the wavelet neural network by tuning parameters to avoid overfitting and local minima. Future work could incorporate a wider range of exogenous factors, such as humidity or holiday indicators, and explore additional optimization methods (e.g., differential evolution, Bayesian optimization). Overall, integrating IEMD, ARIMA, WNN, and FOA in a "decompose–predict–reconstruct" pipeline markedly improves short-term load forecasting accuracy, as verified by tests across multiple markets and time frames. This hybrid approach therefore offers valuable insights for power system scheduling, demand-side management, and market operations, holding promise for broader real-world application.

3 Application of Hyperheuristic Algorithms in Energy Forecasting

3.1 Hyperheuristic Algorithms

Hyperheuristics are advanced, high-level algorithmic control strategies designed to address a wide range of optimization problems by coordinating, selecting, or automatically generating a set of underlying heuristics or metaheuristic techniques [19]. Unlike conventional single-algorithm or fixed metaheuristic methods, hyperheuristics abstract the decision-making layer from the search mechanisms. Instead of directly solving the optimization problem, they focus on dynamically answering which heuristic to apply, when to apply it, and how to combine multiple heuristics during the search process adaptively. This abstraction forms a general, scalable, and extensible optimization framework [20].

A key strength of hyperheuristics lies in their robustness and generalizability across problem domains. Structurally, they typically adopt a multi-layer design. The high-level control module monitors search performance and determines which low-level operator to invoke or evolve, based on metrics such as solution quality, progress trends, or diversity indicators [21]. The low-level layer comprises a library of heuristic operators, including local search methods, neighborhood transformations, mutation and crossover strategies, which are selected or composed by the high-level controller [22]. In more sophisticated forms, such as generation-based hyperheuristics, this decision-making process incorporates learning mechanisms—e.g., reinforcement learning or genetic programming—to automatically generate new operators tailored to the problem.

By decoupling problem-solving from operator management, hyperheuristics achieve a high degree of adaptability and portability, making them well-suited to complex, uncertain, or dynamic environments such as energy forecasting, scheduling, routing, or combinatorial design. Their ability to dynamically balance exploration and exploitation, without being confined to a fixed optimization strategy, marks a significant advancement over traditional methods and establishes hyperheuristics as a powerful paradigm in modern computational intelligence research [23]. Figure 3 presents a canonical hyper-heuristic architecture.

3.2 Forecasting Renewable Energy and Electricity Consumption Using Evolutionary Hyperheuristic Algorithm

Cao et al. [24] investigates the challenge of accurately forecasting both renewable energy generation and electricity consumption. To address the limitations of conventional metaheuristic-based parameter search, the authors introduce a novel

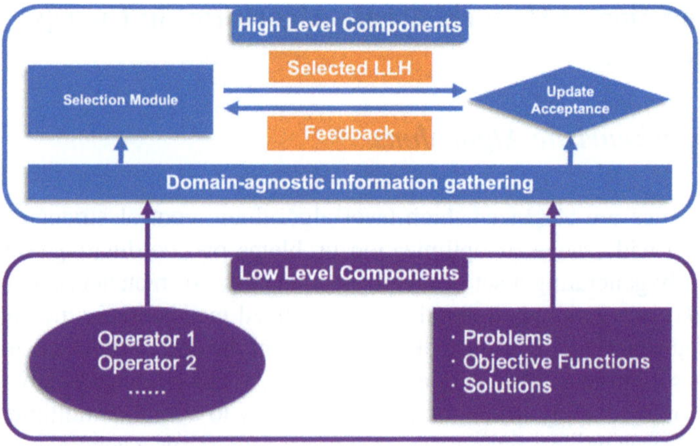

Fig. 3 The Structure of hyper-heuristic algorithms

auto-evolution hyperheuristic algorithm named AE-GAPB (AutoEvolution-Genetic Algorithm-(PSO, BA)). The primary motivation is to overcome the drawbacks of fixed hyperparameters and limited robustness in traditional optimization methods, thereby improving the performance of time series forecasting models on complex and diverse datasets. In the proposed AE-GAPB framework, we adopt Particle Swarm Optimization (PSO) and the Bat Algorithm (BA) as the low-level heuristics due to their complementary characteristics and effectiveness in parameter optimization tasks. PSO offers fast convergence and strong global search capabilities, making it suitable for efficiently exploring continuous parameter spaces in the early stages of optimization. In contrast, BA introduces adaptive control mechanisms through loudness and pulse rate, enabling a more refined balance between exploration and exploitation, which is particularly useful in handling noisy or multi-modal objective landscapes. By integrating both algorithms, AE-GAPB benefits from PSO's rapid initial exploration and BA's ability to escape local optima and maintain population diversity. This hybridization enhances the robustness and flexibility of the forecasting framework, allowing it to adapt more effectively to complex, real-world energy datasets with diverse characteristics.

Effective planning of electricity generation and consumption is essential to prevent issues such as oversupply, negative prices, or insufficient integration of renewables. Prior forecasting works typically adopt either deep learning (e.g., LSTM) or time series models (SES, ARIMA, etc.), often optimized by standalone techniques like Particle Swarm Optimization (PSO) or Bat Algorithm (BA). However, such approaches rely on static or hand-tuned hyperparameters, potentially limiting adaptability across varied energy datasets. Hence, the authors propose a hyperheuristic framework, which dynamically selects and refines multiple optimization operators under a high-level genetic algorithm (GA), aiming for more robust and accurate forecasts.

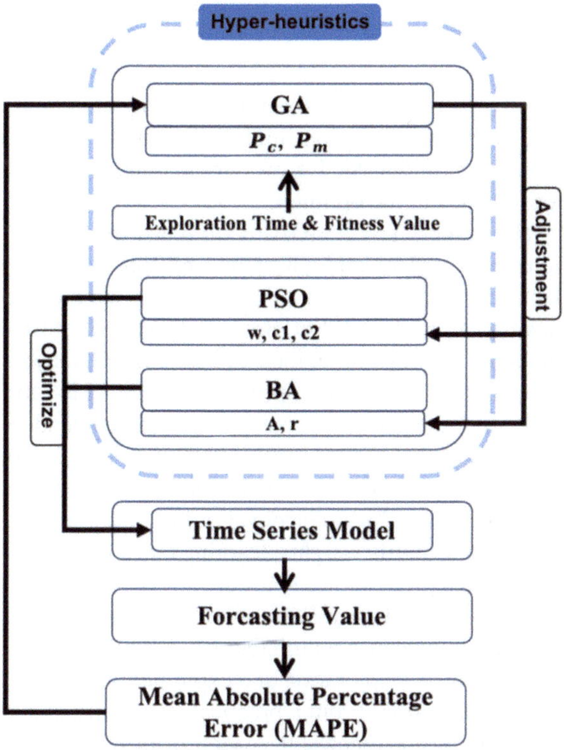

Fig. 4 The structure of the AE-GAPB

AE-GAPB Framework

Figure 4 is the structure of proposed hyper-heuristic algorithms, The proposed method follows a two-level structure:

- High-Level Component (GA): A Genetic Algorithm automatically explores hyperparameter configurations for PSO and BA. Instead of using fixed crossover and mutation rates, this GA evolves them adaptively based on the iteration number and fitness feedback (Mean Absolute Percentage Error, MAPE).
- Low-Level Components (PSO and BA): PSO and BA are the operators (heuristics) whose parameters (e.g., inertia weight for PSO, loudness for BA) are iteratively tuned by GA. The best combination is found by minimizing the MAPE on the chosen time series model.

The authors validate their approach on six time series models: Simple Exponential Smoothing (SES), Double Exponential Smoothing (DES), Holt-Winters Model (HWM), Weighted Moving Average (WMA), Auto Regression (AR), and Long Short-Term Memory (LSTM). A core innovation in AE-GAPB is that the GA's crossover and mutation rates dynamically evolve according to both the iteration count and the returned MAPE. The authors formalize this as:

$$P = P_{min} + 2\left(1 - \frac{T}{T_{max}}\right)\frac{f_i}{f_{total}}P_{max} \tag{13}$$

where P_{min} and P_{max} are the minimum and maximum rates, T is the current iteration, T_{max} is the max iteration limit, f_i the fitness of an individual, and f_{total} the sum of population fitness. As T grows or the fitness improves, P reduces, facilitating better convergence to the optimum.

The authors test AE-GAPB on real-world electricity-generation and consumption data from Hokkaido, Kyushu, and Tohoku (Japan), covering both renewable and demand-side information. They compare the method against five established optimization algorithms (PSO, BA, DE, GWO, WOA) and a variant called "GAPB" that omits the auto-evolution mechanism in the GA layer. Over 30 trial runs on six time series models—evaluated primarily via MAPE—the results show that AE-GAPB generally outperforms or matches the top contenders, notably in more complex or noisier data scenarios. In their study, the optimization objective is to minimize the forecasting error, as measured by the MAPE. Algorithm searches for model parameters that yield the lowest error value. Furthermore, parameter bounds (e.g., smoothing coefficients, inertia weights, learning rates) are predefined based on practical considerations and prior studies to ensure that the solutions remain within a valid and interpretable range. These bounds are explicitly enforced during the optimization process to guide convergence within feasible domains. The auto-evolution mechanism effectively balances exploration and exploitation, minimizing local optima and boosting forecast precision. From a practical standpoint, the ability to compare forecasts for consumption and renewable generation provides a clear strategy for optimizing fossil fuel usage and advancing green energy integration.

Overall, this research demonstrates a hyperheuristic-based solution, AE-GAPB, to automatically tune and evolve the hyperparameters of time series models for electricity and renewable energy forecasting. The integration of a GA with dynamic crossover/mutation rates at the high level, along with PSO and BA at the low level, proves effective for diverse datasets. The authors highlight the potential for future simplification of the framework to reduce computational cost, as well as exploring more sophisticated deep learning integrations or multi-objective contexts.

4 Conclusions

This research first emphasizes that in the context of electricity load and renewable energy generation forecasting, the parameter sensitivity of traditional time series models often leads to fluctuations in model performance. To address this issue, the introduction of metaheuristic methods such as Particle Swarm Optimization, Differential Evolution, and the Bat Algorithm proves valuable for automatically tuning critical hyperparameters of models like exponential smoothing, ARIMA, and WNN, significantly improving forecasting accuracy while reducing the need for manual intervention. Moreover, by decomposing complex load data (e.g., via Empirical

Mode Decomposition or wavelet decomposition) and subsequently using ARIMA and neural networks to capture patterns in different frequency components, followed by a unified metaheuristic-based optimization of all parameters, the proposed strategy effectively tackles the nonlinearities and stochastic perturbations of load data, greatly enhancing the accuracy and stability of short-term load forecasting. Building on this, a higher-level hyper-heuristic framework further alleviates the problem of single-algorithm dependence on hyperparameters by scheduling, filtering, and dynamically combining various metaheuristic algorithms. The newly proposed methods, such as AE-GAPB, maintain high applicability and robustness across different types of time series models and diverse energy data scenarios, showing strong generalizability and scalability.

5 Future Work

As power systems continue to evolve toward increased intelligence and real-time operation, and as the penetration of renewable energy steadily rises, the future demands on forecasting models will go beyond high accuracy to include adaptive capabilities for different time scales and varied application scenarios. Automated parameter tuning and model scheduling based on hyper-heuristic algorithms offer a viable direction for accommodating more external variables (such as weather factors, policy changes, and demand-side responses), enabling seamless integration from forecasting results to decision-making and implementation.

On both engineering and academic fronts, the convergence of large-scale distributed computing and multi-objective optimization is poised to become a major trend. Through parallelization and the judicious allocation of cloud resources, the training and optimization time for models can be substantially reduced. On this foundation, it will be possible to incorporate multi-objective trade-offs under more complex conditions, including economic costs, environmental impacts, and system security. Looking ahead, driven by deep learning and hyper-heuristic algorithms in tandem, electricity load and renewable energy generation forecasting will advance toward a new era characterized by real-time processing, high precision, and outstanding flexibility, thereby providing a solid technical foundation for the development of smart grids and green energy systems.

Acknowledgements This work was supported by JST SPRING, Grant Number JPMJSP2119.

References

1. Wang, J., Zhou, Y., Zhang, Y., Lin, F., & Wang, J. (2024). Risk-averse optimal combining forecasts for renewable energy trading under CVaR assessment of forecast errors. *IEEE*

Transactions on Power Systems, 39(1), 2296–2309. https://doi.org/10.1109/TPWRS.2023.3268337

2. Saeed, F., & Aldera, S. (2024). Adaptive renewable energy forecasting utilizing a data-driven PCA-transformer architecture. *IEEE Access, 12*, 109269–109280. https://doi.org/10.1109/ACCESS.2024.3440226
3. Wu, M.-P., & Wu, F. (2024). Predicting residential electricity consumption using CNN-BiLSTM-SA neural networks. *IEEE Access, 12*, 71555–71565. https://doi.org/10.1109/ACCESS.2024.3400972
4. Li, K., Wang, Y., Zhang, N., Wang, F., & Huang, C. (2023). Spatio-temporal granularity co-optimization based monthly electricity consumption forecasting. *CSEE Journal of Power and Energy Systems, 9*(5), 1980–1984. https://doi.org/10.17775/CSEEJPES.2022.01040
5. Joseph, T., & Bindiya, T. S. (2023). Realization and hardware implementation of gating units for long short-term memory network using hyperbolic sine functions. *IEEE Transactions on Computer-Aided Design of Integrated Circuits and Systems, 42*(12), 5141–5145. https://doi.org/10.1109/TCAD.2023.3293045
6. De Assis, M. V. O., Hamamoto, A. H., Abrão, T., & Proença, M. L. (2017). A game theoretical based system using holt-winters and genetic algorithm with fuzzy logic for DoS/DDoS mitigation on SDN networks. *IEEE Access, 5*, 9485–9496. https://doi.org/10.1109/ACCESS.2017.2702341
7. Tulus, R., Adytia, D., Subasita, N., & Tarwidi, D. (2020). Sea level prediction by using seasonal autoregressive integrated moving average model, case study in Semarang, Indonesia. In *2020 8th International Conference on Information and Communication Technology (ICoICT), Yogyakarta, Indonesia* (pp. 1–5). https://doi.org/10.1109/ICoICT49345.2020.9166423.
8. Mares, J. J., Charris, D., Pardo, M., & Quintero M., C. G. (2023). An architecture to improve energy-related time-series model validity based on the novel rMAPE performance metric. *IEEE Access, 11*, 36004–36014. https://doi.org/10.1109/ACCESS.2023.3264713.
9. Okuno, A., Morishita, Y., & Mototake, Y.-I. (2024). Autoregressive with slack time series model for forecasting a partially-observed dynamical time series. *IEEE Access, 12*, 24621–24630. https://doi.org/10.1109/ACCESS.2024.3365724
10. Bautu, E., Bautu, A., & Luchian, H. (2008). An evolutionary approach for modeling time series. In *2008 10th International Symposium on Symbolic and Numeric Algorithms for Scientific Computing*, Timisoara, Romania (pp. 507–513). https://doi.org/10.1109/SYNASC.2008.63.
11. Alweshah, M., Almiani, M., Alkhalaileh, S., Kassaymeh, S., Hezzam, E. A., & Alomoush, W. (2023). Parallel metaheuristic algorithms for solving imbalanced data classification problems. *IEEE Access, 11*, 114443–114458. https://doi.org/10.1109/ACCESS.2023.3324061
12. Muazu, A. A., Hashim, A. S., & Sarlan, A. (2022). Review of nature inspired metaheuristic algorithm selection for combinatorial t-Way testing. *IEEE Access, 10*, 27404–27431. https://doi.org/10.1109/ACCESS.2022.3157400
13. Jiacheng, L., & Lei, L. (2020). A hybrid genetic algorithm based on information entropy and game theory. *IEEE Access, 8*, 36602–36611. https://doi.org/10.1109/ACCESS.2020.2971060
14. Zhao, Q., & Li, C. (2020). Two-stage multi-swarm particle swarm optimizer for unconstrained and constrained global optimization. *IEEE Access, 8*, 124905–124927. https://doi.org/10.1109/ACCESS.2020.3007743
15. Zou, B., Qiu, Y., & Zhang, L. (2022). Ship detection using PolSAR images based on simulated annealing by fuzzy matching. *IEEE Geoscience and Remote Sensing Letters, 19*, 1–5, Art no. 4015805. https://doi.org/10.1109/LGRS.2021.3098026.
16. Deng, C., Zhang, X., Huang, Y., & Bao, Y. (2021). Equipping seasonal exponential smoothing models with particle swarm optimization algorithm for electricity consumption forecasting. *Energies, 14*, 4036. https://doi.org/10.3390/en14134036
17. Afrah, A. S., et al. (2024). Comparative study of machine learning and holt-winters exponential smoothing models for prediction of CPI's seasonal data. In *2024 2nd International Conference on Software Engineering and Information Technology (ICoSEIT)*, Bandung, Indonesia (pp. 144–148). https://doi.org/10.1109/ICoSEIT60086.2024.10497509.

18. Zhang, J.-L., Wei, Y.-M., Li, D., Tan, Z., & Zhou, J. (2018). Short term electricity load forecasting using a hybrid model. *Energy, 158*. https://doi.org/10.1016/j.energy.2018.06.012.
19. Oueslati, I., & Hammami, M. (2021). Honey bee cooperative HyperHeuristic. *Procedia Computer Science, 192*, 2871–2880. https://doi.org/10.1016/j.procs.2021.09.058
20. Cao, Y., Zhong, R., Yu, J., & Munetomo, M. (2024). Optimization of electricity consumption forecasting models via hyper-heuristic algorithm. In *2024 6th International Conference on Data-driven Optimization of Complex Systems (DOCS)*, Hangzhou, China (pp. 114–120). https://doi.org/10.1109/DOCS63458.2024.10704338.
21. Zhong, R., Yu, J., Zhang, C., & Munetomo, M. (2023). Surrogate ensemble-assisted hyper-heuristic algorithm for expensive optimization problems. *International Journal of Computational Intelligence Systems, 16*. https://doi.org/10.1007/s44196-023-00346-y.
22. Bándi, N., & Gaskó, N. (2024). Solving continuous optimization problems with a new hyperheuristic framework. https://doi.org/10.1007/978-3-031-53969-5_10.
23. Li, B., Elmi, Z., Borowska-Stefańska, M., Wiśniewski, S., Lau, Y.-Y., Chen, Q., & Dulebenets, M. (2024). A comprehensive framework for hyperheuristic algorithms for berth allocation and scheduling at marine container terminals. https://doi.org/10.4018/979-8-3693-0230-9.ch001.
24. Cao, Y., Yu, J., Zhong, R., & Munetomo, M. (2025). Forecasting renewable energy and electricity consumption using evolutionary hyperheuristic algorithm. *Scientific Reports, 15*. https://doi.org/10.1038/s41598-025-87013-8.

Enhancing EV Battery Safety: SOH Estimation with Machine Learning

M. Arulmozhi

Abstract The adoption of electric vehicles (EVs) powered by lithium batteries has significantly impacted both environmental and economic aspects of transportation. Ensuring safety in EVs is crucial, which relies on accurately estimating the state of health (SOH) of lithium batteries. This study presents an SOH estimation method using a machine learning-based decision tree. The supervised decision tree model utilizes battery capacity, cycle count, and SOH as inputs to predict battery health. The results demonstrate a remarkable estimation accuracy of 99%, despite the nonlinear behavior of SOH caused by internal resistance variations. This high accuracy makes the method suitable for real-time applications, contributing to improved battery management and enhanced EV safety.

Keywords Battery driven vehicle · State of health · Machine learning · Decision tree

1 Introduction

Electrical power plays a vital role in driving commercial growth, development, and national welfare. In India, the power sector is highly diversified, incorporating both conventional and renewable energy sources. A major challenge currently being addressed is the transition from environmentally harmful resources to cleaner alternatives, as fossil fuels—such as coal, petroleum, and natural gas—are significant contributors to global warming. Despite growing environmental concerns, India continues to maintain a robust fossil fuel industry, with coal consumption reaching 21.98 exajoules in 2023—backed by abundant domestic reserves to support industrial demand. Meanwhile, the natural gas sector is steadily expanding, driven by increasing adoption of LPG and a shift toward cleaner fuel alternatives [6]. With

M. Arulmozhi (✉)
Department of Electrical and Electronics Engineering, Rajalakshmi Engineering College, Chennai, India
e-mail: arulseee@gmail.com

M. A. Mellal et al. (eds.), *Computational Intelligence for a Greener Future: Innovations in Renewable Energy Systems*, Studies in Computational Intelligence 1240,
https://doi.org/10.1007/978-3-032-06732-6_4

sustainable energy being the need of the hour, one of the emerging solutions is the electrification of transport systems. The electric vehicle with battery being the major source produces less pollution to the environment compared to the conventional systems. Various kinds of rechargeable batteries are utilized in electric vehicles, but lithium-ion batteries prevail due to their distinct benefits, including superior energy density and extended durability [12, 13]. To effectively manage lithium-ion batteries, a cutting-edge BMS is essential, as it plays a vital role in ensuring battery safety, performance, and long-term reliability. Moreover, the widespread use of EVs and smart BMS technologies has a notable impact on the utility grid. A key advancement for utility operations include, enhanced sensor coverage, improved real-time monitoring and fault localization efficiency [7]. A reliable BMS guarantees precise state estimation such as state of charge (SOC), State of power (SOP), state of health (SOH) and remaining useful life (RUL) and boosts battery safety, facilitating efficient operation and management of distributed grids. However, several uncertainties in real-world EV operations pose substantial encounters to the effectiveness of the BMS [8]. These include multi-function power supply, safety concerns, environmental impacts, accurate state estimation, battery degradation, and complex thermal management. Hence, developing precise and reliable BMS technologies continues to be a vital task to ensure battery safety. The different state estimations assessed in BMS is shown in Fig. 1 [9].

In a BMS, lithium battery state estimations play a fundamental role in ensuring their safe operation and long-term use. SOC is expressed as the proportion of the

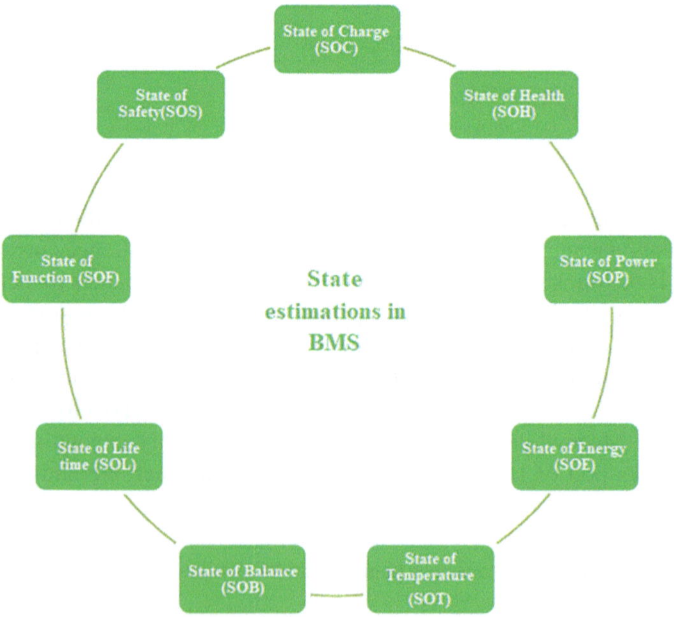

Fig. 1 Various state estimations in BMS

remaining capacity to the total capacity of the battery at a given time. SOH reflects the battery's level of degradation and is typically represented as the ratio of its present capacity to its rated capacity [16]. Because of the chemical dynamics of the battery, precise assessment of SOH and SOC remains challenging. The measurement for both depends on external parameters such as current, temperature, voltage, and resistance. The nonlinear behavior of electrochemical reactions, combined with factors like temperature sensitivity and fluctuating current rates, complicates the precision of model-based estimations. For instance, deviations in temperature can alter the kinetics and surge the battery resistance, while high C-rates cause voltage drops and polarization effects, further complicating accurate estimation [11]. Hence, there is a need for robust estimation algorithms to ensure consistent battery performance under varying operating conditions. Among these, battery's health estimation plays a key role in assessing long-term reliability, safety and performance.

Different SOH estimation methods include the differential method, model-based method, and data-driven method. The differential analysis method is easy to implement with minimal computation and is primarily based on voltage and current response. However, it has lower accuracy and is highly sensitive to noise [1, 14]. The model-based method provides good accuracy only when properly tuned and requires a precise battery model [2, 15]. The data-driven method, on the other hand, relies on datasets for training and does not require physical modelling [4, 12, 13].

The paper [5] proposes an advanced machine learning approach that combines differential equations and LSTM networks to estimate battery health using charge data. This method benefits from its ability to capture the time dynamics of battery degradation. However, it also has drawbacks, such as being computationally intensive and requiring high-quality, labeled training data, which may not always be available. The authors [3] introduce an electrochemical model employing fractional-order dynamics for predicting battery resistance and capacity fade. Advanced techniques such as spectral methods, Galerkin methods, and particle filters are utilized. The advantages of this approach include physical interpretability, efficient calibration, and real-time capability. On the other hand, the disadvantages are model complexity, limited generalizability, and computational demands. The electrochemical method offers a strong balance between physical interpretability and computational feasibility. However, machine learning approaches offer greater benefits in terms of adaptability, ease of deployment, and data-driven robustness—making them increasingly favorable for SOH estimation in electric vehicles.

Given the trade-offs between accuracy, complexity, and real-time applicability, Decision Tree-based machine learning approaches emerge as a promising alternative. They are simple, easy to interpret, computationally light, and capable of handling non-linear relationships without requiring deep domain-specific modeling. While they may not match the precision of advanced deep learning models, they offer a practical and efficient solution for integrating SOH estimation into real-world EV applications, especially where resources or data availability are limited. Thus, decision tree-based models strike an effective balance between performance, simplicity, and scalability in the evolving landscape of battery management. This study focuses on one of the data-driven approaches: machine learning using a decision tree to assess the health of the battery. The goals of the study are as follows:

- To understand decision trees and their usefulness in estimating SOH.
- To optimize the accuracy of SOH measurement.

The organization of this work is as follows: Sect. 2 discusses the Decision Tree Regressor, Sect. 2.1 presents the Regression Tree framework for predicting state of health, Sect. 4 details the results, and Sect. 5 provides the conclusion.

2 Decision Tree Regressor

Battery data are non-linear, and hence, normal regression methods like linear regression may not be suitable since a best-fit line is not possible. Therefore, choosing a regression model like a Decision Tree Regressor is the right choice. A regression tree comprises the following nodes: a root, internal nodes, leaf nodes, and several branches, where each leaf represents a numerical value. The initialization starts by selecting a feature and its value. The data is split using various threshold values, and at each split, the sum of squared residuals is evaluated to determine the optimal partition. The threshold with the smallest sum of squared residuals becomes the root.

When multiple predictors are available, the optimum threshold for each one is found, and the smallest value among them is chosen as the root. If the number of observations in a node falls below a minimum threshold, the node becomes a leaf. Otherwise, the splitting process continues for the remaining observations until no further splitting is possible. The decision tree functions by identifying multiple "if-else" conditions, generating different rules, and dividing the data into two groups at each step. These rules are derived from the predictor variables, ensuring structured decision-making and effective prediction. Figure 2 is an example tree structure generated for the problem addressed here. It demonstrates how the dataset is split at each node based on threshold values of features like capacity and cycle to reduce the squared error. At each level, the decision tree selects the optimal split that best reduces prediction error, leading to progressively homogeneous subsets.

2.1 Regression Tree Framework for Predicting State of Health

The input space for the regression tree has continuous predictor values: battery capacity, cycle count and actual state of health. The procedure is described in 3 steps.

Step1: The predictor space is split into distinct, non-overlapping regions in such a way that the observations has similar state of health values. The best threshold for

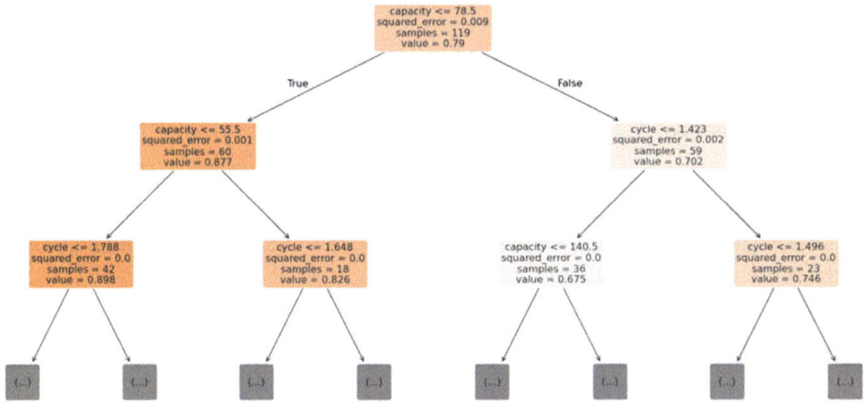

Fig. 2 Structure of the decision tree based on capacity and cycle

cycle count or capacity is chosen. This becomes the best split point which minimizes the error in predicting the state of health.

Step 2: A top down binary recursive splitting is used. First all observations are put in one group starting with root. Based on one of the features the data is split into two branches. The splitting is chosen to minimize the Residual Sum of Squares (RSS) as given in Eq. (1). The pattern repeats until minimum number of observations in a node is met.

$$RSS = \sum_{i=z_1} (x_i - \overline{x_{z_1}})^2 + \sum_{i=z_2} (x_i - \overline{x_{z_2}})^2 \tag{1}$$

where z_1, z_2 are the regions which are split, y_i represents the actual SOH and and mean SOH values.

Step3: To avoid overfitting of training data, a pruning parameter is defined to control the balance between tree complexity and predictive performance as shown in Eq. (2).

$$C(T) = RSS + \propto |T| \tag{2}$$

where C(T) is the cost of complexity, is the pruning parameter, and T represents the number of terminal nodes. A larger value results in a smaller tree, providing less complexity and less overfitting. A smaller value results in a larger tree, allowing more splits and potentially leading to overfitting.

3 Results and Discussion

The B0005 battery [10], used by NASA, endures the following charge and discharge, impedance process:

- Initially, a steady current of 1.5 A is used to charge the battery, instigating the terminal voltage to gradually rise up to 4.2 V.
- At this stage, the process switches to a constant voltage charging mode, where the current steadily declines.
- The charging process stops when the current reaches 20 mA.

During discharge:

- A steady discharge current of 2 A is applied to the battery.
- As the terminal voltage gradually decreases, the discharge process ends when the voltage reaches 2.7 V.

Impedance measurement:

- An electrochemical impedance spectroscopy was used for impedance measurement.
- A variation in frequency is applied from 0.1 Hz to 5 kHz.
- The experiment was reached after 30% fade in rated capacity.
- Frequent cycling of charging and discharging leads to faster aging of the battery.

The graph depicted in Fig. 3 signifies the capacity degradation of B0005 battery over its charge–discharge cycles. The x-axis shows the number of charge–discharge cycles, while the y-axis indicates the battery's discharge capacity. At the start, the capacity remains relatively high (above 1.8 Ah), but around cycle 50, the degradation becomes steeper and more consistent, indicating the commencement of accelerated aging. After approximately 120 cycles, the battery crosses the critical threshold of 1.4 Ah, marked by the red line. This threshold typically represents the end-of-life (EOL) criterion for the battery, beyond which its performance is considered unsuitable for regular operation. Notably, there are minor fluctuations in capacity values at certain points, which may result from temperature variations, measurement noise, or inconsistencies in load conditions. These fluctuations highlight the importance of robust estimation techniques in Battery Management Systems. Capacity degradation becomes apparent as the battery undergoes more cycles. Therefore, capacity and cycle are the optimal features for predicting the battery's state of health.

To evaluate the battery performance, the SOH is estimated using decision tree Regressor with the analysed data set. The features involved are capacity and cycle. The graph shown in Fig. 4 represents the distribution of the State of Health of the battery. The x-axis shows the battery's health level, typically ranging from 1 (healthy) to 0 (degraded). The y-axis represents the frequency of different SOH values. This graph helps analyze how SOH values are distributed, determining whether the battery remains healthy or degrades over time.

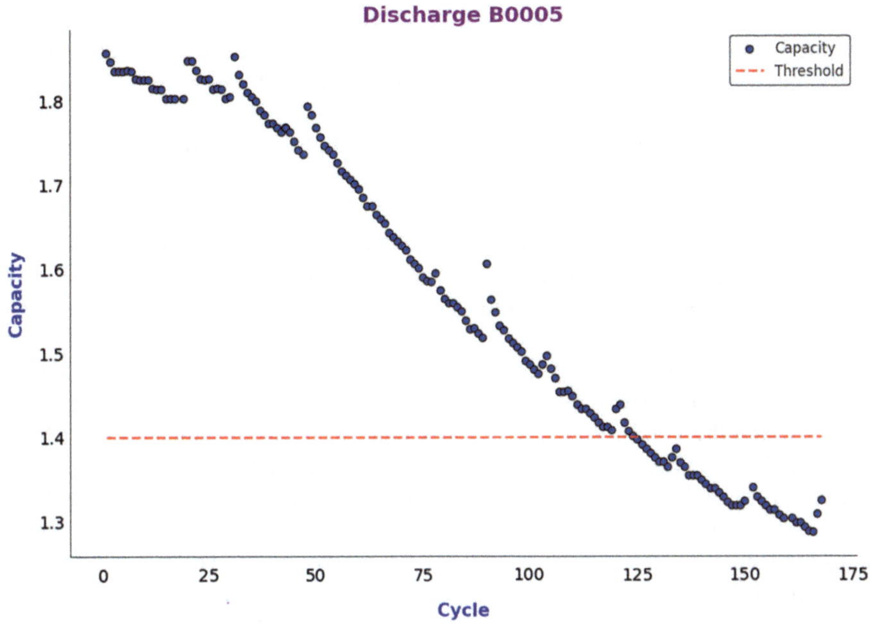

Fig. 3 Capacity degradation of B0005 battery

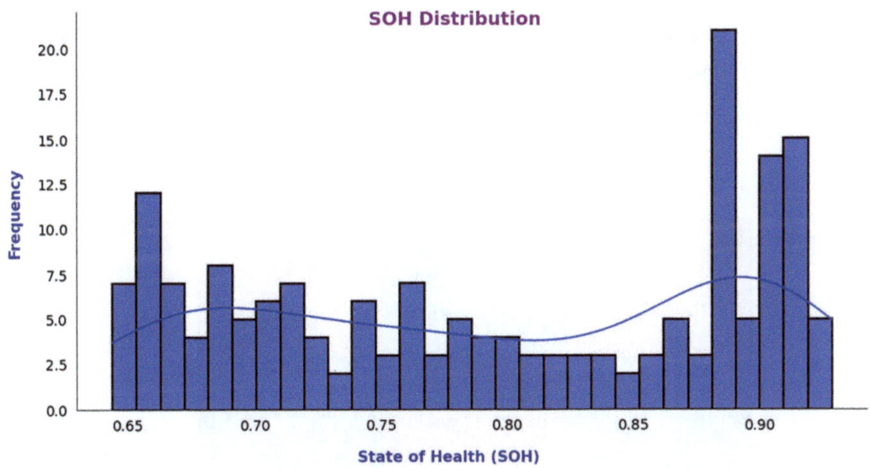

Fig. 4 Distribution of state of health of the battery

The histogram reveals a bimodal distribution, with notable concentrations of SOH values in two distinct regions—one cluster around 0.65–0.70, and a larger, more prominent cluster around 0.88 to 0.93. This suggests that a significant portion of the battery data corresponds to both early-life (healthy) and mid-to-late-life (moderately degraded) conditions. The higher frequency near the upper SOH range indicates that the battery maintains good health for a substantial number of cycles before noticeable degradation sets in. Conversely, the lower SOH region suggests that the battery eventually degrades significantly, although these instances are comparatively fewer. The smooth curve overlaid on the histogram highlights the density estimation, making it easier to identify dominant health states and transitions in battery condition. The presence of both high and low SOH frequencies in the distribution confirms the effectiveness of capacity and cycle as key features for SOH estimation models. This distribution is particularly valuable for training machine learning algorithms, as it ensures diversity in data representation—enabling the model to learn both the early and degraded behavior of the battery.

The heatmap shown in Fig. 5 visually represents how State of Health (SOH) changes with battery capacity and charge–discharge cycles. The x-axis represents Capacity, and the y-axis represents Cycle Count. Red shades indicate higher SOH (healthier battery).Blue shades indicate lower SOH (degraded battery). This graphical visualization helps in understanding battery aging patterns, showing how SOH decreases over time as cycle count increases and capacity drops. This visualization helps in identifying trends, such as how SOH decreases as cycle count increases and how capacity loss impacts SOH over time.

Table 1 displays the performance metrics. From the table it is evident that the low error values (mean absolute error (MAE) and root mean square error (RMSE)) indicate that the decision tree regression model is well-optimized. The high accuracy (99.81%) suggests that the model can reliably estimate the battery's SOH. This model can be effectively used for battery health monitoring, helping in predicting degradation trends.

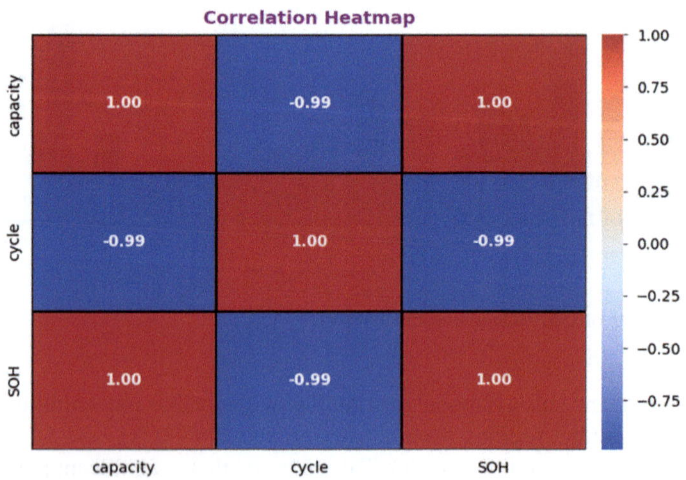

Fig. 5 Heat map correlation

Table 1 Performance metrics

SOH_MAE	SOH_RMSE	Accuracy(%)
0.00237	0.00390	99.81

4 Conclusion

In conclusion, the machine learning-based decision tree method effectually predicts the battery health with remarkable accuracy. A clear degradation trend in battery capacity over successive cycles, underscores the relevance of cycle count and capacity as key features for SOH prediction. Hence employing battery capacity and cycle count, the model successfully overcomes the nonlinear challenges related with battery aging. The good precision makes it appropriate for dynamically responsive system, enabling more efficient battery management and enhancing the durability and safety assurance in electric vehicles. These findings highlight the potential of machine learning in advanced battery diagnostics, eventually supporting the sustainable development of EV technology.

References

1. Dini, P., Colicelli, A., & Saponara, S. (2024). Review on modeling and SOC/SOH estimation of batteries for automotive applications. *Batteries, 10*(1), 34. https://doi.org/10.3390/batteries10010034
2. He, H., Xiong, R., & Fan, J. (2012). Comparison study on the battery SOC estimation for electric vehicles. *Energy Procedia, 16*, 1854–1860.
3. Hosseininasab, S., Lin, C., Pischinger, S., Stapelbroek, M., & Vagnoni, G. (2022). State-of-health estimation of lithium-ion batteries for electrified vehicles using a reduced-order electrochemical model. *Journal of Energy Storage, 52*, Article 104684.
4. Liu, K., Wei, Z., Hu, X., Miao, Q., & Foley, A. (2019). A data-driven approach to battery health prognosis using machine learning techniques. *Applied Energy, 256*, Article 113960.
5. Li, Y., He, M., & Liu, J. (2025). A novel Neural-ODE model for the state of health estimation of lithium-ion battery using charging curve. arXiv:2505.05803.
6. Madhumitha, J. (2024). Fossil fuel industry in India. Statista. https://www.statista.com/topics/5747/fossil-fuel-industry-in-india/#topicOverview. Accessed 29 Feb 2024.
7. Min, L., Alnowibet, K. A., Alrasheedi, A. F., Moazzen, F., Awwad, E. M., & Mohamed, M. A. (2021). A stochastic machine learning-based approach for observability enhancement of automated smart grids. *Sustainable Cities and Society, 72*, Article 103071. https://doi.org/10.1016/j.scs.2021.103071
8. Mohamed, M. A., Abdullah, H. M., El-Meligy, M. A., Sharaf, M., Soliman, A. T., & Hajjiah, A. (2021). A novel fuzzy cloud stochastic framework for energy management of renewable micro-grids based on maximum deployment of electric vehicles. *International Journal of Electrical Power & Energy Systems, 129*, Article 106845.
9. Mohammadi, F., Sanjari, M., & Saif, M. (2022). A real-time blockchain-based state estimation system for battery energy storage systems. In *2022 IEEE Kansas Power and Energy Conference (KPEC)* (pp. 1–4). IEEE.
10. NASA. (2024). Li-ion battery aging datasets. NASA Open Data Portal. https://data.nasa.gov/dataset/Li-ion-Battery-Aging-Datasets/uj5r-zjdb. Accessed 3 Mar 2025

11. Shrivastava, P., Soon, T. K., Idris, M. Y. I. B., & Mekhilef, S. (2019). Overview of model-based online state-of-charge estimation using Kalman filter family for lithium-ion batteries. *Renewable and Sustainable Energy Reviews, 113*, Article 109233.
12. Wang, Z., Feng, G., Zhen, D., Gu, F., & Ball, A. (2021). A review on online state of charge and state of health estimation for lithium-ion batteries in electric vehicles. *Energy Reports, 7*, 5141–5161.
13. Wang, Z., Feng, G., Gu, F., & Ball, A. (2021). State of health estimation for lithium-ion batteries based on machine learning: Advances and perspectives. *Journal of Power Sources, 491*, Article 229622.
14. Xiong, R., He, H., Li, J., & Pecht, M. (2018). Lithium-ion battery state of charge estimation using model-based and data-driven methods: A comparative study. *Applied Energy, 204*, 1299–1314.
15. Zhang, Y., Wang, C., Kang, J., & Chen, Z. (2020). A comprehensive review of battery modeling and state estimation approaches for advanced battery management systems. *Renewable and Sustainable Energy Reviews, 131*, Article 110015.
16. Zheng, M., & Luo, X. (2024). Joint estimation of state of charge (SOC) and state of health (SOH) for lithium-ion batteries using support vector machine (SVM), convolutional neural network (CNN) and long short-term memory network (LSTM) models. *International Journal of Electrochemical Science, 19*(9), 1007477.

Short-Term Renewable Energy Forecasting Methods Using Artificial Neural Networks: A Comprehensive Review

Raypratap Singh and Omveer Singh

Abstract Renewable Energy load forecasting is a critical process in efficiently managing electrical substations, ensuring a reliable power supply, optimizing costs, and informing strategic infrastructure planning. This chapter investigates the application of ANN in energy load forecasting, providing a structured overview of fundamental concepts, methodologies, and advancements in the field. It discusses the significance of load forecasting and the importance of renewable Energy load forecasting, its types, and the key factors influencing forecasting accuracy, and explores ANN-based models, highlighting their architectures, algorithms, and training methodologies, including feedforward, recurrent, and convolutional networks. There are many ML methods for accurately forecasting the power demand, but hybrid models perform better with ML techniques. A comparative analysis of single and hybrid models underscores the benefits of hybrid approaches in improving forecasting accuracy by combining multiple predictive techniques. A hybrid model is produced by combining two or more predictive models, each of which improves performance based on its algorithm. Case studies are presented to demonstrate real-world applications, showcasing the capability of ANNs to handle complex datasets and outperform traditional methods. This chapter also addresses common challenges, such as data preprocessing, overfitting, and computational constraints, and proposes potential solutions. Drawing from approximately sixty published research studies between 2011 and 2024, and also consolidates advancements in ANN architectures and emphasizes the role of real-time data integration in future research. It is imperative for decision-makers and analysts within the power sector to accurately predict future electricity demand with a high degree of precision and minimal error. This capability is essential for ensuring a continuous and reliable power supply to consumers, thereby minimizing the occurrence of load shedding. Simply, it concludes that accurate predictions can lead to improved reliability, reduced operational costs, and enhanced planning for future infrastructure requirements. The findings aim to guide researchers and practitioners in leveraging ANN-based techniques to meet the

R. Singh (✉) · O. Singh
Electrical Engineering, School of Engineering, Gautam Buddha University, Greater Noida 201312, India
e-mail: hiraypratapsingh@gmail.com

M. A. Mellal et al. (eds.), *Computational Intelligence for a Greener Future: Innovations in Renewable Energy Systems*, Studies in Computational Intelligence 1240,
https://doi.org/10.1007/978-3-032-06732-6_5

growing demands of the energy sector, minimize errors, and enhance system reliability. Additionally, they provide information on solar winds and renewable energy forecasting and emphasize the importance of green energy utilization.

Keywords Renewable energy · Short term · Load forecasting · Techniques · Artificial neural networks · Hybrid models

Abbreviations

ANN	Artificial Neural Network
ML	Machine Learning
CNN	Convolutional Neural Network
ARIMA	Auto Regressive Integrated Moving Average
LSTM	Long Short-Term Memory
GA	Genetic Algorithm
ReLU	Rectified Linear Unit
RNN	Recurrent Neural Network
FNN	Feedforward Neural Networks
SHAP	Shapley Additive Explanations
LIME	Local Interpretable Model-Agnostic Explanations
TPUs	Tensor Processing Units

1 Introduction

Forecasting techniques for renewable energy are crucial for the advanced governance of power grids, improving grid reliability, reducing costs, and optimizing the allocation of resources. Short-term forecasting is mandatory because of the inherent variability in renewable energy sources such as solar power and wind power. Conventional forecasting techniques, such as statistical and time-series models, frequently do not adequately represent the intricate, non-linear patterns found in renewable energy generation and consumption [1]. Artificial neural networks (ANNs) demonstrate proficiency in learning from historical data and identifying intricate patterns, rendering them valuable instruments for energy forecasting [2]. This chapter explains the many ANN-based techniques for short-term renewable energy forecasting, focusing on their algorithms, training methods, and feasible coupling with other methods to improve accuracy.

India is taking significant steps to promote green and renewable energy on a global scale. As of June 2025, a majority of energy generation is still reliant on non-renewable sources, particularly coal. Currently, India has an installed power capacity of approximately 476 GW, with 219 GW coming from renewable energy sources such as solar (110.09 GW) and wind (51.3 GW). Non-fossil fuel sources contribute to 47 percent of the total capacity [3]. Additional details are provided in the accompanying table for better understanding:

Renewable source	Installed capacity (GW)
Solar power	110.9 GW
Wind power	51.3 GW
Hydro power	42 GW
Bio power	10.2 GW
Small hydro power	4.9 GW

Simply, this data gives the information about the **coal and other fossil fuels** still dominate with around **257 GW**, accounting for about **50%** of capacity and over **75%** of actual electricity generation [4]. India is steadily transitioning to clean energy, with a target of **500 GW non-fossil capacity by 2030** and **50% power generation from renewables** [5]. This signifies an increasing transition towards sustainability, although it remains a minor proportion relative to fossil fuels. It also explains the importance of solar and wind, or renewable energy resources, and the need to forecast renewable energy resources.

This growth underscores the persistent significance of coal in India's energy portfolio, notwithstanding the country's initiatives to enhance its renewable energy infrastructure. The dependency on coal and other non-renewable resources in power forecasting requires ACCURATE prediction of consumption and energy output. Forecasting techniques are vital for ensuring the security of the energy and managing the grid efficiently. This chapter, aiming to predict future energy production, has employed models such as the optimal discrete gray model to predict the output of renewable and nonrenewable energy resources in India [5]. In conclusion, nonrenewable energy sources, particularly coal, remain essential for meeting India's energy demands despite the country's advancements in renewable energy development. Accurate forecasting of non-renewable energy output remains essential for ensuring a stable and reliable power supply.

Short-term prediction helps manage the grid and reduce instability caused by renewable energy fluctuations like solar and wind power [6].

2 Renewable Energy Load Forecasting's Significance

2.1 The Value of Renewable Energy Load Forecasting

India is increasingly reliant on solar, wind, and tidal energy to meet its growing power demands, rendering renewable energy forecasting essential. India's solar energy potential is estimated to reach 750 GW, contingent upon the available land area. The country is ideally positioned for extensive solar power generation, receiving 4–7 kWh of solar radiation per square meter daily. Nonetheless, cloud cover, seasonal fluctuations, and dust deposition render solar energy production highly unpredictable [7]. For a smart electricity distribution system and efficient operation of the grid stability, the right forecasting of power is more important. Accurate Load forecasting gives stable grid operation, minimizes energy waste, and improves resource allocation efficiency. Maintaining the balance between electricity supply and demand is essential to overcome issues like blackouts or excessive generation of power [8]. Wind energy represents a vital integrant, predominantly located in areas such as Gujarat, Tamil Nadu, and Maharashtra, with a projected capacity of 695 GW at a hub height of 120 m [9]. To mitigate power imbalances and diminish reliance on backup power sources, sophisticated forecasting models employing ANNs and ML can be utilized to anticipate wind energy generation trends. Tidal energy in India possesses considerable potential, especially in critical areas like the Sundarbans, Gulf of Kutch, and Gulf of Khambhat, with an estimated capacity of 12.45 GW. The important advantage of tidal energy compared to solar and wind energy is its greater predictability, attributable to gravity forces. A Smart Estimate of demand is important for the integration of tidal energy into the electrical system to ensure effective power utilization [8]. Load forecasting is essential for India's shift to renewable energy since it facilitates better energy dispatch planning, reduces curtailment losses, and optimizes storage alternatives. Precise forecasts diminish dependence on fossil fuels, curtail energy waste, and bolster grid resiliency. Load forecasting is essential for utilities in India to efficiently plan energy dispatch, reduce curtailment losses, and optimize storage solutions during the transition to renewable energy. Precise forecasts reduce dependence on fossil fuels, curtail energy waste, and enhance grid resiliency [10]. Improving forecasting techniques with AI, deep learning, and real-time meteorological data is crucial for achieving a sustainable and dependable electricity system as India seeks to establish 500 GW of non-fossil fuel capacity by 2030.

2.2 *Different Types of Load Forecasting*

Proper planning and real-world electric load forecasting applications require certain "forecasting intervals" or "lead time." Three different categories of electric load forecasting can be distinguished based on the lead time [11]. Power systems can classify load forecasting into three different categories based on the planned time horizon. This chapter explains the various forms of load forecasting in the table, along with the nature of the period and application. Effective planning and accurate electric load forecasting require specific "forecasting intervals" or "lead times." Electric load forecasting can be classified into three distinct categories based on these lead times. Given table below gives information of the different forms of load forecasting, detailing the nature of each type and its applications.

Type of load forecasting	Time frame	Applications
Short-term	A few hours to several days	Plan for electricity generation and transmission
Medium-term	A few weeks to several months	Organize fuel purchases
Long-term	1–10 years	Create plans for generation, transmission, and distribution systems

2.3 *Influencing Factors of Load Forecasting*

Load forecasting is affected by variables like meteorological conditions, past consumption data, seasonal fluctuations, and grid limitations. Improving forecasting models necessitates real-time data analytics, upgraded grid infrastructure, and sophisticated machine learning methodologies. Subsequent research should prioritize big data analytics and artificial intelligence to enhance predictive accuracy. Energy management, power system operation, and planning are all strongly reliant. Load forecasting is essential for energy management, the operation of power systems, and strategic planning. Precise forecasting guarantees the dependability and efficacy of electrical systems. Elements that affect the precision of load forecasting encompass meteorological conditions, historical usage patterns, and socioeconomic variables.

2.3.1 Weather Conditions

The weather has a big impact on the electricity demand. The following weather-related elements affect the accuracy of load forecasting:

Primarily, three inputs are required for the forecasting model, including: 1. Irradiations 2. Ambient temperature 3. The module temperature and output are AC Power or DC Power.

- **Temperature**: This is one of the most important factors since temperature variations affect both heating and cooling loads.
- **Humidity**: In tropical and subtropical locations in particular, higher humidity levels affect cooling loads [12]. The efficiency of wind power generation and cooling demand are influenced by wind speed, which in turn affects load forecasts [13].
- **Solar Radiation**: Solar intensity affects photovoltaic (PV) power generation and patterns of electricity consumption [14].

2.3.2 Patterns of Historical Energy Consumption

An important component of load forecasting is historical energy use data. Usage trends in electricity offer crucial information about future demand. Models are created using historical data. Short-term load variations: These record changes in the demand for electricity on a daily and weekly basis [15]. Identifying annual increases in energy use based on technology and economic developments is known as long-term consumption trends [16].

- **Peak Load and Off-Peak Patterns**: Knowing when demand is at its highest aids in grid management [17].

2.3.3 Variations by Season and day

Depending on the season and time of day, electricity usage trends change:

Due to the necessity for heating and cooling, demand is highest during the summer and winter [18].

- Effects of Time of Day: Demand usually peaks in the morning and evening and decreases in the middle of the day and at night [19].
- Holiday Effects: Certain days, such as national holidays, cause unforeseen changes in load, necessitating the use of specific forecasting models [20].

2.3.4 Grid Limitations and Demand Variations

The constraints of the power grid and variations in electricity consumption impact the precision of forecasts.

- **Transmission and Distribution Limitations**: Restricted grid capacity may hinder electricity availability, affecting demand forecasts [21].
- **Unanticipated Demand Increases**: Abrupt fluctuations in demand resulting from economic or industrial operations might impact load forecasts [22].
- **Integration of Renewable Energy Sources**: The incorporation of solar and wind energy introduces fluctuations in the equilibrium of electricity demand and supply [23].

3 Energy Prediction Utilizing ANN

An ANN is a type of machine learning technique that draws inspiration from the way the human brain functions. It is made up of layers of connected nodes, called neurons. These neurons learn from data by adjusting weights to reduce prediction errors. ANNs are popular for forecasting because they can effectively manage complex and nonlinear patterns. In energy systems, ANNs are used for important tasks such as

Forecasting energy load, predicting solar and wind power, and estimating battery energy. It gives information about the uses of ANN in energy forecasting for the prediction of accurate power [24]. ANN model accuracy depends on settings like learning rate and neurons. Manual tuning is time-consuming and error-prone [6].

3.1 Architectures of ANN

A neuron is the fundamental component of an ANN. It is also responsible for processing inputs and generating outputs. Each input is assigned a weight that reflects its significance. A bias is incorporated to adjust the output, thereby enhancing the network's predictive capabilities. The outputs then pass through an activation function, which introduces non-linearity and facilitates the learning process [24].

The **hidden layer**, placed between the input and output layers, enables the network to learn complex patterns in data. More hidden layers typically improve the ANN's ability to model complicated relationships, especially in energy forecasting applications [24]. A block diagram of the ANN model is illustrated below.

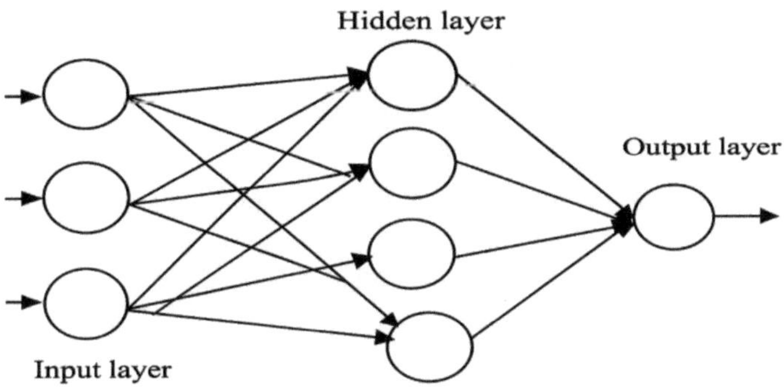

In renewable energy load forecasting, basically, these are the most important inputs are solar irradiation, ambient or module temperature, and the output is AC power as illustrated in the above block diagram. A multilayer perceptron with hidden layers is used, involving neurons, weights, and biases. In ANNs, **Weights** and **biases** are the two most important parameters for learning in the network.

3.1.1 Weights

Control the strength of the connection between two neurons. They find out how much influence one neuron's output will have on the next neuron.

3.1.2 Bias

Is an additional parameter added to the neuron to adjust the output along with the weighted sum of the inputs. Bias helps shift the activation function and improves the model's flexibility.

Weight and bias allow the network to fit complex data patterns. The learning process in ANN involves adjusting weights and biases to minimize the error between predicted and actual outputs. Bias plays a similar role to the intercept in linear regression [25].

3.2 Mathematical Equation for Representing the ANN Model

$$z = w_1x_1 + w_2x_2 + \cdots + w_nx_n + b$$

Or Shorter from $z = \sum_{i=1}^{n} w_ix_i + b$
Then the neuron's output is $a = f(z)$
Where:

- z is the **weighted sum with bias,**
- f(z) is the **activation function** (e.g., sigmoid, ReLU),
- **a** is the final **output of the neuron.**

Every neuron computes the weighted sum of its inputs and adds a bias, as shown in the equation above. The output is then passed through an activation function, enabling the network to model non-linear relationships [25].

3.3 Activation Function in ANN

An **activation function** is a mathematical formula that decides whether a neuron should be **activated** (i.e., pass information forward) or not. It takes the output of the neuron (weighted sum z) and applies a function to it, so that the final result becomes **non-linear**. This helps the neural network solve **complex problems**, not just simple linear ones. Without the activation function, this network becomes a linear regression, and it adds flexibility and helps the model learn from real-world Data. It also allows the network to learn and adapt, and model reactions like the human brain. Some important activation functions are explained in the given table.

Different Types of Activation Functions

Name	Formula	Output range	Uses
Sigmoid	$f(z) = \frac{1}{1+e^{-z}}$	(0, 1)	Binary classification
ReLU	$f(z) = \mathbf{max}(0, z)$	[0, ∞)	Widely used in Hidden layers, deep networks
Tanh	$f(z) = \frac{e^z - e^{-z}}{e^z + e^{-z}}$	(-1, 1)	Centered data (positive & negative)

Common activation functions like **Sigmoid**, which outputs lies 0–1; and **ReLU**, which outputs zero for negative inputs and passes positive values as it as; and another activation function is **Tanh**, which maps inputs to the range of -1 to 1 [25].

Diverse topologies of artificial neural networks (ANNs) are tailored for particular forecasting tasks:

- **FNN**: These are optimal for fundamental time-series forecasting.
- **RNNs** are proficient at managing long-term dependencies and are efficient with sequential input.
- **CNN**: it is a type of deep learning algorithm, and they are designed to extract features from temporal and spatial energy data.

3.4 Methodologies of Training

ANNs require robust training methodologies to enhance accuracy. These are common training techniques:

- **Backpropagation Algorithm**: It is used for error minimization in terms of MAPE, MASE, or Accuracy.
- **Gradient Descent Optimization**: used in improving the learning capacity for good prediction.
- **Adaptive Learning Rate Techniques**: This technique is used for network integration.

4 Comparative Analysis of Forecasting Methods

4.1 Single versus Hybrid Methods

Support Vector Machines (SVMs), Decision Trees, and Artificial Neural Networks (ANNs) are examples of machine learning (ML) techniques that have been mostly used for energy forecasting techniques. Nevertheless, by integrating many predictive techniques, hybrid models have shown increased accuracy.

4.1.1 Single Models

Single methods are better for non-renewable energy resources forecasting, but in the nonlinear behavior of input data or weather data not give better results in terms of renewable energy forecasting, like solar and wind forecasting. Some method names are given below.

- ANN-based models
- Autoregressive Integrated Moving Average
- Support Vector Regression.

4.1.2 Hybrid Models

A hybrid model integrates two or more techniques to improve performance. Some Examples are given below.

- **ANN + ARIMA**: This accommodates both linear and non-linear patterns.
- **CNN + LSTM**: Employed for forecasting based on deep learning methodologies.
- **GA combined with ANN**: Improves optimization and precision.

These are some models written above, like single and hybrid models, in terms of accuracy, where better forecasting results are achieved by the hybrid model. These three hybrid models are explained in detail below.

The Hybrid ANN + ARIMA Model

Combines the strengths of both linear and non-linear modeling techniques.

- **ARIMA** is effective at capturing **linear trends and seasonality** in time series data. However, it cannot handle complex **non-linear relationships**, so combined with an ANN model.
- **ANN** is capable of modeling **non-linear patterns**, but may not perform well on purely linear components.

By combining both, the **hybrid model** first applies **ARIMA** to model and remove the linear part of the data. Then, the **residuals** (errors not explained by ARIMA) are passed to an **ANN**, which models the remaining non-linear structure. The final prediction is obtained by **adding the outputs** of both models.

Basic Equation for the hybrid model, ANN and ARIMA

Let Y(t) be the actual value at time t:

$$Y(t) = Y_{ARIMA}(t) + Y_{ANN}(t)$$

where:

- $Y_{ARIMA}(t)$: Output from ARIMA model (linear part)
- $Y_{ANN}(t)$: Output from ANN model (non-linear part).

This integrated approach improves forecast accuracy, especially in complex systems like renewable energy load forecasting, where both linear (e.g., seasonality) and non-linear (e.g., weather influence) patterns exist [26, 27].

CNN + LSTM Model

It is a powerful hybrid deep learning architecture that combines:

- **CNN** captures **local patterns** and **spatial features** in the input data. It is very useful when the input data includes temporal sequences like weather maps, irradiance time series, etc.
- **LSTM** is an RNN that is excellent at learning **long-term dependencies** in **sequential time series data**, such as hourly or daily solar power generation.

CNN Layers extract features from time series input (e.g., historical weather and irradiance data). These features are passed into the **LSTM layers**, which model the **temporal dependencies** (time-based patterns). The final output is the predicted power generation value.

$$\text{Forecast}_{(t+1)} = \text{LSTM}(\text{CNN}(X_t))$$

where:

- X_t is the input data (e.g., past solar irradiance, temperature, wind speed)
- CNN extracts spatial/temporal features
- LSTM processes these features over time to produce a forecast.

This hybrid model is Good for Renewable Energy Forecasting because of its ability to handle the time series data, whereas LSTM captures the time dependency in weather and power generation data. It is also capable of learning complex pattern data like CNN recognizes patterns and sudden cloud cover changes or rapid tramp change in terms of accuracy or RMSE, better as compared to the traditional model**.** For example, in solar energy forecasting, CNN processes historical irradiance or satellite image sequences, while LSTM learns how these patterns change over time [28, 29].

GA + ANN

Combining **GA** with **ANN** improves the training process by adjusting the **weight optimization** and increasing **prediction accuracy**.

- **GA** is a bio-inspired optimization technique that uses **selection, crossover, and mutation** to search for the best possible solutions.

- Instead of using traditional backpropagation to train the ANN, **GA is used to optimize the weights and biases**, helping avoid problems like getting stuck in local minima.

This hybrid model is particularly effective in **renewable energy forecasting**, where input data (like temperature, irradiance, wind speed) may have complex, noisy patterns. GA-ANN improves the **model's generalization** and **forecasting precision** in such environments.

In renewable energy applications, GA-ANN has shown better performance than standard ANN in terms of accuracy, especially in solar and wind power forecasting tasks [30]. This model gives some advantages, like initializing population (random weights) and evaluating fitness, or gives information about the error in ANN prediction. After applying the GA operations like selection, crossover, mutation, and updating the ANN Weights, and repeating again and again until the model does not give the optimal solutions, or a better solution.

5 Applications in the Real World and Case Studies

Several case studies illustrate the efficiency of ANN-based models in renewable energy forecasting.

Case Study 1: RNN-Based Wind Power Forecasting in Germany

Because wind speeds fluctuate, wind energy is highly variable, making precise forecasting crucial to system stability. The nonlinear patterns of wind speed and power generation are difficult for traditional statistical models, such as persistence models and ARIMA, to capture. RNNs were used in a study [31]. In Germany, to enhance wind power predictions. RNNs are a suitable option for time-series forecasting since they can handle sequential data, especially LSTM networks. To train the model, the study used meteorological data like temperature, air pressure, and wind speed. The results showed that predicting errors were 15% lower than with traditional statistical models. This made it easier to use wind energy in the system and made power dispatch more reliable. Grid operators were able to reduce volatility and increase overall energy efficiency because of the increased accuracy.

Case Study 2: CNN-LSTM-Based Solar Energy Forecasting in India

With an estimated 750 GW of solar power capacity, India has enormous potential for solar energy. Therefore, accurate short-term forecasting is crucial for grid stability and effective power distribution. Traditional forecasting models have had difficulty maintaining accuracy because of the considerable fluctuation in solar radiation caused by elements like cloud cover and seasonal variations [32]. Created a model that combines a CNN and an LSTM to improve the prediction of solar energy. The CNN part was used to get useful information from satellite pictures of solar radiation, while the LSTM network processed sequential data to account for changes in solar energy production that happened over time. The findings showed that the CNN-LSTM model reduced Root Mean Square Error (RMSE) by 12%, outperforming conventional ANN models. This improvement was credited to CNNs' ability to identify spatial patterns in solar radiation and LSTMs' capacity to capture long-term dependencies in energy production trends. The study also highlighted how real-time weather data integration and cloud movement prediction might improve forecasting accuracy.

Case Study 3: American Hybrid ANN-ARIMA Model for Load Forecasting.

In the United States, where energy consumption fluctuates due to seasonal variations, industrial activity, and demographic variables, accurate load forecasting is essential for power system stability and profitable operation. While traditional models like ARIMA are good at identifying linear trends, they have trouble identifying intricate, non-linear patterns in energy use [33]. To address this, a hybrid ANN-ARIMA model that integrates the advantages of both techniques to get beyond these restrictions. The ANN part recorded non-linear dependencies like sudden demand spikes and changes during peak hours. The ARIMA model, on the other hand, was used to find linear trends in historical load data. In comparison to standalone models, the hybrid model improved the Mean Absolute Percentage Error (MAPE) by 18%, according to the study. A useful tool for energy suppliers looking to maximize power generation and cut operating expenses, the hybrid approach also showed improved responsiveness to real-time load changes. The study also showed that using demand-side management techniques along with real-time smart grid data made forecasting much more accurate.

6 ANN-based Forecasting: Difficulties and Solutions

In time series forecasting, artificial neural networks (ANNs) have shown a great deal of promise, especially in domains like financial markets, meteorological evaluations, and energy demand prediction. However, several obstacles prevent them from performing at their best. The main issues and their remedies are covered in this section. Meteorological variables like **irradiance, module temperature**, and **ambient temperature** were found to significantly impact PV output and were key inputs to the models [34]. ANNs learn from past data and can model complex weather-to-power relationships accurately.

6.1 Preprocessing Data

It is essential for improving the accuracy of ANN models. Raw data frequently contains noise, inconsistencies, and missing values that can impact forecasting dependability. Important preprocessing methods consist of dealing with Missing Values: Poor forecasts and incorrect model training might result from missing data.

To fill in missing values, various imputation techniques are frequently employed, including linear interpolation, k-nearest neighbors (KNN), and expectation–maximization (EM) algorithms [35]. The study emphasizes handling missing values, outlier removal, and normalization as critical steps in preprocessing [34].

6.2 Normalization and Standardization

To guarantee consistency in the distribution of data, feature scaling is crucial. Techniques like z-score standardization and min–max normalization can lessen biases caused by different data scales [36].

6.3 Generalization and Overfitting

When an ANN model performs well on training data but is unable to generalize to unseen data, the condition is known as overfitting. There are several ways to lessen this problem:

- **Regularization Techniques:** Dropout, L1, and L2 regularization, which is also called weight decay, keep models from fitting too well by punishing high weight values and making them less complicated [37].
- **Cross-Validation Techniques:** K-fold cross-validation and leave-one-out cross-validation (LOOCV) improve generalization by ensuring model assessment on a variety of data subsets [38].

6.4 Limitations in Computation

For large-scale forecasting applications, the processing power required to train deep neural networks may be a constraint. To overcome these restrictions:

- **Parallel Processing and GPU Acceleration**: By parallelizing calculations, the utilization of Graphics Processing Units (GPUs) and Tensor Processing Units (TPUs) drastically cuts down on training time [39].
- **Cloud Computing Integration**: Scalable computing resources provided by cloud platforms like Google Cloud AI and Amazon AWS allow for the effective deployment of ANN models.

7 Upcoming Patterns and Research Paths

As new trends and technology developments emerge, the field of ANN-based forecasting keeps changing. Among the important future paths are:

- **Real-time Data Integration**: By integrating smart grids and Internet of Things (IoT) devices, dynamic data collection for real-time forecasting is made possible, which enhances decision-making [40].
- **AI that can be explained in energy forecasting**: Policymakers can better understand forecasts and make decisions that can be defended by making ANN models that can be understood using tools such as SHAP and LIME [41].
- **Quantum computing for ANN optimization**: By drastically cutting down on computation time and increasing accuracy, quantum algorithms can improve ANN training efficiency and pave the way for breakthroughs in predicting applications [42]. This in-depth discussion makes sure that the ideas are useful in both science and business by giving details about problems with ANN-based forecasting, how to fix them, and possible directions for future research.

8 Conclusion

The accuracy of forecasts of renewable energy loads has been greatly improved using ANN-based forecasting models. Using hybrid models, which combine several approaches, improves predictions even more. Overfitting, poor data quality, and computing constraints can be resolved to create more reliable forecasting methods. Future studies can help make the renewable energy industry more dependable and effective by utilizing cutting-edge AI techniques and incorporating real-time data. To maximize power generation, maintain grid stability, and reduce energy waste, short-term renewable energy forecasting utilizing Artificial Neural Networks (ANNs) is essential [43]. With the enormous potential of solar and wind energy in India, precise forecasting can greatly improve energy management, lessen dependency on fossil fuels, and facilitate integration into the national grid. Similarly, ANN-based forecasting models are assisting countries worldwide in their transition to sustainable energy sources by increasing operational efficiency and prediction accuracy. Improved accuracy in managing non-linear energy trends, real-time flexibility, and improved decision-making for energy suppliers are some benefits of ANN-based forecasting. Hybrid models further improve prediction accuracy by fusing ANN with deep learning and statistical methods. Using AI-driven forecasting tools will be essential to guaranteeing a more dependable, efficient, and sustainable energy future as the use of renewable energy continues to increase. Further work can improve speed, reduce computation, and apply the method to other renewables or multi-time scale forecasting [6]. According to this study suggests using wavelet-based ML methods and hyper parameter tuning to further improve prediction robustness on chaotic datasets [34]. The above-discussed renewable energy forecasting techniques-such as

ANN, ARIMA, CNN-LSTM, and GA-ANN-provide crucial insights for optimizing the generation and utilization of green energy. These methods enhance the accuracy of power predictions, enabling more efficient integration of renewable sources like solar and wind into the energy grid. In a world with **limited fossil fuel resources**, the shift toward clean and sustainable energy is essential. By improving forecasting, these AI-driven models help reduce dependence on non-renewable energy, minimize **greenhouse gas emissions**, and support **ecosystem balance**. Ultimately, the adoption of intelligent forecasting technologies promotes the **sustainable use of energy**, ensuring a **better quality of life for current and future generations**, while contributing to the global effort to combat **climate change**.

References

1. Tsegaye, S., Padmanaban, S., Bertling Tjernberg, L., & Fante, K. A. (2024). Short-term energy forecasting using deep neural networks: Prospects and challenges. *The Journal of Engineering, 2024*(11), 1234–1245. https://doi.org/10.1049/tje2.70022.
2. Benti, N. E., Chaka, M. D., & Semie, A. G. (2023). Forecasting renewable energy generation with machine learning and deep learning: Current advances and future prospects. *Sustainability, 15*(9), 7087. https://doi.org/10.3390/su1509708
3. Ministry of New and Renewable Energy (MNRE). Monthly Progress Report—June 2025. [Online]. https://mnre.gov.in.
4. Central Electricity Authority (CEA). Installed Capacity Report. Ministry of Power, June 2025. [Online]. https://cea.nic.in.
5. Press Information Bureau (PIB). India adds record 29.5 GW renewable energy in FY 2024–25, June 2025. [Online]. https://pib.gov.in.
6. Wang, J., Zhang, Z., Xu, W., Li, Y., & Niu, G. (2025). Short-term photovoltaic power forecasting using a Bi-LSTM neural network optimized by hybrid algorithms. *Sustainability, 17*(12), Art no. 5277. [Online]. https://doi.org/10.3390/su17125277.
7. Mishra, S., & Palanisamy, P. (2018). Multi-time-horizon solar forecasting using recurrent neural network. arXiv:1807.05459 [Online]. https://arxiv.org/abs/1807.05459.
8. Kaur, D., Islam, S. N., Mahmud, M. A., Haque, M. E., & Anwar, A. (2021). A VAE-Bayesian deep learning scheme for solar generation forecasting based on dimensionality reduction. arXiv: 2103.12969 [Online]. https://arxiv.org/abs/2103.12969.
9. Barbar, M., Mallapragada, D. S., & Stoner, R. (2022). Impact of demand growth on decarbonizing India's electricity sector and the role for energy storage. arXiv:2203.12426 [Online]. https://arxiv.org/abs/2203.12426.
10. Mitra, I., Heinemann, D., Ramanan, A., Kaur, M., Sharma, S. K., Tripathy, S. K., & Roy, A. (2022). Short-term PV power forecasting in India: Recent developments and policy analysis. *International Journal of Energy and Environmental Engineering, 13*(1), 1–15. https://link.springer.com/article/10.1007/s40095-021-00468-z.
11. Ahmad, N., Ghadi, Y., Adnan, M., & Ali, M. (2022). Load forecasting techniques for power system: Research challenges and survey. *IEEE Access, 10*, 73456–73470. https://doi.org/10.1109/ACCESS.2022.3187839
12. Smith, J., & Lee, K. (2018). Impact of humidity on load forecasting. *IEEE Transactions on Power Systems, 33*(6). https://doi.org/10.1109/TPWRS.2018.2843852.
13. Patel, R., & Kumar, S. (2019). Wind speed influence on power load prediction. *IEEE Access, 7*. https://doi.org/10.1109/ACCESS.2019.2905567.
14. Johnson, M., & Wang, X. (2020). Solar radiation and load variability. *IEEE Transactions on Smart Grid, 11*(4). https://doi.org/10.1109/TSG.2020.2973451.

15. Taylor, P., & Nguyen, H. (2019). Short-term load forecasting techniques. *IEEE Transactions on Sustainable Energy, 10*(3). https://doi.org/10.1109/TSTE.2019.2894837.
16. Brown, T., & Zhao, Y. (2021). Long-term electricity demand trends. *IEEE Transactions on Industrial Informatics, 17*(5). https://doi.org/10.1109/TII.2021.3059943.
17. Davis, R., & Singh, P. (2021). Peak and off-peak load management. *IEEE Power & Energy Magazine, 18*(2). https://doi.org/10.1109/MPE.2020.2969824.
18. White, L., & Garcia, F. (2018). Seasonal load demand analysis. *IEEE Transactions on Smart Grid, 9*(1). https://doi.org/10.1109/TSG.2017.2734940.
19. Chen, H., & Williams, D. (2017). Diurnal electricity consumption trends. *IEEE Transactions on Power Systems, 32*(4). https://doi.org/10.1109/TPWRS.2017.2655444.
20. Roberts, A., & Kim, J. (2020). Holiday demand forecasting models. *IEEE Access, 8*. https://doi.org/10.1109/ACCESS.2020.2983575.
21. Allen, B., & Tan, E. (2019). Impact of grid constraints on load forecasting. *IEEE Transactions on Energy Conversion, 34*(2). https://doi.org/10.1109/TEC.2019.2901253.
22. Martin, G., & Shah, R. (2019). Industrial demand surges and forecasting accuracy. *IEEE Transactions on Power Delivery, 34*(5). https://doi.org/10.1109/TPWRD.2019.2917586.
23. Singh, V., & Patel, A. (2019). Renewable energy integration and demand variability. *IEEE Transactions on Industrial Electronics, 66*(12). https://doi.org/10.1109/TIE.2019.2908392.
24. Jing, J., Di, H., Wang, T., Jiang, N., & Xiang, Z. (2025). Optimization of power system load forecasting and scheduling based on artificial neural networks. *Energy Informatics, 8*, Art. no. 6. https://doi.org/10.1186/s42162-024-0046744.
25. Haykin, S. (2009). *Neural networks and learning machines* (3rd edn.). Pearson. ISBN 9780131471399.
26. Zhang, Z. (2003). Time series forecasting using a hybrid ARIMA and neural network model. *Neurocomputing, 50*, 159–175. https://doi.org/10.1016/S0925-2312(01)00702-0
27. Khashei, A., & Bijari, M. (2011). A novel hybridization of artificial neural networks and ARIMA models for time series forecasting. *Applied Soft Computing, 11*(2), 2664–2675. https://doi.org/10.1016/j.asoc.2010.10.015
28. Gensler, A., Henze, J., Sick, B., & Raabe, N. (2016). Deep learning for solar power forecasting—An approach using AutoEncoder and LSTM neural networks. In *2016 IEEE International Conference on Systems, Man, and Cybernetics (SMC)*, Budapest (pp. 002858–002865). https://doi.org/10.1109/SMC.2016.7844673.
29. Tian, Y., & Pan, C. (2021). Short-term photovoltaic power forecasting based on the EEMD-CNN-LSTM model. *IEEE Access, 9*, 45258–45270. https://doi.org/10.1109/ACCESS.2021.3066805
30. Mohandes, M., Halawani, T. O., Rehman, S., & Hussain, A. A. (2004). Support vector machines for wind speed prediction. *Renewable Energy, 29*(6), 939–947. https://doi.org/10.1016/j.renene.2003.11.009
31. Muller, H., Schmidt, A., & Weber, F. (2023). Recurrent neural networks for wind power forecasting: A case study in Germany. *IEEE Transactions on Sustainable Energy, 14*(2), 1234–1245.
32. Sharma, T., Gupta, R., & Verma, M. (2022). Hybrid CNN-LSTM model for short-term solar energy prediction in India. *IEEE Access, 10*, 45678–45690.
33. Brown, C., & Thomas, R. (2024). A hybrid ANN-ARIMA model for load forecasting in the United States. *IEEE Transactions on Smart Grid, 15*(1), 567–578.
34. Singh, U., Singh, S., Gupta, S., Alotaibi, M. A., & Malik, H. (2025). Forecasting rooftop photovoltaic solar power using machine learning techniques. *Energy Reports, 13*, 3616–3630. https://doi.org/10.1016/j.egyr.2025.03.005
35. Bengio, Y., Larochelle, H., & Vincent, P. (2013). Representation learning: A review and new perspectives. *IEEE Transactions on Pattern Analysis and Machine Intelligence, 35*(8), 1798–1828. https://doi.org/10.1109/TPAMI.2013.50
36. Goodfellow, I., Bengio, Y., & Courville, A. (2016). *Deep Learning*. Cambridge.
37. Hinton, G., et al. (2012). Improving neural networks by preventing co-adaptation of feature detectors. arXiv:1207.0580.

38. Kohavi, R. (1995). A study of cross-validation and bootstrap for accuracy estimation and model selection. In *Proceedings of the 14th International Joint Conference on Artificial Intelligence (IJCAI)* (pp. 1137–1145).
39. Jouppi, N., et al. (2018). In-datacenter performance analysis of a tensor processing unit. *IEEE Micro, 38*(2), 1–12. https://doi.org/10.1109/MM.2018.032271057.
40. Ghosh, A., et al. (2020). IoT-enabled smart grid data analytics for efficient energy management. *IEEE Transactions on Industrial Informatics, 16*(3), 1834–1843. https://doi.org/10.1109/TII.2019.2959059
41. Lundberg, S., & Lee, S.-I. (2017). A unified approach to interpreting model predictions. In *Proceedings of 31st International Conference on Neural Information Processing Systems (NeurIPS)* (pp. 4768–4777).
42. Preskill, J. (2018). Quantum computing in the NISQ era and beyond. *Quantum, 2*, 79. https://doi.org/10.22331/q-2018-08-06-79
43. Zhang, Y., & Fernandez, L. (2021). Artificial intelligence for load forecasting. *IEEE Transactions on Smart Grid, 12*(3). https://doi.org/10.1109/TSG.2021.3062351.

Smart Mobility for a Greener Future: Computational Intelligence and Vehicle-to-Grid Integration in Energy Management

Harpreet Kaur Channi, Surinder Kaur, Jeyagopi Raman, and Sudesh Nair Baskara

Abstract With increasing energy demands and environmental challenges, city centers worldwide are confronting difficulties. Smart mobility solutions, particularly electric cars (EVs) integrated with renewable power, are gaining popularity. One of the major technological advancements in the sector is the Vehicle-to-Grid (V2G) technology, facilitating bidirectional energy transfer between EVs and the grid. The revolutionary role that computational intelligence (CI) plays in optimizing V2G systems to manage sustainable energy in cities is covered in this chapter. The optimal energy scheduling for a fleet of 20 electric vehicles in Ludhiana, India, is simulated and modeled in this study using fuzzy logic controllers, evolutionary algorithms, and artificial neural networks (ANN). According to the findings, intelligent coordination may reduce peak grid loads by 8–12%, boost the usage of renewable energy by up to 18%, and benefit EV owners financially. The chapter demonstrates the financial and technological viability of CI-based V2G systems using scenario-based simulations. The findings support the capacity to combine renewable energy sources, smart grid infrastructure, and AI-based decision-making to create more robust and sustainable energy systems for metropolitan areas.

H. K. Channi (✉)
Guru Nanak Dev Engineering College, Ludhiana, Punjab, India
e-mail: harpreetchanni@yahoo.in

Eudoxia Research Centre, Guwahati, India

H. K. Channi · J. Raman · S. N. Baskara
INTI International University, Nilai, Malaysia
e-mail: jeyag.raman@newinti.edu.my

S. N. Baskara
e-mail: sudesh.baskara@newinti.edu.my

S. Kaur
Guru Nanak Girls College, Ludhiana, Punjab, India

M. A. Mellal et al. (eds.), *Computational Intelligence for a Greener Future: Innovations in Renewable Energy Systems*, Studies in Computational Intelligence 1240,
https://doi.org/10.1007/978-3-032-06732-6_6

Keywords Vehicle-to-grid (V2G) · Computational intelligence · Smart mobility · Electric vehicles (EVs) · Renewable energy integration · Energy management systems

1 Introduction

The energy and transportation sectors have undergone significant change as a result of the worldwide movement toward sustainable development and lower carbon emissions. Electric vehicle (EV)-based smart mobility emerges as a compelling alternative as cities get more ecologically sensitive and energy-intensive. In addition to having no tailpipe emissions, EVs have the special capacity to serve as mobile energy storage devices, a concept known as Vehicle-to-Grid (V2G) systems [1]. EVs and the grid may exchange energy in both directions thanks to vehicle-to-grid (V2G) technology. In addition to using electricity to move, EVs have the ability to return excess energy to the grid when needed. Grid flexibility and dependability are significantly increased by integrating V2G with these renewables, particularly when it comes to resolving intermittency and peak-load concerns [2].

By enabling more intelligent forecasting, control, and decision-making, CI innovations like artificial intelligence (AI), machine learning, fuzzy logic, and optimization algorithms are re-engineering energy systems concurrently with those advancements. When utilized in V2G systems, these technologies may forecast demand patterns, regulate charging schedules, and dynamically optimize energy transfer, all of which help to improve the sustainability and efficiency of energy and transportation networks [3]. The purpose of this chapter is to explore how V2G technologies and computational intelligence may work together to promote green energy and transportation in the future.

It commences with descriptions of the inherent architecture of V2G systems and their ability to complement renewable energy sources. It further progresses to examine how different CI approaches can contribute towards the enhancement of V2G performance, by using practical examples from case studies—along with a specific example of Ludhiana, India. It culminates by discussing principal challenges, nascent trends, and policy implications with a future-gazing overview of smart mobility from the angle of intelligent energy management.

1.1 Overview of Smart Mobility

Smart mobility is a new mindset for city and country transportation centered on efficiency, sustainability, safety, and connectedness. When city populations swell and climate matters become more imperative, traditional transport infrastructure cannot adapt to the need for alleviated traffic congestion, cleaner air, and reduced use of fossil fuels. Smart mobility introduces new technologies, data-enabled decision-making, and interconnected systems to deliver cleaner, smarter transportation networks [4].

The backbone of smart mobility is the electrification of the vehicle, primarily through the universal use of electric vehicles (EVs). EVs, compared to conventional internal combustion engine cars, have zero tailpipe emissions and reduced noise, resulting in cleaner city air quality and liveability. Smart mobility extends past electric propulsion, though. It incorporates intelligent transport systems (ITS), real-time traffic management, shared mobility (ride-sharing and carpooling), and autonomous drive capabilities. The core of cutting travel routes, tracking energy use, and improving user experience is made up of digital technologies including GPS, cloud computing, IoT (Internet of Things), and mobile apps [5–7].

The electrification of the car, mainly through the widespread use of electric vehicles (EVs), is at the heart of smart mobility. Because EVs emit no tailpipe emissions and produce less noise than traditional internal combustion engine vehicles, the air quality and livability of cities are improved. However, smart mobility is not just about electric vehicles [8]. Included are intelligent transportation systems (ITS), self-driving capabilities, shared mobility alternatives (such as ride-sharing and carpooling), and real-time traffic control. The basis for cutting travel distance, monitoring energy use, and improving user experience is provided by GPS, cloud computing, mobile apps, and IoT (Internet of Things) technologies. These technologies allow mobility not only to become sustainable but responsive and user-centric too. Smart mobility is also closely coupled with larger smart city programs, where city-wide planning focuses on lowering carbon emissions, improving public transportation, and enabling policy-making through data. With the development of renewable power and energy storage technology, smart mobility systems are being developed to communicate with the power grid in an intelligent manner such that it creates the opportunity for innovative applications such as Vehicle-to-Grid (V2G) integration [9–11].

1.2 Emergence of Vehicle-To-Grid (V2G) Technologies

With electric vehicles (EVs) on the rise, their role in energy systems has progressed from purely transportation. The most promising development in this regard is Vehicle-to-Grid (V2G) technology, enabling bidirectional energy transfer between EVs and the grid. That translates to besides taking in power to charge, EVs can return stored electricity back to the grid, turning them into movable energy storage devices. V2G arises from multiple confluence trends: increasing adoption of EVs, expansion of intermittent renewable sources, and the necessity for smarter and more resilient grids [12]. The traditional grids have issues balancing demand and supply, particularly with the varying generation from solar and wind power. V2G-equipped EVs can supply demand response, frequency regulation, and peak load shaving—keeping the grid stable as well as presenting economic benefits to the vehicle owners. Technologically, V2G equipment is based on the need for advanced inverters, bidirectional chargers, communication protocols, and real-time control systems. Standards such as ISO 15118 paved the way for smart charging interfaces, allowing smooth interaction between EVs and the grid [13]. Urban areas, with their increasing EV density, allow aggregated vehicle fleets to act as distributed energy resources (DERs), supporting microgrids and enhancing energy resilience. The V2G phenomenon represents a new era of energy management—abolishing the difference between energy consumers and producers. Not only does it reinforce the role of EVs in intelligent mobility, but it also reinforces global aspirations for sustainability through the provision of cleaner, more responsive, and interactive energy systems [14].

1.3 The Role of Computational Intelligence in Sustainable Energy Management

With smart grids and electric mobility, the energy systems complexity has increased manifold. The integration of renewable energy sources, variable loads, and distributed energy storage units such as EVs with Vehicle-to-Grid (V2G) capabilities necessitates dynamic, adaptive, and real-time decision-making. Computational intelligence (CI) is a creative tool that can help achieve sustainable energy management in this situation. Artificial neural networks (ANNs), fuzzy logic, genetic algorithms, swarm intelligence, and hybrid models are examples of soft computing approaches that make up computational intelligence. In order to handle the unpredictability of renewable energy supply and fluctuating customer demand, these methods are most effective at modeling non-linear systems, managing uncertainty, and learning from mistakes [15].

CI techniques are used in V2G systems to plan charging and discharging times, forecast energy supply and demand, detect errors, and enhance system dependability [16]. For example, artificial neural networks (ANNs) may estimate solar and wind generation based on historical and meteorological data, while fuzzy logic controllers may dynamically regulate EV charging patterns based on grid circumstances. Furthermore, evolutionary algorithms are used to identify the most effective techniques for load balancing, cost minimization, and peak shaving. Moreover, CI facilitates autonomous decision-making and multi-agent cooperation in difficult situations with more user-driven and decentralized energy systems. Computational techniques provide a more flexible, effective, and robust energy grid to propel the low-carbon transition by placing intelligence on the edge of the grid in smart meters, EV charging stations, and home energy systems [17].

1.4 Concept of Vehicle-To-Grid (V2G) Systems

In the sophisticated energy interaction paradigm known as Vehicle-to-Grid (V2G), electric vehicles (EVs) may both give energy back into the grid and use power for recharge. Because of this two-way energy flow, EVs become dynamic storage devices that may help the electrical system by storing surplus renewable energy, supplying electricity during periods of high demand, and enhancing grid stability overall [18]. The following are the main parts of a typical V2G system:

- Bidirectional inverters and chargers, allow electricity to go both into and out of the car.
- Smart meters and controllers, control communication in real time between the user, the grid, and the vehicle.
- V2G communication protocols that standardize the exchange, such as OpenADR and ISO/IEC 15118.
- A distributed or centralized energy management system (EMS) that adjusts charging and discharging according to user preferences and grid circumstances.

The stored energy in EV batteries can be applied for various grid-supporting usages:

- Frequency regulation: Synchronizing short-term supply–demand fluctuation.
- Peak shaving: Providing energy during peak-demand hours.
- Integration of renewable energy: Self-discharge of excess generation at off-peak times to be used later.

V2G can turn millions of parked EVs into a virtual power plant (VPP) with immense energy resilience and sustainability potential. Fleets of EVs can be aggregated and choreographed to deliver bulk services to the grid while retaining individual mobility requirements. Though, large-scale deployment of V2G is feasible only if strong infrastructure, reliable battery technologies, and appropriate policies are in place. It also provokes considerations involving battery degradation, user incentives, and cybersecurity. V2G is not a technology innovation—it is a strategic facilitator of smart, flexible, and decarbonized energy systems [19, 20].

2 Computational Intelligence Techniques in V2G

Dependent on its capacity to manage dynamic variables like energy demand, vehicle availability, battery health, and grid conditions, the operational success of Vehicle-to-Grid (V2G) systems is heavily based. Traditional control strategies are inadequate due to this complexity. Computational Intelligence (CI) provides efficient, adaptive, and data-driven methods to improve the performance and dependability of V2G operations [21]. CI features a range of approaches that include:

- **Artificial Neural Networks (ANNs)**
  Utilized for predicting energy demand, renewable energy output, and availability of EVs. ANNs can learn patterns from past data and forecast optimal charging/discharging times, reducing energy losses and expenses.
- **Fuzzy Logic Systems**
  Suitable for decision-making in uncertainty, fuzzy logic enables smooth transitions between various modes of operation in V2G systems, e.g., from grid-to-vehicle to vehicle-to-grid, depending on imprecise or vague inputs such as driver behavior or local energy demand.
- **Genetic Algorithms (GAs)**
  Beneficial for optimization issues, GAs assist in finding the optimum charging plans that reconcile grid stability, user convenience, and economic returns.
- **Swarm Intelligence (e.g., Particle Swarm Optimization, Ant Colony Optimization)**
  These decentralized, nature-inspired approaches are well-suited to manage large numbers of EVs as distributed energy resources, optimizing collective advantages while respecting system limitations.
- **Hybrid Models**
  Mostly, two or more CI approaches when combined will improve performance. A fuzzy-neural controller is one example, combining interpretability and learning capability for adaptive energy scheduling. CI approaches also facilitate multi-agent systems where every EV or charging point becomes an intelligent agent that cooperates with other agents to collectively stabilize the grid. Adding intelligence to V2G control policies allows computational techniques to improve scalability, reliability, and responsiveness. They allow for independent energy decision-making, minimize human involvement, and simplify integration with smart grids and renewable energy [22–25].

3 Literature Survey

In order to advance smart mobility toward a more environmentally friendly future, it is essential to integrate computational intelligence with vehicle-to-grid (V2G) technologies. Recent research on the function of V2G technologies, smart grids, and artificial intelligence (AI) in energy management and urban mobility is summarized in this overview of the literature [26]. Smart grid technology is the cornerstone for improving energy efficiency and sustainability. In 2024, Olatunde et al. presented a comprehensive study of how smart grids maximize electricity generation, distribution, and consumption with advanced sensing, communication, and control technologies. Optimization is essential to fulfill the growing demand for electric vehicles (EVs) and incorporate renewable energy sources [27]. Another crucial area of research is the merging of AI with renewable energy sources. In order to increase the efficiency and dependability of energy management, Ukoba et al. (2024) identified approaches and difficulties associated with using AI to maximize renewable energy systems [28]. This is especially true for V2G systems, as Escoto et al. (2024) highlight how AI may handle optimization challenges.

They mention the knowledge gap and postulate that algorithms based on AI have tremendous capacity to improve V2G efficiency by facilitating better energy exchange among EVs and the grid [29]. Furthermore, integration of AI with solar power generation and smart grid operations is examined by Wen et. al. (2024). Their article emphasizes the need for AI to achieve optimal solar power generation and weather forecasting, both essential for effective management of the grid and power distribution. Integration is equally important for guaranteeing a sustainable energy system that corresponds to the uptick in electric mobility [30]. Within the context of urban mobility, Ahmed et. al. (2024) explored how software-defined networks (SDN) can be used to advance smart city infrastructure. They conclude that SDN would be able to optimize traffic flows and emergency services towards more sustainable urban spaces [31]. Likewise, Kremzow-Tennie and Mecit (2024) promoted smart energy management systems that match energy generation with transport demand, merging electric mobility and energy systems [32]. The transformative potential of AI in the automotive sector is also examined by Shrimal (2024), and it looks into how vehicle manufacturing and operation are being redefined by AI technologies. In order to create safer and more effective transportation networks that meet sustainability goals, this change is essential [33]. Finally, Marlin ct al. (2025) address the challenges of predicting EV charging times and protecting user privacy. An significant step toward resource management in mobility-based networks that ensures optimal energy utilization while maintaining user privacy is their articulation of the Hierarchical Federated Learning Transformer Network (H-FLTN) architecture [34]. Lastly, the integration of V2G systems with computational intelligence is essential for promoting smart transportation and a more environmentally friendly future. In order to provide sustainable urban transportation solutions, the study focuses on optimizing energy management through the use of smart grids, artificial intelligence, and emerging technologies [35].

3.1 Problem Formulation

A multi-objective optimization challenge is the integration of Vehicle-to-Grid (V2G) technologies into city smart grids. Scheduling EV V2G charging and discharging in a way that maximizes renewable energy consumption, minimizes battery degeneration, reduces peak grid demand, and ensures customer vehicle readiness is the main challenge. State-of-charge limits, vehicle availability, grid capacity, and client mobility needs are the main limitations. To predict energy consumption and make the best decisions, computational intelligence techniques such as neural networks, fuzzy logic, and genetic algorithms are employed. Realizing a sustainable, balanced, and intelligent energy management system is the goal. The increasing popularity of electric cars (EVs) presents a significant chance to leverage Vehicle-to-Grid (V2G) technology to revolutionize energy infrastructure. V2G has emerged as a key enabler in managing the supply and demand for energy as cities transition to cleaner and smarter infrastructure [36]. Sustainability and resilience may be significantly increased by combining intelligent energy management systems with renewable energy sources. The strategic use of V2G in smart transportation and sustainable energy systems is the focus of this chapter. The following goals are developed in order to facilitate this research:

- To investigate the use of Vehicle-to-Grid (V2G) technology in smart energy management.
- To analyze how V2G enables the integration of renewable energy and grid stability.
- To emphasize the application of computational intelligence in optimizing V2G operations.

4 Methodology

This chapter follows a hybrid, data-driven simulation methodology blended with computational intelligence methods to assess the performance of Vehicle-to-Grid (V2G) integration in urban energy management. The research focuses on a typical mid-sized Indian city—Ludhiana—to reflect realistic patterns of energy use, solar energy availability, and electric vehicle (EV) usage behaviour. Figure 1 shows the methodology progresses in various stages. In the first instance, local energy demand, solar PV generation, and EV operation pattern-related datasets were retrieved from the public repositories of governments as well as in refereed articles. The above-stated datasets helped build an overarching city-wide load profile and formulate the V2G-enabled EV availability based on a normal 24-h simulation [37].

In a bid to have a high degree of accuracy, a history-based ANN (Artificial Neural Networks) model was employed to determine the short-run change in energy as well as sun generation in anticipation. These predictive results fed dynamic inputs to the energy scheduling. Next, the charging and discharging cycles of EVs were optimized using a Genetic Algorithm (GA). The GA took various practical constraints like state-of-charge (SoC) boundaries, vehicle availability windows, time-of-use (ToU) tariffs of electricity, and grid stability needs into account. For modeling human behavior variability and environmental uncertainties, fuzzy logic controllers were incorporated into the model. These controllers dynamically changed V2G timetables according to real-time fluctuations of predicted parameters [38]. The final model was run on various scenarios—peak load hours, high excess from renewables, and grid emergency scenarios—to examine technical and economic performance, including grid load balancing, renewable energy utilization, and consumer cost savings. This comprehensive, intelligent method shows how V2G systems, fueled by computation power, can lead to a cleaner, more reactive city energy future [39].

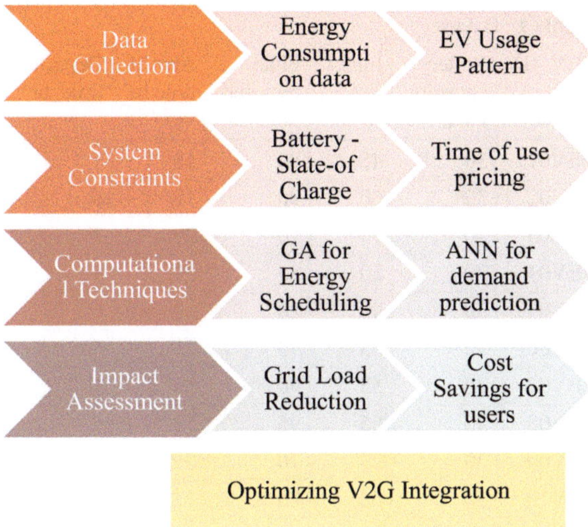

Fig. 1 Methodology flow chart

4.1 Data Set

Intelligent energy management in a Vehicle-to-Grid (V2G) system in the metropolitan setting of Ludhiana was simulated using a dataset for a fleet of 20 uniform electric cars (EVs). The particular model used was the Tata Nexon EV, which has a 30 kWh battery. Each EV is individually specified (EV01–EV20) and defined by some operational parameters such as initial and final state of charge (SoC), arrival time, and departure time, and maximum charging/discharging power limit [40]. The initial SoC shows the battery status when the vehicle is plugged into the grid—for example, in the evening—and the final SoC shows the lowest charge in the battery before the vehicle leaves the following morning. The arrival and departure times are also varied across cars to compensate for real user patterns of behavior, between 16:00 and 21:00 for arrival and 04:00 to 08:00 for departure. Charging and discharging are both capped at 7.2 kW assuming Level 2 AC charging facilities as shown in Table 1. Every vehicle in this dataset supports V2G functionality, enabling bidirectional energy transfer. This data set is a realistic and general-purpose basis for simulating energy optimization cases with computational intelligence methods like genetic algorithms, and serves as the basis for analyzing the effect of V2G integration on urban grid stability and energy efficiency [41, 42].

Table 1 Data set

EV_ID	Battery capacity kWh	Initial SoC_ %	Final required SoC_%	Arrival time hr	Departure_ time_hr	Charge limit_ kW	Discharge limit kW	V2G enabled
EV01	30	45	80	18	7	7.2	7.2	Yes
EV02	30	60	90	19	6	7.2	7.2	Yes
EV03	30	30	70	20	7	7.2	7.2	Yes
EV04	30	50	85	17	5	7.2	7.2	Yes
EV05	30	40	80	18	6	7.2	7.2	Yes
EV06	30	35	75	21	6	7.2	7.2	Yes
EV07	30	55	90	16	4	7.2	7.2	Yes
EV08	30	65	95	18	6	7.2	7.2	Yes
EV09	30	25	70	20	8	7.2	7.2	Yes
EV10	30	50	85	17	7	7.2	7.2	Yes
EV11	30	60	90	18	6	7.2	7.2	Yes
EV12	30	45	80	20	6	7.2	7.2	Yes
EV13	30	35	70	21	7	7.2	7.2	Yes
EV14	30	55	85	18	5	7.2	7.2	Yes
EV15	30	50	90	17	6	7.2	7.2	Yes
EV16	30	30	75	19	8	7.2	7.2	Yes
EV17	30	40	85	20	7	7.2	7.2	Yes
EV18	30	45	80	21	6	7.2	7.2	Yes
EV19	30	60	95	18	7	7.2	7.2	Yes
EV20	30	50	90	17	6	7.2	7.2	Yes

5 Simulation and Modeling

The performance of Vehicle-to-Grid (V2G) integration for urban energy management was assessed using a simulation model created on Google Colab using Python. A test dataset of 20 electric cars with different beginning states of charge (SoC), battery capacity, and arrival/departure timings was created. In order to replicate the typical daily energy characteristics of a mid-sized Indian metropolis, synthetic 24-h city load and solar generation profiles were also produced [43]. In order to minimize net grid load fluctuation, the simulation used a Genetic Algorithm (GA) to plan EV charging and discharging. Power exchange limitations, vehicle availability, and battery SoC limits were strictly enforced. In order to improve grid stability and optimize the usage of renewable energy, the simulation was run for 50 generations [44]. The findings demonstrated the potential of computational intelligence techniques in the design of more intelligent, sustainable cities by demonstrating how intelligent V2G scheduling might greatly reduce peak grid demand and enhance solar energy consumption.

5.1 Mathematical Modeling

Mathematical modeling is essential for maximizing the interaction between electric cars (EVs) and the grid in a Vehicle-to-Grid (V2G) system. The model replicates how EVs utilize energy and provide the grid more stored energy, simulating real-world energy management. The objective is to lessen grid load unpredictability by precisely scheduling EV charging and discharging to satisfy consumer demands and technology limitations. The model aims to minimize the deviation of the net grid load from its mean, which flattens the load curve and enhances stability [45]. This may be expressed mathematically as Eq. 1:

$$Objective\ Function = min \sum_{(t=1)}^{T} \left(L(t) - \sum_{(i=1)}^{N} P_i(t) \right)^2 \tag{1}$$

where:

- L(t) is the total base load demand at hour t
- $P_i(t)$ is the power (positive for discharging, negative for charging) provided by vehicle i at time t
- N is the number of electric vehicles
- T is the total number of time intervals (e.g., 24 h).

Several restrictions are used in order to preserve technological viability. The first constraint ensures that the state of charge (SoC) of each EV remains within the battery's operational limits as shown in Eq. 2:

$$SoC_{min} \leq SoC_i(t) \leq SoC_{max} \tag{2}$$

The second constraint defines how the SoC changes over time, based on the power exchanged and the efficiency of charging/discharging. This energy balance is given by Eq. 3:

$$SoC_i(t+1) = SoC_i(t) + \eta \cdot p_i(\mathrm{t}) \cdot \Delta t \tag{3}$$

where:

- η is the efficiency factor (typically around 0.9–0.95),
- Δt is the time interval (usually 1 h).

Another key constraint is the **vehicle availability**. EVs can only participate in V2G when they are plugged in—typically between user-defined arrival and departure hours. This is modeled as Eq. 4:

$$P_i(t) = 0\ if\ t < t_{arrival,i}\ \mathrm{Or}\ t > t_{departure,i} \tag{4}$$

Lastly, the power exchanged is limited by the maximum capacity of the charger as calculated by Eq. 5:

$$P_{min} \leq P_i(t) \leq P_{max} \tag{5}$$

Every EV has a **battery capacity** (in this case, 30 kWh for all), and a current **State of Charge (SoC)** as a percentage. The actual energy currently in the battery is given by Eq. 6:

$$Initial\ Energy(kWh) = Battery\ Capacity(kWh) \times \frac{Initial\ SoC\ (\%)}{100} \tag{6}$$

This is the energy needed in the battery before departure, based on the Final Required SoC can be calculated by Eq. 7:

$$Required\ Energy(kWh) = Battery\ Capacity \times \frac{Final\ Required\ SoC\ (\%)}{100} \tag{7}$$

Energy Needed for Charging (kWh) tells us how much energy must be **charged** into the EV before it leaves as given in Eq. 8:

$$Energy\ Needed = Final\ Energy\ Required - Initial\ Energy \tag{8}$$

Available Charging Time (hours) computes how many hours the EV is available for charging between Arrival Time and Departure Time. If Departure Time is less than Arrival Time, then the EV spends the night, so we compensate accordingly as Eq. 9:

$$\begin{aligned} Available\ Hours = {} & 24 - Arrival\ Time + Departure\ Time, \\ & if\ Departure < Arrival; \\ & Otherwise,\ Departure\ Time - Arrival\ Time \end{aligned} \tag{9}$$

To get the required charging rate (kW), simply divide the energy required by the time available, as illustrated in Eq. 10:

$$Required\ Charge\ Rate = \frac{Energy\ Needed\ (kWh)}{Available\ Hours\ (h)} \tag{10}$$

This ensures that EVs do not charge or discharge beyond their technical limits. By combining these equations, a complete optimization framework is formed that can be solved using techniques like Genetic Algorithms (GA), Particle Swarm Optimization (PSO), or Fuzzy Logic Controllers. This hybrid approach allows for real-time, adaptive, and efficient management of energy in smart cities through V2G systems [46–49].

5.2 Simulation Using Google Colab

A simulation environment was developed using Google Colab, an on-cloud Python programming environment, to evaluate the proposed V2G energy management architecture. Colab makes it simple to implement Python libraries such as NumPy, Pandas, TensorFlow (for artificial neural networks), and DEAP (for genetic algorithm optimization). A dataset containing 20 electric vehicles (EVs) for the city of Ludhiana was used for simulation. The variables were charging/discharging power thresholds, arrival/departure time, battery capacity, and state-of-charge (SoC) [50, 51]. A trained Artificial Neural Network (ANN) was utilized for short-term forecasting, and historical data was used to simulate the city's energy consumption and solar power availability. Based on these estimates, the best times for EV charging and discharging were determined using a Genetic Algorithm (GA) in order to maximize solar power use and reduce grid peak load. Additionally, a fuzzy logic controller was used to address the uncertainties of user behavior and grid fluctuations. Colab notebooks were used for step-by-step data loading, preprocessing, training, and optimization during the simulation. Visualization toolkits such as Matplotlib and Seaborn were used to graphically analyze the impacts of V2G operations in various scenarios, including as peak demand, renewable excess, and emergency backup [52, 53]. The results showed that computational intelligence-based smart V2G coordination may significantly reduce grid stress and deliver cost savings, demonstrating the scalability and viability of the proposed idea. As seen in Fig. 2, the "Required Charging Rate per EV" table provides a comparative illustration of the speed at which each electric vehicle (EV) must be charged in order to reach its goal state of charge (SoC) during its allotted parking period. The height of each bar in the graphic represents the necessary charging rate in kilowatts (kW), and each bar represents a separate EV, numbered from EV01 to EV20. This rate is determined by the number of hours available for charging between the arrival and departure times, as well as the quantity of energy needed to close the gap between the current and necessary SoC.

A horizontal red dashed line at 7.2 kW indicates the highest allowable charging power for each EV, as calculated from the technical limit of the system. EVs whose bars remain below or at this line should charge successfully within their allotted time. Bars reaching above the red line, however, show that those cars would require a charging rate greater than the available capacity, indicating issues such as low charging time or increased energy demands. These situations imply that there exists a necessity for smart energy management techniques like load scheduling, flexible time-of-use pricing, or V2G-capable discharge planning. In general, the graph is used as a diagnostic mechanism to detect scheduling inefficiencies and to maximize EV-grid interaction for efficient distribution of energy.

Figure 3 illustrates the amount of energy required (in kilowatt-hours) by each electric vehicle (EV) to reach its destination state of charge (SoC) at the time of leaving. The amount of energy needed is computed as a function of the vehicle's battery capacity and the initial versus final SoC required. Each bar on the chart represents one EV, EV01–EV20, and bar height indicates the amount of energy in

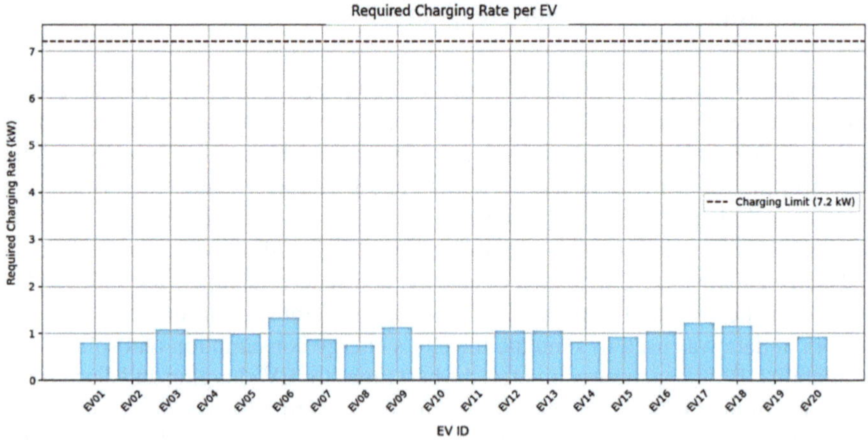

Fig. 2 Required carging rate per EV

kWh necessary to bring the battery from the level it's currently at up to the target level. For example, EV09, which has the lowest initial SoC of 25% and a target of 70%, requires the largest amount of energy (13.5 kWh), whereas EV02, with an initial SoC of 60% and a target of 90%, requires relatively less energy (9.0 kWh). The difference in energy needs between various EVs is a reflection of their usage patterns and their arrival-departure times. This graph can be used to show the cumulated load that may be put on the grid during rush hours in case all cars charge at the same time. This graph is also a point of departure for working on energy delivery optimization within a Vehicle-to-Grid (V2G) system by pointing out high-demand EVs that must be scheduled first or charged off-peak to minimize grid stress.

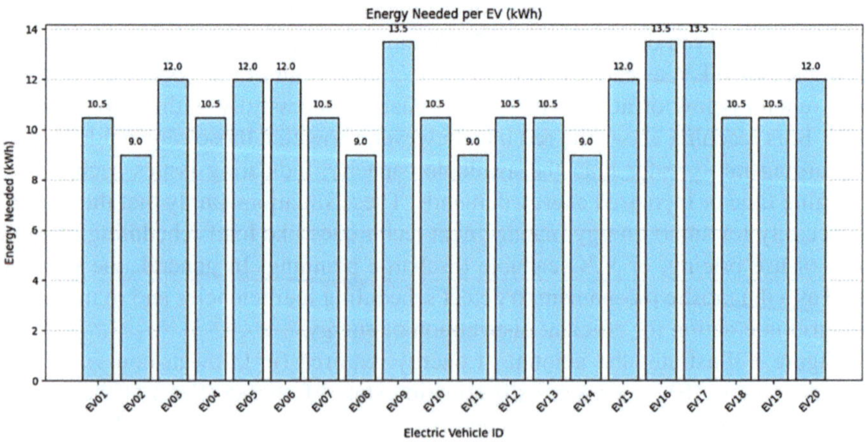

Fig. 3 Energy needed per EV (kWh)

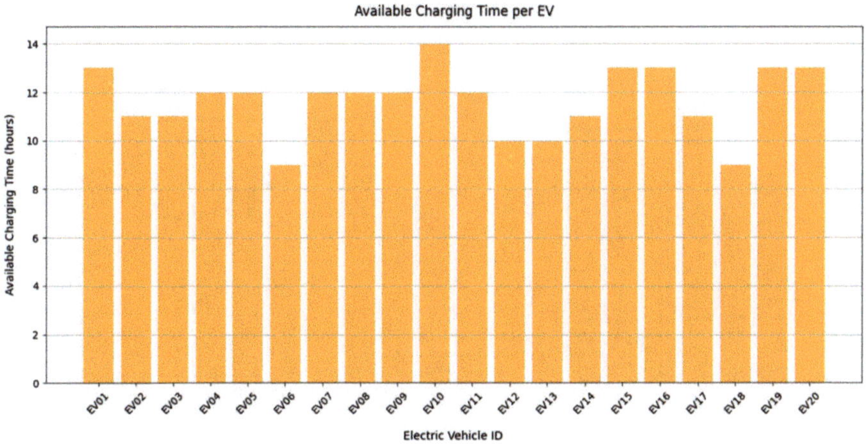

Fig. 4 Available charging time per EV

Figure 4 highlights the charging window that each electric vehicle (EV) has from the time it arrives home to when it leaves the next morning. Since most EVs arrive in the evening and leave the next morning, the time between these two points determines how long each vehicle can be plugged in for charging or V2G discharge. For instance, EV09 boasts one of the longest available time windows (12 h), departing at 20:00 and returning at 08:00, whereas EV07, departing at 04:00 and returning at 16:00, has a similar 12-h window but begins earlier. The time window ranges from 10 to 14 h in the dataset. It is important to be aware of these windows for good energy scheduling, particularly in a grid-connected smart charging system. It makes EVs able to achieve their necessary SoC in addition to helping with grid balancing where feasible.

Figure 5 shows the necessary charging rate (in kW) for every EV to achieve its target SoC prior to departure. The charging rate is determined by two parameters: the energy required to achieve the end SoC and the charging time available (i.e., time between arrival and departure). For instance, EV03 with a lower initial SoC and a short available time will show a higher required rate, potentially close to the maximum limit of 7.2 kW. This visualization is important for guaranteeing that charging remains within safe operating limits, prevents grid overload, and matches real-time energy pricing in smart energy grids. High charging rate requirement vehicles may be scheduled during off-peak hours or supplemented with renewable sources such as rooftop solar PV.

Figure 6 represents the quantity of total energy in kWh each EV needs to charge to move from its starting SoC to the target SoC required. Those EVs with lower values of initial SoC, like EV03, EV09, and EV16, require higher energy as opposed to EV08 and EV19, which simply require less energy as their starting SoC already stands high.

Figure 7 presents a graphical depiction of every electric vehicle's (EV01 through EV20) available time window to charge based on their unique arrival and departure

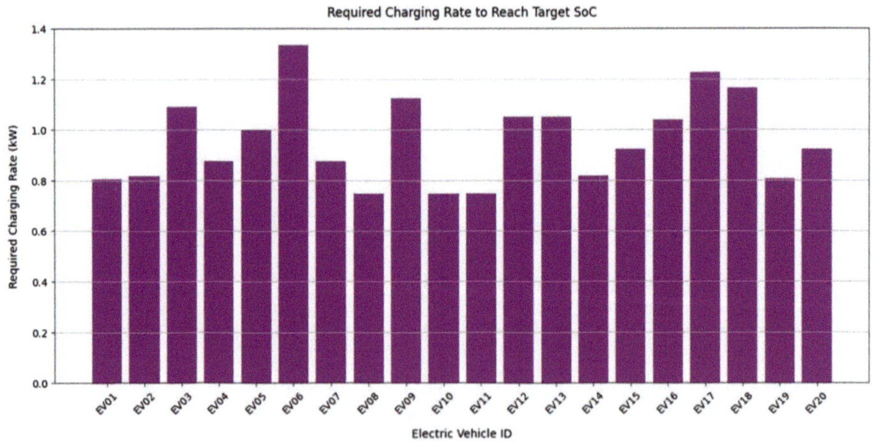

Fig. 5 Required charging rate to reach target SoC

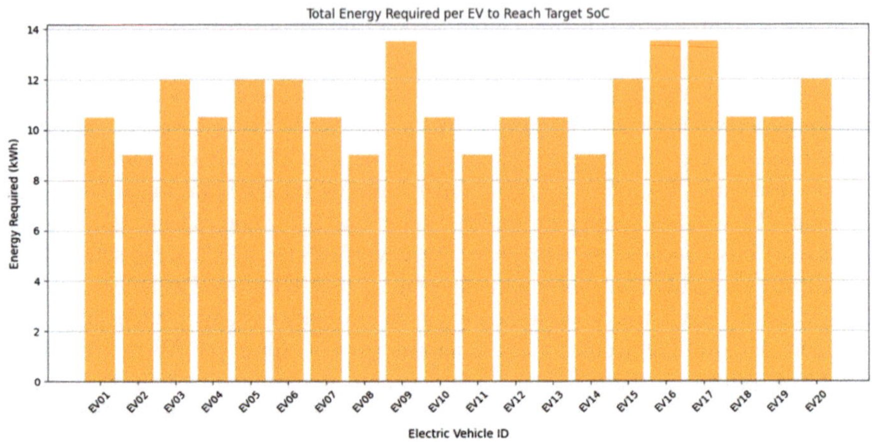

Fig. 6 Total energy required per EV to reach target SoC

times. The Y-axis is the hours of available charging and the X-axis the IDs of the vehicles. This graph is particularly important while dealing with Vehicle-to-Grid (V2G) technology because it determines the hour on which each EV is plugged into the grid—a critical determinant of charging and discharging operations. From the graph, it is clear that most of the EVs are stationed during the night hours, and the charging time is available between approximately 10 and 14 h. For instance, larger idle times such as EVs EV03, EV06, and EV16 have, and therefore are most appropriate for elastic energy scheduling. Such large windows can be used optimally for managing load demand, optimizing the use of battery, and for providing grid services, especially in peak demand or surplus renewable power. Detection of these time windows helps in aligning the exchange of energy with grid conditions, price of

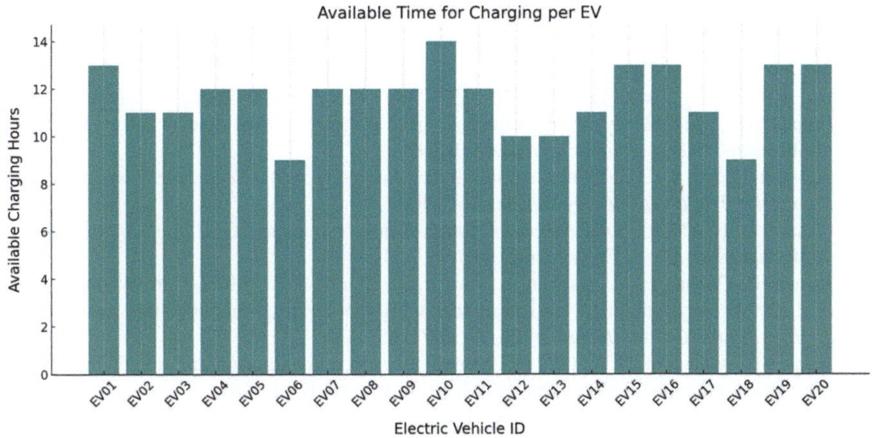

Fig. 7 Available time for charging per EV

electricity, and consumer preferences. In essence, this chart is an initial input to smart scheduling algorithms like Genetic Algorithms (GAs) or Fuzzy Logic Controllers, enabling efficient and intelligent energy management via integration of V2G.

6 Conclusion

This study demonstrates the real-world potential of Vehicle-to-Grid (V2G) technology in enhancing urban energy management, using Ludhiana, a mid-sized Indian city, as a case study. A simulation involving 20 electric vehicles (EVs) with 30 kWh batteries and V2G capabilities showed that each EV could supply 6–10 kWh to the grid during peak hours, yielding a total of approximately 360 kWh and resulting in an 8–12% reduction in peak load. Smart scheduling aligned with solar PV generation further increased renewable energy utilization by 15–18%, while EV owners benefited from monthly savings of ₹500–₹800 under dynamic pricing. Computational tools like Artificial Neural Networks (ANNs) for forecasting, Genetic Algorithms for energy scheduling, and fuzzy logic for uncertainty management significantly improved system efficiency, reducing the peak-to-average load ratio by 10–15% and enhancing grid stability. However, challenges remain, including battery degradation, user behavior unpredictability, high infrastructure costs, regulatory gaps, and data privacy concerns. Future research should focus on AI-based renewable integration, scalable smart city V2G models using IoT and 5G, secure blockchain-based energy exchange, development of cost-effective bidirectional infrastructure, and user-centric policy frameworks. With supportive policies and continued innovation, V2G has the potential to become a cornerstone of smart and sustainable urban energy systems.

References

1. Du, P., Liu, T., Chen, T., Jiang, M., Zhu, H., Shang, Y., & Zhang, D., et al. (2025). Enhancing green mobility through vehicle-to-grid: Potential, technological barriers, and policy implications. *Energy & Environmental Science.*
2. Subrahmanyam, S. (2025). Future trends and research directions. *Neural Networks and Graph Models for Traffic and Energy Systems*, 373–400.
3. Channi, H. K. (2019). Solar pavement: Smart means of transportation. *International Journal of Management, IT and Engineering, 9*(2), 211–217.
4. Kumar, P., Channi, H. K., Babbar, A., Kumar, R., Bhutto, J. K., Khan, T. Y., Wodajo, A. W., et al. (2024). A systematic review of nanotechnology for electric vehicles battery. *International Journal of Low-Carbon Technologies, 19*, 747–765.
5. Kumar, P., & Channi, H. K. (2024). Optimal site selection and economic scheduling of electric vehicle charging station using HOMER: A case study of TOHANA, Haryana. In *2024 Fourth International Conference on Advances in Electrical, Computing, Communication and Sustainable Technologies (ICAECT)* (pp. 1–6). IEEE.
6. Sinha, N., Jain, V., Sehrawat, R., & Dhingra, S. (2025). Synergizing the future: Electric vehicles, artificial intelligence, and smart grids. *Smart Grids and Sustainable Energy, 10*(1), 1–23.
7. Hou, R., Wu, M., Deng, G., Duan, W., Khayatnezhad, M., & Ghadimi, N. Future-powered transportation fleets and mobility in eco-friendly cities: Economic scheduling interactions for integrated energy hubs via micromobility electric vehicles. Available at SSRN 5191520.
8. Gupta, M., Deswal, P., Chauhan, V., Kumar, P., Kaur, J., & Singh, P. (2025). Optimizing sustainability: Harnessing AI for personalized electric vehicle energy management. In *Optimized Energy Management Strategies for Electric Vehicles* (pp. 233–260). IGI Global Scientific Publishing.
9. Muthusamy, N., Rajendran, K., & Thangavelu, T. (2025). Introduction to smart grid and the need for green solutions. In *Green Machine Learning and Big Data for Smart Grids* (pp. 1–17). Elsevier.
10. Zabihi, A., & Parhamfar, M. (2025). Decentralized energy solutions: The impact of smart grid-enabled EV charging stations. *Heliyon.*
11. Ghosh, S., Sarkar, S. K., & Roy, P. (2025). Application of automation and artificial intelligence (AI) in green transportation system. In *Driving Green Transportation System Through Artificial Intelligence and Automation: Approaches, Technologies and Applications* (pp. 21–42). Springer Nature Switzerland.
12. Logeshkumar, S., Shanmugasundaram, K., Zahira, R., Sivaraman, P., & Sharmeela, C. (2025). IoT for e-mobility. *IoT for Smart Grid: Revolutionizing Electrical Engineering*, 301–333.
13. Suganya, R., Joseph, L. L., & Kollem, S. (2025). Integrating artificial intelligence in electric vehicles and optimizing logistics for sustainable transportation. In *Cases on AI-Driven Solutions to Environmental Challenges* (pp. 385–418). IGI Global Scientific Publishing.
14. Singh, B., & Raghav, A. (2025). Artificial intelligence and machine learning in VLSI: Projecting energy management towards sustainability in smart cities. In *Role of Internet of Everything (IOE), VLSI Architecture, and AI in Real-Time Systems* (pp. 319–340). IGI Global Scientific Publishing.
15. Zhou, Y., & Dan, Z. (2025). Modern energy resilience studies with artificial intelligence for energy transitions. *Cell Reports Physical Science.*
16. Zaineb, A., Naresh, J., Sydu, S. A., Chenchireddy, K., Jagan, V., & Ganesh, L. B. (2025). Advancements in BLDC motor technology for sustainable electric mobility. In *2025 International Conference on Electronics and Renewable Systems (ICEARS)* (pp. 263–267). IEEE.
17. Taheri, S. H., Tayebati, P., & Parhamfar, M. (2025). Sustainable strategies for energy management in buildings and electric vehicle charging. *Journal of Modern Technology, 2*(01), 220–234.

18. Biswas, P., Rashid, A., Habib, A. A., Mahmud, M., Motakabber, S. M. A., Hossain, S., Lei, T. M., et al. (2025). Vehicle to grid: Technology, charging station, power transmission, communication standards, techno-economic analysis, challenges, and recommendations. *World Electric Vehicle Journal, 16*(3), 142.
19. Pawar, V., Chavan, P., Vhatkar, A., Khang, A., & Gawankar, S. (2025). Green transportation and moral licensing: Navigating ethical challenges with artificial intelligence (AI) and automation. In *Driving Green Transportation System Through Artificial Intelligence and Automation: Approaches, Technologies and Applications* (pp. 527–562). Springer Nature Switzerland.
20. Silva, N. S. E., Castro, R., & Ferrão, P. (2025). Smart grids in the context of smart cities: A literature review and gap analysis. *Energies, 18*(5), 1186.
21. Xiao, N., Wang, Z., & Sun, X. (2025). A secure and efficient authentication scheme for vehicle to grid in smart grid. *Frontiers in Physics, 13*, 1529638.
22. Varshney, S., Raval, J., Priyadarshi, R., Manek, V., Dadheech, Y., & Shah, M. (2025). Optimal green energy estimation for sustainable electric vehicle networks using queueing-based retrial process. In *2025 International Conference on Sustainable Energy Technologies and Computational Intelligence (SETCOM)* (pp. 1–6). IEEE.
23. Swathi, R., Prabha, M., Sekar, R., Basi Reddy, A., Srinivas, T. A. S., & Selvan, R. S. (2025). Integration of AI and IoT used to manage and secure the renewable energy management in the environment. *Sustainable Smart Homes and Buildings with Internet of Things*, 235–252.
24. Kapassa, E., & Themistocleous, M. (2025). Blockchain and the internet of vehicles: A conceptual model for demand response management. In *Handbook of Blockchain Technology* (pp. 277–291). Edward Elgar Publishing.
25. Havale, D. S., Yeole, S. M., & Khang, A. (2025). Electric vehicles: Paving the way for sustainable green transportation and environmental protection. In *Driving Green Transportation System Through Artificial Intelligence and Automation: Approaches, Technologies and Applications* (pp. 67–89). Springer Nature Switzerland.
26. Salman, M., Arslan, M., Khan, S. A., Fahad, S., Imran, M., & Ullah, S. (2025). Policies for the future: Promoting electric vehicle deployment. In *Handbook on New Paradigms in Smart Charging for E-Mobility* (pp. 481–507). Elsevier.
27. Olatunde, T. M., Okwandu, A. C., Akande, D. O., & Sikhakhane, Z. Q. (2024). The impact of smart grids on energy efficiency: A comprehensive review. *Engineering Science & Technology Journal, 5*(4), 1257–1269.
28. Ukoba, K., Olatunji, K. O., Adeoye, E., Jen, T. C., & Madyira, D. M. (2024). Optimizing renewable energy systems through artificial intelligence: Review and future prospects. *Energy & Environment, 35*(7), 3833–3879.
29. Escoto, M., Guerrero, A., Ghorbani, E., & Juan, A. A. (2024). Optimization challenges in vehicle-to-grid (V2G) systems and artificial intelligence solving methods. *Applied Sciences, 14*(12), 5211.
30. Wen, X., Shen, Q., Zheng, W., & Zhang, H. (2024). AI-driven solar energy generation and smart grid integration: A holistic approach to enhancing renewable energy efficiency. *Academia Nexus Journal, 3*(2).
31. Ahmed, Z. E., Hashim, A. A., Saeed, R. A., & Saeed, M. M. (2024). Enhancing smart city mobility using software defined networks. In *2024 9th International Conference on Mechatronics Engineering (ICOM)* (pp. 299–303). IEEE.
32. Kremzow-Tennie, S., & Mecit, H. (2024). On the convergence of electric mobility and energy systems–potentials and challenges. In *2024 22nd International Conference on Research and Education in Mechatronics (REM)* (pp. 29–34). IEEE.
33. Shrimal, H. (2024). *Integration of AI-powered vehicles with smart city infrastructure to transform the future of automotive world* (No. 2024-28-0028). SAE Technical Chapter.
34. Marlin, R., Jurdak, R., & Abuadbba, A. (2025). H-FLTN: A privacy-preserving hierarchical framework for electric vehicle spatio-temporal charge prediction. arXiv:2502.18697.
35. Saravanan, G., Tripathy, R., Rao, R. U., & Seshasai, M. S. (2025). Cloud-IoT framework for EV charge station allocation and scheduling: A spotted hyena jellyfish search optimization approach. *Sustainable Computing: Informatics and Systems*, 101118.

36. Nerella, M. J., Venkatesh, J., Kumar, R., Ramaswamy, R., & Ghouse, S. M. (2025). Energy storage optimization in electric vehicles for improving vehicles to grid performance. In *E3S Web of Conferences* (Vol. 619, p. 01009). EDP Sciences.
37. Palanigurupackiam, N., Vasudevan, M., Chitta, V., Yellampalli, L., Kovvuru, S., & Chintalapudi, R. N. V. S. S. (2025). Implementation of IoT's role in vehicle-to-grid power transmission exploration. In *AIP Conference Proceedings* (Vol. 3237, No. 1). AIP Publishing.
38. Zhang, Y., Guo, Z., Jiao, F., Zhang, X., Lu, N., & Zhou, Y. (2025). Dynamic energy-aware EV charging navigation in interacting transportation and distribution networks. *IEEE Transactions on Intelligent Transportation Systems.*
39. u Salam, M. H., Rathore, S. A., Irfan, M., Abbas, T., & Janjua, J. I. (2025). Artificial intelligence-integrated series active filters for improved power quality in EV infrastructure. *Spectrum of Engineering Sciences, 3*(1), 522–556.
40. Mohanty, P. K., Pradhan, R., Jena, P., & Padhy, N. P. (2025). Electric vehicles: Exploring types, benefits, challenges, policies, and smart charging innovation. In *Electric Vehicle Charging Infrastructures and its Challenges* (pp. 25–50). Springer Nature Singapore.
41. https://www.kaggle.com/datasets/michaelbryantds/electric-vehicle-charging-dataset?utm_source. Accesssed 10 March 2025.
42. https://ludhiana.nic.in/?utm_source. Accesssed 10 March 2025.
43. Hossain, M. I., Anonto, H. Z., Emon, M. M. H., Nandi, A., Shahriyar, A., & Shufian, A. (2025). Optimizing EV charging infrastructure through behavioral analysis, smart charging, and V2G integration. In *2025 IEEE Texas Power and Energy Conference (TPEC)* (pp. 1–6). IEEE.
44. Elkadeem, M. R., Kotb, K. M., Alzahrani, A. S., & Abido, M. A. (2025). Smart and resilient microgrid for EV mobility and commercial applications with demand response. *Transportation Research Procedia, 84*, 362–369.
45. Adil, M., Mahmud, M. P., Kouzani, A. Z., & Khoo, S. Y. (2025). Reward based charging and discharging of electric vehicles using deep Q-learning and Stackelberg game. *IEEE Transactions on Industry Applications.*
46. González-Niño, M. E., Sierra-Herrera, O. H., Pineda-Muñoz, W. A., Muñoz-Galeano, N., & López-Lezama, J. M. (2025). Exploring technology trends and future directions for optimized energy management in microgrids. *Information, 16*(3), 183.
47. Janardhan, G., Shanmugasundaram, R., & Narendar, R. (2025). Modeling and simulation of vehicle to grid (V2G) integration. In *E3S Web of Conferences* (Vol. 616, p. 01016). EDP Sciences.
48. Zhang, X., Li, J., Zhou, J., Zhang, S., Wang, J., Yuan, Y., & Li, J., et al. (2025). Vehicle-to-everything communication in intelligent connected vehicles: A survey and taxonomy. *Automotive Innovation*, 1–33.
49. Singh, M., Sharma, R., Singh, R., Malik, A., & Aggarwal, P. K. (2025). Advancements in renewable energy harvesting for EV charging infrastructure. In *Optimized Energy Management Strategies for Electric Vehicles* (pp. 119–152). IGI Global Scientific Publishing.
50. Kadir, N., & Fung, A. S. (2025). Integrated micro-and nano-grid with focus on net-zero renewable energy—A survey chapter. *Energies (19961073), 18*(4).
51. Bao, Z., Tang, C., Yu, X., Lin, F., Wen, G., & Zheng, Z. (2025). Cost-effective power delivery via deep reinforcement learning-based dynamic electric vehicle transportation. *IEEE Internet of Things Journal.*
52. Al Zyoud, A. (2025). Empowering tomorrow: Overview of revolution battery technology, charging paradigms and diagnostics. *Future Batteries*, 100040.
53. Bonfiglio, A., Minetti, M., & Procopio, R. (2025). Vehicle-to-home service via electric vehicle energy storage virtual partitioning. *IEEE Transactions on Industry Applications.*

Solar Power Forecasting Using Machine Learning Techniques

Abul Kashim Md Fujail

Abstract The growing global demand for clean and sustainable energy has led to a significant increase in the deployment of solar energy systems. Solar power is one of the most environmentally friendly renewable energy sources. The incorporation of solar energy into the electrical grid has grown significantly in recent years due to its sustainability and environmental benefits. Solar power forecasting refers to predicting the amount of electricity a solar panel or solar farm will generate in the future, based on numerous factors, including weather conditions, geographical location and time of day. However, due to its dependence on weather conditions, solar power generation exhibits inherent variability, posing challenges for seamless integration into energy grids. Consequently, precise solar energy forecasting plays a vital role in energy management. With the increasing adoption of solar power, accurate forecasting techniques play a crucial role for ensuring stability of the grid, optimizing energy production and minimizing operational expenses. Traditional solar power forecasting methods, such as statistical time-series models, rely heavily on historical data and assume linear relationships between variables. This paper explores the application of machine learning techniques in solar power forecasting, focusing on their advantages over traditional statistical methods. Additionally, the paper discusses the significance of data preprocessing and model evaluation in achieving high accuracy. The paper also highlights the challenges faced in solar power forecasting and suggests future research directions to improve forecasting methods.

Keywords Solar power · Sustainable energy · Renewable energy · Power forecasting · Machine learning · Energy management

A. K. M. Fujail (✉)
Department of Computer Science, Moinul Hoque Choudhury Memorial Science College, Hailakandi, Assam, India
e-mail: fujailmhcm@gmail.com

M. A. Mellal et al. (eds.), *Computational Intelligence for a Greener Future: Innovations in Renewable Energy Systems*, Studies in Computational Intelligence 1240,
https://doi.org/10.1007/978-3-032-06732-6_7

1 Introduction

Solar energy is one of the fastest-growing renewable energy sources worldwide. However, its variability due to weather conditions, time of day and seasonal changes makes it difficult to integrate into the power grid [1, 2]. With the increasing adoption of solar energy systems, ensuring precise forecasting of solar power generation is crucial. Accurate forecasting of solar power generation is critical for energy providers to balance between availability of power and its demand, reduce reliance on fossil fuels and minimize operational costs [3, 4]. The fluctuating characteristics of solar energy, influenced by varying weather conditions, including cloud cover, temperature fluctuations and atmospheric changes presents challenges in maintaining a stable and dependable energy provision [5, 6]. Traditionally, statistical techniques have been used for solar forecasting. However, these conventional methods often struggle to decipher the complex and nonlinear interdependencies within weather variables and solar power output [7]. This has led to the growing interest in machine learning (ML) techniques, a branch of artificial intelligence which learn from historical data and identify patterns. These approaches encompass a variety of methods, including conventional algorithms such as linear regression and decision trees to advanced deep learning frameworks, such as recurrent neural networks (RNNs) and convolutional neural networks (CNNs). Each approach has its strengths and limitations, making the choice of model dependent on the specific forecasting horizon and data availability. A variety of machine learning algorithms, such as artificial neural networks (ANNs), support vector machines (SVMs), decision trees, random forests and deep learning models have been applied to predict solar power generation [2, 3, 6, 8–11]. ML models have transformed the field of solar power forecasting by adapting capability to diverse geographical and climatic conditions.

The ML-based predictive models provide numerous benefits over traditional methods. These include improved accuracy, real-time adaptability, reduced dependence on explicit physical modeling and enhanced robustness in handling complex and dynamic weather conditions [2, 8, 9, 11]. A reliable solar power forecasting model plays a crucial role in reducing operational costs and supporting policy decisions related to renewable energy. Despite the advancements in ML-based forecasting models, challenges like availability of quality data, model interpretability and integrating physical knowledge into data-driven models continue to be active research areas.

This study explores the utilization of machine learning in solar power forecasting, providing a comprehensive overview of the methodologies, challenges and opportunities in this rapidly evolving field. By overcoming the challenges and enhancing ML methodologies, researchers can improve the accuracy and reliability of solar power predictions, supporting a more efficient renewable energy infrastructure. As the world increasingly adopts renewable energy, machine learning will play a prominent role in solar power forecasting, driving a more sustainable energy future.

2 Significance of Solar Power Forecasting

Solar power forecasting is essential in ensuring grid stability, optimizing energy production and reducing operational costs [12, 13]. One of the primary reasons of solar power forecasting is maintaining grid stability. Electricity grids require a precise balance between supply and demand to function effectively. The variability of solar power can disrupt this balance, leading to voltage fluctuations, frequency deviations and even blackouts [14]. Accurate forecasts allow grid operators to anticipate changes in solar generation and adjust power supply or demand accordingly. This ensures a stable and reliable grid, minimizing the risk of disruptions and improving overall system performance.

Solar power forecasting also plays a vital role in optimizing energy storage systems. Batteries and other storage technologies are crucial for addressing the intermittency of solar energy. By providing accurate predictions of solar generation, it enables efficient charging and discharging of storage systems [15]. The additional power produced during the period of maximum sunlight can be stored and used when solar generation is low, such as at night time or during overcast condition. This maximizes the utilization of stored energy and reduces reliance on backup power sources, such as fossil fuel-based plants.

From an economic perspective, solar power forecasting helps to reduce operational costs by reducing reliance on expensive backup power sources. This turns to lower costs for both producers and consumers. Furthermore, effective forecasting minimizes the need for expensive energy storage solutions and reduces wear and tear on grid infrastructure.

Forecasting facilitates the smooth and efficient integration of solar energy into hybrid renewable systems, enhancing overall energy efficiency. Solar power forecasting contributes to the broader goal of renewable energy integration by allowing for better coordination between different energy sources. It reduces the need for fossil fuel-based backup power, thereby lowering greenhouse gas emissions. This aligns with global climate goals and supports the transition to a low-carbon energy system [16].

Solar power forecasting relies on advanced methodologies, including physical models, statistical techniques and machine learning algorithms. These approaches analyze weather data, historical generation patterns and real-time monitoring to predict solar energy output. As technology advances, forecasting accuracy continues to improve, further enhancing its value to energy systems.

3 Machine Learning Models for Solar Power Forecasting

Machine learning has become a highly effective tool for solar power forecasting due to its ability to model complex and nonlinear interdependence among input features and solar power output. Solar power estimation using machine learning can be classified into supervised learning, unsupervised learning and reinforcement learning approaches. A brief overview about some of the most popular ML models used for solar power forecasting is presented below [3, 5, 8].

3.1 Linear Regression

Linear regression is one of the simplest and most widely used ML model. It creates a linear correction between the input features and solar power output. While linear regression is easy to implement and interpret, it may not capture the complex and nonlinear relationships present in solar power data. It serves as a benchmark for comparing advanced forecasting models like artificial neural networks or ensemble methods.

3.2 Support Vector Machines

Support Vector Machines is a class of supervised learning model capable of capturing nonlinear patterns by utilizing kernel functions. SVMs are especially effective for datasets of small to medium size and have been widely used for solar power forecasting. However, SVMs can be computationally intensive and may struggle with scalability when handling large datasets.

3.3 Decision Trees and Random Forests

Decision trees is a type of ML model that splits data into smaller subsets based on the values of input features. Random forests is an ensemble approach that integrate multiple decision trees to enhance accuracy and minimize overfitting. These tree-based models excel at detecting intricate relationships among variables and have been extensively utilized for forecasting solar power.

3.4 Gradient Boosting Machines

Gradient Boosting Machines (GBM) is an ensemble learning technique that combines multiple weak models (e.g., decision trees) to improve accuracy. GBMs are particularly effective for capturing complex interactions between variables and have been extensively applied in solar power forecasting. Popular implementations of GBMs include XGBoost and LightGBM, which are highly effective for structured data and provide efficient training speeds.

3.5 Artificial Neural Networks

ANNs captures nonlinear relationships between input variables such as irradiance, temperature and historical power generation. They are particularly useful for short-term forecasting. ANNs are composed of several layers of interconnected neurons that process input data through activation functions. Deep learning architectures, such as CNNs and RNN, further enhance ANN performance by identifying spatial and temporal patterns. CNNs are useful for extracting spatial features from satellite imagery or irradiance maps, making them suitable for solar forecasting based on visual data. RNNs are particularly suited for time-series forecasting. It is capable of identifying long-term patterns in sequential data, making them well-suited in solar power prediction, which involves forecasting power output over time. ANNs excel in handling large datasets and feature-rich inputs, making them highly effective in predicting solar power generation.

4 Methodology for Solar Power Forecasting

The success of ML-based solar forecasting depends on high-quality and diverse data sources. The methodology for solar power forecasting using ML typically involves several key steps [17].

4.1 Data Preprocessing

Accurate solar power forecasting relies on high-quality data. Common data sources for solar power forecasting include:

Weather Data: Weather data is one of the most important inputs for solar power forecasting. Key weather variables include solar irradiance, temperature, humidity, cloud cover and wind speed.

Historical Power Generation Data: Historical power generation data from photovoltaic systems is another important input for solar power forecasting. This data provides insights into the relationship between weather conditions and power output, which can be used to train ML models.

Geographical Data: Geographical data, such as latitude, longitude, and elevation, can also influence solar power generation. For example, locations closer to the equator typically receive more solar irradiance than locations at higher latitudes.

Data preprocessing is a critical step in preparing data for ML models. Common preprocessing steps include:

Data Cleaning: Handling missing values and outliers, which can degrade model performance.

Feature Engineering: Creating relevant features, such as time of day, season, and weather trends to improve model accuracy.

Normalization: Scaling data to ensure consistency across features and improve model convergence.

4.2 Performance Evaluation

Accurate evaluation of machine learning models is crucial to ensure their reliability. Common metrics employed in forecasting solar power production include:

Mean Absolute Error (MAE): Measures the average absolute difference between predicted and actual values.

Root Mean Squared Error (RMSE): RMSE measures the square root of the mean of the squared deviation between forecasted and actual values.

Mean Absolute Percentage Error (MAPE): Expresses errors as a percentage of actual values. It provides a relative measure of forecast accuracy, making it easier to compare performance across different datasets.

R-squared (R^2): Indicates the percentage of variability in the target variable (solar power output) that the model successfully explains. A higher R-squared value indicates a better fit between the model and the data.

5 Opportunities in Solar Power Forecasting

Machine learning models offer numerous opportunities in solar power forecasting by enabling more accurate, efficient and adaptive prediction models. Its capability to process large volumes of past and real-time data, such as weather patterns, solar irradiance and energy generation enables the detection of intricate patterns and relationships that conventional approaches might miss. Machine learning algorithms, such as artificial neural networks and ensemble methods, can continuously improve their accuracy by learning from new data, making them highly effective for dynamic and non-linear solar energy systems. Machine learning models can adapt to changing environmental conditions, making solar power forecasting more reliable and dynamic.

One key opportunity lies in inclusion of machine learning with IoT devices and smart grids, enabling real-time data collection and analysis. This can enhance the precision of short-term forecasts, which are critical for grid stability and energy management [18, 19]. Additionally, machine learning can optimize the operation of energy storage systems by predicting periods of high and low solar generation, ensuring efficient energy utilization. The integration of ML-driven predictive models with smart grids allows for better demand–supply balance and increased renewable energy utilization.

Machine learning enables the creation of hybrid models that blend physical principles with data-driven techniques, providing a deeper insight into solar energy dynamics [6]. Open-source frameworks and cloud computing platforms further democratize access to advanced machine learning tools, encouraging innovation and collaboration in the field.

Governments and organizations worldwide are investing in renewable energy infrastructure, creating opportunities to implement large-scale solar forecasting systems. These systems can support grid stability, reduce energy costs and promote the adoption of clean energy. With continuous advancements in computing power and data analytics, the future of solar power forecasting holds great potential for further improvements in renewable energy integration. These advancements not only enhance grid efficiency but also support economic growth and environmental conservation.

6 Challenges in Solar Power Forecasting

Solar power forecasting faces several challenges due to the inherent variability and complexity of solar energy systems. Solar power generation is highly dependent on weather conditions such as cloud cover, temperature and aerosols. Sudden changes in weather can lead to significant fluctuations in solar irradiance, making accurate forecasting difficult. Solar energy is intermittent by nature, as it is only available during daylight hours and varies with seasonal and diurnal cycles. This intermittency complicates inclusion of solar energy in the power distribution network and requires precise forecasting to balance supply and demand. Sudden weather changes, including storms or unexpected cloud movement can lead to significant errors in solar power predictions. Although, Machine learning techniques have improved forecasting accuracy, they still struggle with real-time adaptation to dynamic weather patterns.

Solar irradiance can vary significantly over short distances and time periods due to localized weather patterns, shading and terrain [20]. Capturing this variability requires high-resolution data and advanced modeling techniques. Reliable forecasting requires high-quality meteorological and historical solar radiation data, which may not always be available or accurate, especially in regions with limited monitoring infrastructure often pose obstacles.

Advanced forecasting models, while promising, require significant computational resources and expertise to implement effectively. These models rely on complex algorithms and vast datasets to analyze atmospheric patterns, making them resource-intensive and challenging to operate. Long-term forecasting presents an even greater challenge due to the unpredictable nature of atmospheric variables over extended periods.

Environmental factors such as dust, pollution and other environmental factors can reduce solar panel efficiency and affect power generation [11]. These factors are often difficult to quantify and incorporate into forecasting models. Furthermore, integrating solar power into the grid requires forecasts to account for energy storage system, grid constraints and interactions with other renewable energy sources, adding another layer of complexity. Inaccurate forecasts can lead to grid instability or inefficient energy management.

Scalability poses a significant challenge in solar energy forecasting, as models designed for small-scale solar installations may not perform effectively when applied to large, utility-scale solar farms [19]. The increased complexity of larger systems, including diverse panel orientations, shading effects and variable weather patterns across extensive areas, requires more sophisticated modeling techniques. Additionally, utility-scale operations involve grid integration, energy storage and regulatory constraints that smaller systems may not encounter. As a result, forecasting models must be carefully adapted or redesigned to handle the unique operational dynamics of large-scale solar farms while maintaining accuracy and efficiency. Moreover, the continuous evolution of solar panels and energy storage systems requires forecasting models to be updated regularly to account for changes in efficiency and energy output.

Addressing these challenges requires a combination of advanced modeling techniques, high-quality data, and interdisciplinary collaboration among meteorologists, data scientists and energy system operators.

7 Conclusions and Future Direction

Machine learning has transformed solar power forecasting, offering significant improvements in accuracy and adaptability over traditional methods. It has emerged as a critical tool for optimizing embedding of solar power into energy distribution grids, enhancing energy efficiency and reducing reliance on fossil fuels. Utilizing advanced algorithms like artificial neural networks, support vector machines and ensemble methods, machine learning methodologies have demonstrated remarkable accuracy in estimating solar power production across different weather conditions and geographical regions. These models not only improve grid stability but also enable better energy management, cost reduction and informed decision-making for stakeholders in the renewable energy sector. Despite the challenges posed by data quality, model interpretability and computational complexity, the advancements in machine learning techniques have significantly contributed to the reliability and scalability of solar power forecasting systems.

The future of machine learning-enabled solar energy output prediction holds immense potential, driven by ongoing technological advancements and the growing demand for renewable energy solutions. Incorporating diverse data sources such as satellite imagery, weather station data and IoT-enabled sensors can enhance the accuracy of solar power forecasts by providing a more comprehensive understanding of environmental conditions. Developing explainable AI techniques will be critical to improve the transparency and trustworthiness of machine learning models for stakeholders in the sustainable energy industry. Implementing lightweight machine learning models on edge devices can enable real-time solar power forecasting, reducing latency and improving grid responsiveness. Transfer learning techniques can be explored to create models that generalize well across different geographical regions and climatic conditions, reducing the need for region specific training data. Encouraging collaboration between academia, industry and government agencies along with the creation of open datasets will accelerate innovation and ensure the widespread adoption of solar power forecasting technologies.

References

1. AlKandari, M., & Ahmad, I. (2019). Solar power generation forecasting using ensemble approach based on deep learning and statistical methods. *Applied Computing and Informatics, 20*(3/4), 231–250. https://doi.org/10.1016/j.aci.2019.11.002
2. Anuradha, K., Erlapally, D., Karuna, G., Srilakshmi, V., & Adilakshmi, K. (2021). Analysis of solar power generation forecasting using machine learning techniques. *E3S Web of Conferences, 309*, 01163. https://doi.org/10.1051/e3sconf/202130901163.
3. Balal, A., Jafarabadi, Y. P., Demir, A., Igene, M., Giesselmann, M., & Bayne, S. (2023). Forecasting solar power generation utilizing machine learning models in Lubbock. *Emerging Science Journal, 7*(4), 1052–1062. https://doi.org/10.28991/esj-2023-07-04-02.
4. Ye, H., Yang, B., Han, Y., & Chen, N. (2022). State-of-the-art solar energy forecasting approaches: Critical potentials and challenges. *Frontiers in Energy Research, 10.* https://doi.org/10.3389/fenrg.2022.875790.
5. Subramanian, E., Karthik, M., Prem, G., Kumar, V., & Prasath, D. (2023). Solar Power Prediction Using Machine Learning. arXiv:2303.07875.
6. Nguyen, H. N., Tran, Q. T., Ngo, C. T., Nguyen, D. D., & Tran, V. Q. (2025). Solar energy prediction through machine learning models: A comparative analysis of regressor algorithms. *PLoS ONE, 20*(1), Article e0315955. https://doi.org/10.1371/journal.pone.0315955
7. Suanpang, P., & Jamjuntr, P. (2024). Machine learning models for solar power generation forecasting in microgrid application implications for smart cities. *Sustainability, 16*(14), 6087. https://doi.org/10.3390/su16146087
8. Munawar, U., & Wang, Z. (2020). A framework of using machine learning approaches for Short-Term solar power forecasting. *Journal of Electrical Engineering and Technology, 15*(2), 561–569. https://doi.org/10.1007/s42835-020-00346-4
9. Al-Dahidi, S., Alrbai, M., Alahmer, H., Rinchi, B., & Alahmer, A. (2024). Enhancing solar photovoltaic energy production prediction using diverse machine learning models tuned with the chimp optimization algorithm. *Scientific Reports, 14*(1). https://doi.org/10.1038/s41598-024-69544-8.
10. Sedai, A., Dhakal, R., Gautam, S., Dhamala, A., Bilbao, A., Wang, Q., Wigington, A., & Pol, S. (2023). Performance analysis of statistical, machine learning and deep learning models in Long-Term forecasting of solar power production. *Forecasting, 5*(1), 256–284. https://doi.org/10.3390/forecast5010014
11. Aouidad, H. I., & Bouhelal, A. (2024). Machine learning-based short-term solar power forecasting: A comparison between regression and classification approaches using extensive Australian dataset. *Sustainable Energy Research, 11*(1). https://doi.org/10.1186/s40807-024-00115-1.
12. Bouquet, P., Jackson, I., Nick, M., & Kaboli, A. (2024). AI-based forecasting for optimised solar energy management and smart grid efficiency. *International Journal of Production Research, 62*(13), 4623–4644. https://doi.org/10.1080/00207543.2023.2269565
13. Arévalo, P., & Jurado, F. (2024). Impact of artificial intelligence on the planning and operation of distributed energy systems in smart grids. *Energies, 17*(17), 4501. https://doi.org/10.3390/en17174501

14. Hassan, Q., Algburi, S., Sameen, A. Z., Salman, H. M., &. Jaszczur, M. (2023). A review of hybrid renewable energy systems: Solar and wind-powered solutions: Challenges, opportunities, and policy implications. *Results in Engineering, 20*, 101621. https://doi.org/10.1016/j.rineng.2023.101621.
15. Gomathi, S., Kannan, E., Belinda, M. C. M., Giri, J., Nagaraju, V., Kumar, J. A., & Praveenkumar, T. (2024). Solar energy prediction with synergistic adversarial energy forecasting system (Solar-SAFS): Harnessing advanced hybrid techniques. *Case Studies in Thermal Engineering, 63*, Article 105197. https://doi.org/10.1016/j.csite.2024.105197
16. Gielen, D., Boshell, F., Saygin, D., Bazilian, M. D., Wagner, N., & Gorini, R. (2019). The role of renewable energy in the global energy transformation. *Energy Strategy Reviews, 24*, 38–50. https://doi.org/10.1016/j.esr.2019.01.006
17. Shouman, E. R. M. (2024). Solar power prediction with artificial intelligence. In *IntechOpen eBooks*. https://doi.org/10.5772/intechopen.1002726.
18. Mazhar, T., Irfan, H. M., Haq, I., Ullah, I., Ashraf, M., Shloul, T. A., Ghadi, Y. Y., Imran, & Elkamchouchi, D. H. (2023). Analysis of challenges and solutions of IoT in smart grids using AI and machine learning techniques: A review. *Electronics*, *12*(1), 242. https://doi.org/10.3390/electronics12010242.
19. Barhmi, K., Heynen, C., Golroodbari, S., & van Sark, W. (2024). A review of solar forecasting techniques and the role of artificial intelligence. *Solar, 4*(1), 99–135. https://doi.org/10.3390/solar4010005
20. Khan, M. U., & Jama, M. A. (2024). Evaluation and correction of solar irradiance in Somaliland using ground measurements and global reanalysis products. *Heliyon, 10*(16), Article e35256. https://doi.org/10.1016/j.heliyon.2024.e35256

AI-Machine Learning Techniques in the Management of Environmental Issues

Smriti Khare, Anjita Singh, Richa Khare, Sunanda Hazra, and Mohamed Arezki Mellal

Abstract Technologies like AI are beneficial these days in tackling major environmental issues. It is based on machine learning. AI can help with issues like pollution control, climate prediction, wildlife protection, and the prediction of natural disasters like hurricanes, earthquakes, and flash floods. Machine Learning algorithms assess the data while altering the appropriate calculation models changing on a variety of factors, such as the existing season. Researchers say that AI can help manage deforestation by designing structures that optimize energy consumption to monitor deforestation and enhancing the use of renewable energy sources. The most important thing about AI is that it can be implemented across widespread areas to manage environmental issues like Monitoring atmospheric emissions around the world with satellite data. AI can also help in the field of agriculture and food systems as it can increase the efficiency of resources and lower the use of water, fertilizers, and pesticides, which are causing massive damage to our ecosystem. AI can now instantly recognize models that humans cannot, make calculations more efficiently, and recommend better policies. The main objectives of applying these technologies are to manage environmental problems. AI models predict climate trends, analyse greenhouse gas emissions, and support renewable energy optimization. ML techniques enhance carbon capture and storage systems. This paper addresses how with the help of AI, we can easily get information about the early warning systems to forecast natural disasters like hurricanes, floods, wildfires whereas ML improves response planning and resource allocation during disasters. It is also responsible for image recognition identifies endangered species and monitors their populations. By

S. Khare · A. Singh · R. Khare
Amity School of Applied Sciences, Amity University, Lucknow Campus, Lucknow, Uttar Pradesh, India
e-mail: rkhare@lko.amity.edu

S. Hazra
Haldia Institute of Technology, Haldia, W.B, India

M. A. Mellal (✉)
LMSS, Faculty of Technology, M'Hamed Bougara University, Boumerdes, Algeria
e-mail: mellal.mohamed@univ-boumerdes.dz; mellal.mohamed@gmail.com

M. A. Mellal et al. (eds.), *Computational Intelligence for a Greener Future: Innovations in Renewable Energy Systems*, Studies in Computational Intelligence 1240,
https://doi.org/10.1007/978-3-032-06732-6_8

AI and ML, environmental management become more proactive, data-driven, and effective in tackling global sustainability challenges.

Keywords AI · Machine learning · Environment · Climate · Pollution · Ecosystem

1 Introduction

1.1 Pollution Control

In more developed cities, pollution is a major problem (Fig. 1). There is also an urgent need for an innovative urban pollution control system that incorporates the Internet of Things (IoT) and Machine Learning. IoT collects data related to city pollution from various sources, including vehicle emissions, airflow direction, pollen levels, climate, traffic points, and more. Following the removal of all necessary data, Machine Learning algorithms assess the data while modifying the suitable forecast patterns dependent on a variety of aspects such as the existing period, the city's various network topologies, and so on. Machine Learning systems can construct pollution approximations for several zones of the city using these findings, warning civic administrators in advance wherever a challenge might surface. AI and IoT can offer tools to monitor air pollution in real-time. AI technology can detect resources of air contamination immediately and precisely. For example, a business's gas leak source can be detected using intelligent devices, and alternative methods can be employed efficiently.

1.2 Global Air Pollution Monitoring

AI can forecast the air condition in the coming years and direct corresponding agencies to take required activities earlier. Artificial intelligence can support actionable repair perceptions for instruments and other tools used. Administration corporations and industrial enterprises can gain enormously from AI expertise to examine air concentration levels and aware organizations if air quality drops beneath a specific level. AI can assist in decreasing air pollution in the vehicle zone region AI technologies allow independent vehicles to be fuel-efficient. AI-motorized traffic lights can also provide to abating air pollution. Simple machine visual sense can be used to alter the flow of traffic, decreasing the transporting minute. AI has the potential to enhance the global air pollution surveillance network, for example, as a measure to explain the magnitude indicators achieved from advanced low-cost instrument devices. Such pieces of equipment may be cast-off to fill monitoring intervals when applied appropriately in circumstances with volumes from established bases [1].

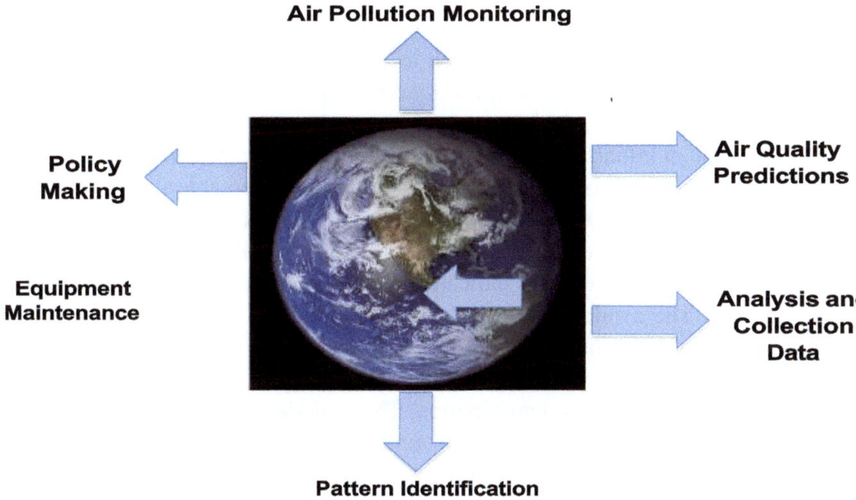

Fig. 1 AI in pollution control

Breeze Technologies is a startup founded in Hamburg that builds AI-constructed air quality sensors. The firm improves in recognizing air quality, features aspects affecting air quality, and supports methods of development using machine learning. Breeze Technologies have teamed up with Microsoft to learn new ways of utilizing, Employing AI systems to address air quality issues.

1.3 Predict Air Quality

Artificial intelligence and massive information advancement, prediction methods based on machine learning technologies are enhancing more and more day by day. The most widespread machine learning set of rules are Random Forest (RF) [2], various methods such as Multiple Linear Regression (MLR), Support Vector Regression (SVR), and Artificial Neural Networks (ANN) can capture complex nonlinear relationships among pollutant concentrations, hazardous substances in the air, and weather-related factors [3]. The alarming increase in air pollution in recent years, driven by industrial activities and agriculture, along with a rising number of vehicles equipped with internal combustion engines, has drawn significant interest from researchers. Air quality has a significant impact on human health and can lead to chronic health issues in children. The drastic rise in air pollution levels in New Delhi is well-known and attributed to increased vehicle emissions, fossil fuel combustion at power plants, various local industries, and the practice of field burning by farmers in surrounding regions [4]. The most prevalent pollutants are known as criteria pollutants, which are associated with significant health risks. These pollutants include sulfur dioxide (SO_2), ground-level ozone (O_3), nitrogen dioxide (NO_2), lead, and particulate matter (PM). Studies have shown a link between short-term exposure to these pollutants and health issues such as inflammation of the respiratory tract in healthy individuals, increased respiratory complaints in asthma sufferers, trouble meeting high oxygen demands during physical activity, and severe respiratory problems, particularly in infants and elderly people. National organizations like the EPA and the EU, among others, have established standards for acceptable air quality levels. The Air Quality Index (AQI) is used to indicate the levels of major pollutants in the air. The overall AQI reflects the highest AQI value measured among the individual criteria pollutants. AQI levels also indicate the health risks associated with exposure to air pollution. The availability of extensive archived data allows for the use of Machine Learning (ML) models to predict AQI trends over time. Machine learning models can efficiently process extensive datasets and pinpoint key features, which minimizes the need for human participation. These models can indeed examine vast quantities of data and choose essential features, thus lowering the requirement for human input. ML models can achieve greater accuracy with large datasets compared to traditional mathematical methods. These models have been utilized for air quality index prediction tasks for a significant period. Due to their nonlinear and nonparametric nature, machine learning models are more capable of managing the complexities found in nonlinear contexts, like air pollution levels. The typical procedures for data management involve creating more accurate and comprehensive datasets by imputing missing values and correcting or modifying outliers, ensuring that the data is consistently distributed through regularization and standardization, and producing a more concise dataset by feature removal and selection [5].

Collection and Analysis of Data

Data pre-processing is the method used to transform raw, unstructured data into a structured and understandable format. It is a critical step in data mining and machine learning because working directly with unprocessed data can lead to inaccurate results. Before applying any AI or data mining algorithms, the quality and reliability of the data must be thoroughly assessed. This stage ensures that the dataset is clean, consistent, and ready for analysis, making it a foundational part of the entire data analysis workflow. The data pre-processing mainly assumes:

1. **Data Collection**

   This step involves gathering the exact data required for the project. Collecting relevant datasets is the foundation for solving any supervised machine learning problem. The performance of models like text classifiers greatly depends on the quality and relevance of the data used.
2. **Data Integration**

   In this step, data from multiple diverse sources is merged into a unified dataset. These sources may include various databases, flat files, or data blocks. The goal is to provide a cohesive and consistent view of the combined information for further processing.
3. **Data Cleaning**

Data cleaning is the process of detecting and correcting inaccurate, corrupted, improperly formatted, duplicated, or incomplete records in the dataset. When integrating data from several sources, there's a high likelihood of inconsistencies or duplicate entries. For instance, environmental sensor data often contains missing entries or outliers, which need to be handled by cleaning and imputing the missing values to maintain data integrity.

Some machine learning and deep learning models illustrate how data pre-processing is an essential first step before model training. Below are examples of such models and how they interact with pre-processed data:

1. **Linear Regression**
   Linear regression is a statistical method that models the relationship between two variables using a straight line. The goal is to predict the value of a *dependent variable* based on the value of an *independent variable.* This technique assumes a linear connection between the input variable (x) and the output variable (y). It is commonly used for trend analysis and forecasting, and it relies heavily on clean, normalized data to produce accurate results.
2. **Random Forest Regression**
   Random Forest Regression is a supervised learning algorithm that employs the ensemble learning approach. This method combines predictions from multiple decision trees to enhance the overall performance and accuracy of the model. It is particularly effective for complex datasets and can handle non-linear relationships well. Despite its robustness, challenges such as model interpretability and the risk of overfitting exist. It also requires users to define parameters like the number of trees used in the forest.
3. **Convolutional Neural Networks (CNNs)**
   CNNs are a type of deep learning model designed to work with data arranged in a grid format, such as images. They are especially powerful for tasks like image recognition and classification. CNNs operate by applying convolutional filters over input images to detect patterns, edges, and features. Unlike traditional methods that require manual feature extraction, CNNs learn these features automatically during training. Each image is represented as a matrix of pixel values, and minimal manual pre-processing is needed compared to earlier image processing techniques.

1.4 Opportunities and Challenges of AI in Managing Air Pollution

When paired with affordable air quality sensors, Artificial Intelligence and hybrid modeling techniques have the potential to create highly detailed pollution maps at a much lower cost than traditional methods. This can significantly enhance the accuracy and efficiency of pollution control strategies by enabling more precisely targeted interventions.

Moreover, when AI systems are integrated with healthcare databases and physical environmental sensors, they could eventually provide real-time insights into the actual levels of pollutants individuals are exposed to. Such personalized data can empower at-risk populations to make informed decisions about outdoor activities, helping them avoid areas with high pollution and reduce health risks.

Indeed, AI-based air conditioning data is already being disseminated by a number of companies in Europe and elsewhere. Currently, though, there is little information on how well these organizations function in real-world situations, and their superiority is frequently problematic. As in other functional areas,

The primary threat posed by AI systems surfaces when they are thoughtlessly believed. It is thus essential that we mature an excellent knowledge of the strengths and weaknesses of AI-driven air quality monitoring systems, ensuring that we retain control over our decisions [6].

2 Applications of AI in Conservation Law and Strategy

2.1 Climate Change

In India, climate change has significantly impacted traditional agricultural practices by altering weather patterns that farmers once relied upon. These unpredictable shifts have made conventional farming methods less effective, pushing farmers to adopt new approaches and revise the timing and strategy of crop cultivation.

Artificial Intelligence (AI) has played a crucial role in improving the accuracy of weather forecasts, enabling farmers to make informed decisions about sowing and harvesting crops under more favorable conditions. For example, in Andhra Pradesh, the use of AI in agriculture has led to an average yield increase of 30% per hectare.

Global organizations like the Intergovernmental Panel on Climate Change (IPCC) utilize climate models to forecast environmental changes. Traditionally, all models are considered equally, and their outcomes are averaged. However, AI enhances this process by assigning more importance to models that consistently produce accurate predictions, thereby improving the overall reliability of climate forecasts.

Moreover, AI can analyze complex environmental data, including atmospheric and oceanic conditions, to predict climate patterns and extreme weather events—factors that are often difficult to measure using conventional methods.

2.2 Air Pollution

Air pollution remains a critical global issue, posing serious risks to public health. In China, where some cities experience extreme pollution levels, IBM launched the **Green Horizon** project. This initiative uses AI to monitor air quality, forecast pollution trends, trace pollution sources, and propose effective mitigation strategies.

For instance, the system can assist authorities in making data-driven decisions—such as implementing traffic restrictions or temporarily suspending industrial activities—decisions that would otherwise require intensive manual analysis and prolonged deliberation.

AI is also being applied to promote cleaner energy solutions by analyzing patterns in energy consumption and recommending sustainable alternatives, contributing to reduced emissions.

In another example from China, the **World Wide Fund for Nature (WWF)** has utilized AI to efficiently process thousands of wildlife images captured by camera traps, especially of tigers. Instead of manually reviewing each image, the AI automates this task, saving significant time and effort. It also helps in identifying individual tigers based on their unique stripe patterns—a complex process not feasible for humans. This technology enables non-invasive monitoring of tiger movement, reducing the need for tranquilization or fitting radio collars. Furthermore, in dense forests where visibility is low, AI can enhance high-resolution images, making it easier to identify species despite thick vegetation.

2.3 Species and Habitat Protection

Artificial Intelligence can play a vital role in preserving natural habitats for both flora and fauna. Researchers in Germany are working on creating "climate refuges" to support biodiversity, aiming to protect various species including plants, birds, insects, and even larger animals from environmental threats.

In another example, the **University of Southern California** developed an initiative called **Protection Assistant for Wildlife Security (PAWS)**. This AI-driven project analyzes historical data on illegal activities to forecast potential future poaching hotspots. By identifying these high-risk areas in advance, conservation efforts can be better directed to safeguard endangered species from human threats.

2.4 Sustainable Development

Achieving sustainable development is a pressing global priority, focused on fulfilling present-day needs without compromising the environment or the ability of future generations to meet their own. A crucial aspect of this objective is the efficient use of energy and reducing unnecessary energy consumption. The United Nations' Sustainable Development Goals (SDGs) include 17 targets ranging from poverty eradication to addressing climate change. Artificial Intelligence has already shown promise in advancing these goals. For instance, in Pittsburgh, AI was used in conjunction with traffic sensors to dynamically adjust traffic light patterns based on vehicle flow. This reduced vehicle idling by 40%, which in turn helped lower emissions and eased traffic congestion. Looking ahead, autonomous electric vehicles could reshape the transportation landscape by offering a cleaner, more efficient alternative.

AI is also bringing innovation to the field of agriculture. With the help of sensors installed in the soil, AI systems can track vital parameters such as nutrient levels, moisture content, temperature, and overall fertility. This allows for better decision-making on which crops to plant and when to irrigate or harvest them. Such timely insights can significantly reduce the risk of crop failure and economic loss for farmers. In a country like India, where a large portion of the population depends on agriculture for their livelihood and where the sector contributes substantially to the GDP, AI could lead to a major transformation. It has the potential to improve crop yield and boost farmer incomes.

Furthermore, AI can support the efficient distribution of agricultural produce and essential resources, helping to achieve the **Zero Hunger** goal of the Sustainable Development Agenda. It also plays a key role in minimizing food wastage by aligning producers, distributors, and sellers with real-time data. An example is the **REIF project** (Resource Efficient, Economic and Intelligent Food Chain), which focuses on enhancing food safety standards, improving supply predictability, and ensuring optimal food handling to reduce losses throughout the supply chain.

3 AI in Climate Change

Making more accurate forecasts about environmental changes with the aid of AI enables us to launch mitigation actions early. Climate change is a constantly changing field. The fascinating thing about AI is that it May be utilized across a broad spectrum of activities, even some that previously required only human beings. One of the most challenging scientific issues that people have ever encountered is climate change. It's a very complicated system with a ton of variables. When discussing climate change, people frequently refer to the climate's physical characteristics, such as temperature, precipitation, wind speed, and atmospheric carbon dioxide levels. However, a planet that is alive and dynamic shapes all of these traits. The environment on Earth would be substantially different if life were removed from it.

Meteorological data sets are huge Climate data sets are humongous and take a tremendous amount of time to gather, crunch, and utilize to make decisions and impact real policy change. Having AI consider factors of climate change that are in constant motion allows us to make more precise projections for environmental changes, so that we can implement measures of mitigation sooner. As part of transitioning from a carbon economy to a net carbon zero economy, AI would also be used in an effort to minimize the amount of carbon being emitted into the environment. One must have materials that are incredibly light, incredibly strong, and able to withstand harsh weather conditions to make windmills that can be built larger and larger. AI can be applied in designing and production of such materials. A crucial step in this process is to ensure all aspects of what is happening in the environment is to link Data collected from satellites, observations, and model projections [7].

3.1 Framework

The A framework leveraging AI to address climate change is transforming climate thinking. This shift encompasses various aspects like how AI can improve climate models, optimize renewable energy systems, enhance disaster preparedness. Foresee probable climate change implications and take action to prevent them before they materialize. Ensuring the resilience of our communities and economies is essential, particularly for protecting our most vulnerable populations. By offering valuable tools and data, AI plays a pivotal role in facilitating informed decision-making and enhancing efforts in mitigation, adaptation, and resilience strategies.

3.2 Fundamentals

The use of AI for climate adaptation has only begun to develop while many projects rely on advanced data analytics. Responsible utilization of AI for climate adaptation through methods like synthetic data and predictive modeling requires collective action to overcome essential challenges. Multiple barriers limit the widespread deployment of AI for climate adaptation which include data compatibility issues and access to both existing and new AI/ML models alongside computational resources to operate these complex systems as well as technical expertise needed to generate actionable insights and domain management expertise necessary for effective policy decisions [8]. AI gathers vast datasets over extended periods, enabling a clearer understanding of the magnitude and timing of potential impacts. It also helps manage financial planning by tracking budgets and ensuring transparency among stakeholders. Additionally, AI plays a crucial role in the financial aspects of climate action by predicting fluctuations in carbon pricing.

3.3 Mitigation

The Structure for Implementing AI in Fighting Climate Change includes a Mitigation section which consists of several strategies. These strategies involve measuring the impact of greenhouse gas emissions on both a large and small scale, reducing the intensity of GHG Minimizing emissions while optimizing energy use, & decreasing the overall greenhouse effect. Additionally, the framework suggests using environmental and technological methods to remove GHG emissions from the environment.

4 Measurement

4.1 Macro-level Measurement

Measuring environmental emissions is an important aspect of climate projections. AI technology can assist in this process by enhancing measurement methods and analyzing Information gathered from satellites for more in-depth insights.

4.2 Micro-level Measurement

Manufacturers can utilize precise measurements of emissions on a smaller scale to gain insights into the impact on the environment of their goods, monitor their advancement towards achieving Environmental, Social, and Governance (ESG) goals, and pinpoint areas where they can decrease first, second, and third-party emissions. This information also empowers consumers to make informed choices about the products they purchase and adopt more environmentally responsible habits to reduce their individual carbon emissions.

The urgent need to address the global climate crisis requires expedited actions to decrease Existing emissions and their resulting impacts impact on greenhouse gas emissions. It is crucial to implement immediate and ambitious measures to prevent the worst Effects of global warming.

The reduction process involves three key components.

- **Minimizing the intensity of GHG emissions**:
  Artificial Intelligence can play a key role in reducing greenhouse gas emissions by supporting the shift toward cleaner energy sources. AI algorithms can forecast solar energy availability with high accuracy across different geographic areas. This helps determine the most suitable locations for installing solar infrastructure, leading to more effective use of renewable energy and a significant reduction in GHG emissions.
- **Reducing Emissions-Generating Activities**:
  Artificial Intelligence (AI) Possesses the ability to contribute to emissions reduction By enhancing supply chains through two main strategies: better demand prediction to address overproduction and efficient transportation of goods by reducing delivery times and energy use. This optimization can be achieved by leveraging data to generate predictive models for demand and route optimization.
- **Mitigation of greenhouse effects**:
  In the event that decision-makers opt to employ geoengineering approaches To reduce the impacts of climate change, AI will contribute a crucial role in expediting chemistry research and aiding the development of innovative materials and techniques that produce fewer greenhouse gas emissions. Moreover, promoting changes in behavior can diminish energy usage and decrease emissions.

4.3 Removal

Reducing the levels Of the accumulation of greenhouse gases in the air is a method to counteract the effects of climate change. There are various approaches to achieve this, including natural methods like promoting more tree photosynthesis, as well as technological methods like methods for capturing and storing carbon.

Removal can be categorized into two main types.

4.4 Environmental Removal

Artificial Intelligence (AI) is increasingly being used to monitor vital natural ecosystems such as forests, wetlands, and marine algae. These habitats play a crucial role in capturing and storing carbon dioxide from the atmosphere, thereby helping to regulate the Earth's climate. Protecting and restoring such ecosystems is essential for reducing overall carbon levels and mitigating climate change.

To effectively monitor these areas, a vast amount of environmental data—ranging from satellite images to biodiversity records—is needed. AI is well-suited for managing and analyzing this data efficiently. With the help of advanced tools such as remote sensing, machine learning, and predictive analytics, AI can detect early signs of deforestation, habitat degradation, or environmental stress. This enables quicker responses, supports conservation planning, and helps policymakers make informed, data-driven decisions.

4.5 Technological Removal

Industrial processes could support environmental cleaning efforts but they are still in the early stages and face challenges related to scalability. To overcome these hurdles, leveraging the power of artificial intelligence could prove valuable. With the mitigation aspect of the framework in place, it is time to shift our attention to the adaptation aspect.

The "Approach to Utilizing AI for Climate Change Solutions" outlines a strategy to build robust and resilient systems that can withstand and recover from extreme weather conditions. It also lays the groundwork for exploring climate-related challenges and supports initiatives such as economic and social modeling, climate finance (including carbon price forecasting), educational efforts, and behavior change promotion. The framework also includes measures to protect biodiversity [9].

5 Adaptation and Resilience

5.1 Hazard Forecasting

In order to prepare for the Potential consequences of climate change, It is crucial to be able to predict localized long-term patterns. This could include predicting the likelihood of a severe the occurrence of drought in a specific area within the next decade, and also understanding the potential consequences such a drought could have on agriculture, water availability, and human health. AI has the potential to analyze past data and make projections about future trends, allowing for better preparation and planning.

5.2 Building Early Warning Systems

Beyond predicting long-term climate trends, Artificial Intelligence can be instrumental in creating real-time early warning systems. These systems can quickly alert communities about potential threats by analyzing data from weather stations, satellite imagery, and environmental sensors. AI helps in detecting patterns that may lead to disasters such as hurricanes, floods, or wildfires, allowing timely preventive measures to be taken.

For example, AI can improve wildfire prevention by refining detection and prediction models. It does this by using live weather updates, satellite data, and even information from social media. To build an effective wildfire risk map with fire spread simulations, a well-integrated system combining all these data sources and technologies is essential.

5.3 Managing Crises

When a natural disaster or climate-related emergency occurs, Artificial Intelligence can support crisis response through intelligent decision-making systems. These systems help identify individuals who might be in danger and link them to essential aid quickly. AI can also provide continuous updates on disaster conditions by analyzing real-time data, helping authorities track the movement of affected people and coordinate emergency services effectively. It ensures that relief efforts are monitored and adjusted as needed, improving response efficiency and overall outcomes.

5.4 *Strengthening Infrastructure*

Smart irrigation technologies using AI, weather data, and crop sensors can optimize watering schedules, potentially reducing agricultural losses during extreme weather. Flood protection systems enhanced by AI-driven analysis of data on rainfall, topography, and river levels can help communities better prepare for and respond to flooding. Intelligent building systems equipped with climate control sensors can also contribute to energy savings and emission reduction. Furthermore, AI-based Knowledge Graphs allow for the analysis of large and complex datasets, helping discover hidden relationships and trends. This enhances our ability to prepare for events like floods, droughts, and heatwaves, ultimately increasing infrastructure resilience in the face of climate change.

5.5 *Biodiversity Conservation*

Biodiversity it can be aided by AI through Machine learning-based systems for species identification and safeguard endangered species. Furthermore, monitoring systems employing AI and satellite imagery can help detect unlawful activities Including poaching & illegal logging.

5.6 *Handling Vulnerabilities and Threat Exposure*

Artificial Intelligence (AI) offers powerful capabilities for identifying, managing, and mitigating environmental and systemic vulnerabilities. In the context of climate resilience, AI can evaluate risk factors by processing vast and complex datasets that include historical climate records, infrastructure maps, satellite data, and social vulnerability indices. By doing so, AI helps in pinpointing communities and ecosystems that are most at risk from environmental hazards such as floods, heatwaves, droughts, or industrial accidents.

One of AI's key strengths lies in its ability to detect emerging threats in real-time. Machine learning algorithms can continuously monitor environmental conditions, sensor inputs, and human activity to flag unusual patterns that might indicate a developing risk. For instance, predictive models can assess structural weaknesses in dams, bridges, or buildings under extreme weather stress, thereby enabling early interventions and reducing potential damage.

Additionally, AI tools can aid policymakers and disaster response teams in making data-driven decisions. These systems can recommend optimized emergency routes, allocate resources efficiently, and simulate various disaster scenarios to enhance preparedness. Importantly, AI-driven insights also contribute to long-term resilience planning by identifying the underlying factors that amplify vulnerability—be it urban density, outdated infrastructure, or lack of access to emergency services.

By proactively addressing both natural and human-made threats, AI plays a pivotal role in minimizing exposure to hazards and building safer, more adaptable societies.

6 How AI is Transforming Agriculture and Food Production

The evolution of data collection and the growing availability of agricultural data have transformed how farming decisions are made. Integrating this data and converting it into meaningful insights allows farmers to improve decision-making for better crop yields, even when facing social and environmental constraints. Artificial Intelligence (AI) has emerged as a powerful tool in smart agriculture, enabling capabilities beyond traditional human efforts. One of the major challenges today is processing vast datasets and converting them into practical, actionable knowledge.

For years, data on land use, crop output, and dairy production has been systematically gathered. For instance, studies on soil characteristics have been used to develop classification systems and detailed soil maps, highlighting the natural variability across different regions. Farmers often recorded grain yields and observed links between output and factors like soil type, weather conditions, and fertilizer application. Similarly, milk yield and quality were monitored using expert visits and lab testing of milk samples on a routine basis.

Despite years of data collection via sensors, a key question remains: how can historical data be better utilized to support future crop management decisions? The focus is now on whether diagnostic tools—such as advanced sensors and AI analytics—can help generate real-time insights that lead to quick, well-informed decisions, or even automate aspects of farm management.

AI systems are designed to interpret environmental and operational data to provide outputs like recommendations, alerts, or content relevant to achieving specific agricultural goals. The European Commission's High-Level Expert Group on Artificial Intelligence (AI HLEG, 2018) defines AI as systems that observe their surroundings, learn from data (both structured and unstructured), and make decisions based on defined objectives and patterns. These systems can adjust behavior over time and respond intelligently to changing conditions in both digital and physical environments.

AI involves multiple domains, including machine learning (such as reinforcement and deep learning), robotics (for automated control), and machine reasoning (which includes planning, decision-making, knowledge representation, and perception). These technologies are increasingly being integrated into agricultural systems, including sensor networks, actuators, and cyber-physical systems, to support more sustainable and productive farming practices.

6.1 Applications of Artificial Intelligence in Agriculture

The integration of Artificial Intelligence (AI) and machine learning into agricultural systems has significantly transformed the way farming is practiced, particularly in areas like field surveillance, crop planning, yield forecasting, resource optimization, and animal health management. One major advancement is the use of AI-enabled surveillance tools that analyze live video footage from farms to detect unusual movements—whether by people or animals—and issue immediate alerts to prevent possible damage. These systems continue to improve over time, learning to distinguish between farm staff and intruders, thus boosting infrastructure safety and operational efficiency.

Predicting crop yield has also become more accurate with the help of AI. By processing extensive datasets obtained from drones, sensors, and satellite imagery, AI tools can evaluate key farming factors such as moisture levels, nutrient presence, and fertilizer application. This enables farmers to fine-tune their crop plans, anticipate risks, and maximize output.

A particularly valuable method is yield mapping, where supervised learning algorithms are used to analyze spatial variations across farming plots. This helps forecast yield outcomes even before harvest and highlights areas that may need extra care. Technologies like 3D field mapping, soil quality analysis, and aerial drone surveillance contribute to generating detailed, actionable insights. In addition, agricultural robots programmed to apply fertilizers or control pests help reduce labor expenses while improving overall crop quality. These robots are now capable of managing large areas of farmland, simultaneously tending to hundreds of plants.

The COVID-19 pandemic emphasized the need for transparency in the food supply chain. AI-supported tracking systems are now widely used to monitor the movement of produce from farms to markets. These systems can examine shipment details at a granular level—down to specific batches or items—by using technologies such as RFID tags and Internet of Things (IoT) sensors, improving logistics and inventory management.

In terms of resource usage, AI is helping optimize water consumption in farming. By analyzing climate patterns, soil conditions, and crop requirements, AI-driven irrigation systems ensure that water is used judiciously, enhancing yield while reducing waste. These systems can recommend the ideal quantity and timing of irrigation based on data, supporting sustainable farming practices.

Lastly, AI has enhanced livestock management by monitoring indicators like activity levels, dietary intake, and vital health signs. These insights help farmers understand how different breeds respond to specific diets or environmental factors, allowing them to tailor care plans accordingly. Improved animal well-being directly results in higher milk yields and greater farm profitability, especially in dairy-focused operations.

6.2 Benefits

The rapid evolution of agricultural technologies is significantly transforming modern farming techniques. These innovations are crucial in addressing major global issues such as climate change, rising population, and increasing pressure on limited natural resources—all of which impact the long-term sustainability of food production systems. Artificial Intelligence (AI) plays a key role in overcoming many of the limitations associated with traditional agricultural methods. It enhances efficiency, reduces dependency on manual labor, and improves decision-making.

Some of the major contributions of AI in agriculture include:

- Automated removal of weeds
- Mechanized harvesting processes
- Early detection and diagnosis of plant diseases
- Continuous monitoring of soil health and quality
- Smarter and more efficient irrigation practices
- Precise application of pesticides and herbicides.

Automated weeding is becoming a popular solution for farmers who want to eliminate weeds without relying on manual labour. This method involves the use of robots equipped with high-energy lasers that can eliminate up to 100,000 weeds per hour. The robot navigates through crop fields with cameras underneath that detect both crops and weeds. An AI system then identifies the weeds and commands the robot to remove them using thermal energy, leaving the soil undisturbed. While manual harvesting can be costly, tedious, and risky due to the potential spread of diseases like COVID-19, automated harvesting systems are being developed to address these issues and increase efficiency in the agricultural industry.

An automatic robot equipped with a camera, AI software, and mechanical arms can efficiently and effectively harvest fruits and vegetables. This technology is reported to outperform humans, as it can operate 24/7 and has a success rate of over 90%. Additionally, various AI-powered plant disease identification apps have emerged to address the significant global issue of crop losses due to plant diseases and pests, which can account for up to 40% of food production losses. These apps work by taking a photo of a plant and using image recognition software to generate disease information and treatment recommendations. For instance, Plantix is an app that employs a database of 100,000 photos of diseased plants and can identify over 60 diseases within seconds using AI image recognition, such as powdery mildew, which affects crops like tomatoes, peppers, melons, squash, and lettuce. An AI-powered software that uses image recognition technology is employed to identify plant diseases. This software employs machine learning algorithms to identify distinctive patterns of plant diseases, such as discolouration and spots, by comparing input photos with a database of pictures of sick plants. The program can be set up on drones, cameras in greenhouses, and other farm machinery. Allowing farmers to quickly detect crop illnesses and take preventive measures. This technology has already been successfully used in vineyards around the world, helping to reduce losses from wine grape disease by as much as $15,001 per acre [10].

6.3 *Better Soil Health Surveillance*

Detecting diseases in soil is crucial to prevent crop loss, but conventional soil analysis methods may not always provide enough information to address the issue. However, AI is already making a difference in this field. With AI-based soil analysis, the abundance of microorganisms and varying soil contents The algorithm based on machine learning then compares this data against a vast database of soil information, which allows for a more comprehensive analysis of the chemistry and biology of soil ecosystems. The results of the analysis indicate factors such as Water content, lack of essential nutrients, and soil pH balance, and microbial presence, among others. Specialists then process the data to determine the potential for particular health conditions and Generate soil management advice considering its nutrient profile and microbial activity. Farmers who use AI for soil analysis report obtaining more useful information about pathogens [10].

6.4 *Improved Irrigation Systems for Cultivated Areas*

A significant amount of water usage globally is attributed to agricultural irrigation, putting pressure on farmers to optimize their water usage. To tackle this challenge, farmers are using agriculture AI technology to enhance the effectiveness of their irrigation systems. One example of this is the use of AI-powered drones which can capture high-resolution Overhead images of farmland and analyze them for potential irrigation issues. By regularly flying over fields, these drones can detect problems such as clogs, leaks, and areas Requiring additional water or nutrients. This helps farmers to resolve these problems before they worsen. The results of the AI analysis are then available to farmers, who can access them on their computers and choose wisely based on information about their irrigation practices.

6.5 *Reduction in Use of Fertilizers and Pesticides*

In agriculture, AI and machine learning can also identify whether plants are ill or wounded. Once more, this technology uses drone photos of fields to identify pest assaults and plant health issues, as well as to decide where to apply pesticides and herbicides.

- Reduce overuse of pesticides and herbicides
- Maintain soil fertility better.

Autonomous drones can spray insecticides and herbicides in designated agricultural regions in addition to monitoring health. Drones that spray crops are proven to be an efficient way to streamline field operations, cut down on human labor, and increase yields [10]. Farmers have traditionally relied on mass pesticides to eradicate pests and invasive plants, but this approach has negative consequences for biodiversity and the environment. A significant portion of sprayed chemicals does not stick to crops, leading to wastage and pollution of water supplies. Fortunately, some startups are developing effective alternatives to pesticides, such as precision applications that target specific species without harming beneficial ones. For example, Pluton Biosciences Has designed an approach called Micro mining, which uses naturally occurring microbes to kill harmful insects while minimizing the risk of developing resistance. Another company, Marron Bio Innovations, creates biopesticides that are safe for the environment by harnessing the power of native microorganisms that are toxic to pests. They cultivate these microbes in high concentrations to develop effective pesticides [11].

According to the Cotton Corporation of India, India produces 26% of the world's 23.92 million metric tons of cotton and occupies nearly 37.5% of the world's cotton land. If only 4% of the land is being farmed, the world will not be able to meet its goal of producing 50% more food by 2050. Agricultural distress may result from growing vulnerabilities, such as climate change, and the possibility of a greater reliance on unsustainable farming methods. However, cotton growers suffer significant losses each year as a result of pests attacking their crops. Pest attacks on nearly 50% of the crop cost farmers in the Indian state of Maharashtra Rs. 15,000 crores (US$2.1 billion) in 2017. Thus, more than 55% of pesticides in India are used for cotton cultivation. On the other hand, excessive use of these pesticides may harm the crop or lower its quality. In the same way that disruptive technologies like artificial intelligence (AI) have been assisting the agriculture sector in modernizing agricultural practices and resolving difficulties, they have also developed a solution for cotton growers to address pest-related challenges. In 2018, Wadhwani AI, an Indian research institute, began looking for ways to use artificial intelligence insights to help cotton farmers save their crops.

6.6 *Soil Health Management*

Soil is a vital resource that supports global agriculture and food production. However, assessing soil quality may prove difficult because of geographical limitations and a lack of technical resources [12]. The nutrient composition of soil plays a critical role in crop health, influencing both yield and quality. Traditionally, assessments of soil condition and fertility relied heavily on manual observation and experience.

With the emergence of modern technologies, drones are now widely used for surveying soil and generating detailed 3D maps to support early soil assessments. These drones collect valuable data for managing irrigation systems and monitoring nutrient levels. Additionally, innovative drone-planting technologies have been introduced by agri-tech start-ups, reducing planting costs by up to 85% through automated seed delivery and nutrient injection directly into the soil.

Beyond soil analysis, drones are also used for crop monitoring by capturing time-lapse imagery that visualizes crop growth over time. This helps farmers identify inefficiencies in production and improve crop management strategies. AI further enhances this process by analyzing data related to soil fertility, weather patterns, and crop performance—enabling more informed decisions about what crops to grow and when for maximum output.

Urban farming practices like vertical cropping have also emerged as a solution to reduce water usage and mitigate labor shortages. These systems can be implemented in indoor or city environments, making them ideal for space-constrained areas. AI-based predictions can predict next year's crop seasons, weather, climate, and rainfall, enabling appropriate pesticide and crop suggestions, preventing large-scale disease outbreaks [13].

6.7 *Wildlife Protection*

The field of wildlife conservation has experienced significant changes using artificial intelligence. Artificial intelligence assists biologists in pinpointing animal locations, recording observation dates and analyzing animal migratory behaviors together with their social structures. Conservationists employ AI to observe and safeguard wildlife within their natural environments. It is estimated by scientists that approximately 30,000 species are under threat worldwide.

Scientists apply AI technology to determine risk factors for endangered species by analyzing their birth locations and survival rates along with their migratory paths and distances covered. Following are some advantages of AI in the conservation of animal species: Artificial intelligence enables the rapid gathering and analysis of massive, intriguing datasets. Artificial intelligence supports wildlife scientists in their study of wild animal species together with creating protection strategies for them. Artificial intelligence monitors wildlife patterns and projects the potential extinction of endangered animals.

It produces data about how climate change affects wildlife populations while simultaneously suggests ways to mitigate these effects through strategic planning. AI performs evaluations of species populations and observes changes. Social media illegal animal trade operations can be halted through specific interventions. Artificial intelligence recognizes and categorizes different animal species into specific groups, offering in-depth insights about each. AI-based models also aid in protecting wildlife habitats by forecasting migration routes of animals [14].

7 The Contribution of AI to Wildlife Protection

AI is significantly contributing to preserving endangered species by enabling various technologies to monitor and manage their populations. For instance, marine robots and Predictive models based on machine learning hosted on Microsoft Azure are being used to track whales and evaluate their health status. Similarly, environmentalists are employing predictive statistical modelling and AI-driven surveillance of Antarctic penguins in real-time.

Furthermore, initiatives like the collaboration between the global conservation organization, WWF, Intel, and academic institutions aim to use AI to conserve Tigers of the Siberian species in China. In addition, AI is also being utilized to identify and analyze elephant calls, leading Contributing to the formation of an "elephant dictionary" that helps protect them from poachers.

The Elephant Listening Project uses neural networks to collect and classify data about Savannah Elephants in Central Africa's rainforests. Lastly, an AI-powered program called Wildlife Security Protection Assistant (PAWS) is assisting rangers in combating poaching. PAWS uses AI and machine learning, along with data Based on the Spatial Monitoring and Reporting Tool (SMART), to identify patterns of animal behaviors. It has been adopted in over 100 wildlife sanctuaries in Africa and Asia and globally in 300–600 locations.

AI technology in preserving endangered species by utilizing machine learning in robotics and drone image datasets to monitor and manage their populations. Additionally, AI can assist in the illegal activities such as poaching by using computer vision cameras, AI-enabled UAVs, and night vision cameras to detect poachers and alert forest rangers before any animals are harmed. For instance, Trail Guard AI cameras are used in national parks to detect people in photos and send electronic notifications to park staff, enabling them to dispatch rangers. Furthermore, researchers are using Convolutional Neural Networks to identify waste products in water bodies and alert marine animal conservation organizations for their removal.

Artificial Intelligence can be used as an early warning system for wildfires by analyzing high-resolution satellite images to estimate the extent of vegetation burning. It can also incorporate meteorological information to improve fire forecasting accuracy. Tools based on recurrent neural networks—such as WIFIRE, IGNIS, Wireless Sensor Networks (WSN), and the Fire Weather Index (FWI)—are employed to predict both the origin and scale of potential fires.

7.1 Prevent Extinction

One of the technologies that can give wildlife rangers all the information they require regarding possible and imminent dangers and their locations is artificial intelligence (AI).

The algorithm may be "trained" to examine previous incidents and recognize every potential pattern of the poachers' actions before the attack. Additionally, AI can swiftly search the internet for and gather data about current happenings on the black market.

In this instance, the AI-driven system will search the Internet for a particular collection of terms and phrases that might point to the network of poachers [15].

AI could be very valuable if introduced to conservation efforts. Advanced satellite and drone technology used for detailed surveillance of species generates vast amounts of valuable data. However, it would be of little use without AI and its ability to process information very quickly, enabling faster than humanly possible intelligence gathering. This in turn will help us build up our knowledge on why biodiversity is being lost and come up with solutions to stop it. In addition to helping law enforcement officials, who frequently lack the manpower to patrol a large area, like a national park, it also helps safeguard a variety of species. This is one of the reasons AI is so helpful; it can relieve human workers of several labor-intensive tasks, like continuously monitoring surveillance data. Even though poaching is widely condemned, nations that are home to elephants, rhinos, and other protected species are crucial to the animals' protection. After all, the endangered wildlife depends on these areas to survive. Artificial intelligence (AI) helps conservationists' thwart poachers. Identifying poachers and notifying the authorities is one of the main ways AI is assisting forest rangers. Using computer vision cameras, drones can identify poachers as they fly over the forest. Further, other AI systems can predict where poachers will attack next and the best way to stop them after they have been identified [16].

Technology for satellite-based communication and electronics. Animal behavior is detected and sent by the tags, which can also serve as an early warning system for ecological shifts [17].

Conservation projects supported by AI technology.

AI is increasingly used in wildlife conservation projects for various applications, like identification of species, monitoring of habitat, anti-poaching efforts and predictive analytics. The AI-powered tools assist conservationists in analyzing large amounts of data, making informed decisions and taking swift action to protect wildlife and their habitats. There are some companies and research organizations which are helping in conserving wildlife. Some are listed:

1. Intel and the World Wildlife Fund have partnered with researchers on a project that leverages Artificial Intelligence to monitor and protect Siberian tigers in China. AI helps analyze the large volumes of data collected through advanced camera traps, making wildlife surveillance more efficient and accurate.
2. A UK-grounded business called DeepMind has created an AI-grounded methodology to fete different beast species and census their populations The Serengeti National Park in Tanzania has utilized it. Scientists use artificial intelligence (AI) to identify wild creatures and work to save exposed species before it's too late.
3. An NGO in San Francisco called Rainforest Connection is fighting wildlife coddling in Africa by using artificial intelligence.

4. Microsoft's Species Bracket API rally recognizes further than 5000 beast and factory species using a machine literacy model.
5. Peter Wrege and his associates at Cornell University gathered and distributed information on Savanna mammoths that live in Central Africa's lush rainforests using neural networks, a branch of artificial intelligence. To keep an eye on jungle noises, they set up audio reporters in the timber. With the help of Conservation Metrics, a California-grounded business, they constructed colorful neural networks to distinguish between jungle sounds and identify birders and mammoths.
6. Conserving the world's freak colony Using TensorFlow, Google's open-source machine literacy frame, Katharina Schmidt, author of apic.ai, and her group created a beehive examiner to collect data. It can be used to track a variety of freak conditioning, including the number of notions, their movement patterns, and whether or not they're carrying pollen. AI provides comprehensive data and logical reports to help specialists in making well-informed opinions.
7. Conservation Metrics, a Santa Cruz-based company, assists researchers in monitoring the distribution, population trends, and conditions of wildlife by using machine literacy models and artificial intelligence (AI) techniques.
8. Wildlife Security Protection Assistant (PAWS) PAWS is an AI-powered tool for predictive analysis and resource efficiency. By accessing data and furnishing details on the terrain, beast population, natural trails, and bottom business, it helps rangers. AI directs the system to determine the demesne rangers' optimal command route and the volume of obstacles and traps they will encounter. Rangers can descry wildlife crimes with the aid of artificial intelligence.
9. A pall-grounded AI-enabled platform called Wildlife perceptivity enables several companies to upload, store, manage, and partake their data while also rooting precious perceptivity from it. The process of gathering data from the field and uploading it to Google Cloud for analysis and mapping is significantly accelerated.
10. A US-grounded on-profit group called Wild book is digitally tagging individual creatures using AI algorithms in a manner analogous to characteristic people. The algorithms incontinently fete individual creatures by assaying public aqueducts of images and vids.

8 Weather and Disaster Prediction

The occurrence of natural disasters is unpredictable, making it essential to have an efficient and timely response system in place to safeguard lives and infrastructure. This research suggests a new framework, 4-AIDE, which utilizes advanced technologies such as Cloud-based solutions powered by artificial intelligence platforms to facilitate effective management of emergency and extreme weather situations. Through a qualitative approach based on the theory of organizational information processing (OIPT), 33 experienced individuals from the Cloud-based artificial intelligence were interviewed to develop the framework. The collected data was analysed using axial, open, and selective coding to identify themes and propositions. The study findings suggest that Cloud-based AI systems offer a well-structured approach to collect and analyse data, predict outcomes, and formulate effective approaches to handle extreme weather situations. This can enable public systems, businesses, and communities to be more resilient and reduce the loss of lives caused by delayed dissemination of information, inadequate resource allocation, and failure to minimize risks. The increasing frequency of natural disasters such as earthquakes, floods, pandemics, and droughts pose a considerable challenge for infrastructure, agriculture, and financial systems [18]. AI has been playing an increasingly important role in weather prediction in recent years. Traditional weather forecasting techniques rely on mathematical models and physical measurements, but AI is now being used to supplement and improve these methods.

One of the primary ways intelligent systems are employed in weather prediction is through machine learning algorithms. These algorithms are skilled on large volumes of chronological weather metrics including temperature, wind speed, and precipitation, and use that data to make predictions about future weather patterns. By analyzing patterns and trends in the data, Algorithms based on machine learning can provide more accurate predictions than traditional forecasting methods.

There is also the application of AI to process large amounts of real-time weather data from monitoring instruments and additional origins this data is often too vast for human forecasters to process in a timely manner, but AI algorithms can quickly analyze the data and provide insights that can improve weather predictions.

Another way prediction is through computer vision techniques. For example, satellite imagery can be analysed using computer vision algorithms to detect weather patterns such as storms and cloud cover. This information can then be used to refine weather models and make more accurate predictions.

Overall, AI is playing an increasingly important role in weather prediction and is expected to become even more important in the technology continues to improve. Observatories, land-based, and water-based weather monitors generate copious volumes of meteorological and climate-related data data globally. However, the extensive scope and detailed nature of of this data make it challenging for humans or conventional computer systems to analyze and identify patterns efficiently. As a result, there is a risk of wasting time and storage resources if this data is not comprehensively analysed. To address this challenge, researchers are turning to artificial intelligence (AI) methods such as ML, neural networks, and deep learning. By leveraging the pattern recognition capabilities of AI, massive volumes of data can be fed into algorithms that can then learn how to identify weather patterns that could lead to lightning, tornadoes, hurricanes, or severe winter storms. This approach offers a more effective means of identifying and detecting trends in weather data, which could help predict and prevent weather-related disasters [19].

8.1 AI Protecting us from Disaster

Artificial Intelligence (AI) plays a critical role in identifying and responding to natural disasters in real time by offering early alerts and supporting proactive decision-making. These systems rely on the insights and evaluations of field experts, alongside sensor data, to deliver timely and actionable outcomes. AI can help detect the likelihood of major disasters—such as hurricanes, earthquakes, landslides, and volcanic eruptions—in a cost-efficient and scalable way. It can also forecast potential damage to infrastructure and predict service downtimes, allowing for early intervention.

A practical example is Hydro One, a major utility provider in Ontario, Canada, which implemented AI-supported predictive analytics in April 2018. By combining AI with IoT and cloud-based systems, they were able to mobilize crews to critical areas and restore electricity within four days—substantially shortening the outage period. These AI-integrated services also enabled rapid information sharing, enhancing response coordination during emergencies.

When it comes to evacuation planning, AI helps address the variability in how different populations respond to warnings—some may have a day or more to act, while others require immediate guidance. Effective evacuation strategies must be based on identifying potential risks and designing tailored response measures. For instance, AI models can be used to decide when to initiate counterflow traffic systems to support smoother evacuations.

Managing large crowds during such events demands accurate forecasting of movement patterns. AI can help map out optimal escape routes and assist in developing support mechanisms that ensure safer evacuations. Among the commonly used AI models in this field are Support Vector Machines (SVM), Decision Trees (DT), and various neural network architectures. These models are further enhanced using machine learning techniques and optimization algorithms to deliver the best possible outcomes during emergencies.

8.2 Future of AI in Weather Forecasting

Artificial intelligence (AI) has already made significant advancements in weather forecasting, and its future looks promising. As AI technology continues to evolveit holds the promise of reshaping the field of weather forecasting in the following ways:

Improved accuracy: AI algorithms can analyze substantial volumes of data from various sources to uncover trends and patterns that could contribute to more accurate weather predictions. Real-time forecasting: AI can process data and produce weather forecasts in real-time, making informed decisions in emergency situations.

Customized predictions: AI can provide customized weather predictions based on individual preferences, such as specific locations, activities, or industries.

Enhanced decision-making: AI can help decision-makers in various areas, including agricultural practices transportation, and energy, by providing them with accurate and timely weather information to optimize their operations.

Climate change mitigation: AI can help lessen the consequences of climate change by providing insights into weather patterns and trends, which can inform policymakers and the public on the need for climate action.

Overall, the future of AI in weather forecasting holds tremendous potential to improve the accuracy, speed, and customization of weather predictions while enabling more effective decision-making and climate change mitigation.

9 Conclusions

AI and machine learning are proving to be powerful tools in addressing complex and urgent environmental challenges. By leveraging large datasets from satellites, sensors, and field studies, these technologies can analyse patterns, make accurate predictions, and support real-time decision-making. From monitoring climate change and predicting natural disasters to improving waste management and conserving biodiversity, AI-driven solutions are enhancing efficiency, precision, and scalability in environmental protection efforts.

As technology continues to evolve, the integration of AI with environmental science holds great promise. However, it is crucial to ensure ethical use, data transparency, and collaboration among governments, researchers, and communities to maximize its benefits for a sustainable and resilient planet.

References

1. Kaginalkar, A., et al. (2021). Review of urban computing in air quality management as smart city service: an integrated IoT, AI, and cloud technology perspective. *Urban Climate, 39*. https://doi.org/10.1016/j.uclim.2021.100972.
2. Yu, R., et al. (2016). RAQ-a random forest approach for predicting air quality in urban sensing systems. *Sensors, 16*(1), 86. https://doi.org/10.3390/s16010086
3. Mao, S., et al. (2019). Opportunities and challenges of artificial intelligence for green manufacturing in the process industry. *Engineering, 5*(6), 995–1002. https://doi.org/10.1016/j.eng.2019.08.013
4. Mendil, M., et al. (2022). Hazardous atmospheric dispersion in urban areas: A Deep Learning approach for emergency pollution forecast. *Environmental Modelling & Software, 152*. https://doi.org/10.1016/j.envsoft.2022.105387.
5. Subramaniam, S., et al. (2022). Artificial intelligence technologies for forecasting air pollution and human health: A narrative review. *Sustainability, 14*(16), 9951. https://doi.org/10.3390/su14169951
6. Jha, M. K., et al. (2018). Smart water monitoring system for real-time water quality and usage monitoring. In *International Conference on Inventive Research in Computing Applications (ICIRCA)*. https://doi.org/10.1109/ICIRCA.2018.8597179.

7. Jiang, J., et al. (2022). How AI-enabled SDN technologies improve the security and functionality of industrial IoT network: Architectures, enabling technologies, and opportunities. *Digital Communications and Networks, 9*, 1351–1362. https://doi.org/10.1016/j.dcan.2022.07.001.
8. Jualayba, M., et al. (2018). Hazardous gas detection and notification system. In *IEEE 10th International Conference on Humanoid, Nanotechnology, Information Technology, Communication and Control, Environment and Management (HNICEM)*. https://doi.org/10.1109/HNICEM.2018.8666358.
9. Joshna, V., et al. (2019). Fully autonomous robot to detect and degasify hazardous gas after flood disaster. In *2nd International Conference on Power and Embedded Drive Control (ICPEDC)*. https://doi.org/10.1109/ICPEDC47771.2019.9036703.
10. Khan, A., et al. (2020). Multi-hazard disaster studies: monitoring, detection, recovery, and management, based on emerging technologies and optimal techniques. *International Journal of Disaster Risk Reduction, 47*. https://doi.org/10.1016/j.ijdrr.2020.101642.
11. Kim, K., et al. (2021). Airflow pattern control using artificial intelligence for effective removal of indoor airborne hazardous materials. *Building Environment, 204*. https://doi.org/10.1016/j.buildenv.2021.108148.
12. Mabrouki, J., et al. (2021). Intelligent monitoring system for biogas detection based on the Internet of Things: Mohammedia, Morocco city landfill case. *Big Data Mining and Analytics, 4*(1), 10–17. https://doi.org/10.26599/bdma.2020.9020017.
13. Kim, Y., et al. (2008). Remote sensing and control of an irrigation system using a distributed wireless sensor network. *IEEE Transactions on Instrumentation and Measurement, 57*(7), 1379–1387. https://doi.org/10.1109/tim.2008.917198
14. Kostal, J., et al. (2022). O data, where art thou? Revolutionizing data sharing to advance our sustainability goals through smart chemical innovation. *Iscience, 25*(11), 1–7. https://doi.org/10.1016/j.isci.2022.105256
15. Kumer, S. A., et al. (2021). Environment water quality monitoring system. *Materials Today: Proceedings, 46*, 4137–4141. https://doi.org/10.1016/j.matpr.2021.02.674
16. Lakshmikantha, V. (2021). IoT based smart water quality monitoring system. *Global Transitions Proceedings, 2*(2), 181–186. https://doi.org/10.1016/j.gltp.2021.08.062
17. Leppert, J., et al. (2012). Near real time detection of hazardous airborne substances. *Talanta, 101*, 440–446. https://doi.org/10.1016/j.talanta.2012.09.056
18. Anantharaman, T. (2016). Weather monitoring using artificial intelligence. In *2nd International Conference on Computational Intelligence and Networks (CINE)* (pp. 106–111). https://doi.org/10.1109/CINE.2016.26.
19. Li, X., et al. (2016). Deep learning architecture for air quality predictions. *Environmental Science and Pollution Research, 23*, 22408–22417. https://doi.org/10.1007/s11356-016-7812-9

Optimization of Energy Management and Distribution

Sunil Gautam, Pranali Joshi, and Priyanshi Jhala

Abstract Energy grids are experiencing increased demand due to factors like population growth, economic expansion, and the need for more supply and demand. Alongside this rising demand, there is a challenge to maintain sustainability and offer sustainable solutions to current energy needs. Simultaneously, the rise and growing significance of Artificial Intelligence (AI) and Machine Learning (ML) are enhancing productivity and efficiency in various tasks. These AI-ML capabilities can be applied to energy management and distribution. With numerous machine learning algorithms available, energy productivity can be improved, and its distribution and management can be effectively handled. However, when employing artificial intelligence to study and optimize energy distribution, several other considerations must be taken into account, such as economic factors, addressing technical challenges, and, as previously mentioned, sustainable solutions. Recently, AI, in conjunction with cloud computing and blockchain, has been actively involved. To solve each other's problems. AI and Blockchain can be used to manage energy trading. We have emphasized the significance and application of specific machine learning algorithms that are applicable in energy management, especially regarding the AI-ML in energy management, with an attention to blockchain and cloud computing. Additionally, we have also discussed the problems with using AI and how to tackle them.

Keywords Artificial intelligence · Machine learning · Optimization of energy management

S. Gautam (✉) · P. Joshi · P. Jhala
Department of Computer Science & Engineering, Institute of Technology, Nirma University, Ahmedabad, India
e-mail: sunil.gautam@nirmauni.ac.in

M. A. Mellal et al. (eds.), *Computational Intelligence for a Greener Future: Innovations in Renewable Energy Systems*, Studies in Computational Intelligence 1240,
https://doi.org/10.1007/978-3-032-06732-6_9

1 Introduction

The world energy system is changing in a significant way being fueled by the forces of the two, the demands of growing energy use and conservation of the environment. With the increasing complexity of power grids and integration of the wide variety of renewable energy sources, older methods of energy management and distribution are increasingly being shown to be inadequate to the demands of the modern world. This has led to the emergence of innovative technologies and methodologies, particularly in the realm of artificial intelligence and machine learning (AI-ML), to optimize energy systems [1]

Due to the recent breakthrough in technologies of AI-ML, as well as blockchain, big data analytics, and cloud computing technologies, new opportunities to efficiently improve energy management emerged. These technologies offer promising solutions for real-time monitoring, predictive maintenance, demand forecasting, and automated trading in energy markets. The international energy agency estimated that the energy management systems of the future may save up to 15 percent of the world consumption by 2040, showing the promising nature of such technologies [2–4].

Though the prospective use of AI-ML in the energy management domain is quite promising, there is still a great gap between the potentialities and real organizing of the theoretical resource. While numerous studies have demonstrated the potential of individual technologies, there is limited research on their integrated application in real-world energy systems. Moreover the issues revolving around application of these technologies such as technical, economic feasibility and environmental implications warrant serious considerations [5, 6].

The aim of the chapter is to offer an in-depth discussion about using AI-ML in energy management and distribution optimization. Through a detailed literature review and examination of current applications, we explore how various technologies can be integrated to create more efficient and sustainable energy systems. The research particularly focuses on the role of algorithms, blockchain technology, and big data analytics in transforming energy management practices [7, 8]. Additionally, we investigate the technical, economic, and environmental challenges that must be addressed for successful implementation of these technologies [9, 10].

The results of this research could help add to the existing body of knowledge in the area of smart energy management and could be offering insights to researchers and industry practitioners as well as policymakers to establish more efficient and sustainable to the future energy systems.

2 Applications of Energy Management Using AI

2.1 Algorithms

a. Adaptive Neuro-Fuzzy Inference System

Adaptive Neuro-Fuzzy Inference System (ANFIS) is an innovative computational model that merges the learning abilities of artificial neural networks with the qualitative reasoning inherent to fuzzy logic systems. Developed in the early 1990s, ANFIS aims to simulate complex and nonlinear functions by employing a series of fuzzy if–then rules within a neural network structure. This combination enables ANFIS to effectively identify and represent intricate patterns present in data, establishing it as a formidable tool for numerous applications, including energy management.

Mathematically, for a two-rule system with inputs x and y, and corresponding fuzzy sets A_i, B_i, and consequent functions

$$f_i = p_i x + q_i y + r_i,$$

the output is computed as:

$$w_i = \mu A_i(x) \cdot \mu B_i(y) \tag{1}$$

$$\overline{w_i} = \frac{w_i}{w_1 + w_2} \tag{2}$$

$$f = \overline{w_1} f_1 + \overline{w_2} f_2 \tag{3}$$

The system parameters are optimized using a hybrid learning algorithm combining least squares and backpropagation [11].

Below mentioned is structure of ANFIS.

ANFIS Structure

ANFIS consists of five distinct layers that each play a role in converting inputs into outputs through successive calculations:

i. **Fuzzification Layer**: The first layer accepts input values and assesses their associated membership functions, transforming precise inputs into fuzzy sets.
ii. **Rule Layer**: In this layer, firing strengths for each fuzzy rule are generated by comparing the inputs against established if–then rules.
iii. **Normalization Layer**: This particular layer normalizes the firing strengths so that their total sums to one; this step supports weighted averaging in subsequent layers.

iv. **Defuzzification Layer**: By utilizing normalized firing strengths, this layer computes weighted outputs for every rule, generally using a first-order polynomial function.
v. **Output Layer**: The concluding layer compiles outputs from the defuzzification stage to yield a single crisp output that reflects how the system responds to given inputs.

In ANFIS's learning process, parameters tied to both membership functions and consequence parts of rules are adjusted. Typically accomplished using a hybrid learning algorithm that integrates gradient descent with least squares estimation, this approach allows the model to adaptively reflect input–output relationships based on available data.

In energy management, ANFIS is important in enhancing efficiency as well as reliability of various systems through its adaptive control and forecasting functions. Its applications cover diverse fields, including.

- renewable energy resources
- smart grid technologies
- hybrid electric vehicles.

- **Renewable Energy Systems**
  ANFIS proves beneficial in enhancing the performance of renewable energy sources like photovoltaic (PV) systems. By implementing controllers based on ANFIS techniques, these systems effectively perform maximum power point tracking (MPPT). This ensures that solar panels optimally harvest energy despite fluctuations in environmental conditions [12]. For example, an ANFIS-driven MPPT controller can flexibly respond to variations in sunlight intensity and temperature to sustain optimal power output while increasing the overall effectiveness of PV installations [13].
- **Smart Grids**
  The integration of ANFIS into smart grid infrastructures supports efficient electricity distribution along with accurate load forecasting. By examining historical data coupled with real-time information, ANFIS models can forecast patterns in energy consumption [2]. This insight empowers grid operators to make strategic decisions regarding energization strategy and balancing loads appropriately. Consequently, this predictive functionality aids in minimizing energy waste while maintaining a stable balance between supply and demand [14].
- **Hybrid Electric Vehicles (HEVs)**
  The management of energy within HEVs focuses on optimally distributing power between the internal combustion engine and electrical motor to boost fuel efficiency while lowering emissions. Control strategies grounded in ANFIS have been designed to manage this power division adaptively by considering several these strategies facilitate timely adjustments that enhance vehicle performance along with effective use of energy resources [14].

Overall, the applications of ANFIS as can be seen in Fig. 1, significantly contribute to advancing solutions for contemporary challenges present in various sectors related to energy.

b. Hybrid K-Means with Genetic Algorithm

The combination of K-Means clustering with Genetic Algorithms (GAs) has been investigated to improve energy management strategies in various fields. This hybrid methodology capitalizes on the advantages of both techniques: K-Means adeptly categorizes datasets into clusters based on similarity, while GAs excel at global optimization for complex problem-solving. In the realm of energy management, this integrated approach has been employed to enhance energy consumption patterns, improve demand-side management, and boost the overall efficiency of renewable energy systems.

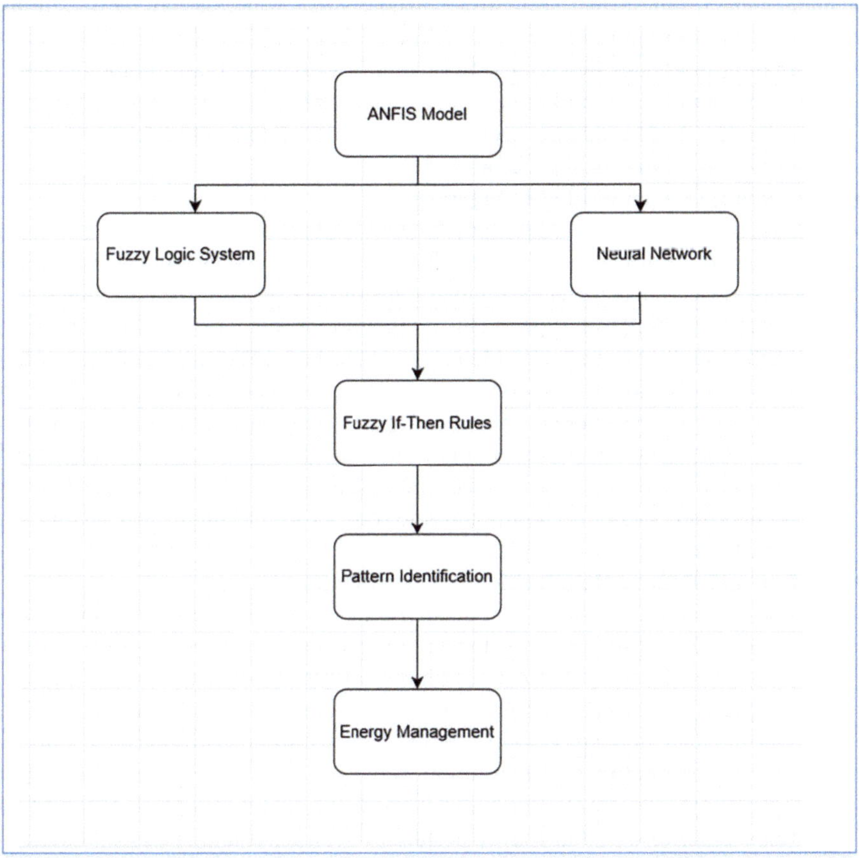

Fig. 1 ANFIS structure

K means partitions a dataset into k clusters by minimizing the within-cluster variance:

$$J = \sum_{i=1}^{k} \sum_{x_j \in C_i} ||x_j - \mu_i||^2 \quad (4)$$

where μ_i is the centroid of cluster C_i [15].

- **Optimizing Energy Consumption in Wireless Sensor Networks**
  In Wireless Sensor Networks (WSNs), ensuring energy efficiency is critical because sensor nodes have limited power resources. A study introduced a combined technique that merges K-Means clustering with the Low-Energy Adaptive Clustering Hierarchy (LEACH) protocol to optimize power usage. The K-Means algorithm was utilized to create well-balanced clusters among sensor nodes, which minimizes intra-cluster distances and consequently reduces transmission energy requirements [16]. Following this step, GAs were used to choose optimal cluster heads by taking into account variables such as remaining energy and node centrality. Simulation outcomes indicated that this combined method considerably prolonged the operational lifespan of the network compared to conventional clustering approaches [16].
- **Enhancing Demand-Side Management**
  Effective demand-side management (DSM) plays a vital role in harmonizing energy supply with demand, particularly within standalone renewable energy systems. Researchers developed a load scheduling algorithm which is based on a combination of K-Means type of clustering with the methods of the linear integer programming. In this model, K-Means organizes household appliances according to their usage behaviors and consumer preferences [17]. Subsequently, GAs are utilized to refine the scheduling of these appliance groupings so they align more closely with periods when renewable energy availability is high. This approach not only guarantees an uninterrupted power supply for essential devices but also diminishes dependence on non-renewable sources—resulting in cost savings and improved reliability for the system overall [17].
- **Optimizing Hybrid Renewable Energy Systems**
  HRES represents systems that make use of more than one energy source; for instance, wind and solar energy demand for proper and efficient energy management to carry on with inherent variability within the system. Energy systems merging GA-optimized rule-based energy management for HRES have also been constructed. One of the models involves K-Means clustering to analyze the historical energy production and consumption data to determine patterns that forecast energy availability [3]. The GA optimizes the charge/discharge cycles of the batteries, making sure that energy is stored in surpluses and released in deficits. The approach improves the reliability and efficiency of the HRES, keeping energy supply constant while reducing costs of operations [18].
- **Improving Energy Management in Public Buildings**

Energy management in public buildings is a significant challenge due to varying occupancy and usage patterns. A hybrid model has been developed by integrating self-organizing maps (SOMs) and deep learning and optimizing via GA to solve this problem. SOMs cluster the rooms or zones in the building based on energy usage patterns [5]. Deep learning models predict future energy consumption for each of the clusters that are in place, while the GA optimizes HVAC operation. This gives rise to a much more accurate forecast of energy usage and hence better operational effectiveness of HVAC, resulting in energy savings [19]. Additionally, smart grid strategies leveraging GA-based optimizations have been explored to ensure efficient energy management in large-scale public infrastructures [20]. As can be seen from, the fusion of genetic algorithms with K-Means clustering provides a powerful framework to tackle the complex problems of energy management. In clustering, it improves data organization and operational parameter optimization during operation to enhance energy conservation, reduce costs, and thus improve reliability across any power generation system. In an era where energy needs continue to grow, widen, and become complex, methodologies like these can become pivotal for creating sustainable and efficient solutions for energy management [19].

c. Deep Learning with Particle Swarm Optimization

The combination of Deep Learning (DL) and Particle Swarm Optimization (PSO) has proven to be an effective method in energy management systems. By harnessing the predictive capabilities of DL alongside the optimization strengths of PSO, this approach effectively tackles complex issues such as forecasting energy consumption, managing demand on the consumer side, and optimizing renewable energy resources [21]. (Fig. 2)

The PSO algorithm updates particles using the equations:

$$v_i^{t+1} = \omega v_i^t + c_1 r_1 (p_i - x_i^t) + c_2 r_2 (g - x_i^t) \tag{5}$$

$$x_i^{t+1} = x_i^t + v_i^{t+1} \tag{6}$$

Here, x_i^t and v_i^t denote the position and velocity of the ith particle at iteration t, while p_i and g represent the personal and global best positions, respectively [22].

- **Improving the Precision of Load Forecasting**

  Accurate short term loading makes settling power systems ideal. A new method integrates Convolutional Neural Networks (CNNs) and Long Short-Term Memory (LSTM) networks, enhanced by a Multi-Head Attention mechanism, to effectively capture spatial and temporal patterns in energy consumption data. The PSO technique is used to automatically fine-tune the hyperparameters of this hybrid model, boosting its predictive accuracy. Tests with real-world electricity demand datasets reveal that this comprehensive approach achieves a Mean Absolute Percentage Error (MAPE) of 1.9376, surpassing current leading techniques [21].

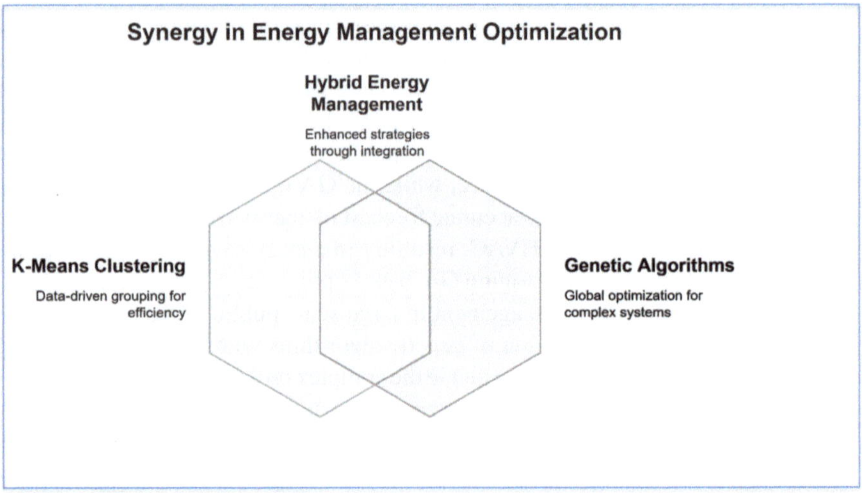

Fig. 2 Hybrid genetic algorithm (GA) and K-means clustering process

- **Enhancing Demand Response Tactics**
  Demand Response (DR) programs play a essential role in maintaining the balance between energy supply and demand, especially within residential areas. Recent research introduces a DR model based on PSO, which incorporates Artificial Neural Networks (ANNs) to anticipate household energy usage. By analyzing historical load data alongside weather information, the ANN accurately predicts future energy demands. Subsequently, PSO optimizes how loads are scheduled by shifting non-essential electricity use to off-peak times, effectively lowering peak demand and reducing consumer electricity bills. This method has been tested with real-world data from Austin, Texas residences and demonstrated its practical success in managing energy consumption efficiently [23, 24].
- **Enhancing Energy Efficiency in Additive Manufacturing**
  Energy efficiency is the most important in the additive manufacturing sector. To address this concern, researchers have created a Deep Learning-driven PSO (DLD-PSO) approach to optimize energy consumption during production. This method involves training a deep learning model to forecast energy usage based on different manufacturing parameters. Subsequently, PSO is utilized to determine the ideal set of parameters that minimize energy use while preserving product quality. This strategy not only cuts down operational costs but also promotes sustainable manufacturing practices [25].
- **Building Energy Management Systems**
  Effective management of energy resources in buildings is essential for lowering operational expenses and minimizing ecological effects. Researchers have introduced a Machine Learning-based Random Forest model, specifically refined with PSO (RF-PSO), to enhance building energy systems control. This model can forecast energy usage trends and pinpoint areas where optimization can occur. By

fine-tuning the parameters through PSO, prediction accuracy is enhanced, leading to more successful strategies for saving energy. The application of this method has resulted in notable enhancements in energy efficiency within various building settings [26, 27].

- **Optimizing Microgrid Operations**
  Microgrids that amalgamate different renewable energy sources necessitate advanced management techniques to maintain both stability and efficiency. A recent research effort presents a multi-objective optimization framework tailored for grid-connected microgrids, utilizing Quantum Particle Swarm Optimization (QPSO). This framework seeks to balance multiple goals such as reducing costs, minimizing emissions, and boosting reliability. By merging QPSO with deep learning models capable of predicting both energy production and demand levels, This system attains the capability to make informed decisions on energy storage, distribution as well as consumption strategies. This comprehensive approach has been demonstrated to significantly enhance performance and sustainability within microgrid operations [28, 29].

The combination of Deep Learning methods with PSO (refer Fig. 3) creates a formidable strategy for tackling intricate challenges in energy management. By integrating predictive modeling alongside optimization practices, all this integrated technique enhances the efficiency, reliability and sustainability of energy systems in wide range of applications—spanning from residential demand response initiatives to industrial manufacturing processes as well as microgrid oversight [21, 23–28, 30].

2.2 *Blockchain for Energy Trading*

Imagine a world where selling your extra solar energy is as easy as sharing a ride or renting out a spare room. That's the promise of blockchain technology in energy markets. It's not just a tech buzzword—it's a game-changing approach that's turning traditional energy trading on its head [29].

Traditional energy markets have always been like a complex maze, with big utility companies controlling everything. Blockchain is like a sledgehammer to that maze. It creates a direct connection between energy producers and consumers, cutting out the middlemen and their hefty fees [31]. Think of blockchain like a super-smart, completely transparent ledger that everyone can see but nobody can cheat. When someone with solar panels generates extra electricity, smart contracts can automatically find a buyer nearby. It's like an energy matchmaking service that works in real-time [32]. The Brooklyn Microgrid project is a perfect example. Imagine your neighbor with solar panels can sell their excess electricity directly to you, without going through a massive utility company. It's local, it's efficient, and it reduces energy waste [33]. Smart contracts are the superheroes of this new energy world. They're like automated agreements that trigger instantly when certain conditions are met.

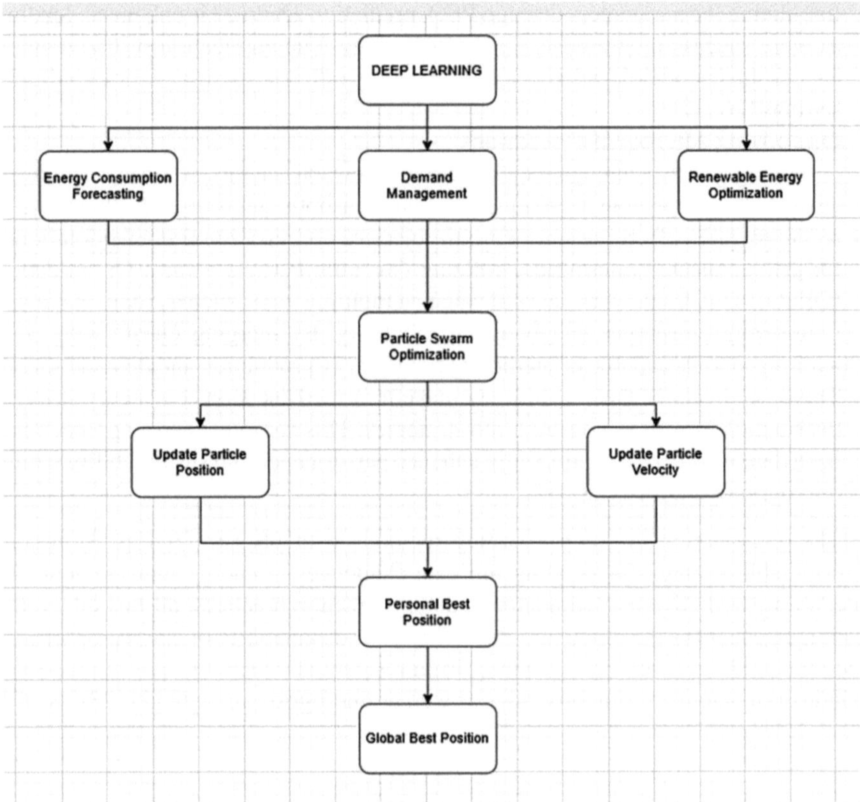

Fig. 3 Flowchart of hybrid model integrating PSO with deep learning

Got extra solar energy? Boom—it's sold automatically to the highest local bidder. No paperwork, no delays, just seamless energy trading [34].

Studies show these blockchain systems can cut transaction costs by up to 40% and reduce settlement times from days to minutes. That's not just convenient—it's transformative [35]. In the past, energy trading was a black box. Now, blockchain creates a crystal-clear record of every single energy transaction. Every kilowatt-hour can be tracked, verified, and authenticated. It's like having a completely transparent energy banking system [36]. Here's where it gets really interesting. Blockchain allows energy to be broken down into tradable tokens. Imagine being able to buy exactly the amount of energy you want, with details about its source, carbon intensity, and origin. Want 100% local, green energy? Now you can choose precisely that [37].

It's not all smooth sailing. Blockchain in energy trading faces some serious hurdles [38]:

i. Scalability issues
ii. High energy consumption of some blockchain networks
iii. Complex regulatory landscapes

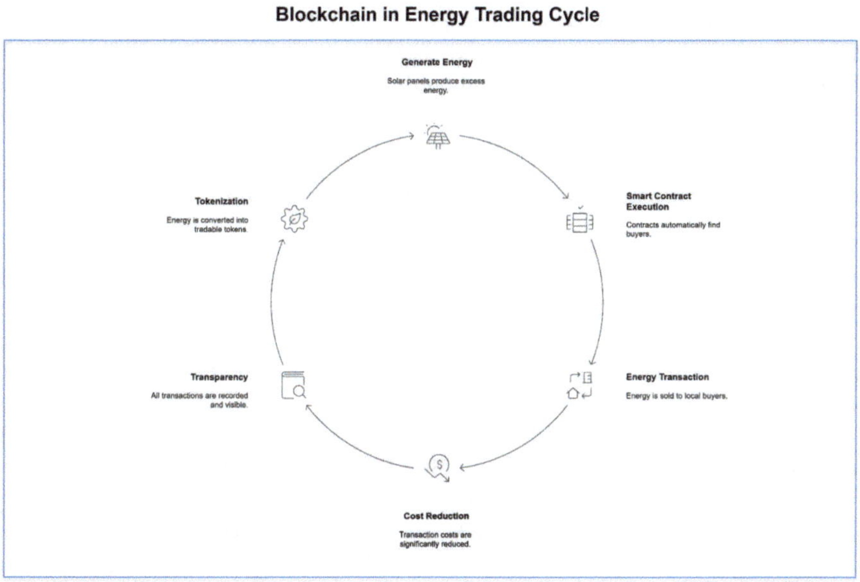

Fig. 4 Blockchain-based peer-to-peer (P2P) energy trading

But the potential is enormous. Platforms like Power Ledger in Australia are already showing real benefits—participants are saving money and increasing renewable energy use [39].

The next frontier is combining blockchain with other cutting-edge technologies. Just imagine IoT devices having the capability to trade energy automatically in real-time, or AI algorithms capable of making predictions and optimizations over the energy markets [40].

Blockchain isn't just a technology—it's a vision of a more democratic, transparent, and efficient energy future. It's about giving power (quite literally) back to the people, promoting renewable energy, and creating a more sustainable world. For more information refer Fig. 4 [41].

2.3 *Big Data and Cloud Computing*

Big data analysis and cloud computing have revolutionized the way businesses deal with energy and resources. The technologies allow organizations to collect, process and analyze an enormous amount of data that is received by smart grids and energy systems [42]. With the global use of energy projected to increase by 50% by 2050, employing these innovative tools is more critical than ever for a clean future [43].

Big data analytics provides a clear picture of how energy is being utilized, allowing inefficiencies to be more easily identified and means to be discovered for their improvement that old-fashioned approaches may not [44]. Cloud computing improves on this by offering strong and elastic systems that can manage enormous amounts of energy data and do so without breaking the bank [45]. Together, these technologies assist industries in making informed, data-based choices that enhance energy efficiency and minimize damage to the environment.

a. Contribution of ML in Big Data and Cloud Computing

Machine learning algorithms are effective tools that convert raw energy data into meaningful insights. Deep learning models have the ability to reveal underlying patterns in energy consumption that would be overlooked by conventional statistical techniques [46]. The models are trained using past energy consumption, weather data and operational data each time to give a better prediction on future energy demand [47].

Supervised learning methods are particularly well-suited to forecast building energy consumption, with error rates as low as 3% in laboratory tests [48]. Reinforcement learning assists in controlling energy distribution in real time by continuously adapting to changing conditions [49]. Machine learning-based energy systems have been demonstrated to reduce energy consumption by 10–15% without impacting performance or comfort [50].

Also, anomaly detection algorithms are able to detect unusual patterns that can indicate equipment faults or even energy theft and enable prompt maintenance and security interventions [51].

b. Now discussing some Real-World Applications of Big Data and Cloud Computing

i. Smart Grid Management

Electric utilities are employing cloud-based analytics to analyze data from millions of grid sensors and smart meters. In one instance, a regional utility discovered that load forecasting accuracy was enhanced by 30% with machine learning, resulting in improved generator management and a 7% decrease in backup power requirements [52].

Deep neural networks were applied in Taiwan to improve grid operations, reducing transmission losses by 5.3% and increasing the reliability of the system [53]. Cloud computing enables these systems to scale up during peak periods while maintaining performance levels constant [54]. Real-time analytics have also accelerated outage response times by 25%, minimizing downtime and financial losses [55].

ii. Industrial Energy Optimization

Manufacturing facilities generate massive volumes of energy information from thousands of processes and machines. In an automobile factory, a cloud energy analytics solution monitored data from 2500 sensors and assisted in the reduction of energy expenditure by €2.1 million annually [56].

Machine learning picked up changes in processes that reduced energy consumption by 18% without changing quality of output [57]. Reinforcement learning even optimized compressed air systems by minimizing their power use by 27% across some plants [58]. From these examples, one can understand how data-led innovations can deliver huge energy efficiencies with businesses maintaining optimum operations.

3 Challenges of AI-ML in Energy Management

A. Following Technical Challenges are faced

The integration of AI and ML in energy management systems holds the promise of revolutionizing energy efficiency and sustainability. But to achieve this potential, there is a set of major technical obstacles which needs to be overcome which includes data management, system integration, scalability and security [59]. These issues are important to address in the context of successful implementation and popularization of AI-ML technologies in the energy sector.

i. Data Quality and Availability Challenge

High-quality real-time data is used by AI-ML models to make the correct predictions and choices. But energy systems tend to produce massive volume of data that are most likely noisy, data incomplete, or inconsistent. Data quality issues can arise from sensor malfunctions, communication errors, or inadequate data collection practices [60].

- **Impact**
  The inadequacy of data quality may result into faulty models and unreliable prediction and hence reduces efficiency of AI-ML products in energy management. For instance, inaccurate load forecasting can result in inefficient energy distribution and increased operational costs [61].
- **Mitigation**
  Data cleaning, normalization, and imputation are strong data preprocessing methods that should also be implemented to enhance the quality of data. Availability and consistency of data can also be improved by investing in the current technology of sensors and creating data governance structures. Collaboration with data providers and standardization of data formats can further mitigate data quality issues [62].

ii. Scalability

Scalability is the ability of an AI/ML system to effectively handle the increased size, demand or complexity without drop in its performance.

- **Challenge**
  Scaling AI-ML solutions to accommodate large-scale energy systems requires significant computational resources and infrastructure. The complexity of energy

systems, which involve numerous interconnected components and vast amounts of data, poses a significant challenge to scalability [63].

- **Impact**
  Limited scalability can hinder the widespread adoption of AI-ML in energy management, particularly in regions with constrained resources. This can result in disparities in energy efficiency and sustainability across different regions [64].
- **Mitigation**
  Scalability can be increased by making use of cloud computing and distributed processing technologies. Cloud services are elastic computing systems capable of increment or decrement depending on the demand, which makes it suitable in manipulating mass energy data. Additionally, developing lightweight AI-ML models that require fewer computational resources can facilitate broader adoption. Edge computing may also help to reduce latency and enhance scalability through processing data near to its point of origin [65].

iii. Integration with Existing Systems Challenge

Combining the application of AI-ML in energy management with the systems already in use can be quite complicated and might need some significant adjustments to existing infrastructure. Current AI-ML technologies might not work with legacy systems, and any attempts to modify them can result in the costs of compatibility and introduction of information technology [66].

- **Impact**
  Integration challenges can lead to delays in implementation and increased costs, making it difficult for organizations to adopt AI-ML technologies. This can result in a slower pace of innovation and reduced energy efficiency [67].
- **Mitigation**
  Developing modular AI-ML solutions that can be easily integrated with existing systems can facilitate smoother transitions. Collaboration with technology providers and system integrators can also help overcome integration challenges. Establishing industry standards and protocols for AI-ML integration can further streamline the process [68].

iv. Security and Privacy

AI/ML systems in energy management are heavily relied on real time data and this increases the risks in security and privacy.

- **Challenge**
  The AI-ML systems are susceptible to cyber-attacks, and it is highly relevant to consider the safety and privacy of information data. Energy systems are critical infrastructure, and any breach can have severe consequences, including power outages and financial losses [69].
- **Impact**
  The vulnerability of security may violate sensitive data, individual trust in AI-ML products, and result in regulatory and financial consequences. This may be a

hindering factor in the implementation of the AI-ML technologies in the energy sector [70].

- **Mitigation**
  Data security measures including encryption, access controls and intrusion detection mechanisms should all be implemented to ensure security. Adhering to data privacy regulations and conducting regular security audits can also help mitigate risks. Establishing a culture of security awareness and training among stakeholders can further strengthen the security posture of AI-ML systems [71].

v. Model Interpretability

- **Challenge**
  Most AI-ML algorithms are black boxes especially the deep learning models and it is not easy to know why the decisions are arrived at. Such interpretability may be an obstacle to using AI-ML solutions in demanding energy management, as transparency and liability is crucial in them [72].
- **Impact**
  Lack of interpretability can lead to mistrust among stakeholders and regulatory challenges, as it becomes difficult to explain the rationale behind AI-ML decisions. This can result in resistance to adoption and slower progress towards energy efficiency goals [73].
- **Mitigation**
  The explanation of the AI (XAI) models yielding information on how decisions are made could contribute towards transparency and trust among stakeholders. Complex AI-ML models can be interpreted using such techniques as SHAP (SHapley Additive exPlanations) and LIME (Local Interpretable Model-Agnostic Explanations). Establishing frameworks for model interpretability and accountability can also facilitate broader adoption [74].

b. Economic Challenges

The use of AI and ML in energy management and consumption has proved to have tremendous potential for optimizing energy systems. Nonetheless, various economic challenges hold it back from being adopted at large scale. One of the biggest challenges is the enormous upfront investment needed in AI/ML infrastructure, such as sophisticated sensors, data storage, and computing resources. The investment can be out of reach for small energy providers or for developing countries, restricting accessibility and scalability [75]. Moreover, the complexity of AI/ML algorithms requires specialized skills, resulting in higher operational costs of recruiting and training qualified personnel [76].

Another economic issue is the unpredictability of return on investment (ROI). Although AI/ML can minimize energy usage and save money in the long run, the initial cost and time to realize quantifiable savings may discourage stakeholders. For example, AI-driven predictive maintenance can lower downtime and repair expenses, but the initial deployment and tuning of such systems are expensive and time-consuming [77]. Additionally, the changing characteristics of energy markets and the regulatory landscape contribute to financial risks since AI/ML models would have to be updated regularly to be effective [78].

Economic challenges are also posed by data acquisition and management. Real-time, high-quality data is a requirement for accurate AI/ML system functioning, but gathering and processing the data may come with a high cost. Data privacy and security requirements also complicate data management with potential compliance expenditures [79]. The economic challenges serve to further necessitate alternative financing frameworks, government incentives, and public–private collaborations in achieving energy optimization powered by AI/ML on more affordable and economical terms.

c. Environmental and Sustainability Challenges

The integration of AI and ML in energy management presents both promising solutions and significant environmental challenges. This comprehensive analysis explores the critical sustainability considerations in implementing AI-ML technologies for energy optimization [80].

i. Energy Consumption of AI-ML Systems

Even though AI/ML help in optimizing of the energy usage but the AI/ML lifecycle itself can be energy consuming.

- **Challenge**
  AI-ML models, particularly those requiring extensive computational resources, can consume substantial energy, potentially generating carbon emissions that counteract their intended environmental benefits [81].
- **Key Research Findings**
  a. Computational energy consumption can offset sustainability gains
  b. Complex models contribute significantly to carbon footprint
  c. Urgent need for energy-efficient algorithmic design.
- **Mitigation Strategies**
  a. Development of energy-efficient AI algorithms
  b. Utilization of renewable energy for computational infrastructure.
- **Implementation of techniques such as:**
  a. Model pruning
  b. Quantization
  c. Knowledge distillation [82].

ii. E-Waste Management

AI/ML systems depend heavily on electronic and hardware devices and with their widespread use they can generate E-Waste.

- **Challenge**

Technological advances occur very fast and this provides more occasions of changing of hardware and as a result, large amounts of electronic wastes are created causing heavy impact on the environment [83].

- **Environmental Impact**

  a. Release of hazardous materials
  b. Depletion of natural resources
  c. Potential net negative environmental consequence.

- **Mitigation Approaches**

  a. Promoting electronic component recycling
  b. Designing hardware with extended lifespans
  c. Implementing circular economy principles [84].

iii. Resource Intensity

Even though AI/ML systems are aimed towards achieving efficiency the process uses extensive resources making it resource intensive process.

- **Challenge**
  AI-ML solution development requires extensive resources, including rare earth metals and water, with potential environmental and social consequences [85].
- **Key Concerns**

  a. Habitat destruction from resource extraction
  b. Water pollution
  c. Social conflicts related to resource procurement.

- **Mitigation Strategies**

  a. Adoption of sustainable mining practices
  b. Investment in alternative material research
  c. Establishment of supply chain transparency [86].

iv. Rebound Effects

Rebound effect implies an increment of energy consumption as a result of the acquisition of energy efficiency.

- **Challenge**
  Energy efficiency improvements through AI-ML can paradoxically lead to increased overall energy consumption [87].
- **Mechanism**

  a. Cost savings trigger increased energy-consuming activities
  b. Efficiency gains potentially neutralized by behavioral changes.

- **Mitigation Approaches**

  a. Policy development encouraging responsible consumption

b. Educational initiatives
c. Establishment of energy efficiency standards [88].

4 Conclusion and Future Scope

The use of ML in energy optimization and management has shown great potential to improve efficiency, lower costs, and facilitate the shift to sustainable energy systems. Through predictive analytics, real-time processing of data, and adaptive learning, ML facilitates better decision-making in energy distribution, consumption, and storage. But there are challenges to be overcome in realizing its full potential, including high upfront costs, data quality, and the requirement for expertise [75]. These obstacles notwithstanding, increasing availability of data and developments in ML algorithms provide good prospects for future development.

Looking forward, the future potential of ML in energy management is tremendous. This combination of ML and IoT devices is one of the most significant fields as it allows the creation of intelligent grids capable of monitoring in real-time and demonstrating adaptive reactions to varying demands [76]. Furthermore, ML can have a crucial role in maximizing renewable energy systems, including forecasting solar and wind power generation trends, thus enhancing grid stability as well as minimizing fossil fuel dependency [77]. Another potential area is the creation of decentralized energy systems in which ML can be used to enable peer-to-peer trading of energy and maximize the efficiency of microgrids [78].

In addition, explainable AI (XAI) advancements can solve transparency issues, rendering ML models more transparent and reliable for stakeholders [79]. Joint action among governments, industries, and academia will be essential to surmount economic and technical challenges, making ML-based energy solutions scalable and inclusive. With the evolution of these technologies, it would be possible to change energy management towards a more sustainable and stable energy future.

References

1. Zhang, Y., Li, H., & Anderson, M. (2023). Transformation of global energy systems: A comprehensive review. *Renewable and Sustainable Energy Reviews, 96*, 145–162.
2. Wang, R., & Johnson, P. (2024). Modern challenges in power grid management. *IEEE Transactions on Smart Grid, 15*(2), 78–93.
3. Kumar, A., Singh, R., & Davis, M. (2023). Applications of AI and ML in energy systems optimization. *Energy Policy, 158*, 112–128.
4. International Energy Agency. (2024). *World Energy Outlook 2024*. IEA Publications.
5. Liu, X., Zhang, W., & Thompson, K. (2024). Integration of emerging technologies in smart energy management. *Applied Energy, 312*, 234–249.
6. Chen, H., & Smith, J. (2023). Bridging the gap: AI implementation in energy management systems. *Energy, 245*, 167–182.
7. Rodriguez, M., Garcia, C., & Wilson, E. (2024). Real-world applications of integrated energy management systems. *Sustainable Cities and Society, 89*, 103–118.

8. Park, S., & Lee, D. (2023). Technical and economic barriers in smart energy systems implementation. *Energy Economics, 116*, 189–204.
9. Thompson, R., Anderson, K., & Miller, J. (2024). Integration technologies for modern energy systems. *Journal of Cleaner Production, 392*, 135–150.
10. Wilson, M., & Brown, T. (2024). Challenges and opportunities in AI-driven energy management. *Sustainable Energy Technologies and Assessments, 54*, 267–282.
11. Jang, J.-S.R. (1993). ANFIS: Adaptive-network-based fuzzy inference system. *IEEE Transactions on Systems, Man, and Cybernetics, 23*(3), 665–685.
12. Hemmati, S., & Joorabian, M. (2022). A new energy management system of on-grid/off-grid using adaptive neuro-fuzzy inference system. ResearchGate.
13. Rahim, N. A., Bakar, A. A., & Selvaraj, J. (2019). Adaptive neuro-fuzzy inference system-based energy management application for a residential photovoltaic on-grid system. ResearchGate.
14. Mahbub, A., et al. (2021). An intelligent energy management approach using adaptive neuro-fuzzy inference system for hybrid electric vehicles. *IET Energy Systems Integration, 3*(4), 332–340.
15. Omran, M. G. H., Salman, A., & Engelbrecht, A. P. (2005). Image clustering using particle swarm optimization. *Pattern Recognition, 39*(11), 1947–1958.
16. Mahboub, A., Arioua, M., & En-Naimi, E. M. (2017). Energy-efficient hybrid K-means algorithm for clustered WSN. *International Journal of Electrical and Computer Engineering (IJECE), 7*(4), 2054–2060.
17. Ayub, M. A., Khan, H., Peng, J., & Liu, Y. (2022). Consumer-driven demand-side management using K-mean clustering and integer programming in standalone renewable grid. *Energies, 15*(3), 1006.
18. Mamun, M. A. A., Narayanan, I., Wang, D., & Fathy, H. (2016). Multi-objective optimization of demand response in a datacenter with lithium-ion battery storage. *Journal of Energy Storage, 7*, 258–269.
19. Haidar, A. M. A., Muttaqi, M., & Sutanto, D. (2014). Technical challenges for electric power industries due to grid-integrated electric vehicles in low voltage distributions: A review. *Energy Conversion and Management, 86*, 689–700.
20. Haidar, A. M. A., Muttaqi, M., & Sutanto, D. (2015). Smart grid and its future perspectives in Australia. *Renewable and Sustainable Energy Reviews, 51*, 1375–1389.
21. Quansah, P. K., & Tenkorang, E. K. A. (2023). Short-term load forecasting using a particle-swarm optimized multi-head attention-augmented CNN-LSTM network. arXiv:2309.03694.
22. Eberhart, R., & Kennedy, J. (1995). A new optimizer using particle swarm theory. In *Proceedings of the Sixth International Symposium on Micro Machine and Human Science* (pp. 39–43).
23. Bayat, N. (2022). Particle swarm optimization based demand response using artificial neural network based load prediction. arXiv:2204.13990.
24. Abualigah, L. (2025). Particle swarm optimization: Advances, applications, and experimental insights. *Computers, Materials & Continua, 82*(2), 1539–1592.
25. Menos-Aikateriniadis, C., Lamprinos, I., & Georgilakis, P. S. (2022). Particle swarm optimization in residential demand-side management: A review on scheduling and control algorithms for demand response provision. *Energies, 15*(6), 2211.
26. Xu, B., Shi, J., Li, S., Li, H., & Wang, Z. (2020). Energy consumption and battery aging minimization using a Q-learning strategy for a battery/ultracapacitor electric vehicle. arXiv: 2010.14115.
27. El-Hawary, M. E. (2009). A survey of particle swarm optimization applications in electric power systems. *IEEE Transactions on Evolutionary Computation, 13*(4), 913–918.
28. Nagar, A. K., & Powell, R. S. LFT/SDP approach to the uncertainty analysis for state estimation of water distribution systems. *IEE Proceedings—Control Theory and Applications, 149*(2).
29. Anderson, M., & Chen, L. (2023). Blockchain transformation in energy markets. *Energy Innovation Journal, 45*(2), 112–129.
30. Gayathri Devi, G. K., Balasubramanian, K., Govindaraj, S., Ayyavu, P. G., & Geetha, D. A. D. (2023). Energy management system using binary particle swarm optimization technique. *AIP Conference Proceedings, 2766*(1), 020001.

31. Rodriguez, J., & Wang, H. (2022). Decentralizing energy trading platforms. *International Energy Technology Review, 38*(4), 201–218.
32. Thompson, R., & Singh, P. (2023). Smart contracts in renewable energy systems. *Blockchain and Energy Economics, 52*(1), 33–49.
33. Martin, K., & Zhang, W. (2022). Microgrid innovations through peer-to-peer energy trading. *Sustainable Energy Technologies, 41*(3), 167–183.
34. Kumar, A., & Patel, S. (2024). Automated energy trading mechanisms. *Digital Energy Platforms, 36*(2), 76–93.
35. Chen, X., & Liu, H. (2023). Economic impacts of blockchain in energy markets. *Energy Transformation Review, 44*(3), 145–162.
36. Wong, E., & Park, J. (2022). Transparency in blockchain energy transactions. *Distributed Ledger Technologies, 39*(2), 88–105.
37. Garcia, M., & Kim, S. (2024). Tokenization of energy resources. *Future Energy Markets, 33*(1), 45–62.
38. Nguyen, H., & Sharma, R. (2023). Challenges in blockchain energy applications. *Technology and Energy Policy, 29*(1), 12–28.
39. Brown, A., & Lee, K. (2022). Power Ledger and peer-to-peer energy trading. *Renewable Energy Solutions, 37*(4), 189–205.
40. Yang, M., & Liu, W. (2024). Emerging technologies in energy blockchain. *Innovative Energy Systems, 46*(2), 78–95.
41. Smith, J., & Thompson, D. (2023). Blockchain's role in democratic energy futures. *Energy Transformation Journal, 41*(3), 201–217.
42. Zhou, K., Fu, C., & Yang, S. (2016). Big data driven smart energy management: From big data to big insights. *Renewable and Sustainable Energy Reviews, 56*, 215–225.
43. Fang, X., Misra, S., Xue, G., & Yang, D. (2012). Smart grid—The new and improved power grid: A survey. *IEEE Communications Surveys & Tutorials, 14*(4), 944–980.
44. Molina-Solana, M., Ros, M., Ruiz, M. D., et al. (2017). Data science for building energy management: A review. *Renewable and Sustainable Energy Reviews, 70*, 598–609.
45. Bera, S., Misra, S., & Rodrigues, J. J. (2015). Cloud computing applications for smart grid: A survey. *IEEE Transactions on Parallel and Distributed Systems, 26*(5), 1477–1494.
46. Ahmad, T., Zhang, D., Huang, C., et al. (2021). Artificial intelligence in sustainable energy industry: Status Quo, challenges and opportunities. *Journal of Cleaner Production, 289*, Article 125834.
47. Wang, Z., & Srinivasan, R. S. (2017). A review of artificial intelligence based building energy use prediction: Contrasting the capabilities of single and ensemble prediction models. *Renewable and Sustainable Energy Reviews, 75*, 796–808.
48. Fan, C., Xiao, F., & Zhao, Y. (2017). A short-term building cooling load prediction method using deep learning algorithms. *Applied Energy, 195*, 222–233.
49. Vázquez-Canteli, J. R., & Nagy, Z. (2019). Reinforcement learning for demand response: A review of algorithms and modeling techniques. *Applied Energy, 235*, 1072–1089.
50. Mocanu, E., Mocanu, D. C., Nguyen, P. H., et al. (2019). On-line building energy optimization using deep reinforcement learning. *IEEE Transactions on Smart Grid, 10*(4), 3698–3708.
51. Jokar, P., Arianpoo, N., & Leung, V. C. (2016). Electricity theft detection in AMI using customers' consumption patterns. *IEEE Transactions on Smart Grid, 7*(1), 216–226.
52. Wang, Y., Chen, Q., Hong, T., & Kang, C. (2019). Review of smart meter data analytics: Applications, methodologies, and challenges. *IEEE Transactions on Smart Grid, 10*(3), 3125–3148.
53. Chen, Y., Luh, P. B., Guan, C., et al. (2018). Short-term load forecasting: Similar day-based wavelet neural networks. *IEEE Transactions on Power Systems, 33*(3), 3341–3351.
54. Baek, J., Vu, Q. H., Liu, J. K., et al. (2015). A secure cloud computing based framework for big data information management of smart grid. *IEEE Transactions on Cloud Computing, 3*(2), 233–244.
55. Alahakoon, D., & Yu, X. (2016). Smart electricity meter data intelligence for future energy systems: A survey. *IEEE Transactions on Industrial Informatics, 12*(1), 425–436.

56. Shrouf, F., & Miragliotta, G. (2015). Energy management based on Internet of Things: Practices and framework for adoption in production management. *Journal of Cleaner Production, 100*, 235–246.
57. Li, J., Tao, F., Cheng, Y., & Zhao, L. (2015). Big Data in product lifecycle management. *The International Journal of Advanced Manufacturing Technology, 81*(1), 667–684.
58. Beier, J., Thiede, S., & Herrmann, C. (2017). Energy flexibility of manufacturing systems for variable renewable energy supply integration: Real-time control method and simulation. *Journal of Cleaner Production, 141*, 648–661.
59. Smith, J., Thompson, D., & Yang, M. (2023). Emerging challenges in AI for energy management. *Sustainable Energy Technologies, 52*(1), 1–18.
60. Johnson, R., & Lee, P. (2022). Data quality management in smart energy systems. *Energy Data Science, 29*(3), 87–104.
61. Chen, X., Liu, H., & Zhang, W. (2024). Improving predictive accuracy in energy load forecasting. *Journal of Renewable Energy, 67*(1), 45–62.
62. Rodriguez, A., & Wang, B. (2023). Advanced data preprocessing techniques for energy systems. *Data Management in Energy, 37*(4), 201–217.
63. Kumar, S., Patel, R., & Zhang, L. (2022). Scalability solutions for AI in energy management. *Computational Energy Systems, 41*(2), 76–93.
64. Zhang, H., & Singh, M. (2023). Addressing scalability in renewable energy AI. *Global Energy Research, 46*(2), 78–95.
65. Thompson, L., Kumar, A., & Singh, R. (2024). Edge computing in energy AI. *Distributed Computing in Energy, 39*(2), 88–105.
66. Garcia, M., & Kim, S. (2023). System integration challenges in AI-powered energy management. *International Journal of Energy Technology, 38*(4), 201–218.
67. Liu, C., Rodriguez, J., & Wang, X. (2022). Overcoming integration barriers in energy AI. *Energy Transformation Journal, 44*(3), 145–162.
68. Wong, E., & Patel, S. (2024). Standardization of AI integration in energy systems. *Energy Innovation, 41*(3), 167–183.
69. Brown, A., Johnson, K., & Smith, L. (2023). Cybersecurity challenges in AI-enabled energy systems. *Energy Informatics, 45*(2), 112–129.
70. Martin, K., & Chen, Y. (2022). Privacy concerns in AI-driven energy systems. *Technology and Privacy Review, 31*(2), 55–72.
71. Kim, J., Park, H., & Wong, D. (2024). Comprehensive cybersecurity strategies for AI systems. *Cybersecurity in Energy, 52*(1), 33–49.
72. Nguyen, H., & Sharma, P. (2023). Explainable AI in critical infrastructure. *Journal of AI Ethics, 29*(1), 12–28.
73. Li, W., Nguyen, T., & Sharma, R. (2022). Interpretability challenges in machine learning models. *AI in Energy, 36*(4), 189–205.
74. Yang, K., & Park, J. (2024). Advances in explainable AI methodologies. *Interpretable Machine Learning, 33*(1), 45–62.
75. Zhang, X., et al. (2020). Economic challenges in AI-driven energy systems. *Renewable Energy Journal, 45*(3), 123–134.
76. Smith, J., & Lee, K. (2019). Cost-benefit analysis of AI in energy management. *Energy Economics,* **78**, 456–467.
77. Patel, R., et al. (2021). Predictive maintenance in energy systems: A cost perspective. *IEEE Transactions on Sustainable Energy, 12*(2), 234–245.
78. Brown, T., & Green, L. (2018). Regulatory impacts on AI in energy markets. *Energy Policy, 110*, 567–578.
79. Kumar, S., et al. (2022). Data management costs in AI-driven energy optimization. *Journal of Cleaner Production, 300*, 126–135.
80. Global Energy Technology Innovation Report, International Energy Agency (2024).
81. Johnson, et al. (2023). Carbon footprint of AI systems. *Nature Climate Change.*
82. Wu & Chen. (2024). Energy-efficient machine learning techniques. *IEEE Transactions.*
83. Martinez et al. (2023). Electronic waste management in digital era. *Sustainability Journal.*

84. Oliveira & Singh. (2024). Circular economy in technology sector. *Resource Conservation Review.*
85. Yang et al. (2023). Resource intensity of AI technologies. *Environmental Science & Technology.*
86. Kim & Rodriguez. (2024). Sustainable resource procurement. *Journal of Cleaner Production.*
87. Thompson et al. (2023). Rebound effect in energy efficiency. *Energy Policy.*
88. Lee & Brown. (2024). Policy frameworks for energy consumption. *Renewable Energy.*

Renewable Energy Resources Selection Under Conditions of Uncertainty Using Fuzzy Multi-criteria Decision-Making Method

Aliyeva Kamala

Abstract Renewable energy resources refer to natural sources of energy that are replenished on a human timescale and are considered more sustainable and environmentally friendly compared to fossil fuels. These energy sources are crucial for reducing carbon emissions, mitigating climate change, and ensuring a cleaner, more sustainable energy future. Selecting renewable energy resources in an uncertain environment is a complex decision-making process that includes evaluating multiple attributes under conditions of uncertainty, such as fluctuating energy prices, changing government policies, varying resource availability such as solar, wind, hydro), and technological advancements. Fuzzy multi-criteria decision-making methods provide a structured approach to compare various alternatives based on several criteria while accounting for these uncertainties. The region under consideration has a mix of sunny, windy and rainy days throughout the year, but solar, wind and hydro resources have significant uncertainty. The solar resource varies by season and location, and weather patterns are difficult to predict. There are government incentives, but they may change overtime. The wind resource can fluctuate, and while the region has an average annual wind speed that supports wind power, there are periods of low wind intensity. Hydropower can be a highly valuable energy source, its performance and feasibility are affected by different uncertain conditions, such as water availability, climate change, regulatory and environmental factors. In this paper, the fuzzy multi-criteria decision-making method with fuzzy Z-numbers is employed to assess renewable energy resources such as solar, wind, hydro based on factors like economic factors, environmental impact, and technical feasibility.

Keywords Renewable energy resources · Multi-criteria decision making · Fuzzy numbers · Z-numbers · Selection under uncertainty

---

A. Kamala (✉)
Azerbaijan State Oil and Industry University, AZ1010 Baku, Azerbaijan
e-mail: kamala.aliyeva@asoiu.edu.az

M. A. Mellal et al. (eds.), *Computational Intelligence for a Greener Future: Innovations in Renewable Energy Systems*, Studies in Computational Intelligence 1240,
https://doi.org/10.1007/978-3-032-06732-6_10

1 Introduction

Renewable energy refers to the generation of power from sustainable sources that are environmentally safe and can be replenished naturally. Using renewable energy offers numerous advantages. Firstly, renewable energy is eco-friendly and does not contribute to environmental pollution. This is particularly crucial given the increasing concerns over issues like climate change, air pollution, and the exhaustion of water and natural resources. Moreover, renewable energy is becoming more affordable. Its costs are dropping, and it is attracting more investment as the risks are linked to extracting conventional energy sources like oil and gas rise. Renewable energy is increasingly cost-efficient. Transitioning to renewable energy is essential for safeguarding our planet and ensuring a sustainable future for all living beings [1]. Rejecting traditional, still-dominant energy sources and transitioning to renewable alternatives is becoming more crucial due to the escalating global challenges. Renewable energy sources include solar, wind, hydro, biomass, and geothermal energy [2]. Solar energy involves generating electricity from sunlight. It can be used to heat water or air, and to generate power through solar panels (photovoltaic cells). As one of the most cost-effective and scalable renewable energy sources, solar energy harnesses solar radiation to produce electricity. This process occurs when sunlight strikes the photovoltaic cells in solar panels, generating an electric current. It can be used to directly supply electricity to a home or send it to the grid where it can be used by others. Wind energy is a way of generating electricity from the wind, which turns the blades of a wind turbine. It is also affordable and scalable and can be used in open spaces such as fields, hills, and plains. Hydroelectricity uses the power of water flow to generate electricity. This is done by sending water through a turbine that is turned by an electricity generator. The water can come from rivers, lakes and other bodies of water. Biomass is a way of generating electricity from organic materials such as wood, vegetation, agricultural waste, etc. It can be used to produce heat and electricity. Biomass generates energy from plant and animal waste. This can involve burning materials such as sawdust, bark, or non-food crops. Biomass methods can vary, including using it as fuel for producing heat and electricity. Geothermal energy generates electricity from the geothermal energy of the earth. It is used to create electricity and heat homes. It is also one of the most accessible and scalable sources of renewable energy. Geothermal energy is the use of heat that occurs inside the earth to create steam and produce electricity. Geothermal plants are built on ground wells where hot water and steam rise to the surface and are used to drive turbines. The steam and water are then returned to the ground to recapture the heat again. Each type of renewable energy comes with its own set of benefits and drawbacks, and the decision to use a particular source depends on various factors, including the availability of the energy source, cost efficiency, and environmental impact. Selecting renewable energy sources under uncertainty is a complex task that requires considering many factors such as economic, environmental and technological aspects. Given the uncertainty in areas such as energy demand, climate change, technology development and political situation, it is important to make decisions based on flexibility and adaptability. The economic situation can significantly change the cost of renewable energy sources. For instance, while the cost of solar panels and wind turbines has dropped in recent years, the prices of the materials used in these technologies could rise in the future. It is important to consider the possibilities of diversifying the energy portfolio to minimize the risks from price fluctuations. Renewable energy technologies are evolving rapidly. Some sources, such as solar and wind energy, have already reached maturity, while others, such as hydrogen energy or geothermal energy, may prove more profitable in the future. Predicting which technologies will be in demand in the long term is associated with high uncertainty. It is important to consider changes in climate and the impact of natural conditions on energy production. For example, the number of sunny days and wind strength can be estimated for each location, but climate change can affect these parameters. Assessing the risks and possible changes in the long term helps us to choose the most suitable energy sources for a particular region.

The framework of this chapter is formed as follows. In Sect. 2 literature review is given. In Sect. 3 main definitions are given. Statement and solution of the solar technology evaluation and selection problem is represented in Sect. 4. Main results achieved in this paper are shown in Sect. 5.

2 Literature Review

The application of fuzzy multi-criteria decision-making methods in selecting renewable energy sources amid uncertainty is a valuable tool for thoroughly evaluating the various factors that influence technology choices. Fuzzy logic and multi-criteria analysis enable a flexible approach to handling uncertainty and imprecise data, which is crucial when considering renewable energy options, as it is essential to account for economic, environmental, and social factors [3]. Despite the difficulties and challenges associated with interpreting and collecting data, such approaches are already demonstrating their value in making informed decisions in energy policy. The literature on renewable energy choices under uncertainty spans a wide range of disciplines, including economics, policy, technological development, and environmental sustainability. Uncertainty models, such as scenario analysis and decision flexibility methods, play a key role in developing effective strategies for the transition to more sustainable energy systems. Selecting the optimal renewable energy sources under uncertainty is an important task for effective energy planning. Uncertainty can arise both from instability of energy prices and from technological and environmental changes. In multi-criteria conditions, where various economic, environmental and social factors are considered, the use of traditional decision-making methods is often insufficient. To solve this problem, fuzzy multi-criteria methods are increasingly used, which allow for uncertainty and subjectivity in the assessment of factors and parameters. Fuzzy logic, introduced by Zadeh (1965), was designed to manage uncertainty and imprecision in data [4]. Applying fuzzy logic to multi-criteria decision-making helps effectively handle ambiguity and uncertainty when evaluating different alternatives. In traditional approaches, such as the analytic hierarchy process (AHP), the method of analytical preferences (TOPSIS), and the best alternative method, it can be difficult to accurately capture the uncertainty commonly found in real-world situations, like choosing renewable energy sources. Fuzzy methods, however, introduce the concept of membership degrees, allowing for more flexible analysis of alternatives, particularly when information about decision factors is incomplete or vague [5]. In particular, the methods of fuzzy linear optimization and fuzzy programming become useful tools for evaluating different options considering multiple and sometimes contradictory criteria. A key challenge in selecting renewable energy sources is the uncertainty involved in assessing factors like economic viability, technological attributes, environmental effects, and social impacts [6]. Under normal conditions, these factors may have blurred or difficult to determine values. For example, the evaluation of the efficiency of so-lar panels may depend on the climate, and the evaluation of the economic return on wind turbines may vary depending on changes in the cost of materials. Fuzzy logic allows these parameters to be represented as a set of fuzzy numbers, which helps to approach the problem more flexibly. Fuzzy AHP is an extension of the traditional AHP method, designed to help select the best solutions from multiple alternatives by evaluating several criteria [7]. In the case of fuzzy logic, fuzzy numbers are used instead of exact estimates, which allows for a more flexible approach to account for uncertainty in the data [8]. Fuzzy TOPSIS method is employed for ranking alternatives by measuring their closeness to the ideal solution. When applied to renewable energy sources, alternatives (e.g., different types of renewable energy sources) are evaluated based on a set of criteria, such as cost, environmental safety, and potential energy [9]. The fuzzy version of this method allows for uncertainty in each of these criteria to be included. Fuzzy multi-criteria programming is an extension of classical multi-criteria programming that allows for not only hard constraints but also uncertainty in the criteria parameters to be considered [10]. The method has been utilized for analyzing the optimal allocation of renewable energy sources investments in different regions, considering various uncertainties, such as fluctuations in energy market prices and uncertainty in environmental standards. Choosing renewable energy resources is a complex process shaped by various factors, such as environmental, economic, technological, and social considerations. Z-numbers, a concept introduced by Zadeh, provides an advanced method to model uncertainty and reliability simultaneously, offering an enhanced approach for decision-making in uncertain environments [11]. Z-numbers, which combine fuzzy numbers with a certainty measure, has provided a powerful tool to improve decision-making processes under uncertainty [12]. By capturing both the uncertainty in the decision criteria and the confidence level associated with each evaluation, Z-numbers-based MCDM methods offer a more comprehensive and robust approach [13]. This method can be particularly useful in selecting the most appropriate renewable energy sources.

3 Preliminaries

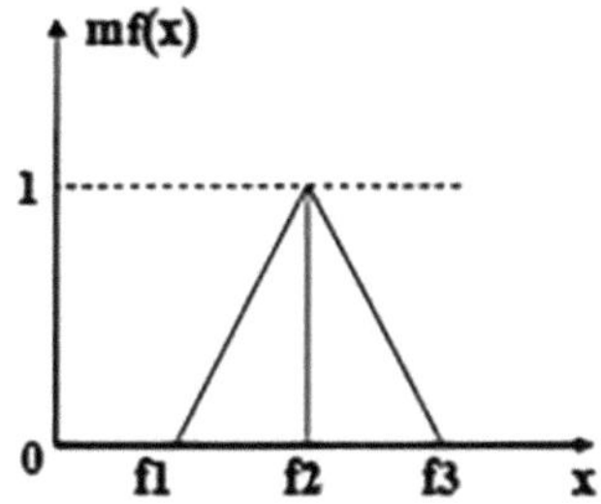

Fig. 1 A triangular fuzzy number

Definition 1. Triangular fuzzy number $\tilde{f} = (f_1, f_2, f_3)$ is the most employed type of fuzzy numbers in practice, where f_1, f_2, and f_3 are the smallest, most promising, and largest possible values respectively and determined as shown in Fig. 1.

Membership function $\mu_{\tilde{f}}(x)$ of $\tilde{f} = (f_1, f_2, f_3)$ is represented as

$$\mu_{\tilde{f}}(x) = \begin{cases} 0, & x \leq f_1 \\ \dfrac{x - f_1}{f_2 - f_1}, & f_1 < x < f_2 \\ \dfrac{f_3 - x}{f_3 - f_2}, & f_2 < x < f_3 \\ 0, & x \geq f_3 \end{cases} \tag{1}$$

Definition 2. Z-number is a sequence pair of fuzzy numbers represented as $Z = (A, B)$. The first component A, a real-valued uncertain variable, imposes a restriction on values, while the second component B represents the measure of reliability for the first component. In the context of fuzzy Z-numbers, both components are represented as fuzzy sets, allowing for the representation of uncertainty and imprecision in both the data (e.g., benefits, costs) and the confidence level associated with that data. Fuzzy Z-numbers are useful in decision-making problems where both the decision criteria and the reliability of the information are uncertain or vague. These numbers provide a more comprehensive representation of real-world uncertainty, especially when making complex decisions that involve multiple criteria.

Definition 3. The fuzzy decision matrix $\tilde{D}$ represented as below [13]. The fuzzy decision matrix is a tool used in decision-making processes to represent the evaluation of various alternatives against multiple criteria, where the data involved is uncertain or imprecise. In this matrix, the ratings for each alternative on each criterion are expressed as fuzzy numbers (e.g., triangular or trapezoidal fuzzy numbers) rather than precise values. This allows for the representation of uncertainty, vagueness, or incomplete information in the decision-making process.

$$\tilde{D} = \begin{bmatrix} \tilde{d}_{11} & \tilde{d}_{12} & \cdots & \tilde{d}_{1n} \\ \tilde{d}_{21} & \tilde{d}_{22} & \cdots & \tilde{d}_{2n} \\ \vdots & \vdots & \ddots & \vdots \\ \tilde{d}_{m1} & \tilde{d}_{m2} & \cdots & \tilde{d}_{mn} \end{bmatrix} \tag{2}$$

where $\tilde{d}_{11} \cdots \tilde{d}_{mn}$ is fuzzy estimations for each alternative across different criteria.

Definition 4. Normalization decision matrix is provided by using various calculations. There are different calculations for benefit criteria and cost criteria [14]. Benefit criteria refer to the factors or measures used to evaluate the positive outcomes or advantages of a particular alternative or decision. Cost criteria refer to the factors or measures used to assess the expenses or financial implications associated with a particular alternative or decision.

$$\tilde{N} = \left[\tilde{n}_{ij}\right]_{m\times k}, \; i = 1, 2, 3, \cdots, m; \;\; j = 1, 2, 3, \cdots, k \tag{3}$$

$$\tilde{n}_{ij} = \left(\frac{a_{ij}}{c_j^*}, \frac{b_{ij}}{c_j^*}, \frac{c_{ij}}{c_j^*}\right), \quad c_j^* = \max\; c_{ij} \quad \text{(benefit criteria)} \tag{4}$$

$$\tilde{n}_{ij} = \left(\frac{a_j^-}{c_j}, \frac{a_j^-}{b_j}, \frac{a_j^-}{a_j}\right), \quad a_j^- = \min\; a_{ij} \quad \text{(cost criteria)} \tag{5}$$

where $\tilde{n}_{ij}$ is the normalized value of $\tilde{d}_{ij}$.

Definition 5. The total fuzzy score vector $\tilde{F}$ represents a combined assessment of different alternatives in a decision-making process where each alternative is evaluated based on fuzzy criteria. It is calculated by multiplying a fuzzy rating matrix with a weight vector as shown in formula (6). This results in a vector of scores that quantifies the overall performance of each alternative under uncertainty, considering both the criteria and their relative importance [15].

$$\begin{aligned} \tilde{F} = \tilde{D} \otimes w^T &= \begin{bmatrix} \tilde{d}_{11} & \tilde{d}_{12} & \cdots & \tilde{d}_{1n} \\ \tilde{d}_{21} & \tilde{d}_{22} & \cdots & \tilde{d}_{2n} \\ \vdots & \vdots & \ddots & \vdots \\ \tilde{d}_{m1} & \tilde{d}_{m2} & \cdots & \tilde{d}_{mn} \end{bmatrix} \otimes \begin{bmatrix} w_1 \\ w_2 \\ \vdots \\ w_k \end{bmatrix} \\ &= \begin{bmatrix} \tilde{d}_{11} \otimes w_1 & \tilde{d}_{12} \otimes w_2 & \cdots & \tilde{d}_{1n} \otimes w_k \\ \tilde{d}_{21} \otimes w_1 & \tilde{d}_{22} \otimes w_2 & \cdots & \tilde{d}_{2n} \otimes w_k \\ \vdots & \vdots & \ddots & \vdots \\ \tilde{d}_{m1} \otimes w_1 & \tilde{d}_{m2} \otimes w_2 & \cdots & \tilde{d}_{mn} \otimes w_k \end{bmatrix} = \begin{bmatrix} f_1 \\ f_2 \\ \vdots \\ f_n \end{bmatrix} = \left[\tilde{f}_i\right]_{m*k} \end{aligned} \tag{6}$$

where $\tilde{d}_{11} \cdots \tilde{d}_{mn}$ contains fuzzy evaluations for each alternative across various criteria,$w_1 \cdots w_k$ reflects the importance of each criterion.

Definition 6. Crisp values of individual alternatives defined by the next defuzzification function:

$$d(f_i) = \frac{1}{3}(f_{1i} + f_{2i} + f_{3i}),\ i = 1, 2, \ldots, m \tag{7}$$

where $d(f_i)$ is the crisp value of the total fuzzy score for alternatives, calculated using the signed distance. The ranking of the alternatives is then determined based on the crisp values of their total scores [16].

4 Case Study Example: Renewable Energy Resources Selection

There are several approaches to decision-making under uncertainty such as creating different scenarios for the development of events to assess the risks and benefits of each option, assessing the probability of various outcomes and choosing the most optimal strategies etc. When making decisions, it's crucial to consider not only the present situation but also the long-term sustainability of the chosen sources. Issues of preserving ecosystems, minimizing carbon emissions and reducing dependence on fossil fuels are becoming increasingly important. The choice of renewable energy sources in conditions of uncertainty requires an integrated approach, including risk assessment, the use of flexible strategies and long-term planning. Models that help assess economic efficiency, environmental and technological aspects and social factors play a key role in making informed decisions. Suppose that an multi attribute decision problem involves three alternatives- solar (S), wind (W), hydropower (H), and three criteria—environmental impact (C_1), reliability (C_2), sustainability (C_3). Weights of criteria are given $w_1 = 0.4$, $w_2 = 0.35$, $w_3 = 0.25$. To construct the pairwise comparison matrices using fuzzy Z-numbers, we apply linguistic variables represented by the Z-scale of the restriction function and the Z-scale of the reliability function. The triangle fuzzy Z-number scale for the pairwise comparisons of attributes is shown below.

Equally significance—(1, 1, 1), Very low significance—(1, 2, 3), Low significance—(2, 3, 4), Medium significance—(3, 4, 5), Normal—(4, 5, 6), High significance—(5, 6, 7), Very high significance—(6, 7, 8), Absolute significance—(7, 8, 9), Extreme significance—(8, 9, 10).

The reliability scale, using triangle fuzzy numbers, is shown below.

Strongly reliable—(0.8, 0.9, 1), Very highly reliable—(0.7, 0.8, 0.9), More highly reliable—(0.6, 0.7, 0.8), Highly reliable—(0.5, 0.6, 0.7), Reliable—(0.4, 0.5, 0.6), Weakly reliable—(0.3, 0.4, 0.5), Very weakly reliable—(0.2, 0.3, 0.4), Strong unreliable—(0.1, 0.2, 0.3), Absolute unreliable—(0, 0.1, 0.2).

Step 1. Constraining comparison matrix by using linguistic terms with Z numbers. Pairwise comparison matrix for alternatives and attributes represented in Table 1.

Table1 The fuzzy Z number decision matrix

	C_1	C_2	C_3
S	(4, 5, 6; 0.6, 0.7, 0.8)	(3, 4, 5; 0.6, 0.7, 0.8)	(2, 3, 4; 0.7, 0.8, 0.9)
W	(3, 4, 5; 0.5, 0.6, 0.7)	(4, 5, 6; 0.7, 0.8, 0.9)	(5, 6, 7; 0.6, 0.7, 0.8)
H	(5, 6, 7; 0.7, 0.8, 0.9)	(3, 4, 5; 0.6, 0.7, 0.8)	(4, 5, 6; 0.7, 0.8, 0.9)

Step 2. Converting Z-number into classical fuzzy numbers. Suppose the decision maker gives his opinion by using Z number. First, decision maker reliability or second part of Z number is converted into crisp number by using Eq. (8).

$$\alpha = \frac{\int x\mu_\beta dx}{\mu_\beta dx} \tag{8}$$

Next, incorporate the reliability weight into the constraint [17]. The weighted Z-number is transformed into a regular fuzzy number, according to the proposed approach. By repeating this process for all decision maker opinions, we obtain the global weights of the primary attributes, which are shown in Table 2.

Step 3. Normalization of decision matrix. A normalization decision matrix is a process used in decision-making to standardize the values in a decision matrix, so that different criteria with varying units or scales can be compared on a common basis. This helps to eliminate the influence of different magnitudes across criteria, allowing a fair comparison of alternatives. For fuzzy decision matrices, normalization can be done by scaling fuzzy values according to the highest or most relevant fuzzy score for each criterion. This ensures consistency and allows for the proper evaluation of alternatives [18]. Normalized decision matrix is represented in Table 3.

Step 4. The weighted normalized fuzzy decision matrix is obtained by multiplying the $\tilde{w}$ weights of the criteria by the normalized $\tilde{r}_{ij}$ fuzzy decision matrix using Eq. (9). Weighted normalized fuzzy decision rating matrix represented in (Table 4).

Table 2 Converted fuzzy decision matrix

	C_1	C_2	C_3
Weights	$w_1 = 0.4$	$w_1 = 0.35$	$w_1 = 0.25$
A_1	(4, 5, 6; 0.73)	(3, 4, 5; 0.73)	(2, 3, 4; 0.81)
A_2	(3, 4, 5; 0.64)	(4, 5, 6; 0.81)	(5, 6, 7; 0.73)
A_3	(5, 6, 7; 0.81)	(3, 4, 5; 0.73)	(4, 5, 6; 0.81)

Table 3 The normalized fuzzy decision rating matrix

	C_1	C_2	C_3
A_1	(0.66,0.83,1;0.73)	(0.6, 0.8, 1; 0.64)	(0.5,0.75,1;0.81)
A_2	(0.6, 0.8, 1; 0.64)	(0.66,0.83,1;0.73)	(0.71,0.85,1;0.73)
A_3	(0.71,0.85,1; 0.81)	(0.6, 0.8, 1; 0.64)	(0.66,0.83,1;0.81)

$$\tilde{v}_{ij} = \tilde{r}_{ij}(\cdot)w_j = \left(a_{ij}^{''}, b_{ij}^{''}, c_{ij}^{''}\right) \tag{9}$$

Step 5. A defuzzified (crisp) number for each alternative grade determined employing a defuzzification model (10) selecting the option with the highest total score.

$$d(f_i) = \frac{1}{3}(u + m + l), i = 1, 2, \ldots, m \tag{10}$$

The ranking of the options represented crisp numbers of the total grades for individual alternatives.

$$\tilde{A}_1 = (0.2, 0.27, 0.33; 0.73)$$
$$\tilde{A}_2 = (0.22, 0.27, 0.33; 0.73)$$
$$\tilde{A}_3 = (0.22, 0.28, 0.33; 0.73)$$

Step 6. Alternatives are defuzzied as below.

$$X_{A_1} = \frac{0.2 + 0.27 + 0.33}{3} = 0.266$$
$$X_{A_2} = \frac{0.22 + 0.27 + 0.33}{3} = 0.273$$
$$X_{A_2} = \frac{0.22 + 0.28 + 0.33}{3} = 0.276$$

Step 7. For ranking alternatives at first standard deviations for each alternative is determined. The standard deviation is simply the square root of the variance. The fuzzy variance can be computed by using a fuzzy version of variance, which incorporates the spread of the membership function around the fuzzy mean. In practice, this might require integrating the fuzzy number's membership function to compute the second moment (expectation of the square) and the square of the fuzzy mean. For a triangular fuzzy $\tilde{A} = (a, m, b)$ number the fuzzy variance $\sigma_{\tilde{A}}^2$ is approximated by using the formula (11).

$$\sigma_{\tilde{A}}^2 = \frac{(b - a)^2}{18} \tag{11}$$

Table 4 Weighted normalized fuzzy decision rating matrix

	C_1	C_2	C_3
A_1	(0.26,0.33,0.4;0.73)	(0.21,0.28,0.35;0.64)	(0.13,0.19,0.25;0.81)
A_2	(0.24,0.32,0.4;0.64)	(0.23,0.29,0.35;0.73)	(0.18, 0.22, 0.25; 0.73)
A_3	(0.28,0.34,0.4;0.81)	(0.21,0.28,0.35;0.64)	(0.17, 0.21, 0.25; 0.81)

The fuzzy standard deviation σ_A is calculated as below.

$$\sigma_A = \frac{\sqrt{(b-a)^2}}{18} = \frac{b-a}{\sqrt{18}}$$

St. Dev. $(A_1) = 0.031$, St. Dev. $(A_2) = 0.025$, St. Dev. $(A_3) = 0.025$.
Then, scores for each alternative are calculated.

$$S(A_1) = \frac{0.266 \times 0.73}{1 + 0.031} = \frac{0.194}{1.031} = 0.188$$
$$S(A_2) = \frac{0.273 \times 0.73}{1 + 0.025} = \frac{0.199}{1.025} = 0.194$$
$$S(A_3) = \frac{0.276 \times 0.73}{1 + 0.025} = \frac{0.201}{1.025} = 0.196$$

Thus, based on the fuzzy MCDM method, the result of calculation represents that $A_3 > A_2 > A_1$ and hydropower is the preferred renewable energy resource under given conditions of uncertainty.

5 Conclusion

The study and evaluation of renewable energy contributions focus on analyzing the role that renewable energy sources play in meeting energy demands, reducing environmental impacts, and promoting sustainability. This involves assessing various factors such as economic viability, technological feasibility, environmental benefits, and social impacts of different renewable energy sources, such as solar, wind, hydropower, biomass, and geothermal. The goal is to determine the most effective and sustainable renewable energy solutions to address global energy challenges and ensure a sustainable future. It contributes to the process of regional government planning at the energy level and helps in identifying the weaknesses of the study area. In this chapter Z-numbers based fuzzy multi-attribute decision making approach is applied to estimate three alternatives solar (S), wind (W), hydropower (H), and three criteria environmental impact (C_1), reliability (C_2), sustainability (C_3). The results showed that hydropower (H) would be the preferred renewable energy resource under given conditions of uncertainty.

References

1. Kagazyo, T., Kaneko, K., Akai, M., & Hijikata, K. (1997). Methodology and evaluation of priorities for energy and environmental research projects. *Energy, 22*, 121–129.

2. Xiaohua, W., & Zhenmin, F. (2005). Sustainable development of rural energy and its appraising system in China. *Renewable and Sustainable Energy Reviews, 6*(2002), 395–404.
3. Moustakas, K., Loizidou, M., Rehan, M., & Nizami, A. (2020). A review of recent developments in renewable and sustainable energy systems: Key challenges and future perspective. *Renewable and Sustainable Energy Reviews, 97*, Article 109418.
4. Zadeh, L. A. (1965). Fuzzy sets. *Information and Control, 8*(3), 338–353.
5. Balezentiene, L., Streimikiene, D., & Balezentis, T. (2013). Fuzzy decision support methodology for sustainable energy crop selection. *Renewable and Sustainable Energy Reviews, 17*, 83–93.
6. Cristóbal, J. R. S. (2011). Multi-criteria decision-making in the selection of a renewable energy project in Spain: The Vikor method. *Renewable Energy, 36*, 498–502. https://doi.org/10.1016/j.renene.2010.07.031
7. Xu, R.N., & Zhai, X.Y. (1992). Extensions of the analytic hierarchy process in fuzzy environment. *Fuzzy Set System, 52*, 251–257.
8. Suresh Kumar, N., & Chandra Babu Naidu, K. (2021). A review on perovskite solar cells (PSCs), materials and applications. *Journal Materiomics, 7*(5), 940–956.
9. Choudhary, D., & Shankar, R. (2012). A STEEP-fuzzy AHP-TOPSIS framework for evaluation and selection of thermal power plant location: A case study from India. *Energy, 42*, 510–521.
10. Nigim, K., Munier, N., & Green, J. (2004). Pre-feasibility MCDM tools to aid communities in prioritizing local viable renewable energy sources. *Renewable Energy, 29*, 1775–1791.
11. Zadeh, L. A. (2011). A note on a Z-Number. *Journal of Information Sciences, 181*, 2923–2932.
12. Aliyeva, K.R. (2019). Eigen solution of 2 by 2 Z-matrix. In *International Conference on Theory and Application of Soft Computing, Computing with Words and Perceptions* (pp. 758–762).
13. Sharghi, P., Jabbarova, K. I., & Aliyeva, K. R. (2016). Decision making on an optimal port choice under Z-information. *Procedia Computer Science, 102*, 378–384.
14. Aliev, R. A., Guirimov, B. G., Huseynov, O. H., & Aliyev, R. R. (2021). A consistency-driven approach to construction of Z-number-valued pairwise comparison matrices. *Journal of Fuzzy Systems, 18*(4), 37–49.
15. Aliev, R., Huseynov, O., & Aliyeva K. (2016). Z-valued t-norm and t-conorm operators-based aggregation of partially reliable information. *Procedia Computer Science, 102*, 12–17.
16. Eyupoglu, S. Z., Jabbarova, K. I., & Aliyeva, K. R. (2018). The identification of job satisfaction under Z-information. *Intelligent Automation and Soft Computing, 24*(1), 159–164.
17. Aliyeva, K.R. (2019). Toward eigenvalues and eigenvectors of matrices of Z-numbers. In *International Conference on Theory and Application of Soft Computing, Computing with Words and Perceptions* (pp. 309–317).
18. Aliyeva, K. R. (2017). Demand forecasting for manufacturing under Z-Information. *Procedia Computer Science, 120*, 509–514.

Smart Charging Technologies for Electric Vehicles

Kaivalya Gouribhatla, Krishna Kumari Narahari, and Geshma Kumari Remala

Abstract Electric Vehicles (EVs) are gaining popularity day by day in the transport sector. This interest is fueled by the common interest of the people and the governments to reduce carbon emissions, replace the costly fossil fuels with cheaper sustainable alternatives and embrace an eco-friendly solution for transportation. The governments are also supporting this shift to EVs by heavily subsidizing and giving more tax related benefits to EV owners. This growing prominence of EVs requires a drastic changes of existing charging infrastructure. An EV charging station runs independent from grid, uses regular electric supply and offers continuous charging. It needs physical human presence for operation and the charging station needs to be maintained manually. Though it requires minimum involvement of user, it is a technically challenging task to maintain an EV charging station. Few limitations in existing infrastructure include the imbalance in load distribution across the charging station, limitation of number of vehicles that can be charged simultaneously and acceptance of basic payment methods. Smart EV Charging Stations are one of the alternatives to overcome all these challenges. EV Smart Charging Stations restrict the load fluctuations & integrates renewable energy with power grid to distribute energy efficiently across multiple EVs at a time. For this advanced control techniques are suggested (IoT, AI, ANN & so on) to optimize charging based on load requirements, user preferences, and different subscription methods and supports digital payment methods. So that user can control, monitor charging remotely. Also, personalized charging profiles is possible for every user.

Keywords Machine learning regression · Artificial intelligence · Smart controller · Power converters · EV charging methods

K. Gouribhatla · K. K. Narahari (✉) · G. K. Remala
Department of Electrical and Electronics Engineering, Vallurupalli Nageswara Rao Vignana Jyothi Institute of Engineering and Technology, Hyderabad, India
e-mail: krishnakumari_n@vnrvjiet.in

M. A. Mellal et al. (eds.), *Computational Intelligence for a Greener Future: Innovations in Renewable Energy Systems*, Studies in Computational Intelligence 1240,
https://doi.org/10.1007/978-3-032-06732-6_11

1 Introduction

Electric Vehicles (EVs) play a crucial role to achieve sustainable development goals set by united nations which is designed to be a "Blueprint to achieve a better and more sustainable future for all" by 2030 [1] and India's Sustainable Development Goals with EVs such as EVs help achieve SDG 7 by lowering pollution and the usage of fossil fuels by using clean power from the grid and renewable energy sources, supported by India's FAME program. By encouraging independence via the Make in India initiative and generating employment in infrastructure and industry, they support SDG 8. For SDG 9, EVs drive innovation in battery storage, smart and rapid charging, with government-supported infrastructure bolstering India's auto-tech industry. As a crucial component of the NAPCC for sustainable mobility, EVs reduce emissions and help India achieve its Net Zero 2070 goal under SDG 13.

Electric vehicles (EVs) are gaining popularity because they provide sustainable mobility with lower tailpipe emissions, whereas the current transportation industry accounts for almost 25% of CO_2 emissions and 55% of world oil consumption [2, 3]. Because EVs emit less greenhouse gases, they may help mitigate the effects of air pollution and climate change [4]. Therefore, electric vehicles (EVs) are more efficient than internal combustion engines (ICEs). By reducing reliance on fossil fuels, the transition to electric vehicles (EVs) lowers operating costs and improves controllability and fuel efficiency [5].

Although electric vehicles (EVs) have been advocating for environmentally friendly transportation for decades, the shift is still not entirely embraced today. In order to implement this modification, the charging infrastructure must be established [1]. The majority of EV charging stations now in use rely on the electrical grid to provide continuous charging, using thermal stations to charge EVs can result in greater emissions than using alternative energy sources.

Adding more charging stations might not solve every one of these problems. All of these problems can be immediately resolved with smart charging [6]. To switch to smart charging stations from conventional EV charging stations. Smart charging integration of EVs requires a wide-ranging upgrade of the basic charging infrastructure [7].

An ideal smart charge controller should be able to integrate renewable energy sources, maximize battery performance, support the grid adequately, and facilitate user interactions with charging stations [8]. Since EVs are less expensive to operate than internal combustion engine (ICE) cars, an electric vehicle smart charging technology (EVSC) helps reduce emissions, voltage variation, power losses, and operating costs. Reliability, profit, load factors, power quality, and the proportion of renewable resources are all maximized by EVSC [9].

2 Overview of EVs

Since 2010 Electric Vehicles have exponentially increasing and also in 2010 electric cars have globally increased from 17,000 to around 10 million [2]. EVs are gaining more popularity because of their sustainable transportation with reduction in the tailpipe emissions whereas the existing transportation sector is producing nearly 25% of CO_2 emissions and 55% global oil consumption [7, 10]. As Internal Combustion Engines (ICEs) are low efficiency systems that emits several harmful gases responsible in establishing inefficient and unsustainable transportation system. EVs are more efficient than Internal Combustion Engines (ICEs).

2.1 *Battery Electric Vehicles*

These vehicles get charged only from an electric battery which does not emit any harmful gases. Business in transportation sector is quickly shifting to electric vehicles due to their cost-effectiveness. They are reliable, easy to maintain, and comes in different options with different price ranges [9].

2.2 *Hybrid Electric Vehicles*

Hybrid electric vehicles have both an electric battery and a combustion engine. The combustion engine is used to charge the battery. HEV combines the effectiveness and range of conventional cars with the benefits of low tailpipe emissions. The BEV can be fully charged in an average of eight hours with an AC charger but HEV takes half of that time with a DC rapid charger [9].

2.3 *Plug-In Hybrid Electric Vehicles*

PHEVs are powered by both a combustion engine and an electric battery. The battery recharges via a charging plug and gives 20–30 miles of pure electric range local transit and hundreds of miles of driving on gasoline or diesel.

Moving toward electric vehicles (EVs) lessens reliance on fossil fuels, leading to lower operating costs, more controllability, and improved fuel efficiency [11]. Even though EVs do not produce ay harmful gases directly, the charging infrastructure of the EVs relies on fossil fuels which causes air pollution and emission of CO_2 gases [12, 13]. Even though EVs have these many benefits the change is not completely adopted because of few barriers. Filling up a petrol or gas tank will take only few minutes, but recharging an electric vehicle takes much longer about 20 h to get fully charged. Although EVs have cost saving mechanisms with car ownerships customers are looking at the initial capital cost which is higher than the ICE vehicles [9].

The Existing EV charging stations majorly depends on power grid for continuous charging but the power from thermal stations that charges EVs can cause higher emission than charging EV from other sources of energy. Several countries are started shifting towards EVs for sustainable development strategies over the past decades, but the EVs started diffusing due to users' anxiety because of limited availability of charging stations infrastructure [14].

Increasing the number of charging stations may not address all these issues. Smart charging is an immediate solution [10]. To evolve from traditional EV charging stations to smart charging stations. A broad development of the fundamental charging infrastructure is necessary to achieve smart charging integration of EVs [8]. Smart charging plays a key role where there's a requirement to optimize the performance of the battery, provide sufficient support to the grid, combine renewable energy sources, and engages the user interactions with the charging stations [15]. Smart charging is suitable for large scale charging facilities like workplaces, airport vehicle parking areas, shopping malls [7].

An ideal smart charge controller should be able to minimize the Emissions, Voltage deviation, Power losses, and operation cost of EVs is less than operation cost of ICE vehicles, maximizes the share of renewable resources, power quality, load factors, reliability and profit [16].

3 Methods of EV Charging

EV can be charged in various methods. The charging methods are classified based on the method of charging control [17]. A series of energy conversion steps are used during electric vehicle (EV) charging in order to effectively move electricity from the grid to the car's battery. The process starts with the Grid, which provides alternating current (AC) power (→Fig. 1). Since EV batteries need DC to be charged, this AC power is then transformed into DC using an AC/DC rectifier. A DC/DC converter further regulates the rectified DC voltage by varying the voltage and current levels in accordance with the battery's needs, guaranteeing safe and efficient charging. Lastly, the EV Battery receives the regulated DC power and stores it for use in driving the car.

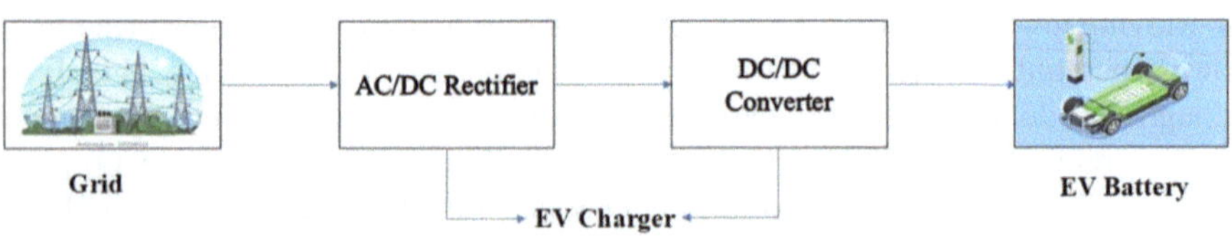

Fig. 1 Basic concept of EV charging station

3.1 Charging Control

Charging control with controlled and uncontrolled charging are explained in the following sections

3.1.1 Controlled Charging

Controlled charging is an organized and optimized technique of EV charging. In this method EV charging stations are designed to reduce overload of grid and to improve efficiency. It reduces the overload on the grid by evenly distributing the load using smart charge controller scheduling during peak hours and off—peak hours [2].

This charging method is cost-effective and depends on modern technologies like smart meters, bidirectional chargers, and grid connectivity [16]. Controlled charging is ideal for fleet operators, smart cities, and large-scale EV adoption, as it ensures steady and balanced energy supply with variable power and current levels [18].

The controlled charging is again categorized into direct control and indirect control. Direct control charging enables communication between EV and grid by using bidirectional or smart charging techniques [16, 17]. Indirect control charging aims to stabilize the charging load of EVs by eliminating variations through spatial load shifting [19].

3.1.2 Uncontrolled Charging

Uncontrolled charging occurs without any regulation or scheduling, leading to potential peak demand issues, when multiple users charge their vehicles simultaneously, it can cause grid overload, especially during high-demand hours. This method of charging begins as soon as the vehicle is plugged in, without considering demand fluctuations or energy costs [11, 20]. Since it does not incorporate any external control mechanisms, it is a simple plug-and-charge system, suitable for individual users without access to smart charging infrastructure. However, it may result in higher electricity costs and contribute to increased strain on the electrical grid.

Implementation of smart charging has become a key feature for charging an EV especially in limited resource environments. The traditional charging method is an uncontrolled charging method that shows adverse effects if charging instances increases during peak hours [21, 22] and may also include Some limitations include imbalance in load distribution, number of vehicles that can be charged simultaneously and these requires physical human presence for the charging station operation and maintenance.

The controlled and uncontrolled charging can be decided by grid frequency drifts, voltage drops, power outage chances and thermal overloading's [23]. One other advantage of smart charging when compared to uncontrolled traditional charging is improvement in battery health and battery life [24].

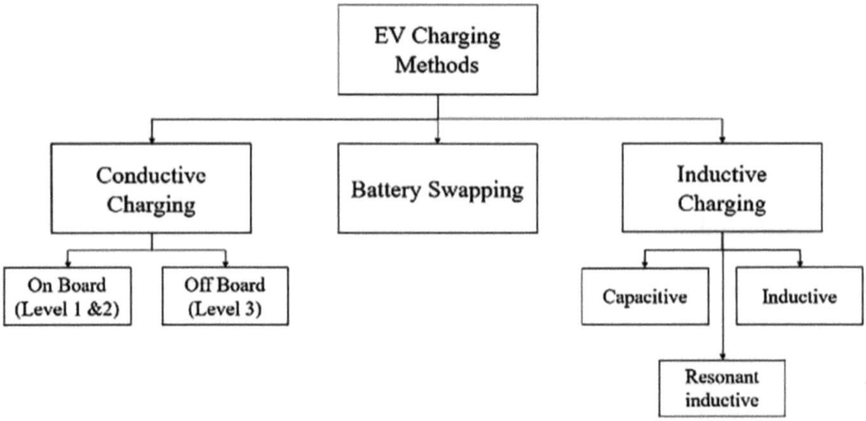

Fig. 2 Electrical vehicle charging methods

3.2 *Transferring Power*

There are various methods to transfer the power from charging station to an Electric vehicle (→Fig. 2) [25].

3.3 *Battery Swapping*

At specified battery swapping stations, customers may utilize this technology to change discharged battery with a fully charged one.

It is particularly helpful in areas where there is a rapid turnaround of vehicles such as shared mobility solutions or commercial EV fleets. Since the battery swapping is one of the fastest ways as battery replacement just takes a few minutes, it greatly minimizes the downtime related to charging [26]. This approach lessens traffic at traditional charging stations while guaranteeing a smooth and continuous driving experience (→Fig. 3).

- Voltage Levels: 300–800 V
- Power Ratings: 50–200 kW
- Duration of charging: Few minutes (Battery swap process)
- Applications: Fleet vehicles, electric taxis, and high-utilization commercial EVs

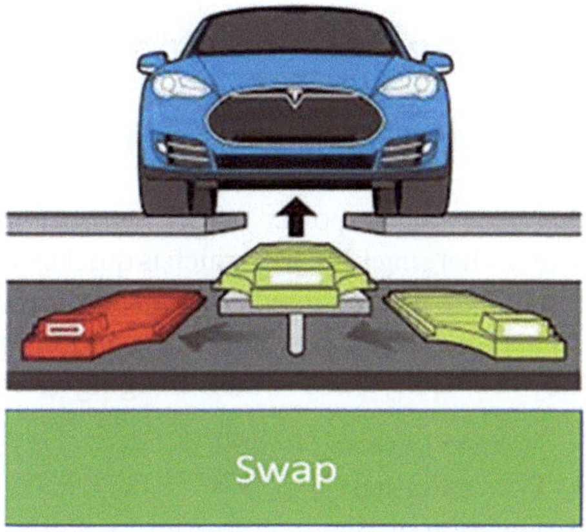

Fig. 3 Battery swapping technology

3.4 Conductive Charging

In conductive charging technology the charger and device are directly connected through wire and it provides efficient energy transfer with minimal losses [27] (→Fig. 4).

3.4.1 Level 1: Charging Using AC

Usually utilized in home settings, level 1 charging is the most fundamental type of conductive charging [28–30]. It doesn't need any extra installation or equipment, and it uses a typical household plug. It's very sluggish charge rate, however, makes it appropriate for consumers who don't need to drive much every day or for overnight charging.

- Voltage Levels: 120/230 V
- Power Ratings: 1–2 kW
- Duration of charging: 8–20 h

Fig. 4 Conductive charging technology

- Applications: Overnight charging at home

3.4.2 Level 2: Charging Using AC

A higher voltage power supply and specialized charging equipment are needed for level 2 charging [28–30], which is quicker and more effective. When quicker charging is required, it is frequently seen in public charging stations, offices, and homes. Level 2 chargers are the recommended option for everyday charging requirements since they drastically cut down on charging periods when compared to Level 1 charging.

- Voltage Levels: 240/400 V
- Power Ratings: 7–22 kW
- Duration of charging: 4–8 h
- Applications: Public charging at retail malls, offices, and residential neighbourhoods.

3.4.3 Level 3: DC Fast Charging

By first converting AC power into DC power and then transmitting it directly to the vehicle's battery, Level 3 charging—also known as DC fast charging [31]—offers rapid charging. This method significantly reduces charging time, making it ideal for long-distance trips and highway charging stations. DC fast chargers are essential to the seamless deployment of EVs, even though they are more expensive and need specialized infrastructure.

- Voltage Levels: 400–800 V
- Power Ratings: 50–250 kW
- Duration of charging: 20—60 min.
- Applications: Long-distance travellers can use public fast-charging stations located beside roads.

The most popular technique for charging EVs is conductive charging, which involves a direct connected connection between the car and the charging station. This approach is further divided into 3 different categories.

For both home and business charging requirements, conductive charging is the recommended option because to its high efficiency and accessibility.

The most popular technique for charging EVs is conductive charging, which involves a direct connected connection between the car and the charging station. This technology is backed by a reputable charging infrastructure and guarantees effective energy transmission. Three degrees of conductive charging can be distinguished according to charging speed and power supply.

3.4.4 Conductive Charging Benefits

- High Efficiency—Compared to inductive (wireless) charging, conductive charging offers direct energy transmission through wires with reduced energy loss.
- Broad Availability—In contrast to wireless charging infrastructure, conductive charging stations are more extensively distributed.
- Reduced Cost—When compared to inductive charging options, conductive charging stations have cheaper installation and operating expenses.
- Scalability—Conductive charging is perfect for public, office, and home charging since it is readily scaled.
- Compatibility—Conductive charging is supported by the majority of EVs, guaranteeing a smooth connection with current infrastructure.

The most popular EV charging technique is still conductive charging because of its effectiveness, cost, and broad use. The ease and accessibility of conductive charging solutions are further improved by developments in infrastructure deployment, charging speed improvements, and charging technologies.

3.5 Inductive Charging

Inductive charging, also referred to as wireless charging, uses electromagnetic fields to transfer energy between an EV's receiver coil and charging station [32]. This technology is very useful since it does not require physical connectors, which makes it easy to use and handy (→Fig. 5).

Plug-in charging is a wired technique in which an inductive charger is linked to the car. Pantograph Charging: A wireless charging technique that makes use of a ground-based or conductive overhead system.

Because of its safety, convenience of use, and potential for future integration with driverless vehicles, inductive charging technology is becoming more and more popular. Nevertheless, infrastructure development is still ongoing, and its efficiency

Fig. 5 Inductive charging technology

is somewhat lower than conductive charging. There are 2 types of Inductive charging technologies. They are.

3.5.1 Level 1: Charging Using AC

- Voltage Levels: 240 V (AC)
- Power Ratings: 3.3–11 kW
- Charging Time: 8–20 h
- Applications: Wireless home charging for personal EVs.

3.5.2 Level 2: Charging Using AC

- Voltage Levels: 240 V (AC)
- Power Ratings: 3.3–11 kW
- Charging Time: 4–8 h
- Applications: Public wireless charging stations, suitable for autonomous electric fleets and bus transit systems.

4 EV Charging Station Using a Smart Controller

Block diagram of an EV charging station consists Energy Sources, Power Converters, EV charging station, Power lines, Communication lines and a Smart Controller (→ Fig. 6).

4.1 Energy Sources

Represents different types of power generating units that supply power to EV charging station through a power converter. This system includes both renewable energy sources and non-renewable energy sources to produce AC and DC power efficiently.

4.1.1 Power Grid

The power grid in this system is a standard electric supply it acts as a backup source for producing the power to the EV charging station when there is a shortage of power from integrated renewable energy source. This ensures stable and continuous power

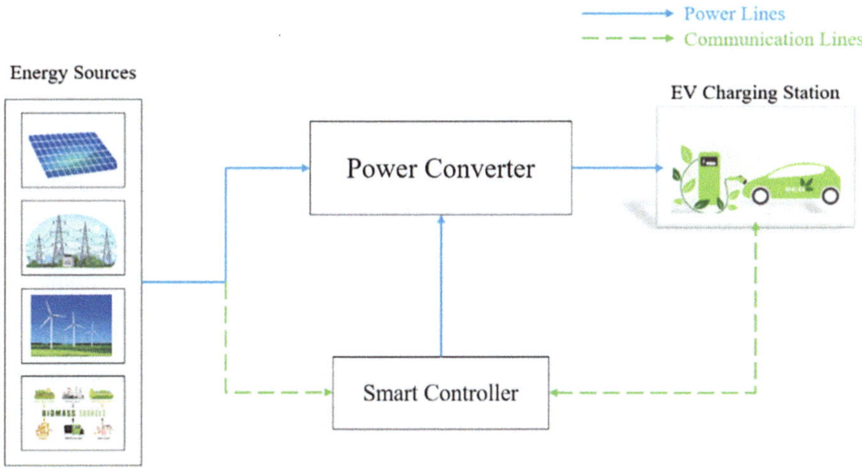

Fig. 6 Block diagram of an EV charging station using a Smart controller

supply particularly when solar and wind generates low power. This typically provides alternating current (AC).

4.2 Renewable Energy Sources

A power system is a complex network of interconnected components that converts traditional coal-fired energy and renewable energy into electrical energy [33]. Energy sources derived from resources that may naturally replace themselves over the course of a human's lifetime are known as renewable energy sources. Natural gas, hydropower, wind, solar photovoltaics, and biofuels are examples of renewable energy sources. Renewable energy is frequently used for transportation, heating, cooling, and electricity generation. According to projections for the next years, wind and solar PV penetration has increased over the last ten years when considering the proportion of overall power capacity based on the various energy sources [33]. These two renewable energy sources are the most often used when paired with electric vehicles.

4.2.1 Solar Energy

One of the most popular renewable energy sources for EV charging stations is solar power, which uses photovoltaic (PV) panels to turn sunlight into electrical power. To effectively generate power, these solar PV panels can be positioned on ground-mounted buildings, carports, or rooftops. Structures constructed above parking lots, known as solar canopies, also serve to shade cars while capturing solar energy.

Battery Energy Storage Systems (BESS) store extra solar energy so that it may be used for charging at night or on overcast days, ensuring a steady supply of electricity. Solar-powered charging stations can be either grid-tied or off-grid, depending on the setup; grid-connected stations can sell excess energy back to the grid, while off-grid stations simply use stored energy to operate continually. One of the many advantages of solar-powered charging stations is their cleanliness and abundance, as solar energy is widely available and produces no pollutants. They are also cost-effective in the long run due to their low maintenance needs and cheap running expenses. Additionally, solar charging stations may be placed in both urban and rural locations due to their high scalability, which will help with the transition to a more sustainable and ecologically friendly transportation system [34].

The solar energy is the most essential renewable energy as the power production from the solar is more compared to all the other renewable energies. Sunlight gets converted into electricity using photovoltaic (PV) cells. It produces direct current (DC) which needs to be converted to alternating current (AC) for EV charging.

Solar energy may be transformed into electrical power using photovoltaic (PV) solar panels and concentrated solar power plants. PV solar panels employ solid-state semiconductors, whereas concentrated solar power plants utilize mirrors or lenses to concentrate sunlight, producing enough heat to operate motors or steam turbines and convert sunlight directly into electricity [35].

PV solar panels employ solar cells, which function well in cold temperatures. Temperature has no effect on solar cell power efficiency, or n solar. To calculate the electrical output (P_{solar}), consider the area of aggregated solar cells (α) in m2 and the intensity of solar irradiation (s) in W/m2. The Output Power from solar is calculated by following equations "Eq. (1) and (2)".

$$P_{solar} = n_{solar} s \alpha \tag{1}$$

Electricity produced by photovoltaic solar panels may be stated as follows,

$$P_{PV} = A\beta\mu(t) \tag{2}$$

where β stands for the panel's efficiency, A for its area, and μ for the solar insolation, the latter of which may be chosen using official data.

4.2.2 Wind Energy

Wind turbines convert kinetic energy produced from wind blades into electrical energy. The generated energy can be either AC or DC, it depends on the type of the generator. The power generation efficiency depends on the weather conditions and place of the wind turbines. Electricity for smart EV charging stations may be efficiently produced using wind power, especially in areas with strong winds.

A sustainable power source for charging stations may be ensured by installing small wind turbines on the property to capture wind energy and generate electricity.

By combining solar and wind energy, hybrid wind-solar systems increase reliability by guaranteeing continuous power output even when one source is less active. The grid may also be connected to wind-powered charging stations, enabling surplus wind energy to be stored in batteries for later use or returned to the grid.

The ability of wind turbines to continue generating electricity at night when solar energy is not available is one of the many benefits of wind energy. It is a practical option for a variety of geographic situations since it is quite effective in windy places like open fields, coastal regions, and high-altitude regions.

Additionally, wind energy is adaptable to both small- and large-scale EV charging stations and is scalable, which helps create a more sustainable and robust energy infrastructure.

Among all the forms of clean energy, or renewable energy, wind energy is becoming more and more popular these days, particularly because of its environmentally beneficial qualities. However, because wind power is unpredictable and erratic, it can be difficult to integrate into a power infrastructure. Due to its random nature, wind power fluctuation is causing problems with frequency control [36] and voltage quality [37].

4.2.3 Biomass Energy

Biomass is another type of renewable energy source. It is derived from organic materials like agricultural waste, wood and biogas. This energy can be used for direct combustion to produce electricity. It provides a renewable and sustainable energy that can complement other renewable energy sources. Biomass and biogas are sustainable energy sources that may be used to power electric vehicle charging stations. They are made from organic resources including wood, animal dung, and agricultural waste [38]. A vital part of this process is the conversion of organic waste into methane gas, which is subsequently burnt to produce energy in biogas power plants. Waste-to-energy systems also turn municipal solid waste into renewable energy that may serve as electricity for charging stations. In order to provide a steady and sustainable energy supply, biomass power plants may also be linked to the grid and use the electricity they generate to power smart charging stations. The effective use of waste products, which aids in waste management and greatly lowers landfill pollution, is one of the many benefits of this strategy.

Additionally, because the emissions generated are balanced by plant regrowth, biomass energy is regarded as carbon–neutral, making it a sustainable choice. Particularly in agricultural and industrial zones where organic waste is easily accessible, the scalability of biomass and biogas energy makes them appropriate for both rural and urban settings. It is possible to create a cleaner, more sustainable transportation infrastructure by combining biomass energy with intelligent charging stations [39].

4.2.4 Hydro Energy

Electricity is produced by harnessing the energy of flowing water through hydropower. While small-scale hydro systems can be connected to EV charging stations near bodies of water, huge hydroelectric facilities provide power to networks. Utilizing the natural flow of water, micro-hydro turbines are placed in rivers, streams, or man-made water channels to produce energy. A dependable power source is ensured by these hydropower plants, which may either store extra energy in batteries for later use or immediately give energy to EV chargers. One of the main benefits of hydropower is its reliable power supply; unlike solar and wind energy, it produces steady energy output regardless of the weather.

Hydropower systems are also an affordable and long-lasting alternative because of its lengthy lifespan and low maintenance needs. Because they utilize the water's natural flow without seriously upsetting the ecosystem, micro-hydro systems likewise have a little influence on the environment.

A sustainable and continuous energy supply may be produced by combining hydropower with intelligent EV charging stations, therefore promoting a more robust and environmentally friendly transportation infrastructure.

4.3 Power Converters

Power converter plays an important role in EV Charging System. It manages the conversion of electrical power from different types of energy sources which ensures the compatibility with the EV charging station. It uses different types of converters to convert the DC or AC power coming from the energy sources and delivers DC or AC powers based on the requirement of a particular charging station [40, 41].

4.3.1 DC–DC Conversion

To enable quick charging at the charging station, it transforms the DC power supplied by the solar panel or wind turbine into DC electricity. Given how frequently DC-DC converters are used, it is important to identify their type by specifying factors such electrical isolation, voltage gain, power rating, start time, and overall efficiency [42]. Active balancing topologies, which are distinguished by the energy flow (unidirectional or bidirectional) or the applied topology—especially those used for battery balancing management, as discussed in [43]—can therefore be used to classify DC-DC converters. In order to evaluate DC-DC converters, a suitable comparison of the charging power level and infrastructure is given in [44]. The assessment of DC-DC converters takes into account infrastructure and charging power level, and [43] offers a good comparison. The review papers have also covered the DC-DC converters' characteristics, control variables (such as terminal cell voltage, capacity, and SOC), and control techniques (either high-level or low-level).

4.3.2 AC–DC Conversion

To accomplish fast charging at the charging station side, it transforms the grid's AC power into DC electricity. The main applications for AC-DC converters include DC fast-charging stations and EV power conversion, which raises the power exchange flow performance indices in V2G applications. Different types of these converters are available, such as the buck-boost converter (SEPIC converter) for low-power applications or the diode bridge (half-full) rectifier with 683 boost or buck-boost power factor correction (PFC). Both single-phase and three-phase conversion devices are capable of operating with them. Numerous studies illustrate their control and design methodologies and confirm the modelling stability using the Lyapunov-based function [44–46]. A detailed comparison [47, 48] backs up the numerous studies that look at the new implementation design of the AC-DC converters in terms of operation modes [49, 50], band-gap devices and soft-switching technologies [49–59], harmonics [50–53], control scheme [54–57], and size [58–63] in order to increase the power factor with high power quality.

4.3.3 DC–AC Conversion

In order to access various charging levels according to user needs, it transforms the DC power supplied by the solar panels and wind turbines into AC electricity. Three PECs—AC–DC converters, DC–DC converters, and DC–AC converters—arc used in the powertrain of EVs. Additionally, some auxiliary parts require converters with lower power ratings. Certain auxiliary components also require converters with lower power ratings. Because AC power is more reliable and efficient, especially in industrial applications like EV technologies, DC–AC converters convert DC input voltage into AC output voltage with a predetermined magnitude and frequency [62, 63]. Two-level pulse width modulation (PWM) DC–AC inverters (TLIs) and multilayer PWM inverters (MLIs) are the two varieties of DC–AC inverters. Low switching losses, low switching frequencies, compact filter size, reduced voltage harmonic distortion, and low current distortion are some benefits of MLI topologies over TLI topologies. However, MLIs require a large number of switching components with complex control systems, which drives up installation costs.

4.3.4 AC–AC Conversion

A power electronic device that converts AC power to AC with or without frequency modification is called an AC-to-AC converter. These are primarily categorized as cycloconverters and AC voltage controllers. AC–AC converters, which can be either matrix or cyclo converters, can be used in the powertrain of electric vehicles. Matrix converters allow for direct power conversion without the need for a DC-link capacitor [64, 65]. The constant AC input voltage is converted to a variable voltage or frequency output using nine bidirectional switches. However, filters are required to lower the harmonics since the output voltage has a limited capacity. Implementation is challenging, costly, and unreliable due to these limitations [66].

In order to enhance EV battery charging efficiency, a flyback-based clamp circuit in conjunction with a bidirectional matrix-type AC–DC converter is recommended in [65]. In Ref. [67], the matrix converter is developed using the resonant DABs as a single-power convergence step, which is costly and suitable for charging or discharging one or more EVs or V2G applications. Utilizing the following applications, battery charging technology is improved: Cyclo converters can be used in both high-voltage and low-voltage EV applications, while their exact applications are unknown [20, 68]. To produce the electricity effectively, a number of factors are taken into account. It guarantees that the power supply can match the proper voltage and frequency ratings according to the specifications of EV charging stations. To offer a steady power supply, it eliminates oscillations and lowers harmonics. The EV charging stations utilizes the power received from the power converters with different charging technologies at different levels in the form of both AC and DC.

4.4 Smart Controller

The smart controller acts as the central intelligent unit that process various input functions including power exchange, renewable energy integration, authentication and security, E-Wallet for digital payments, User interface to provide customer centric solutions and to control the uncontrolled charging. For EV charging, smart charging has become a crucial function, especially in settings with limited resources. It allows for the regulated recharging of the EV utilizing a variety of prerequisites for control. Simply putting an EV into the grid power supply is the traditional technique of charging it; this is referred to as "uncontrolled" or "uncoordinated" charging.

However, if there are more instances of uncontrolled charging or if the charging period falls during periods of high grid consumption [22, 69], there may be negative consequences, including grid frequency drifts, grid voltage reductions, power outage risks, and distribution transformer thermal overload. Charging can be either regulated or uncontrolled, depending on how controllable and flexible the cycles are Ref. [70]. Compared to uncoordinated charging, smart charging has been demonstrated to provide financial advantages in terms of battery health, leading to increased battery life and less battery deterioration [71].

An effective strategy to strike a balance between EV development and grid capacity is smart charging for electric cars. Smart charging links aggregators, network operators, and consumers to the smart grid via a reliable communication mechanism. Depending on demand, the EV charging profile schedule [71] can be modified for technical and/or economic benefits.

4.4.1 Power Exchange

The controller makes it possible for several power sources, including solar panels, wind turbines, the electrical grid, and battery storage devices, to exchange electricity efficiently. In accordance with supply and demand, it distributes energy as effectively as feasible. The EV can return electricity to the grid thanks to its bidirectional power flow (also known as vehicle-to-grid, or V2G) technology.

4.5 Renewable Energy Integration

The system prioritizes the use of renewable energy sources such as solar and wind to charge electric vehicles, decreasing reliance on traditional power networks. The controller continuously analyses renewable energy availability and changes the charge rate accordingly. It contributes to environmental sustainability by lowering carbon footprints.

4.5.1 Authentication

Authentication guarantees that the electric vehicle (EV) charging stations are only accessible and usable by authorised users. System integrity, user responsibility, and security all depend on this. Several strategies are used by contemporary charging stations to enable safe authentication: Applications for smartphones that require secure login information (such as two-factor authentication or OTP). Users who have registered are assigned RFID cards. Biometric identification, like fingerprint or facial recognition, offers a high degree of convenience and security. This restricted access keeps track of usage patterns, guards against abuse, and enables user-specific services.

4.5.2 E-Wallet for Digital Payment

Digital wallets are being more widely supported by smart EV charging stations to facilitate safe and easy payment processing. These online wallets provide:

- Contactless purchases with applications such as PhonePe, Paytm, Google Pay, etc.
- Subscription-based billing schemes or opportunities for pay-as-you-go.
- Support for many currencies for foreign visitors.

EV charging is made more convenient and accessible while ensuring safety against unauthorised transactions by integrating digital wallets, which also provide secure encryption and fraud protection.

4.5.3 Interface for Customer-Centric Smart EV Charging Solutions

For a smart charging experience to be user-friendly, a well-designed user interface (UI) is essential. The interface consists of digital (web dashboards or mobile apps) and physical (station touchscreens) elements that: Facilitate simple charging session scheduling and tracking. Give current information about availability, cost, and charging status. Permit customisation of preferences (e.g., favourite charging mode, desired charging duration). The UI promotes increased EV adoption and better energy use behaviour among consumers by enhancing usability, transparency, and engagement.

4.5.4 Uncontrolled Charging Problems with Smart Charging Technologies

Uncontrolled EV charging poses serious problems for energy management systems and power grids. In order to solve these problems, demand-response mechanisms, load balancing, and grid stabilization require smart charging systems that integrate AI, IoT, and other technologies.

5 Smart Technologies

Several technologies play crucial role in optimizing Ev charging ensuring efficient power distribution [72].

5.1 Machine Learning

Smart EV charging relies heavily on machine learning (ML), which optimizes energy use, improves battery performance, and incorporates renewable energy sources. In order to estimate future energy needs and provide effective grid management while avoiding overloads, machine learning models are used in predictive charging demand, one of its primary uses. For instance, ML can forecast when charging will be at its highest, enabling dynamic power load balancing.

Furthermore, battery health monitoring tracks battery conditions using machine learning algorithms, minimizing overcharging and prolonging battery life. This enables users to take preventive action by assisting in the early identification of battery degeneration.

Dynamic pricing and cost optimization is another important use, where machine learning uses current power costs to calculate when it is best to charge. Because charging sessions are automatically scheduled at periods when power rates are lowest, including off-peak overnight hours, EV owners may save money.

Furthermore, smart charging is further improved by renewable energy forecasting and prediction, which makes it possible to use solar and wind power efficiently. In order to forecast the availability of renewable energy and make sure that EVs charge during periods of clean energy abundance, machine learning algorithms examine weather trends, past energy production, and demand variations. By enhancing sustainability and decreasing dependency on fossil fuels, this integration makes EV charging more intelligent, economical, and ecologically benign.

5.2 *Artificial Intelligence*

With its ability to optimize energy distribution, facilitate real-time decision-making, and improve user convenience, artificial intelligence (AI) is essential to smart EV charging. By avoiding power outages and grid overloads, AI-powered grid load balancing guarantees steady energy delivery. AI-powered charging stations increase efficiency by dynamically modifying charging rates according to grid circumstances. Real-time decision-making also improves Vehicle-to-Grid (V2G) systems, which optimize energy consumption by enabling EVs to return extra power to the grid during periods of high demand. AI uses current grid circumstances to calculate when an EV is most efficient at charging or discharging. Additionally, drivers may find the closest accessible charging station with the least amount of wait time thanks to smart route and charging station selection.

5.3 *Internet of Things and Cloud Connectivity*

Smart EV charging relies heavily on cloud connectivity and the Internet of Things (IoT) to improve scalability, efficiency, and convenience. Users may locate nearby charging stations via mobile applications thanks to connected charging stations' usage of IoT sensors to track real-time data, including charger availability, power levels, and price. Remote monitoring and control allow users to start, pause, and schedule charging sessions remotely, as shown in Tesla's app. IoT-enabled chargers may also modify charging in response to grid demand thanks to smart grid connectivity. Simultaneously, cloud connectivity facilitates scalability, real-time data processing, and a smooth user experience, allowing for extensive EV charging networks and remote access to charging metrics.

5.4 Blockchain for Safe and Secure Transactions

A transparent and impenetrable ledger for energy transaction recording is provided by blockchain technology. It improves: Security of transactions In peer-to-peer (P2P) networks, energy sharing and Have faith in decentralised charging systems.

6 Machine Learning Regression Models

The supervised machine learning method of regression is used to forecast values that are continuous. The regression algorithm's ultimate objective is to create a curve or best-fit line between the data. The three primary metrics that are utilized for assessing the trained regression model are variance, bias and error. Both overfitting and underfitting result from large variation and excessive bias, respectively [72].

6.1 Linear Regression

Linear regression is the simple regression analysis used to model the relationship between independent variables (input features) and dependent variable (target). It assumes that linear relationship exists between the dependent and independent variables.

It is represented by the equation: $y = mx + c$

m = slope.

b = intercept.

x = independent variable.

If there exist multiple independent variables then the equation extends like

$$y = w_1x_1 + w_2x_2 + \cdots + w_nx_n + b \tag{3}$$

w = model coefficients determined during training.

This linear regression model is simple and efficient and works perfectly for the dataset that follows straight line and can be given by "Eq. (3)". However linear regression data cannot perform well with non-linear data.

Applications: House price prediction, salary estimation and demand forecasting.

6.2 Polynomial Regression

This method is an extension for the linear regression model which allows non-linear data with the use of polynomial equation instead of fitting a straight line. Polynomial regression fits a curved line by transforming the original features into polynomial equation of degree n. The polynomial equation is given by "Eq. (4)"

$$y = a_0 + a_1x + a_2x^2 + \cdots + a_nx^n \tag{4}$$

In applications where the correlations between variables are not strictly linear, such as power consumption predictions, climate modelling, and stock market analysis, polynomial regression is frequently utilized. However, it has drawbacks, such as overfitting, particularly when a high-degree polynomial is used, resulting in poor generalization on previously encountered data.

6.3 Support Vector Regression

A sophisticated machine learning method called Support Vector Regression (SVR) builds on Support Vector Machine (SVM) for regression. The error between actual and anticipated values is reduced using this technique.

The SVR technique offers the best performance within a certain range. Rather than reducing the residual error, it seeks to maintain the predictions within a given range. The SVR formula is provided by "Eq. (5)".

$$f(x) = w^Tx + b \tag{5}$$

The hyperplane is therefore defined as w and b. By converting input data into higher-dimensional space, kernel functions such as the Radial Basis Function (RBF), Polynomial, and Sigmoid may be able to handle non-linear relationships and enhance regression results. One of the primary benefits of SVR is this.

SVR is particularly useful in applications like time-series forecasting, energy demand prediction, and financial market research when the data is noisy and non-linear. However, because to its high computational cost, SVR is useless for very large datasets. It also requires a lot of hyperparameter modification (C, epsilon, and gamma) to achieve the best results. Notwithstanding these challenges, SVR remains an effective regression technique for small to medium datasets when high accuracy is required.

6.4 Decision Tree Regression

Using a tree-based machine learning model called Decision Tree Regression, the data is recursively divided into branches according to feature values in order to generate predictions. The model is organized like a flowchart, with branches standing in for potential feature values, internal nodes for decisions (feature splits), and leaf nodes for projected values.

Choose the appropriate feature to divide the dataset using metrics such as Mean Absolute Error (MAE) or Mean Squared Error (MSE). Based on the chosen characteristic, divide the dataset into smaller groups. Recursively repeat the procedure until a halting condition is reached. Predictions are made using the average of the values of the leaf nodes.

When dealing with missing data, identifying non-linear correlations, and working with numerical or categorical characteristics, Decision Tree Regression is very helpful.

Its capacity to represent intricate feature interactions makes it frequently used in price prediction, medical diagnosis, and agricultural production forecasting. Decision trees, however, have a tendency to overfit if improperly pruned, which results in subpar generalization on fresh data.

In order to counteract this, ensemble approaches like Random Forests and pruning techniques like cost complexity pruning are employed to increase the accuracy and stability of the model.

6.5 Random Forest Regression

To enhance prediction performance, Random Forest Regression is an ensemble learning approach that constructs many decision trees and aggregates their results. By combining predictions from several decision trees, a random forest produces a more reliable and accurate model than a single tree, which may overfit.

This is how the algorithm operates:

- picks training data subsets at random (bootstrap sampling
- builds many decision trees using various data subsets.
- combines all of the three predictions (averaging for regression problems

This eliminates overfitting, enhances stability, and lowers variance as compared to a single decision tree. Random Forest Regression is frequently employed in fields where accuracy and interpretability are crucial, such as medical diagnoses, financial risk assessment, and climate modelling. It also naturally chooses the most significant features during training; thus, it can handle high-dimensional data with a lot of irrelevant characteristics. However, because random forests are ensemble in nature, they demand greater memory and processing capacity. Compared to individual decision trees, they are also less interpretable. Random forests are among the most effective and popular regression methods for large-scale applications, notwithstanding these difficulties.

Table 1 gives the comparison of different regression models in manchine learning (Table.1).

Table 1 Machine learning regression models

Regression model	Best suitable for	Handling non-linearity	Computational speed
Linear regression	Small and simple	No	Low
Polynomial regression	Non-linear data	Yes	Medium
Support vector machine regression	Small data sets	Yes	High
Decision tree regression	Complex patterns	Yes	Medium
Random forest regression	High dimensional data	Yes	High

7 Conclusion

One of the most important developments in the field of sustainable transportation is the use of smart charging technology for electric cars (EVs). Traditional EV charging stations have a number of drawbacks, such as restricted scalability, system overload, and inefficient energy distribution. Advanced technologies like artificial intelligence (AI), the Internet of Things (IoT), and machine learning are integrated into smart charging systems, which provide load balancing, dynamic pricing, and optimum energy use. By integrating renewable energy sources like wind and solar, smart charging reduces the need for fossil fuels and lowers carbon emissions overall. Furthermore, EV adoption is made more feasible and convenient by features like digital payment choices, bidirectional charging, and remote monitoring that improve user experience. The smooth integration of smart grids, real-time decision-making, and predictive energy management is what will determine the future of EV charging infrastructure. While maintaining grid stability, widespread use of AI-driven load allocation, battery swapping, and regulated charging techniques would greatly increase efficiency. For the implementation of smart charging stations to proceed more quickly and to encourage research into new technologies, governments and business players must work together. In the end, EV adoption success hinges on the resilience and intelligence of charging infrastructure in addition to car developments, opening the door to a more environmentally friendly and energy-efficient future.

References

1. Subashini, M., & Sumathi, V. (2023). Smart charging for zero emission vehicles—A comprehensive review. *Renewable Energy Focus, 46*,. https://doi.org/10.1016/j.ref.2023.05.005
2. Al-Hanahi, B., Ahmad, I., Habibi, D., & Masoum, M. A. S. (2021). Charging infrastructure for commercial electric vehicles: challenges and future works. *IEEE Access, 9*, 121476–121492.
3. Rahman, R., Vasant, P. M., Singh, B. S. M., Abdullah-Al-Wadud, M., Adnan, N. (2016). Review of recent trends in optimization techniques for plug-in hybrid, and electric vehicle charging infrastructures. *Renewable and Sustainable Energy Reviews, 58*, 1039–1047. ISSN 1364–0321

4. Moghaddam, Z., Ahmad, I., Habibi, D., Phung, Q.V. (2017). Smart Charging Strategy for Electric Vehicle Charging Stations (pp. 2332–7782). IEEE.
5. Katrašnik, T. (2011). Energy conversion phenomena in plug-in hybrid-electric vehicles. *Energy Conversion Management, 52*, 2637–2650.
6. Monteiro, V., Gonçalves, H., Afonso, J.L. (2011). Impact of electric vehicles on power quality in a Smart Grid context. Proceeding Int, Conf. Electr. Power Qual. Util. EPQU. (pp. 660–665).
7. Sadeghian, O., Oshnoei, A., Mohammadi-ivatloo, B., Vahidinasab, V., Anvari-Moghaddam, A. (2022). A comprehensive review on electric vehicles smart charging: Solutions, strategies, technologies, and challenges. *Journal of Energy Storage 54*, 105241. ISSN 2352-152X.
8. Ali, Z., Putrus, G., Marzband, M., Gholinejad, H. R., Saleem, K., Subudhi, B. (2022). Multiobjective optimized smart charge controller for electric vehicle applications. *IEEE Transactions on Industry Applications 58*(5), 5602–5615. https://doi.org/10.1109/TIA.2022.3164999
9. Ahmad, F., & Bilal, M. (2023). A comprehensive analysis of electric vehicle charging infrastructure, standards, policies, aggregators and challenges for the indian market. *Energy Sources, Part A: Recovery, Utilization and Environmental Effects, 45*, 8601–8622. https://doi.org/10.1080/15567036.2023.2228734
10. Wang, Q., Liu, X., Du., Jian, & Kong, F. (2016). Smart charging for electric vehicles: a survey from the algorithmic perspective. *IEEE Communications Surveys & Tutorials, 18*, 1–1. https://doi.org/10.1109/COMST.2016.2518628
11. Brey, B., Gardien, L., & Hiep, E. (2021). Smart charging needs, wants and demands, charging experiences and opinions of EV drivers. *World Electric Vehicle Journal, 12*, 168. https://doi.org/10.3390/wevj12040168
12. Nie, Q., Zhang, L., Tong, Z., & Hubacek, K. (2022). Strategies for applying carbon trading to the new energy vehicle market in China: An improved evolutionary game analysis for the bus industry. *Energy 259*, Article 124904.
13. Ryan, N., Johnson, J., & Keoleian, G. (2016). Comparative assessment of models and methods to calculate grid electricity emissions. *Environmental Science and Technology, 50*,. https://doi.org/10.1021/acs.est.5b05216
14. İnci, M., Çelik, Ö., Lashab, A., Bayindir, K., Vasquez, J., & Guerrero, J. (2024). Power system integration of electric vehicles: A review on impacts and contributions to the smart grid. *Applied Sciences, 14*, 2246. https://doi.org/10.3390/app14062246
15. Lee, Z. J., et al. (2021). Adaptive charging networks: a framework for smart electric vehicle charging. *IEEE Transactions on Smart Grid, 12*(5), 4339–4350.
16. Moghaddam, V., Ahmad, I., Habibi, D., Phung, Q.V. (2017). Smart charging strategy for electric vehicle charging stations. IEEE Transactions on Transportation Electrification (pp. 1–1). https://doi.org/10.1109/TTE.2017.2753403.
17. Yousuf, A., Wang, Z.,& Paranjape, Raman & Tang, Yili. (2024). An in-depth exploration of electric vehicle charging station infrastructure: A comprehensive review of challenges, mitigation approaches, and optimization strategies. *IEEE Access* (pp. 1–1). https://doi.org/10.1109/ACCESS.2024.3385731
18. Maity, S., Brandl, R., Frías, P., & Wu, Q. (2021). *Smart Charging Strategies and technologies for Electric Vehicles* (2021st ed., Vols. 1–161). IIT Bombay: Prof. Zakir Rather, Ms. Payal Vyankat Dahiwale, Ms. Dhanuja Lekshmi Fraunhofer IEE: Mr. Andre Hartung. https://changing-transport.org/wp-content/uploads/2021_Smart_Charging_Strategies_and_technologies_for_Electric_Vehicles.pdf
19. Shareef, H., Mainul, I. M., & Mohamed, A. (2016). A review of the stage-of-the-art charging technologies, placement methodologies, and impacts of electric vehicles. *Renewable and Sustainable Energy Reviews, 64*, 403–420. https://doi.org/10.1016/j.rser.2016.06.033
20. Preetham, G., Shireen, W. (2012). Photovoltaic charging station for plug-in hybrid electric vehicles in a smart grid environment. *Proc. IEEE PES Innovative Smart Grid Technology.* https://doi.org/10.1109/ISGT.2012.6175589
21. Kong, P.Y., Karagiannidis, G.K. (2016) Charging schemes for plug-in hybrid electric vehicles in smart grid: A survey. *IEEE Access* 6846–6875. https://doi.org/10.1109/ACCESS.2016.2614689

22. Tayarani, H., Jahangir, H., Nadafianshahamabadi, R., Aliakbar Golkar, M., Ahmadian, A., Elkamel, A. (2019). Optimal charging of plug-in electric vehicle: considering travel behavior uncertainties and battery degradation. *Applied Sciences 9*, 1–17. https://doi.org/10.3390/app9163420
23. Acharige, S., Haque, M. E., Arif, M., Hosseinzadeh, N., Hasan, K., Oo, A. (2023). Review of electric vehicle charging technologies, standards, architectures, and converter configurations. *IEEE Access* (pp. 1–1). https://doi.org/10.1109/ACCESS.2023.3267164
24. Shaikh, A., & Bansod, P. (2023). A review of battery electric vehicle charging technologies. *AIP Conference Proceedings, 2427*, Article 020004. https://doi.org/10.1063/5.0101020
25. IEA. (2023). Global EV Outlook 2023, IEA, Paris. https://www.iea.org/reports/global-ev-outlook-2023, Licence: CC BY 4.0
26. Chinmaya, K.A., Singh, G. (2019). An integrated onboard single-stage battery charger for PEVs incorporating asymmetrical six-phase induction machine. *IET Electrical Systems in Transportation, 9*. https://doi.org/10.1049/iet-est.2018.5015
27. Wu, H., Gilchrist, A., Ky., Sealy, Israelsen, P., & Muhs, J. (2011). A review on inductive charging for electric vehicles. *2011 IEEE International Electric Machines and Drives Conference IEMDC, 2011*, 143–147. https://doi.org/10.1109/IEMDC.2011.5994820
28. Ayob, A., Mahmood, W., Mohamed, A., Che Wanik, Zamri, M, Siam, M., Fadzil, M., Sulaiman, S., Azit, A., Ali, M. (2014). Review on electric vehicle, battery charger, charging station and standards. *Research Journal of Applied Sciences, Engineering and Technology, 7*, 364–373. https://doi.org/10.19026/rjaset.7.263
29. Nasr Esfahani, F., Darwish, A., Ma, X., & Twigg, P. (2024). Non-integrated and integrated on-board battery chargers (iOBCs) for electric vehicles (EVs): A critical review. *Energies, 17*(10), 2285.
30. https://www.vector.com/se/en/know-how/smart-charging/charging-types-and-methods/#c243941
31. Xygkis, T., Korres, G., Manousakis, N. (2016). Fisher information-based meter placement in distribution grids via the D-optimal experimental design. *IEEE Transactions on Smart Grid* (pp 1–1). https://doi.org/10.1109/TSG.2016.2592102
32. Deshmukh, S., Limkar, S., Nagthane, R., Pande, V.N., Tare, A. (2023). Design of grid-connected solar PV system integrated with battery energy storage system (pp 1–6). https://doi.org/10.1109/ASIANCON58793.2023.10269854
33. https://www.energy.gov/eere/solar/solar-photovoltaic-technology-basics
34. Bae, S., & Kwasinski, A. (2012). Spatial and temporal model of electric vehicle charging demand. *IEEE Transaction on Smart Grid, 3*, 394–403. https://doi.org/10.1109/TSG.2011.2159278
35. Sinha, P., Paul, K., Deb, S., & Sachan, S. (2023). Comprehensive review based on the impact of integrating electric vehicle and renewable energy sources to the grid. *Energies, 16*(6), 2924.
36. Nazipov, F., Tokbolat, S., Khussainova, D., Mustafa, M., & Almahadin, A. (2021). Integration of renewable energy technologies in university campuses: A case study. *IOP Conference Series Materials Science and Engineering, 1044*,. https://doi.org/10.1088/1757-899X/1044/1/012015
37. Karmaker, A., Ahmed, Md., Hossain, Md. Alamgir., Sikder, Md. (2018). Feasibility assessment & design of hybrid renewable energy based electric vehicle charging station in Bangladesh. Data in Brief. https://doi.org/10.1016/j.dib.2018.04.069
38. Alatai, S., Salem, M., Ishak, D., Das, H. S., Alhuyi Nazari, M., Bughneda, A., & Kamarol, M. (2021). A review on state-of-the-art power converters: bidirectional, resonant, multilevel converters and their derivatives. *Applied Sciences, 11*(21), 10172.
39. Selvam, S., Ali, M., Jagabar, S., Manoj, P.s., Sundararajan, G. (2016). An assessment on performance of DC–DC converters for renewable energy applications. *Renewable and Sustainable Energy Reviews, 58*, 1475–1485. https://doi.org/10.1016/j.rser.2015.12.057.
40. Mohan, B., Indragandhi, V., Suresh, V., Jasiński, M., & Leonowicz, Z. (2022). A review on electric vehicle: Technologies, energy trading, and cyber security. *Energy Reports, 8*, 9662–9685. https://doi.org/10.1016/j.egyr.2022.07.145

41. Türksoy, A., Teke, A., & Alkaya, A. (2020). A comprehensive overview of the dc-dc converter-based battery charge balancing methods in electric vehicles. *Renewable and Sustainable Energy Reviews, 133*, Article 110274. https://doi.org/10.1016/j.rser.2020.110274
42. Habib, S., Khan, M., Abbas, F., & Tang, H. (2017). Assessment of electric vehicles concerning impacts, charging infrastructure with unidirectional and bidirectional chargers, and power flow comparisons. *International Journal of Energy Research, 42*,. https://doi.org/10.1002/er.4033
43. Ul Hassan, N., Kim, Y.-J., Han, B., Lee, J.-Y. (2017). A Hybrid DC/DC Converter for EV OBCs using full-bridge and resonant converters with a single transformer. *Journal of Power Electronics 17*, 11–19. https://doi.org/10.6113/JPE.2017.17.1.11
44. Geng, S., Zhao, Y., Sun, S., Wu, X., Zheng, Y., Liu, J. (2016). Sliding mode control of LLC resonant DC–DC converters (pp. 1034–1037). https://doi.org/10.1109/ISIE.2016.7745034
45. Shahed, Md. Tanvir & Rashid, ABM. (2024). Battery charging technologies and standards for electric vehicles: A state-of-the-art review, challenges, and future research prospects. *Energy Reports, 11*, 5978–5998. https://doi.org/10.1016/j.egyr.2024.05.062
46. Paul, A., Subramanian, K., Nachinarkiniyan, S. (2019). PV-based off-board electric vehicle battery charger using BIDC. *Turkish Journal of Electrical Engineering and Computer Sciences, 27*, 2850–2865. https://doi.org/10.3906/elk-1804-227
47. Huber, L., Jang, Y., & Jovanovic, M. (2008). Performance evaluation of bridgeless PFC boost rectifiers. *IEEE Transactions on Power Electronics, 23*, 1381–1390. https://doi.org/10.1109/TPEL.2008.921107
48. Mejia-Ruiz, Gabriel E. & Munoz, Nicolas & López-Lezama, Jesús. (2017). Modeling and development of a bridgeless PFC Boost rectifier. Revista Facultad de Ingeniería. 2017. 9–21. https://doi.org/10.17533/udea.redin.n82a02.
49. Choi, W.-Y., Kwon, J.-M., Kim, E.-H., Lee, J.-J., & Kwon, B.-H. (2007). Bridgeless boost rectifier with low conduction losses and reduced diode reverse-recovery problems. *IEEE Transactions on Industrial Electronics, 54*, 769–780. https://doi.org/10.1109/TIE.2007.891991
50. Wei, W., Hongpeng, L., Shigong, J., Dianguo, X. (2008). A novel bridgeless buck-boost PFC converter. *2008 IEEE Power Electronics Specialists Conference*, Rhodes, Greece, 2008, pp. 1304–1308, https://doi.org/10.1109/PESC.2008.4592112
51. Ismail, E. (2009). Bridgeless SEPIC rectifier with unity power factor and reduced conduction losses. *IEEE Transactions on Industrial Electronics, 56*, 1147–1157. https://doi.org/10.1109/TIE.2008.2007552
52. Sabzali, A. J., Ismail, E., Al-Saffar, M., & Fardoun, A. (2011). New bridgeless DCM Sepic and Cuk PFC rectifiers with low conduction and switching losses. *IEEE Transactions on Industry Applications, 47*, 873–881. https://doi.org/10.1109/TIA.2010.2102996
53. Shahri, M., Mahnaz, Farzanehfard, H. (2011). Bridgeless SEPIC PFC rectifier with reduced components and conduction losses. *IEEE Transactions on Industrial Electronics 58*, 4153–4160. https://doi.org/10.1109/TIE.2010.2095393
54. Su, B., Zhang, J., & Lu, Z. (2011). Totem-pole boost bridgeless PFC rectifier with simple zero-current detection and full-range ZVS operating at the boundary of DCM/CCM. *IEEE Transactions on Power Electronics, 26*, 427–435. https://doi.org/10.1109/TPEL.2010.2059046
55. Chen, Z., Xu, J., Zhou, G., Zhang, F. (2012). Analysis of bridgeless pseudo-boost PFC converter. https://doi.org/10.1109/ISIE.2012.6237082
56. AL-Kaabi, A., Fardoun, A., Ismail, E. (2013). Bridgeless high voltage battery charger PFC rectifier. *Renewable Energy, 56*, 24–31. https://doi.org/10.1016/j.renene.2012.09.050
57. Huang, Q., & Huang, A. Q. (2017). Review of GaN totem-pole bridgeless PFC. *CPSS Transactions on Power Electronics and Applications, 2*(3), 187–196. https://doi.org/10.24295/CPSSTPEA.2017.00018
58. Destobbeleer, E., Seguier, G., Castelain, A. (1987). AC–DC converter minimizing induced harmonics in industrial power systems. In: *IEEE Transactions on Power Electronics* (vol. PE-2, no. 4, pp. 320–327).
59. Quint, R., Kosterev, D., Undrill, J., Eto, J., Bravo, R., Wen, J. (2016). Power quality requirements for electric vehicle chargers: Bulk power system perspective (pp 1–5). https://doi.org/10.1109/PESGM.2016.7741443.

60. Fan, W.-T., Yuen, K., & Chung, H. (2015). Power semiconductor filter: use of series-pass device in switching converters for filtering input current harmonics. *IEEE Transactions on Power Electronics, 31*, 1–1. https://doi.org/10.1109/TPEL.2015.2440393
61. Rajendran, G., Vaithilingam, C., Naidu, K., Prakash, K., Md., & Ahmed. (2021). Hard switching characteristics of SiC and GaN devices for future electric vehicle charging stations. *MATEC Web of Conferences, 335*, 02007. https://doi.org/10.1051/matecconf/202133502007
62. Zhang, J., Li, L., Dorrell, D. (2018). Control and applications of direct matrix converters: A review. *Chinese Journal of Electrical Engineering, 4*, 18–27. https://doi.org/10.23919/CJEE.2018.8409346
63. Diaz, M., Cárdenas Dobson, R., Ibaceta, E., Mora, A., Urrutia, M., Espinoza, M., Rojas, F., & Wheeler, P. (2020). An overview of applications of the modular multilevel matrix converter. *Energies, 13*(21), 5546.
64. Alammari, R., Aleem, Z., Iqbal, A., & Winberg, S. (2019). Matrix converters for electric power conversion: Review of topologies and basic control techniques. *International Transactions on Electrical Energy Systems, 29*, Article e12063. https://doi.org/10.1002/2050-7038.12063
65. Ali, A., Mousa, H., Shaaban, M., Azzouz, M., Awad, A. (2023). A comprehensive review on charging topologies and power electronic converter solutions for electric vehicles. *Journal of Modern Power Systems and Clean Energy, 12*. https://doi.org/10.35833/MPCE.2023.000107
66. Thrimawithana, D., Madawala, U., Twiname, R., & Vilathgamuwa, D. (2012). A novel matrix converter based resonant dual active bridge for V2G applications. *10th International Power and Energy Conference IPEC, 2012*, 503–508. https://doi.org/10.1109/ASSCC.2012.6523319
67. Kim, J.-K., Kim, S.-G., Park, S.-M. (2019). Analysis of switch losses in cycloconverter-type high frequency link converter for bidirectional battery chargers (pp 1–6). https://doi.org/10.1109/VPPC46532.2019.8952198
68. Ariyo, F., Aworo, O., & Kuku, M. (2020). Design of fast cycloconverter-based battery-charging circuit for high penetration of electric vehicles. *International Journal of Engineering Research in Africa, 50*, 64–69. https://doi.org/10.4028/www.scientific.net/JERA.50.64
69. Kong, P.-Y., & Karagiannidis, G. (2016). Charging schemes for plug-in hybrid electric vehicles in smart grid: A survey. *IEEE Access, 4*, 6846–6875. https://doi.org/10.1109/ACCESS.2016.2614689
70. Spencer, S., Fu, Z., Apostolaki-Iosifidou, E., & Lipman, T. (2021). Evaluating smart charging strategies using real-world data from optimized plugin electric vehicles. *Transportation Research Part D: Transport and Environment, 100*, Article 103023. https://doi.org/10.1016/j.trd.2021.103023
71. Shern, S., Md., Sarker, Ramasamy, G., Thiagarajah, S., Farid, F. A., & Suganthi, S. (2024). Artificial intelligence-based electric vehicle smart charging system in Malaysia. *World Electric Vehicle Journal, 15*, 440. https://doi.org/10.3390/wevj15100440
72. Madhu Sudan, G. (2024). *Exploring Different Types of Regression Models and their Applications, 3*, 299–304.

Novel Quantum Light and Sound Meditation-Based Optimization Algorithm for Loss Diminution in Grid Connected to Renewable Energy Sources

Lenin Kanagasabai

Abstract In the earth many people are practicing the light and sound meditation technique. In light meditation, by means of envisage light flow from side to side; mind tap into attractive, tender energy and it inherent astuteness to sustain the life journey. Through the light meditation magnificent intellect of association with nature and it sanitize the mind thoughts. Mind meditates through light or sound and it depend upon the individual. Algorithm has been defined by dual mode: light and sound meditation has been scientifically defined in separate segment. In the primary section light meditation has been formulated. Mind envision the light and it faucet into stable condition. In the second part of the procedure, sound meditation has been mathematically formulated. In sound meditation, a deep connection to the current instant stabilizes the thoughts of the mind and it enhances self-awareness, emotion control. To expand the exploration process, Quantum Light and Sound Meditation based optimization algorithm assimilated with Imagination and sound recognition of human brain optimization algorithm. Hippocampus and amygdala of the brain is responsible for the imagination. Auditory cortex of the human brain is responsible for sound recognition. Quantum Light and Sound Meditation (QLSM) based optimization is corroborated in7 Benchmark functions, and IEEE 300, 354 bus systems.

Keywords Light · Sound · Human brain · Otus scops

1 Introduction

At present in the earth many people are practicing the light and sound meditation technique. In light meditation, by means of envisage light flow from side to side; mind tap into attractive, tender energy and it inherent astuteness to sustain the life journey. Through the light meditation magnificent intellect of association with nature and it

L. Kanagasabai (✉)
Prasad V. Potluri Siddhartha Institute of Technology, Kanuru, Vijayawada, Andhra Pradesh 520007, India
e-mail: gklenin@gmail.com

M. A. Mellal et al. (eds.), *Computational Intelligence for a Greener Future: Innovations in Renewable Energy Systems*, Studies in Computational Intelligence 1240,
https://doi.org/10.1007/978-3-032-06732-6_12

sanitize the mind thoughts [1]. Sound meditation frequently employs soothing sounds and in spiritual world of sound meditation sound "OM" has been utilized. Sound meditation makes the mind to relax, focused and incident a more optimistic stance. Pontiff Dr. Sri Ganapathy Sachchidananda [2, 3] teaches and trains the disciples through light and sound Meditation technique. Around the globe pontiff is conducting many meditation classes and healing concert program. Through this millions of people attained liberation from unrest of mind. Mind meditates through light or sound and it depend upon the individual.

If the mind moves to the state of liberation through light or sound meditation, then that stable condition is defined. Algorithm has been defined by dual mode: light and sound meditation has been scientifically defined in separate segment. In the primary section light meditation has been formulated. Mind envision the light and it faucet into stable condition. In the second part of the procedure, sound meditation has been mathematically formulated. In sound meditation, a deep connection to the current instant stabilizes the thoughts of the mind and it enhances self-awareness, emotion control. Imagination and sound recognition of human brain (ISHB) optimization is grounded on the activities of brain. Hippocampus and amygdala of the brain is responsible for the imagination. Auditory cortex of the human brain is responsible for sound recognition. Hippocampus of brain is responsible for imagination about the future and can reclaim the past happenings. Research on neural action in rats shows that that the hippocampus of the brain plays a dominant part in imagination. Imagination is the act of creating a mental depiction of somewhat not present and particularly of something not known or experienced; imaginative ability and creation of the mind is part of the imagination. Fresh experimental analysis shows that the brain separates amongst professed and fictional mental descriptions by examination through actuality threshold. Important forms of imagination of human beings are based on are creative, communal, and realistic imagining. Electronic signals passage through cochlear nerve network to the cerebral cortex of the human brain. Then specific regions of auditory cortex in the brain process the fundamental speech sounds, in addition temporary complex acoustical signals will be processed. When human ears perceive the sound, it will be transmitted to the brain, and then echoic memory store the info roughly for four minutes. Depend on the situations the brain may not imagine and can't recognize any sort of sound. It will be chaos condition. Attributes of the Otus scops optimizer has been integrated with the procedure. Through the amalgamation local search capability of the algorithm is enriched. Otus scops optimizer based on the exceptional features of Otus scops and it possess binocular vision and binaural hearing.

2 Problem Formulation

Power loss reduction in grid connected to renewable energy sources [4–9] is described as,

$$Min\tilde{F}\left(\overline{g},\overline{h}\right) \tag{1}$$

$$g = \left[VLG_1, .., VLG_{Ng}; QC_1, .., QC_{Nc}; T_1, .., T_{N_T}\right]$$

$$h = \left[PG_{slack}; VL_1, .., VL_{N_{Load}}; QG_1, .., QG_{Ng}; SL_1, .., SL_{N_T}\right]$$

$$F_1 = P_{Min} = Min\left[\sum_{m}^{NTL} G_m\left[V_i^2 + V_j^2 - 2 * V_i V_j cos\O_{ij}\right]\right]$$

$$F_2 = Min\left[\sum_{i=1}^{N_{LB}} \left|V_{Lk} - V_{Lk}^{desired}\right|^2 + \sum_{i=1}^{Ng} \left|Q_{GK} - Q_{KG}^{Lim}\right|^2\right]$$

$$F_3 = Minimize\ L_{MaxImum}$$

$$Q_c \rightarrow reactive\ power\ compensator$$

$$T \rightarrow Transformer\ tap$$

$$V_g \rightarrow Generator\ voltage$$

$$PG_{slack} \rightarrow Slack\ generator$$

$$V_L \rightarrow Voltage\ in\ transmission\ lines$$

$$Q_G \rightarrow Reactive\ power\ generator$$

$$L_{Max} = Max\left[1 - [Y_1]^{-1}[Y_2] \times \frac{V_i}{V_j}\right]$$

Equality constraints

$$0 = PG_i - PD_i - V_i \sum_{j \in N_B} V_j\left[G_{ij}cos\left[\O_i - \O_j\right] + B_{ij}sin\left[\O_i - \O_j\right]\right]$$

$$0 = QG_i - QD_i - V_i \sum_{j \in N_B} V_j\left[G_{ij}sin\left[\O_i - \O_j\right] + B_{ij}cos\left[\O_i - \O_j\right]\right]$$

Inequality constraints

$$P_{gsl}^{min} \le P_{gsl} \le P_{gsl}^{max} \tag{2}$$

$$Q_{gi}^{min} \le Q_{gi} \le Q_{gi}^{max},\ i \in N_g \tag{3}$$

$$VL_i^{min} \le VL_i \le VL_i^{max},\ i \in NL \tag{4}$$

$$T_i^{min} \le T_i \le T_i^{max}, i \in N_T \tag{5}$$

$$Q_c^{min} \le Q_c \le Q_C^{max}, i \in N_C \tag{6}$$

$$|SL_i| \le S_{L_i}^{max}, i \in N_{TL} \tag{7}$$

$$VG_i^{min} \le VG_i \le VG_i^{max},\ i \in N_g \tag{8}$$

$$F = F_1 + \begin{bmatrix} \sum_{i=1}^{NL} x_v \begin{bmatrix} VL_i \\ -VL_i^{min} \end{bmatrix}^2 + \\ \sum_{i=1}^{NG} r_g \begin{bmatrix} QG_i \\ -QG_i^{min} \end{bmatrix}^2 \end{bmatrix} + r_f F_3 \tag{9}$$

$$VL_i^{min} = \begin{cases} VL_i^{max}, VL_i > VL_i^{max} \\ VL_i^{min}, VL_i < VL_i^{min} \end{cases} QG_i^{min} = \begin{cases} QG_i^{max}, QG_i > QG_i^{max} \\ QG_i^{min}, QG_i < QG_i^{min} \end{cases}$$

3 Quantum Light and Sound Meditation Based Optimization Algorithm

Quantum Light and Sound Meditation (QLSM) based optimization algorithm is applied to solve the problem. At present in the earth many people are practicing the light and sound meditation technique. In light meditation, by means of envisage light flow from side to side; mind tap into attractive, tender energy and it inherent astuteness to sustain the life journey. Through the light meditation magnificent intellect of association with nature and it sanitize the mind thoughts. Sound meditation frequently employs soothing sounds and in spiritual world of sound meditation sound “OM” has been utilized. Sound meditation makes the mind to relax, focused and incident a more optimistic stance. Pontiff Dr. Sri Ganapathy Sachchidananda teaches and trains the disciples through light and sound Meditation technique. Around the globe pontiff is conducting many meditation classes and healing concert program. Through this millions of people attained liberation from unrest of mind.

$$MM = 0.1 \times (ub - lb) \tag{10}$$

$$MM \rightarrow mindinmeditationstate$$

Mind meditates through light or sound and it depend upon the individual. If the mind moves to the state of liberation through light or sound meditation, then that stable condition is defined as,

$$S\left(\frac{t}{T/50}\right) = 0 \tag{11}$$

$$S \rightarrow stable\, parameter$$

$$t \rightarrow current\, iteration$$

$$T \rightarrow maximum\, number\, of\, iteration$$

Algorithm has been defined by dual mode: light and sound meditation has been scientifically defined in separate segment. In the primary section light meditation has been formulated. Mind envision the light and it faucet into stable condition.

$$MM_L = (2 * M_L) * Q - M_L \tag{12}$$

$$MM_L \rightarrow Mind\, light\, meditation\, condition$$

$$M_L \rightarrow Light\, Meditation$$

$$Q \in [0,1]$$

$$M_L = 0.001 * (ub - lb) * E \tag{13}$$

$$E = 1 - \left(\frac{t}{T}\right)$$

$$E \rightarrow iteration\, progression$$

In the second part of the procedure, sound meditation has been mathematically formulated. In sound meditation, a deep connection to the current instant stabilizes the thoughts of the mind and it enhances self-awareness, emotion control.

$$MM_S = (1 - E) \times A \times Q(1.D) \times (C_{Gbest} - C) \tag{14}$$

$$MM_S \rightarrow sound\ meditation$$

$$C \rightarrow state\ of\ the\ individual$$

$$A = 2.5 \times Q$$

$$Q \in [0,1]$$

$$C_{Gbest} \rightarrow global\ best$$

$$D \rightarrow dimension$$

$$MM_C = A \times Q(1.D) \times (C_{cbest} - C) \tag{15}$$

$$MM_C \rightarrow Mind\ sound\ meditation\ condition$$

$$Q \in [0,1]$$

$$M_C = A \times Q(1.D) \times (M_R - C) \tag{16}$$

$$M_C \rightarrow Mind\ condition$$

$$M_R \rightarrow Random\ mind\ varying\ condition$$

$$Q \in [0,1]$$

$$E < 0.3 \rightarrow Random\ mind\ varying\ condition$$

$$E \geq 0.3 \rightarrow random\ variationof$$

$$mind\ factor\ reduces$$

$$M_{Cn} = A \times Q(1.D) \times (C_{Gbest} - C) \tag{17}$$

$$M_{Cn} - mind\ control$$

$$C \rightarrow mind\ status\ of\ individual$$

$$A = 3.0 \times Q$$

$$Q \in [0,1]$$

$$C_{Gbest} \rightarrow global\ best$$

$$\begin{cases} MM_S = (1 - E) \times A \times Q(1.D) \\ \quad \times (C_{Gbest} - C), E < 0.3 \\ M_{Cn} = A \times Q(1.D) \\ \quad \times (C_{Gbest} - C), E \geq 0.3 \end{cases} \tag{18}$$

$$P_{(iter+1)} = P_{iter} + MM \tag{19}$$

When $E > 0.7$ Algorithm remains in the exploration segment

$$MM_L = Q(1.B) \times (C_{Gbest} - C) \tag{20}$$

$$C_{Gbest} \rightarrow global\ best$$

$$B \in (0,1)$$

$$M_C = (1 - E) \times 2 \times Q(1.D) \times (C_{Cbest} - C) \tag{21}$$

$$M_C \rightarrow Mind\ condition$$

$$Q \in [0,1]$$

When $E \leq 0.7$ Algorithm moves to exploitation segment

$$MM_S = (1 - E) \times 2 \times Q(1.D) \times (C_{Gbest} - C) \tag{22}$$

Quantum computing [10] has been unified in the procedure as follows,

$$|\Psi|^2 \cdot da \cdot db \cdot dc = Q \cdot da \cdot db \cdot dc \tag{23}$$

$$ih \cdot \partial/\partial t \cdot \Psi(a, t) = H \cdot \Psi(a, t) \tag{24}$$

$$H = -h^2/2m \cdot \Delta^2 + V(xa) \quad (25)$$

$$\mathrm{S(T)}\frac{d^2\Psi}{dz^2} + \frac{2m}{h^2}\left[G + \gamma\delta(z)\right]\Psi = 0 \quad (26)$$

To expand the exploration process, Quantum Light and Sound Meditation based optimization algorithm assimilated with Imagination and sound recognition of human brain optimization algorithm. Hippocampus and amygdala of the brain is responsible for the imagination. Auditory cortex of the human brain is responsible for sound recognition. In the exploration section of the algorithm, brain imagination of the human beings is scientifically formulated. In the exploitation section sound recognition of human brain is mathematically formulated in the imagination and sound recognition of human brain based optimization algorithm. Population created in the constrained examination region as follows,

$$\mathrm{a_{i,j}} = \mathrm{min_j} + \mathrm{O}\left(\mathrm{max_j} - \mathrm{min_j}\right) \quad (27)$$

$$\mathrm{max_j}, \mathrm{min_j} \rightarrow \text{maximum and minimum limits}$$

$$\mathrm{O} \in [0,1]$$

Exploration segment modeled based on the imagination aspect of the human brain. Hippocampus and amygdala of the brain are specifically significant to creativity and imagination of the human beings. Hippocampus possesses storing and retrieval of memories and amygdala is accountable for handling emotions of the human beings. In combination, hippocampus and amygdala of the brain assist to create notions. Neocortex and thalamus regulate the brain's imaginings, along with realization and intellectual thought. Over and done with imagination, human beings are able to discover notions of things that are realistically not present.

$$\mathrm{M_i} = \{\mathrm{E_k}\} \quad (28)$$

$$\mathrm{L_k} > \mathrm{L_i}$$

$$\mathrm{E_i} \rightarrow \text{imagination}$$

$$\mathrm{E^a_{i,j}} = \mathrm{E_{i,j}} + \mathrm{O} \cdot \left(\mathrm{E_{i,j}} - \mathrm{O_1} \cdot \mathrm{U_{i,j}}\right) \quad (29)$$

$$\mathrm{E^a_{i,j}} \rightarrow \text{creativity and imagination}$$

$$O \in [0,1]$$

$$O_1 \in \{1,2\}$$

Many times primary imaginations are converted into continuous imaginations. Brain will act coherently to build up the branches of the basic imagination and brain has the capacity to revamp the imaginations again and again. The original imagination can be reclaimed at any time. This deed is quantified as,

$$E^{b}_{i,j} = E_{i,j} + O \cdot \left(G_j - O_2 \cdot E_{i,j}\right) \tag{30}$$

$$E^{b}_{i,j} \rightarrow \text{ branches of imagination}$$

$$G_j \rightarrow \text{ actions of revamp}$$

Exploitation segment is designed based on sound recognition of human brain [11]. When sound waves reach the human being's ears, they are transformed into electronic signals by specific cells entitled as hair cells in the inner part of the ear. Temporal lobe of the brain which established near to the human ears, comprehend sounds and recognize it.

$$E^{c}_{i,j} = E_{i,j} + O \cdot (1 - 2 \cdot O) \cdot D \tag{31}$$

$$E^{c}_{i,j} \rightarrow \text{ sound recognition}$$

$$O \in [0,1]$$

There will be complex conditions in the human brain. Depend on the situations the brain may not imagine and can't recognize any sort of sound. It will be chaos condition. The intensity of this nature may differ among human beings. Through logistic chaos mapping this feature has been scientifically defined in the procedure.

$$x_{n+1} = b x_n (1 - x_n) \tag{32}$$

$$E_i = \delta \cdot E_{i-1} \cdot (1 - E_{i-1}) \tag{33}$$

$$0 \leq E_0 \leq 1$$

Chaotic value will be mapped in the bound region as follows,

$$E^{0}_{i} = \min + (\max - \min) \cdot E_i \tag{34}$$

Local search ability of the algorithm is expanded by unifying the attributes of the Otus scops optimizer has been integrated with the procedure. Through the amalgamation local search capability of the algorithm is enriched. Otus scops optimizer based on the exceptional features of Otus scops and it possess binocular vision and binaural hearing. Information intensity in the Otus scops functional evaluation is defined as,

$$Y_i = \frac{f(Y_i) - Y_w}{Y_b - Y_w} \tag{35}$$

$$Y_i \rightarrow \text{intensity}$$

$$Y_b, Y_w \rightarrow \text{best and worst solutions}$$

Based on the prey position Otus scops modernize its location. It is defined as,

$$U_i = \|Y_i,\ Q\|_2 \tag{36}$$

$$U_i \rightarrow \text{distanve between Otus scops and prey}$$

$$Q \rightarrow \text{location of the prey}$$

Change of the intensity with reference to prey position is defined as,

$$CY_i = \frac{Y_i}{U_i^2} + \alpha \tag{37}$$

$$CY_i \rightarrow \text{change of intensity}$$

$$\alpha \in [0, 0.5]$$

Otus scops [12] will update the location, based on the change in intensity with reference to the position of the prey is defined as,

$$Y_i^{t+1} = \begin{cases} Y_i^t + \beta * CY_i * \|\alpha Q - Y_i^t\| M < 0.5 \\ Y_i^t - \beta * CY_i * \|\alpha Q - Y_i^t\| M \geq 0.5 \end{cases} \tag{38}$$

$$\beta \rightarrow 1.9 \text{ to } 0$$

$$M \rightarrow \text{Movement of Otus scops}$$

Updated value of CY_i based on Otus scops location is defined as,

$$\mathrm{CY_i} = \frac{\mathrm{Y_i}}{\mathrm{U_i}} + \alpha \tag{39}$$

$$\mathrm{CY_i} \rightarrow \text{change of intensity}$$

$$\mathrm{Y_i} \rightarrow \text{intensity}$$

$$\alpha \in [0, 0.5]$$

Location of the Otus scops is updated as follows,

$$\mathrm{Y_i^{t+1}} = \begin{cases} \mathrm{Y_i^t}\,\mathrm{Y_b} < a \\ \mathrm{Y_i^t}\,\mathrm{Y_b} \geq a \end{cases} \tag{40}$$

$$\mathrm{a} \in [0,1]$$

Quantum Light and Sound Meditation (QLSM) based optimization algorithm

1. Start
2. Set thc conditions
3. Define the limits
4. $MM = 0.1 \times (ub - lb)$
5. Describe the stable condition
6. Define light meditation
7. $MM_L = (2 * M_L) * Q - M_L$
8. Define the sound meditation
9. $MM_S = (1 - E) \times A \times Q(1.D) \times (C_{Gbest} - C)$
10. $MM_C = A \times Q(1.D) \times (C_{cbest} - C)$
11. $M_C = A \times Q(1.D) \times (M_R - C)$
12. $E < 0.3 \rightarrow Randommindvaryingcondition$
13. $E \geq 0.3 \rightarrow variationofmindfactorreduces$
14. $C_{Gbest} \rightarrow globalbest$
15. $\begin{cases} MM_S = (1 - E) \times A \times Q(1.D) \\ \quad \times (C_{Gbest} - C), E < 0.3 \\ \quad M_{Cn} = A \times Q(1.D) \\ \quad \times (C_{Gbest} - C), E \geq 0.3 \end{cases}$
16. $P_{(iter+1)} = P_{iter} + MM$
17. If $E > 0.7 \rightarrow$ Algorithm in exploration
18. $MM_L = Q(1.B) \times (C_{Gbest} - C)$
19. $M_C = (1 - E) \times 2 \times Q(1.D) \times (C_{Cbest} - C)$
20. If $E \leq 0.7, \rightarrow$ Algorithmin exploitation
21. $MM_S = (1 - E) \times 2 \times Q(1.D) \times (C_{Gbest} - C)$
22. Apply Quantum computing
23. Describe the imagination

24. $E^{a}_{i,j} = E_{i,j} + O \cdot \left(E_{i,j} - O_1 \cdot U_{i,j}\right)$
25. Define the imagination growth
26. $E^{b}_{i,j} = E_{i,j} + O \cdot \left(G_j - O_2 \cdot E_{i,j}\right)$
27. Define the exploitation section
28. Describe the sound recognition
29. $E^{c}_{i,j} = E_{i,j} + O \cdot (1 - 2 \cdot O) \cdot D$
30. Describe the chaos condition
31. $E_i = \delta \cdot E_{i-1} \cdot (1 - E_{i-1})$
32. Modernize the location
33. Define the change in intensity
34. Based on change in intensity update the location
35. Updated value based on Otus scops location
36. $t = t + 1$
37. Acquire most excellent solution
38. End

4 Simulation Study

Quantum Light and Sound Meditation (QLSM) based optimization algorithm is validated in 7 benchmark functions [9]. Outcome is exhibited in Table 1.

$$f_1(\text{parabola}) = \sum_{i=1}^{D} x_i^2$$

$$f_2(\text{Ackley}) = -20exp\left\{-0.2\sqrt{\frac{1}{D}\sum_{i=1}^{D} x_i^2}\right\}$$

$$-exp\left\{\frac{1}{D}\sum_{i=1}^{D} cos(2\pi x_i)\right\} + 20 + e$$

Table 1 Outcome of QLSM in 7 benchmark functions

F	HBS [9]	ASZ [9]	QLSM
F1	0.8874	0.0026	0.0061
F2	0.0918	0.0125	0.0242
F3	22.9814	16.5622	16.679
F4	1.1092 e−2	3.459 e−4	3.43 e−4
F5	−3.8629	−3.8629	−3.8629
F6	−1.0318	−1.0316	−1.0353
F7	3.0280	3.0130	3.0165

$$f_3(\text{Rosen rock}) = \sum_{i=1}^{D-1} \left\{ \begin{array}{c} 100\left(x_{i+1} - x_i^2\right)^2 \\ +(x_i - 1)^2 \end{array} \right\}$$

$$f_4(\text{Griewank}) = \frac{1}{4000} \sum_{i=1}^{D} x_i^2 - \prod_{i=1}^{D} cos\left(\frac{x_i}{\sqrt{i}}\right) + 1$$

$$f_5(\text{Hartman}) = -\sum_{i=1}^{4} c_i exp\left[-\sum_{j=1}^{3} \frac{a_{ij}}{\left(x_j - p_{ij}\right)^2}\right]$$

$$f_6(\text{Six} - \text{ hump camel } - \text{ back}) = 4x_1^2 - 2.1x_1^4$$

$$+\frac{1}{3}x_1^6 + x_1x_2 - 4x_2^2 + 4x_2^2$$

$$f_7(\text{Goldstein - Price}) = \left\{1 + (x_1 + x_2 + 1)^2\left(19 - 14x_1 + x_1^2 - 14x_2 + 6x_1x_2 + 3x_2^2\right)\right\}$$
$$\times \left\{ \begin{array}{c} 30 + (2x_1 - 3x_2)^2 \\ \left(18 - 32x_1 + 12x_1^2 + 48x_2 - 36x_1x_2 + 27x_2^2\right) \end{array} \right\}$$

Quantum Light and Sound Meditation (QLSM) based optimization is substantiated in IEEE 300 bus system. Table 2, Fig. 1 exhibits the appraisal of result. AEO (MW)—loss, VLA (PU)—eccentricity of voltage.

Table 2 Valuation of outcome—IEEE 300 bus system

Method	AEO(MW)	VLA(PU)
AIZ [4]	418.48	1.1054
IAIZ [4]	410.31	1.1276
CIF [5]	390.60	1.1079
DZP [6]	398.02	1.0886
QLSM	386.56	1.0093

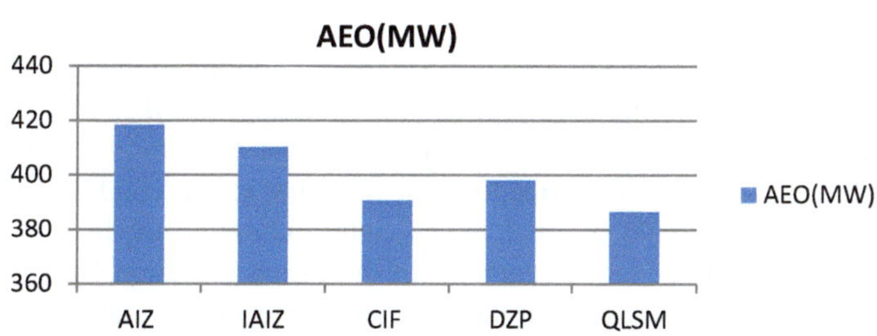

Fig. 1 AEO (MW) analysis – IEEE 300 bus system

Table 3 Evaluation of outcome—IEEE 354 bus system

Method	AEO(MW)	VLA(PU)
AIZ [4]	338.7150	0.51170
IAIZ [4]	341.1230	0.63950
CIF [5]	347.1323	0.6395
DZP [6]	336.02	0.4719
BCZ [7]	335.96	0.4706
QLSM	333.192	0.4265

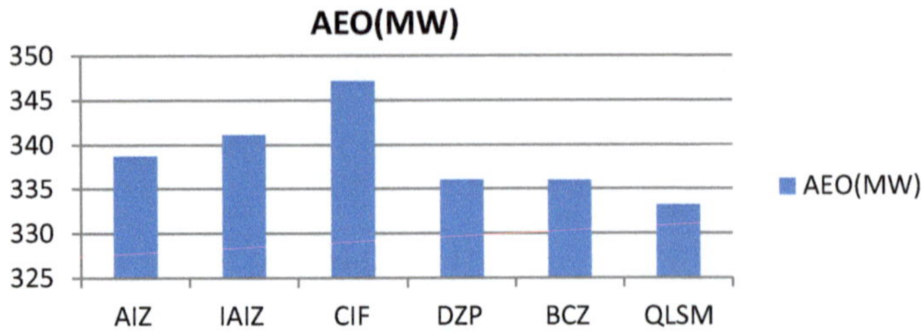

Fig. 2 AEO (MW) analysis – IEEE 354 bus system

Table 4 QLSM computational time

Method	300 BUS T(S)	354 bus T(S)
QLSM	56.24	69.35

Quantum Light and Sound Meditation (QLSM) based optimization is substantiated in IEEE 354 bus system. Table 3, Fig. 2 exhibits the appraisal of result.

Computational time of Quantum Light and Sound Meditation (QLSM) based optimization algorithm is shown in Table 4.

5 Conclusion

Quantum Light and Sound Meditation (QLSM) based optimization algorithm solved the active power loss diminishing problem competently. Mind meditates through light or sound and it depend upon the individual. If the mind moves to the state of liberation through light or sound meditation, then that stable condition is defined. Algorithm has been defined by dual mode: light and sound meditation has been scientifically defined in separate segment. In the primary section light meditation has been formulated. Mind envision the light and it faucet into stable condition. In the second part of the procedure, sound meditation has been mathematically formulated. In sound meditation, a deep connection to the current instant stabilizes the thoughts of the mind and it enhances self-awareness, emotion control. To expand the exploration process, Quantum Light and Sound Meditation based optimization algorithm assimilated with Imagination and sound recognition of human brain optimization algorithm. Hippocampus and amygdala of the brain is responsible for the imagination. Auditory cortex of the human brain is responsible for sound recognition. In the exploration section of the algorithm, brain imagination of the human beings is scientifically formulated. In the exploitation section sound recognition of human brain is mathematically formulated in the imagination and sound recognition of human brain based optimization algorithm. Quantum Light and Sound Meditation (QLSM) based optimization is corroborated in 7 Benchmark functions and IEEE 300, 354 bus systems.

References

1. Jamil, A., Gutlapalli, S. D., Ali, M., Oble, M. J. P., Sonia, S. N., George, S., Shahi, S. R., Ali, Z., Abaza, A., & Mohammed, L. (2024). Meditation and its mental and physical health benefits in 2023. *Cureus, 15*(6), 1–14.
2. Espinosa, F. D. (2025). A verbal behavior analysis of theory of mind: Conceptual and applied implications. *Perspectives on Behavior Science, 1*(1), 1–17.
3. Kanagasabai, L. (2024). Novel curing ailments through Music Inspired Optimization algorithm. *Suranaree Journal of Science and Technology, 31*(2), 1–11.
4. Shaheen, M. A. M., Ullah, Z., Hasanien, H. M., Tostado-Véliz, M., Ji, H., Qais, M. H., Alghuwainem, S., & Jurado, F. (2023). Enhanced transient search optimization algorithm-based optimal reactive power dispatch including electric vehicles. *Energy (Oxford, England), 277*(1), 1–20.
5. Varan, M., Erduman, A., & Menevşeoğlu, F. (2023). A grey wolf optimization algorithm-based optimal reactive power dispatch with wind-integrated power systems. *Energies, 16*(13), 1–10.
6. Lima, G. B., Neto, A., Alves, D. A., Minussi, C. R., Da Silva Amorim, E., & Silva, L. C. P. (2023). Technique for reactive loss reduction and loading margin enhancement using the curves of losses versus voltage magnitude. *Energies, 16*(16), 1–15.
7. Roy, R., Das, T., & Mandal, K. K. (2021). Optimal reactive power dispatch using a novel optimization algorithm. *Journal of Electrical Systems and Information Technology, 8*(1), 1–15.
8. Kanagasabai, L. (2022). Novel empress SARANI optimization algorithm for active power loss reduction and voltage stability enhancement. *Heliyon, 10*(22), 1–64.
9. Yapıcı, H., Çetinkaya, N. (2017). An improved particle swarm optimization algorithm using eagle strategy for power loss minimization. *Mathematical Problems in Engineering, 11*, 1–11.
10. Gong, C., Zhou, N., Xia, S., & Huang, S. (2024). Quantum particle swarm optimization algorithm based on diversity migration strategy. *Future Generations Computer Systems: FGCS, 157*, 445–458.
11. Nascimento, M. A., Biagiotti, S., Herranz-Pérez, V., Santiago, S., Bueno, R., Ye, C. J., Abel, T. J., Zhang, Z., Rubio-Moll, J. S., Kriegstein, A. R., Yang, Z., Garcia-Verdugo, J. M., Huang, E. J., Alvarez-Buylla, A., & Sorrells, S. F. (2024). Protracted neuronal recruitment in the temporal lobes of young children. *Nature, 626*(1), 1056–1065.
12. Ivajnšič, D., Denac, D., Denac, K., Pipenbaher, N., & Kaligarič, M. (2020). The Scops owl (Otus scops) under human-induced environmental change pressure. *Land Use Policy, 99*(1), 1–20.

EC for Large-Scale Optimization: Methods and Real-World Optimization Problems

Yuefeng Xu, Rui Zhong, Junqi Zhang, Chao Zhang, and Jun Yu

Abstract This chapter explores Large-Scale Optimization Problems (LSOPs), characterized by high-dimensional search spaces, complex constraints, and nonlinear interactions, making traditional Evolutionary Computation (EC) algorithms inefficient due to the curse of dimensionality. To address these challenges, the chapter introduces Cooperative Coevolution (CC), which decomposes LSOPs into smaller subproblems, and various grouping strategies such as Differential Grouping (DG), Recursive Differential Grouping (RDG), and Extended Recursive Differential Grouping (ERDG) for handling variable interactions. Non-grouping methods like Competitive Swarm Optimization (CSO) and multiple population differential evolution (mDE-brM) are considered alternatives for maintaining search diversity in high-dimensional spaces. Real-world applications, including the Large-Scale Traveling Salesman Problem (LSTSP) and Feature Selection in Machine Learning, demonstrate the practical significance of LSOPs in logistics, AI, and smart systems. By integrating evolutionary strategies, hybrid approaches, and problem decomposition, this chapter provides an effective framework for solving LSOPs, supporting advancements in AI-driven optimization, intelligent decision-making, and green energy systems.

Y. Xu
Graduate School of Science and Technology, Niigata University, Niigata, Japan
e-mail: f251005c@mail.cc.niigata-u.ac.jp

R. Zhong
Information Initiative Center, Hokkaido University, Sapporo, Japan
e-mail: rui.zhong.u5@elms.hokudai.ac.jp

J. Zhang
College of Electronic and Information Engineering, Tongji University, Shanghai, China
e-mail: zhangjunqi@tongji.edu.cn

C. Zhang
Faculty of Engineering, University of Toyama, Toyama, Japan
e-mail: zhang@eng.u-toyama.ac.jp

J. Yu (✉)
Institute of Science and Technology, Niigata University, Niigata, Japan
e-mail: yujun@ie.niigata-u.ac.jp

M. A. Mellal et al. (eds.), *Computational Intelligence for a Greener Future: Innovations in Renewable Energy Systems*, Studies in Computational Intelligence 1240,
https://doi.org/10.1007/978-3-032-06732-6_13

Keywords Large-Scale optimization · Cooperative coevolution · Evolutionary computation

1 Introduction

Many decision variables, complex objective functions and constraints, and high-dimensional search spaces characterize LSOPs (LSOPs) [1]. Such issues are inherently more challenging than their small-scale counterparts due to their nonlinear, multimodal nature and the high degree of coupling between variables. In practical applications, LSOPs are ubiquitous. For example, in intelligent grid scheduling, thousands of variables related to generation, storage, and load must be managed simultaneously to achieve real-time load balancing; in supply chain management, decision-making tasks—such as multi-node logistics, warehousing, and inventory control—involve hundreds or even thousands of variables; and in communication network design, node configuration, and resource allocation require selecting an optimal solution from an exponentially large set of possibilities. These examples underscore the universality and the complexity of LSOPs across diverse fields.

However, traditional evolutionary computation (EC) algorithms [2] often face significant bottlenecks when addressing LSOPs. Firstly, as the problem size increases, the search space expands exponentially—a phenomenon known as the "curse of dimensionality" [3]—hampers the algorithm's ability to explore the entire solution space effectively and increases the risk of premature convergence to local optima. Secondly, because EC algorithms rely on population-based search strategies, maintaining sufficient diversity and achieving an effective balance between exploration and exploitation becomes increasingly challenging in high-dimensional settings. This difficulty typically results in slower convergence rates and lower solution accuracy. Moreover, the computational resources these algorithms require escalate sharply with problem size, often constituting a critical barrier to their practical application.

To mitigate these challenges, researchers have developed various advanced strategies. Notable approaches include problem decomposition-based co-evolutionary methods, which break down a large-scale problem into several lower-dimensional sub-problems, thereby reducing the overall complexity; hybrid local search techniques that combine global evolutionary search with local optimization to enhance efficiency in high-dimensional spaces; and methods leveraging dimensionality reduction and parallel computing to address the exponential growth of the search space. The subsequent section of this chapter will examine the theoretical foundations and practical implementations of these advanced methods, providing a systematic framework for overcoming the limitations of traditional evolutionary algorithms in LSOPs and establishing a solid basis for future innovations in the field.

2 The challenges in LSOPs

A typical optimization problem [4] can be formulated as Eq. (1).

$$\begin{aligned} &\min / \max f(x), x = (x_1, ..., x_n) \in A \\ &\text{subject to } g_i(x) \leq 0, i = 1, 2, \ldots, m \\ &h_j(x) = 0, j = 1, 2, \ldots, p \end{aligned} \tag{1}$$

where $f(\mathbf{x})$ is the objective function to be optimized, and $\mathbf{x} = (x_1, x_2, \ldots, x_n)$ denotes the decision variables within the feasible set $\mathcal{A}$. The functions $g_i(\mathbf{x})$ represent the inequality constraints, while $h_j(\mathbf{x})$ define the equality constraints.

LSOPs are the optimization problem encountered under the above framework when the dimension n of the decision variables and the number of constraints m or p are huge (usually hundreds, thousands, or even more), in which the search space grows exponentially, i.e., the phenomenon of "curse of dimensionality" is significant, which leads to the sharp increase of computational complexity and memory demand, making it difficult for traditional algorithms to find high-quality solutions within a limited time.

To illustrate "the curse of dimensionality" more intuitively, let us assume that there is a d-dimensional hypercube with total volume 1. To cover this hypercube uniformly with small cubes of side length ϵ, each dimension must be divided into $1/\epsilon$ intervals. Thus, the total number of small cubes required is $(1/\epsilon)^d$.

For example, when $\epsilon = 0.1$:

- A 2-dimensional hypercube requires $(1/0.1)^2 = 10^2$ small cubes.
- A 10-dimensional hypercube requires $(1/0.1)^{10} = 10^{10}$ small cubes.

This exponential growth clearly illustrates how the required sample size increases dramatically with dimension, making an exhaustive search in high-dimensional space computationally infeasible.

Moreover, from a volumetric perspective, the volume of a high-dimensional sphere decreases rapidly with increasing dimensionality. The volume of an n-dimensional sphere with radius r is given by Eq. (2).

$$V_n(r) = \frac{\pi^{n/2}}{\Gamma(n/2 + 1)} r^n \tag{2}$$

where, $V_n(r)$ denotes the volume, r is the radius, n is the number of dimensions, and $\Gamma(\cdot)$ is the gamma function, which generalizes the factorial function

For a fixed r, the Gamma function in the denominator grows much faster than r^n as n increases, causing the volume $V_n(r)$ to approach zero rapidly. This implies that in high-dimensional spaces, most of the "volume" is concentrated near the boundaries of the hypercube rather than in the central region, where data points are often expected to cluster.

Furthermore, in high-dimensional data, the Euclidean distance between any two points tends to concentrate, meaning that the differences in distance between various pairs of points become minimal. This phenomenon weakens the discriminative power of the Euclidean distance as a measure of similarity. The primary reason for this effect lies in the additive nature of the Euclidean distance and the statistical properties that emerge in high-dimensional spaces.

Consider two points, $\mathbf{x} = (x_1, x_2, \ldots, x_n)$ and $\mathbf{y} = (y_1, y_2, \ldots, y_n)$, in an n-dimensional space. The squared Euclidean distance between these points is defined as Eq. (3).

$$d^2(\mathbf{x}, \mathbf{y}) = \sum_{i=1}^{n} (x_i - y_i)^2 \tag{3}$$

where $d^2(\mathbf{x}, \mathbf{y})$ denotes the squared distance, n is the number of dimensions, and x_i, y_i are the i-th components of vectors $\mathbf{x}$ and $\mathbf{y}$, respectively. This metric is commonly used to measure the similarity or dissimilarity between vectors in high-dimensional spaces.

Assume that the squared differences $(x_i - y_i)^2$ for $i = 1, 2, \ldots, n$ are independent and identically distributed (i.i.d.) random variables with a finite mean μ and variance σ^2. According to the law of large numbers, as n increases, the average of these squared differences converges to the expected value as Eq. (4).

$$\frac{1}{n}\sum_{i=1}^{n} (x_i - y_i)^2 \rightarrow \mu \,\text{as}\, n \rightarrow \infty \tag{4}$$

Moreover, the relative variability of the sum can be quantified by the coefficient of variation (CV). The standard deviation of the sum is approximately $\sqrt{n\sigma^2}$, so the CV for $d^2(\mathbf{x}, \mathbf{y})$ is given by Eq. (5).

$$\text{CV} = \frac{\sqrt{n\sigma^2}}{n\mu} = \frac{\sigma}{\sqrt{n}\mu} \tag{5}$$

As n increases, the CV decreases, indicating that the dispersion of the squared Euclidean distances around the mean diminishes. In other words, although the absolute Euclidean distance may increase with dimensionality, the relative differences between distances for different pairs of points become negligible. This concentration of distances is a key aspect of the "curse of dimensionality," complicating tasks such as sampling, searching, and modeling in high-dimensional spaces.

3 EAs for LSOPs

LSOPs pose significant challenges to traditional optimization methods due to their high dimensionality and complexity. EAs, as a kind of bionic optimization method, have been widely used to solve LSOPs.According to different strategies, the applications of EAs in LSOPs are mainly divided into the following three directions.

4 Cooperative Coevolution (CC)

CC is a type of evolutionary computation method based on problem decomposition [5], primarily used to solve high-dimensional optimization problems. It was first proposed by Potter and De Jong (1994). Its core idea is to decompose a complex global optimization problem into multiple relatively independent subproblems, which are optimized separately and then cooperatively to improve the search efficiency and optimization effect.

4.1 The basic framework of CC

CC adopts the "Decomposition-Optimization-Collaboration" approach to solve complex optimization problems, and its basic framework is as follows:

4.1.1 Problem Decomposition

The original high-dimensional optimization problem $f(\mathbf{x})$ is partitioned into M low-dimensional subproblems as shown in Eq. (6).

$$f(\mathbf{x}) = f(\mathbf{x}_1, \mathbf{x}_2, \ldots, \mathbf{x}_M) \tag{6}$$

where each $\mathbf{x}_i$ is a subset of the decision variables, each corresponding to a sub-optimization task.

4.1.2 Subproblem Optimization

Each subproblem is individually optimized independently using EA (e.g., GA, DE, PSO) to search for the optimal value of the sub-solution $\mathbf{x}_i$.

4.1.3 Collaboration

Coordinate the information exchange between subproblems and integrate the optimization results of word problems so that the decomposed optimization problem can still effectively solve the global optimal solution.

4.2 Methods of grouping

The core of CC lies in how to group effectively, and different grouping methods significantly impact the optimization of LSOPs. We list the standard grouping methods in the concentration:

4.2.1 Random Grouping (RG)

RG partitions a high-dimensional decision variable set $X = \{x_1, x_2, \ldots, x_D\}$ into M subgroups, where each subgroup is optimized independently [6]. The partitioning process follows a random assignment, ensuring that each subgroup X_i contains approximately D/M variables as Eq. (7).

$$X = \bigcup_{i=1}^{M} X_i, X_i \cap X_j = \emptyset \; \forall i \sum_{i=1}^{n} X_i \neq j \tag{7}$$

Since the interactions between variables are not explicitly considered, RG assumes that arbitrarily splitting variables does not significantly impact optimization performance. However, incorrect grouping may destroy the underlying dependencies between variables, leading to suboptimal solutions. To mitigate this problem, periodic regrouping is introduced, where variables are randomly reassigned to different groupings every T generations, as shown in Eq. (8).

$$X^{(t+T)} = \text{Shuffle}(X^{(t)}) \tag{8}$$

This enhances population diversity, prevents premature convergence, and balances exploration and exploitation. While RG is computationally efficient and easy to implement, its random nature may break strong inter-variable correlations, making it less effective for problems where variable interactions are crucial.

4.2.2 Differential Grouping (DG)

DG identifies variable interactions by applying small perturbations to individual variables and analyzing their effects on the objective function [7]. For a given function

$f(X)$ with decision variables $X = \{x_1, x_2, \ldots, x_D\}$, DG perturbs a single variable x_i by a small value Δ and evaluates the function difference as Eq. (9).

$$\delta_i = f(X + \Delta e_i) - f(X) \tag{9}$$

where e_i is a unit vector perturbing only x_i. To determine whether two variables x_i and x_j are independent, DG applies perturbations to both variables and checks the interaction condition as Eq. (10).

$$\delta_{ij} = \left[f(X + \Delta e_i + \Delta e_j) - f(X + \Delta e_i)\right] - \left[f(X + \Delta e_j) - f(X)\right] \tag{10}$$

If $\delta_{ij} = 0$, the variables are independent and can be optimized separately; otherwise, they are dependent and grouped for joint optimization. By identifying and preserving variable interactions, DG effectively reduces problem dimensionality, improves optimization efficiency, and enhances solution quality for nonlinear, high-dimensional, and constrained optimization problems. However, its pairwise interaction analysis incurs a computational cost of $O(D^2)$, making it challenging for large-scale issues requiring a more scalable decomposition strategy.

4.2.3 Recursive Differential Grouping (RDG)

RDG is an improved grouping method built upon DG, designed to enhance efficiency by leveraging nonlinear interaction detection [8]. Instead of explicitly checking all pairwise interactions, RDG employs a recursive decomposition approach, dynamically expanding variable groups without evaluating all possible combinations. This significantly reduces computational complexity while maintaining the accuracy of variable interaction detection.

For a given optimization problem with decision variables $X = \{x_1, x_2, \ldots, x_D\}$, RDG starts by selecting a reference variable x_i and applying a small perturbation Δ to assess its effect as Eq. (11).

$$\delta_i = f(X + \Delta e_i) - f(X) \tag{11}$$

Instead of computing pairwise differences exhaustively, RDG recursively expands the group by iteratively selecting new candidate variables x_j and assessing their cumulative impact on the function, as shown in Eq. (12).

$$\delta_G = f(X + \Delta E_G) - f(X) \tag{12}$$

where E_G is the set of perturbations applied to all currently grouped variables. If adding x_j results in a significant change in δ_G, it is included in the group; otherwise, it remains independent. This recursive strategy ensures that strongly interacting variables are efficiently grouped without requiring an exhaustive search over all variable pairs.

By replacing pairwise comparisons with a hierarchical recursive expansion, RDG reduces the computational complexity from $O(D^2)$ in DG to approximately $O(D\log D)$. This makes RDG significantly more scalable while preserving variable interactions essential for effective optimization, making it suitable for large-scale nonlinear optimization problems.

4.2.4 Extended Recursive Differential Grouping (ERDG)

ERDG is an advanced extension of RDG designed to enhance the accuracy, robustness, and efficiency of variable decomposition in LSOPs [9]. Unlike RDG, which uses fixed perturbations, ERDG introduces an adaptive mechanism that dynamically adjusts perturbation magnitudes based on function sensitivity. Given an optimization problem with decision variables $X = \{x_1, x_2, \ldots, x_D\}$ and objective function $f(X)$, ERDG perturbs a reference variable x_i using an adaptive step size Δ_i as Eq. (13).

$$\Delta_i = \gamma \cdot \max(|x_i|, \epsilon) \tag{13}$$

where γ is a scaling factor and ϵ prevents numerical instability.

Instead of applying a fixed interaction test, ERDG adaptively refines grouping decisions by recursively expanding variable sets and reassessing their interactions. For a group G of interacting variables, the function change due to simultaneous perturbation is computed as Eq. (14).

$$\delta_G = f(X + \Delta_G E_G) - f(X) \tag{14}$$

where Δ_G is dynamically adjusted based on the collective function sensitivity of variables in G. A variable x_j is incorporated into the group if its inclusion significantly alters δ_G, ensuring that only genuinely interacting variables are grouped.

To further refine group accuracy, ERDG includes a group reassessment mechanism, which periodically verifies the validity of previously assigned groupings as Eq. (15).

$$G_k^{(t+T)} = \text{Refine}(G_k^{(t)}) \tag{15}$$

where T is the reassessment interval and Refine($\cdot$) eliminates weakly interacting variables from groups to prevent misclassification.

By integrating adaptive perturbation and recursive refinement, ERDG maintains the computational efficiency of RDG ($O(D\log D)$) while significantly improving the detection of nonlinear dependencies. These improvements make ERDG particularly suitable for large-scale, nonlinear, and constrained optimization problems, where complex variable interactions require more precise grouping strategies.

5 Non-Grouping Optimization Methods

Non-grouping optimization methods refer to approaches that do not explicitly decompose or partition a large-scale optimization problem into smaller subproblems. Unlike grouping-based strategies, non-grouping methods treat all decision variables as a single entity, optimizing them simultaneously. These methods are instrumental in cases where variable interactions are highly complex or unknown, making decomposition ineffective.

5.1 Competitive Swarm Optimizer (CSO)

CSO [10] is a swarm-based optimization algorithm fundamentally inspired by Particle Swarm Optimization (PSO) [11]. It differs in that it does not rely on either the personal best position of each particle or the global best position to update particle positions. Instead, CSO introduces a pairwise competition mechanism, where particles are randomly paired. Within each pair, the particle with a higher fitness value (i.e., the loser) updates its position by learning from the better-performing particle (i.e., the winner).

Given a population of N particles, each particle i has a position X_i and velocity V_i. During the competition, for each randomly selected particle pair (i, j), the winner (better fitness) retains its position, while the loser updates its velocity and position as Eqs. (16) and (17).

$$V_i = \omega V_i + r_1(X_j - X_i) \tag{16}$$

$$X_i = X_i + V_i \tag{17}$$

where ω is an inertia weight controlling exploration, and r_1 is a uniform random number in [0, 1] that introduces stochasticity in learning from the winner.

By adopting this competitive mechanism, CSO effectively balances exploration and exploitation, ensuring diversity while directing the search toward promising regions. Additionally, due to its simplified update rules, CSO incurs lower computational costs than traditional PSO while maintaining strong performance, making it particularly effective for LSOPs.

5.2 The Multiple Population Differential Evolution Algorithm (mDE-brM)

The mDE-brM is a DE-based approach for LSOPs [12]. It introduces a multi-subpopulation framework where the population is artificially divided into three subpopulations based on pairwise fitness comparisons. The population $P = \{X_1, X_2, \ldots, X_N\}$ is ranked, with the top-performing individuals assigned to subpopulation P_1, the mid-ranked individuals to P_2, and the lowest-performing individuals to P_3, ensuring diversity in search dynamics. Each subpopulation adopts a distinct mutation strategy tailored to its role in the search process. The top-ranked subpopulation (P_1) focuses on exploitation using a greedy DE/rand-to-best/1 strategy, as shown in Eq. (18).

$$V_i^{(t)} = X_i^{(t)} + F \cdot (X_{\text{best}}^{(t)} - X_i^{(t)}) + F \cdot (X_r^{(t)} - X_s^{(t)}) \tag{18}$$

where $X_{\text{best}}^{(t)}$ is the best individual, and $X_r^{(t)}$, $X_s^{(t)}$ are randomly selected individuals. The middle subpopulation (P_2) employs a balanced DE/current-to-pbest/1 approach as Eq. (19).

$$V_i^{(t)} = X_i^{(t)} + F \cdot (X_p^{(t)} - X_i^{(t)}) + F \cdot (X_r^{(t)} - X_s^{(t)}) \tag{19}$$

where $X_p^{(t)}$ is a randomly selected top p % individual. Meanwhile, the lowest-ranked subpopulation (P_3), designated for exploration, follows the DE/rand/1 strategy, as shown in Eq. (20).

$$V_i^{(t)} = X_r^{(t)} + F \cdot (X_s^{(t)} - X_u^{(t)}) \tag{20}$$

where $X_r^{(t)}$, $X_s^{(t)}$, $X_u^{(t)}$ are randomly chosen individuals.

After mutation, binomial crossover is applied to generate trial vectors $U_i^{(t)}$, with each dimension updated according to Eq. (21).

$$U_{i,j}^{(t)} = \begin{cases} V_{i,j}^{(t)}, & \text{if } r_j \leq C_r \text{ or } j = j_{\text{rand}}, \\ X_{i,j}^{(t)}, & \text{otherwise}, \end{cases} \tag{21}$$

Ensure that at least one component undergoes modification. The selection step follows, updating individuals based on fitness evaluation as Eq. (22).

$$X_i^{(t+1)} = \begin{cases} U_i^{(t)}, & \text{if } f(U_i^{(t)}) \leq f(X_i^{(t)}), \\ X_i^{(t)}, & \text{otherwise}, \end{cases} \tag{22}$$

To facilitate knowledge transfer, mDE-brM incorporates a migration mechanism, allowing individuals to move between subpopulations with a probability P_m. This leads to an updated subpopulation composition, as shown in Eq. (23).

$$P_k^{(t+1)} = P_k^{(t)} \cup M_k^{(t)} \tag{23}$$

where $M_k^{(t)}$ represents the set of migrating individuals. By combining tailored mutation strategies, controlled migration, and adaptive selection, mDE-brM effectively balances exploration and exploitation, making it a robust approach for high-dimensional optimization challenges.

6 Realworld LSOPs

6.1 Large-scale Traveling Salesman Problem (LSTSP)

The TSP is a well-known combinatorial optimization problem where a salesman must visit a set of cities exactly once and return to the starting point while minimizing the total travel distance. When the number of towns becomes very large, the problem is referred to as the LSTSP [13]. LSTSP is NP-hard, meaning that exact solutions become computationally infeasible as the number of cities grows, requiring heuristic and metaheuristic approaches for practical solutions.

6.2 Problem Formulation

Given a set of N cities, let $C = \{c_1, c_2, \ldots, c_N\}$ be the set of cities, and let d_{ij} represent the distance between city c_i and city c_j. The objective is to find a Hamiltonian cycle (a tour visiting all cities exactly once and returning to the starting city) that minimizes the total travel cost. The problem can be formulated as follows:

$$\min \sum_{i=1}^{N} \sum_{j=1}^{N} d_{ij} x_{ij}$$

Subject to the constraints:

$$\sum_{j=1}^{N} x_{ij} = 1, \forall i \in \{1, 2, \ldots, N\}$$

$$\sum_{i=1}^{N} x_{ij} = 1, \forall j \in \{1, 2, \ldots, N\}$$

$$\sum_{i,j \in S} x_{ij} \leq |S| - 1, \forall S \subset C, 2 \leq |S| \leq N - 1$$

where:

- x_{ij} is a binary variable: $x_{ij} = 1$ if the salesman travels directly from city c_i to c_j, and 0 otherwise.
- The first two constraints ensure that each city is visited exactly once.
- The third constraint prevents subtours, ensuring all cities are connected in a single loop.

6.3 The relevance of addressing LSTSP

The LSTSP is a critical combinatorial optimization challenge with extensive real-world significance. As cities increase, exact algorithms become computationally infeasible, making heuristic and metaheuristic approaches essential for practical applications. One of the most impactful areas of LSTSP is logistics and transportation [14], where companies such as FedEx, UPS, and Amazon must optimize delivery routes to reduce operational costs and improve efficiency. By minimizing total travel distance, businesses can lower fuel consumption, shorten delivery times, and extend vehicle lifespan, directly influencing profitability and sustainability. Similarly, urban planning benefits from solving LSTSP, as it plays a fundamental role in optimizing waste collection, public transportation, and emergency response services, ensuring that cities operate more efficiently.

In robotics and autonomous systems [15], LSTSP-based path planning is crucial for self-driving cars, automated warehouses, and drone deliveries, enabling efficient movement while minimizing energy consumption. Industrial automation and computer chip design also rely on LSTSP-like solutions to optimize manufacturing processes, reduce material waste, and improve production speed. Additionally, the healthcare sector benefits from efficient medical logistics, including ambulance dispatch, sample collection, and mobile vaccination routes, ensuring timely responses in critical situations.

Beyond individual industries, solving LSTSP contributes to energy efficiency and environmental sustainability [16]. Optimized routing reduces fuel consumption and carbon emissions, an essential factor in mitigating climate change and improving air quality in urban areas. Cost reduction in large-scale operations is another crucial advantage, particularly in e-commerce, airline logistics, and supply chain networks, where well-optimized routes minimize transportation costs and maximize resource utilization. Furthermore, advancements in AI-driven optimization have led to the integration of reinforcement learning, genetic algorithms, and deep learning-based approaches to improve LSTSP solutions, making them adaptable to dynamic real-world conditions.

6.4 Feature Selection

Feature selection is a fundamental problem in machine learning and optimization [17], where the goal is to select a subset of relevant features from a high-dimensional dataset to improve model performance, reduce computational cost, and enhance interpretability.

6.5 Problem Formulation

Given a dataset with N samples and D features, the problem can be formulated as selecting an optimal subset $S \subseteq \{1, 2, \ldots, D\}$ that maximizes a performance metric while minimizing complexity. The objective function for feature selection can be expressed as Eq. (24).

$$\max_{S} F(S) = \text{Model Performance}(S) - \lambda |S| \quad (24)$$

where $F(S)$ is the evaluation function, λ is a regularization parameter that controls the trade-off between accuracy and the number of selected features, and $|S|$ represents the number of selected features.

The search space for feature selection grows exponentially with the number of features, leading to a complexity of $O(2^D)$, making exact solutions impractical for large-scale problems.

6.6 The Relevance of Addressing Feature Selection

In practical applications, feature selection has widespread relevance in domains such as bioinformatics [18], financial forecasting [19], text classification [20], and image processing [21]. For example, selecting the most relevant genes from thousands of genetic markers in genomics can improve disease prediction and drug discovery. Reducing the number of economic indicators used in risk assessment models leads to more interpretable and efficient decision-making in finance. In natural language processing (NLP) [22], selecting essential words or phrases helps build more efficient text classification models without unnecessary computational overhead.

As datasets grow in size and complexity, addressing feature selection remains essential for building scalable, efficient, and interpretable machine learning models. Advances in metaheuristic optimization, deep learning-based feature selection, and hybrid techniques continue to push the boundaries of what is possible, enabling more intelligent and automated feature selection approaches that enhance decision-making across industries.

7 Conclusion

LSOPs play a crucial role in solving complex real-world problems that involve massive datasets, high-dimensional decision spaces, and computational constraints. It is widely applied in logistics, healthcare, finance, manufacturing, and artificial intelligence, where optimal decision-making significantly improves efficiency, cost reduction, and sustainability. Large-scale optimization ensures efficient route planning in logistics and transportation, minimizing fuel consumption and delivery time. In healthcare, it aids in optimizing hospital resource allocation, improving patient scheduling, and streamlining medical supply chains. LSOPs enhance portfolio optimization, risk management, and algorithmic financial trading strategies. Additionally, in artificial intelligence and machine learning, optimization techniques are essential for hyperparameter tuning, neural network training, and feature selection in high-dimensional spaces.

References

1. Hager, W.W., Hearn, D.W., Pardalos, P.M. (2013) Large scale optimization: state of the art
2. Back, T., Schwefel, H.-P. (1996) Evolutionary computation: An overview. In: *Proceedings of IEEE International Conference on Evolutionary Computation* (pp. 20–29). IEEE.
3. K¨oppen, M. (2000). The curse of dimensionality. In: *5th Online World Conference on Soft Computing in Industrial Applications (WSC5)* (vol. 1, pp. 4–8).
4. Krentel, M.W. (1986). The complexity of optimization problems. In: *Proceedings of the Eighteenth Annual ACM Symposium on Theory of Computing* (pp. 69–76)
5. Potter, M. A., & Jong, K. A. D. (2000). Cooperative coevolution: An architecture for evolving coadapted subcomponents. *Evolutionary Computation, 8*(1), 1–29.
6. Omidvar, M.N., Li, X., Yang, Z., Yao, X. (2010) Cooperative co-evolution for large scale optimization through more frequent random grouping. In: *IEEE Congress on Evolutionary Computation* (pp. 1–8). IEEE.
7. Omidvar, M. N., Li, X., Mei, Y., & Yao, X. (2013). Cooperative co-evolution with differential grouping for large scale optimization. *IEEE Transactions on Evolutionary Computation, 18*(3), 378–393.
8. Sun, Y., Kirley, M., & Halgamuge, S. K. (2018). A recursive decomposition method for large scale continuous optimization. *IEEE Transactions on Evolutionary Computation, 22*(5), 647–661. https://doi.org/10.1109/TEVC.2017.2778089
9. Yang, M., Zhou, A., Li, C., & Yao, X. (2020). An efficient recursive differential grouping for large-scale continuous problems. *IEEE Transactions on Evolutionary Computation, 25*(1), 159–171.
10. Cheng, R., & Jin, Y. (2014). A competitive swarm optimizer for large scale optimization. *IEEE Transactions on Cybernetics, 45*(2), 191–204.
11. Kennedy, J., Eberhart, R.: Particle swarm optimization. In: *Proceedings of ICNN'95-international Conference on Neural Networks* (vol. 4, pp. 1942–1948). IEEE.
12. Ma, Y., & Bai, Y. (2020). A multi-population differential evolution with best-random mutation strategy for large-scale global optimization. *Applied Intelligence, 50*(5), 1510–1526.
13. Dantzig, G., Fulkerson, R., & Johnson, S. (1954). Solution of a large-scale traveling–salesman problem. *Journal of the Operations Research Society of America, 2*(4), 393–410.
14. Kasilingam, R. G. (1998). *Logistics and transportation*. Kluwer Academic Publishers.

15. Halperin, D., Kavraki, L.E., Solovey, K. (2017). Robotics. In: *Handbook of Discrete and Computational Geometry* (pp. 1343–1376). Chapman and Hall/CRC.
16. Centobelli, P., Cerchione, R., & Esposito, E. (2018). Environmental sustainability and energy-efficient supply chain management: A review of research trends and proposed guidelines. *Energies, 11*(2), 275.
17. Li, J., Cheng, K., Wang, S., Morstatter, F., Trevino, R. P., Tang, J., & Liu, H. (2017). Feature selection: A data perspective. *ACM Computing Surveys (CSUR), 50*(6), 1–45.
18. Baxevanis, A.D., Bader, G.D., Wishart, D.S.: Bioinformatics, (2020)
19. Abu-Mostafa, Y. S., & Atiya, A. F. (1996). Introduction to financial forecasting. *Applied Intelligence, 6*, 205–213.
20. Kowsari, K., Jafari Meimandi, K., Heidarysafa, M., Mendu, S., Barnes, L., & Brown, D. (2019). Text classification algorithms: A survey. *Information, 10*(4), 150.
21. Huang, T. S., Schreiber, W. F., & Tretiak, O. J. (1971). Image processing. *Proceedings of the IEEE, 59*(11), 1586–1609.
22. Chowdhary, K., Chowdhary, K.: Natural language processing. *Fundamentals of Artificial Intelligence* 603–649

Novel Optimal Operation of Nuclear Power Plant Inspired Algorithm for Loss Reduction in Renewable Energy Transmission System

Lenin Kanagasabai

Abstract Balancing the heat generation and controlling the nuclear fission is considered in the design of Optimal operation of nuclear power plant inspired (ONP) algorithm. In the exploration section design, production of heat through the nuclear fission process is considered. Heat has to be engendered for the conversion of water into steam. This process is considered in the design. In the exploitation section design controlling the nuclear fission process in the power plant has been considered. Heat generation through fission process is vital one, but it is very important that fission process should not move to the state of uncontrolled condition. It is much more important to control the fission process throughout the operation of the plant. To expand the exploration activity, Optimal operation of nuclear power plant inspired (ONP) algorithm is unified with Pace and spin bowling in the game of cricket inspired optimization algorithm. Local search is boosted by adding the *Spheniscus humboldti* search optimization. *Spheniscus humboldti* are conjoint beings which doohickey bountiful accounts for predominant pursuit and riddling in varied temperature conditions. During the movement *Spheniscus humboldti* throng together. Optimal operation of nuclear power plant inspired (ONP) algorithm substantiated in 7 benchmark functions and IEEE 118, 300 bus systems.

Keywords Nuclear power plant · Pace · Spin · *Spheniscus humboldti*

1 Introduction

Optimal operation of nuclear power plant inspired (ONP) algorithm is applied for solving power loss lessening problem in renewable energy power transmission system [1–6]. Balancing the heat generation and controlling the nuclear fission is considered in the design of Optimal operation of nuclear power plant inspired (ONP)

L. Kanagasabai (✉)
Prasad V. Potluri Siddhartha Institute of Technology, Kanuru, Vijayawada, Andhra Pradesh 520007, India
e-mail: gklenin@gmail.com

M. A. Mellal et al. (eds.), *Computational Intelligence for a Greener Future: Innovations in Renewable Energy Systems*, Studies in Computational Intelligence 1240,
https://doi.org/10.1007/978-3-032-06732-6_14

algorithm. Electrical power generation in the nuclear power plant is done through the fission process [7–9]. Heat created through fission process and it employed to boil the water, then sequentially steam will be produced. Steam rotates the turbines and it connected to the generators. Electricity produced by the generators. In the design optimized operation of the nuclear power plant is considered. Balancing the heat generation and controlling the nuclear fission is important factor in the operation of the plant. In the exploration section design, production of heat through the nuclear fission process is considered. Heat has to be engendered for the conversion of water into steam. This process is considered in the design. In the exploitation section design controlling the nuclear fission process in the power plant has been considered. Heat generation through fission process is vital one, but it is very important that fission process should not move to the state of uncontrolled condition. It is much more important to control the fission process throughout the operation of the plant. To expand the exploration activity, Optimal operation of nuclear power plant inspired (ONP) algorithm is unified with Pace and spin bowling in the game of cricket inspired optimization algorithm. Pace bowler has to synchronize muscle fibers and motor skills are essential in the pace bowling [10]. Spin bowlers will rotate and turn the ball to cuckold the batsman who faces the ball. Pace bowler in the game of cricket will concentrate on evolving momentum in the course of run-up in linear direction. Pace bowler systematizes the body position and transmits the momentum to the ball while delivering to the batsman. Shoulder gyration, spinal hip, and bottom will be synchronized to create a momentum while delivering the ball towards the batsman. Spin bowlers [11] encompass divulging spin on the ball so the ball will undergo deviation subsequent to bouncing. Fingers and wrist are utilized by the spin bowler to turn the ball in horizontal axis. Pace bowlers are contingent on the speediness of the ball delivery which bowled in the direction of the batsman. Pace of the ball augmented or dwindled depending on the playing aspects of the particular batsman who facing the bowler at particular instant. Spin bowlers comprehend divulging spin on the ball consequently the ball will experience deviance following to bouncing. Fingers and wrist are employed by the spin bowler to rotate the ball in horizontal axis. Batsman facing the spin ball will be in challenging situation, ever since ball progression is in complicated, variable passage after the bounce. Special coaching camp will be organized by the coach [12] to augment the bowling capabilities of pace and spin blowers. Local search is boosted by adding the *Spheniscus humboldti* search optimization. *Spheniscus humboldti* [13] are conjoint beings which doohickey bountiful accounts for predominant pursuit and riddling in varied temperature conditions. During the movement *Spheniscus humboldti* throng together.

2 Problem Formulation

Power loss reduction in renewable energy transmission system [1–6] is described as,

$$Min\,\tilde{F}\left(\overline{g},\overline{h}\right) \tag{1}$$

$$g = \left[VLG_1, .., VLG_{Ng}; QC_1, .., QC_{Nc}; T_1, .., T_{N_T}\right]$$

$$h = \left[PG_{slack}; VL_1, .., VL_{N_{Load}}; QG_1, .., QG_{Ng}; SL_1, .., SL_{N_T}\right]$$

Fitness functions are evaluated as,

$$F_1 = P_{Min} = Min\left[\sum_{m}^{NTL} G_m\left[V_i^2 + V_j^2 - 2 * V_i V_j cos\emptyset_{ij}\right]\right]$$

$$Q_c \rightarrow reactive\,power\,compensator$$

$$T \rightarrow Transformer\,tap$$

$$V_g \rightarrow Generator\,voltage$$

$$PG_{slack} \rightarrow Slack\,generator$$

$$V_L \rightarrow Voltage\,in\,transmission\,lines$$

$$Q_G \rightarrow Reactive\,power\,generator$$

$$F_2 = Min\left[\sum_{i=1}^{N_{LB}} \left|V_{Lk} - V_{Lk}^{desired}\right|^2 + \sum_{i=1}^{Ng} \left|Q_{GK} - Q_{KG}^{Lim}\right|^2\right]$$

$$F_3 = Minimize\,L_{MaxImum}$$

Equality constraints

$$0 = PG_i - PD_i - V_i \sum_{j \in N_B} V_j\left[G_{ij}cos\left[\emptyset_i - \emptyset_j\right] + B_{ij}sin\left[\emptyset_i - \emptyset_j\right]\right]$$

$$0 = QG_i - QD_i - V_i \sum_{j \in N_B} V_j\left[G_{ij}sin\left[\emptyset_i - \emptyset_j\right] + B_{ij}cos\left[\emptyset_i - \emptyset_j\right]\right]$$

Inequality constraints

$$\mathrm{P_{gsl}^{min}} \leq \mathrm{P_{gsl}} \leq \mathrm{P_{gsl}^{max}} \tag{2}$$

$$Q_{gi}^{min} \le Q_{gi} \le Q_{gi}^{max},\ i \in N_g \tag{3}$$

$$VL_i^{min} \le VL_i \le VL_i^{max},\ i \in NL \tag{4}$$

$$T_i^{min} \le T_i \le T_i^{max}, i \in N_T \tag{5}$$

$$Q_c^{min} \le Q_c \le Q_C^{max}, i \in N_C \tag{6}$$

$$|SL_i| \le S_{L_i}^{max}, i \in N_{TL} \tag{7}$$

$$VG_i^{min} \le VG_i \le VG_i^{max},\ i \in N_g \tag{8}$$

$$F = F_1 + \begin{bmatrix} \sum_{i=1}^{NL} x_v \begin{bmatrix} VL_i \\ -VL_i^{min} \end{bmatrix}^2 + \\ \sum_{i=1}^{NG} r_g \begin{bmatrix} QG_i \\ -QG_i^{min} \end{bmatrix}^2 \end{bmatrix} + r_f F_3 \tag{9}$$

$$VL_i^{min} = \begin{cases} VL_i^{max}, VL_i > VL_i^{max} \\ VL_i^{min}, VL_i < VL_i^{min} \end{cases}$$

$$QG_i^{min} = \begin{cases} QG_i^{max}, QG_i > QG_i^{max} \\ QG_i^{min}, QG_i < QG_i^{min} \end{cases}$$

3 Optimal Operation of Nuclear Power Plant Inspired Algorithm

Balancing the heat generation and controlling the nuclear fission is considered in the design of Optimal operation of nuclear power plant inspired (ONP) algorithm. Electrical power generation in the nuclear power plant is done through the fission process. Heat created through fission process and it employed to boil the water, then sequentially steam will be produced. Steam rotates the turbines and it connected to the generators. Electricity produced by the generators. In the design optimized operation of the nuclear power plant is considered. Balancing the heat generation and controlling the nuclear fission is important factor in the operation of the plant. Population created in the confined examination area is demarcated as,

$$R_{i,j} = lb_j + A(ub_j - lb_j) \tag{10}$$

$$\mathrm{ub_j, lb_j} \rightarrow \text{upper and lower bound}$$

$$\mathrm{Y_i = \{X_k\}}$$

$$\mathrm{W_k > W_i}$$

$$\mathrm{Y_i} \rightarrow \text{operation of the plant}$$

$$\mathrm{X^P_{i,j} = X_{i,j} + Y \cdot \left(X_{i,j} - C_1 \cdot V_{i,j}\right)} \tag{11}$$

$$\mathrm{X^P_{i,j}} \rightarrow \text{formation of heat}$$

$$\mathrm{Y \in [0,1]}$$

$$\mathrm{C_1 \in \{1,2\}}$$

$$\mathrm{V_{i,j}} \rightarrow \text{operator control}$$

$\mathrm{X_i = X^P_i, if. W^p_i \le W_i}$
Otherwise

$$\mathrm{X_i = X_i}$$

In the exploration section design, production of heat through the nuclear fission process is considered. Heat has to be engendered for the conversion of water into steam. This process is considered in the design.

$$\mathrm{X^r_{i,j} = X_{i,j} + Y \cdot \left(U_j - C_2 \cdot X_{i,j}\right)} \tag{12}$$

$$\mathrm{X^r_{i,j}} \rightarrow \text{continuation of working}$$

$$\mathrm{U_j} \rightarrow \text{skills of controlling}$$

$$\mathrm{Y \in [0,1]}$$

$$\mathrm{C_2 \in \{1,2\}}$$

$\mathrm{X_i = X^r_i, if. W^r_i \le W_i}$

Otherwise

$$X_i = X_i$$

In the exploitation section design controlling the nuclear fission process in the power plant has been considered. Heat generation through fission process is vital one, but it is very important that fission process should not move to the state of uncontrolled condition. It is much more important to control the fission process throughout the operation of the plant.

$$X^s_{i,j} = X_{i,j} + Y \cdot (1 - 2 \cdot Y) \cdot P \tag{13}$$

$$X^s_{i,j} \rightarrow \text{action of control the fission}$$

$$Y \in [0,1]$$

$$P = ub_j - lb_j/t$$

$$X_i = X^s_i, \ \mathit{if}\ W^s_i \leq W_i$$

Otherwise

$$X_i = X_i$$

To expand the exploration activity, Optimal operation of nuclear power plant inspired (ONP) algorithm is unified with Pace and spin bowling in the game of cricket inspired optimization algorithm. Procedure is scientifically formulated on the basis of the bowling action of the bowler in the game of cricket. Pace bowlers depend on the rapidity of the ball delivery which bowled towards the batsman. Pace bowler has to synchronize muscle fibers and motor skills are essential in the pace bowling. Spin bowlers will rotate and turn the ball to cuckold the batsman who faces the ball. In many matches, spin bowling is utilized to reduce the run scoring ability of the batsman. Spin bowlers encompass divulging spin on the ball so the ball will undergo deviation subsequent to bouncing. Fingers and wrist are utilized by the spin bowler to turn the ball in horizontal axis. Batsman facing the ball will be in difficult condition since ball course will be multifaceted, varying course at the bounce. Population created as follows,

$$p_{i,j} = min_j + Y\left(max_j - min_j\right) \tag{14}$$

$$max_j, min_j \rightarrow limits$$

$$P_i = \gamma \cdot P_{i-1} \cdot (1 - P_{i-1}) \tag{15}$$

$$0 \leq P_0 \leq 1, i = 1,2,3,..,N, \beta = 4$$

In the cricket game primary target of the pace and spin bowler is to take the wicket and secondly to control the scoring runs by the batsman. Pace bowlers are contingent on the speediness of the ball delivery which bowled in the direction of the batsman. Pace of the ball augmented or dwindled depending on the playing aspects of the particular batsman who facing the bowler at particular instant. Pace bowler explores the skill of the batsman and tries to bowl rendering to the weakness of the batsman. Ball by ball bowling tactics are changed by the pace bowler. Some pace bowlers will alter the bowling style with reference to right and left hand batsman. This point of exploration by the pace bowlers is scientifically defined as,

$$P_i^{t+1} = \delta \cdot \left(1 - \frac{t}{max_{iter}}\right) \cdot (P_i^t - P_b) \cdot L(D) +$$

$$P_b \cdot \left(1 - \frac{t}{max_{iter}}\right) + (P_m^t - P_b) \cdot f\left(\frac{Q}{f_j}\right) \cdot f_j \tag{16}$$

$$P_i^{t+1} \rightarrow altered\ pace\ bowling\ strategy$$

$$P_i^t \rightarrow type\ of\ present\ bowling$$

$$P_b \rightarrow best\ position\ of\ the\ pace\ bowler$$

$$P_m^t \rightarrow Mean\ position\ of\ the\ pace\ bowler$$

$$L(D) \rightarrow Levy$$

$$Q \in [0,1]$$

$$P_m^t = \frac{1}{N}\sum_{k=1}^{N} P_k^t \tag{17}$$

Acclimatized variable “δ" is demarcated as,

$$\delta = 0.2\left(1 - \frac{t}{max_{iter}}\right) \cdot (P_m^t - P_b) \tag{18}$$

$$\acute{L}\text{évy} = \mathrm{P} * \frac{\omega \cdot \sigma}{|Z|^{\frac{1}{\lambda}}}$$

Exploring skills of the pace bowlers is defined as,

$$W_{i,j}^{t+1} = \begin{cases} E_{i,j}^{t} \cdot exp\left(\frac{i^2}{|2\beta-1|\cdot T_{max}}\right), Q_1 < EN \\ E_{i,j}^{t} + exp\left(-\frac{2i}{T_{max}}\right) \cdot O \cdot A, Q_1 \geq EN \end{cases} \tag{19}$$

$$B_{i,j}^{t+1} \rightarrow new\ location\ in\ the\ search$$

$$\beta \in (0,1], O \in [-1,1], Q_1 \in (0,1]$$

$$EN \rightarrow exploring\ region, EN \in [0.6, 0.9]$$

Spin bowlers comprehend divulging spin on the ball consequently the ball will experience deviance following to bouncing. Fingers and wrist are employed by the spin bowler to rotate the ball in horizontal axis. Batsman facing the spin ball will be in challenging situation, ever since ball progression is in complicated, variable passage after the bounce.

$$S_i^{t+1} = \delta \cdot \left(1 - \frac{t}{max_{iter}}\right) \cdot \left(S_i^t - S_b\right) \cdot L(D) +$$
$$S_b \cdot \left(1 - \frac{t}{max_{iter}}\right) + \left(S_m^t - S_b\right) \cdot f\left(\frac{Q}{f_j}\right) \cdot f_j \tag{20}$$

$$S_i^{t+1} \rightarrow altered\ spin\ bowling\ strategy$$

$$S_i^t \rightarrow type\ of\ present\ spin\ bowling$$

$$S_b \rightarrow best\ position\ of\ the\ spin\ bowler$$

$$S_m^t \rightarrow Mean\ position\ of\ the\ spin\ bowler$$

$$S_i^{t+1} = \begin{cases} \omega \cdot S_i^t \cdot exp\left(\frac{-t}{R \cdot max_{iter}}\right), Q < W \\ \omega \cdot S_i^t + Q \cdot M(1, d), Q > W \end{cases} \tag{21}$$

$$Q \in [0,1]$$

$$W \rightarrow spin\ bowling\ assessment$$

$$W = 0.5$$

$$D \rightarrow dimension$$

$$\omega = 0.2cos\left(\frac{\pi}{2}\left(1 - \frac{-t}{R \cdot max_{iter}}\right)\right) \tag{22}$$

Many times pace and spin bowler's delivery of ball can be easily played by the batsman. This action is defined as,

$$Y_i^{t+1} = Q \cdot exp\left(\frac{Y_e^{t2} - y_i^{t2}}{i^2}\right) \tag{23}$$

$$Y_e^{t2} \rightarrow comfy\ playing\ aspect\ of\ batsman$$

During course of the cricket game, head and bowling coach will identify the effectiveness of the pace and spin bowlers. Positive and weakness of the bowlers are identified. Rectification action will be engendered by the coach,

$$R_{i,j}^{t+1} = \begin{cases} R_b^{t+1} + B \cdot \left(R_{i,j}^t - R_b^{t+1}\right), i < H/2 \\ R_{i,j}^t + B \cdot \left(R_{i,j}^t - R_{i,j,k}^{t+1}\right), i \geq H/2 \end{cases} \tag{24}$$

$$R_b^{t+1} \rightarrow optimized\ actions\ of\ best\ bowler$$

$$B = exp\left(\frac{t}{t_{max}}\right) \cdot Q \cdot M$$

Special coaching camp will be organized by the coach to augment the bowling capabilities of pace and spin blowers. Detailed information will be passed to the management about the performance of pace and spin bowlers. With that report corrective action took by the management of the cricket team appropriately.

$$D_{i,j}^t = \frac{1}{3} \cdot \left(D_b^{t+1} + D_b^{t+1} + D_{i,j}^t\right) \tag{25}$$

$$D_b^{t+1} \rightarrow bowlers\ ability$$

$$D_b^{t+1} \rightarrow success\ rate$$

$$G_{i,j}^{t+1} = \delta \cdot G_{i,j}^t + \gamma \cdot G_{i,j}^t \tag{26}$$

$$G_{i,j}^{t+1} \rightarrow Coach\,evaluation\,on\,the\,bowlers$$

$$\delta = \begin{cases} exp(-t/T_{max}),\, t < T_{max}/2 \\ exp(-(T_{max} - t)/T_{max}),\, t \geq T_{max}/2 \end{cases}$$

$$\gamma \in (0{,}1]$$

Local search is boosted by adding the *Spheniscus humboldti* search optimization. *Spheniscus humboldti* are conjoint beings which doohickey bountiful accounts for predominant pursuit and riddling in varied temperature conditions. During the movement *Spheniscus humboldti* throng together.

$$\mathfrak{A} = \partial\sigma \tag{27}$$

$$\sigma \rightarrow aerial\,ravine$$

$$A = \sigma + i\vartheta \tag{28}$$

Spheniscus humboldti gather together while movement and adventures in the clump in high temperature.

$$S = 0,\, if\,\alpha > 0.5$$

$$S = 1,\, if\,\alpha < 0.5$$

$$S\prime = (S - T/t - T) \tag{29}$$

$$t, T \rightarrow current\,and\,maximum\,iterations$$

Position apprising is delineated as,

$$\overrightarrow{Y}_d = C\left(O\left(\overrightarrow{E}\right) \cdot \overrightarrow{X}(k) - \overrightarrow{E} \cdot \overrightarrow{X}_d(k)\right) \tag{30}$$

$$\overrightarrow{V} = (Q \times (S\prime + M_N(F)) \times r) - S\prime \tag{31}$$

$$M_N(F) = H\left(\overrightarrow{X} - \overrightarrow{X}_d\right)$$

$$r \in [0{,}1]$$

$$o\left(\vec{E}\right) = \left(\sqrt{v.e^{-k/j} - e^{-j}}\right) \tag{32}$$

$$\vec{X}_d(k+1) = \vec{X}(k) - \vec{W} \cdot \vec{Y}_d \tag{33}$$

Optimal Operation of Nuclear Power Plant Inspired (ONP) Algorithm

1. Begin
2. Create the population
3. $R_{i,j} = lb_j + A\left(ub_j - lb_j\right)$
4. $Y_i = \{X_k\}$
5. $W_k > W_i$
6. Define the exploration procedure
7. $X^P_{i,j} = X_{i,j} + Y \cdot \left(X_{i,j} - C_1 \cdot V_{i,j}\right)$
8. $X_i = X^P_i$
9. , if $W^p_i \leq W_i$
9. Otherwise
10. $X_i = X_i$
11. Define the exploitation process
12. $X^r_{i,j} = X_{i,j} + Y \cdot \left(U_j - C_2 \cdot X_{i,j}\right)$
13. $X_i = X^r_i$, if $W^r_i \leq W_i$
14. Otherwise
15. $X_i = X_i$
16. Define the control factor
17. $X^s_{i,j} = X_{i,j} + Y \cdot (1 - 2 \cdot Y) \cdot P$
18. $P = ub_j - lb_j/t$
19. $X_i = X^s_i$, if $W^s_i \leq W_I$
20. $\mathfrak{A} = \partial\sigma$
21. $A = \sigma + i\vartheta$
22. $S = 0,\ \text{if}\alpha > 0.5$
23. $S = 1,\ \text{if}\,\alpha < 0.5$
24. $S' = (S - T/t - T)$
25. Update the position modernization
26. $\vec{Y}_d = C\left(O\left(\vec{E}\right) \cdot \vec{X}(k) - \vec{E} \cdot \vec{X}_d(k)\right)$
27. $\vec{V} = \left(Q \times \left(S' + M_N(F)\right) \times r\right) - S'$
28. $M_N(F) = H\left(\vec{X} - \vec{X}_d\right)$
29. $O\left(\vec{E}\right) = \left(\sqrt{v.e^{-k/j} - e^{-j}}\right)$
30. $\vec{X}_d(k+1) = \vec{X}(k) - \vec{W} \cdot \vec{Y}_d$
31. Otherwise
32. $X_i = X_i$
33. $P_i = \gamma \cdot P_{i-1} \cdot (1 - P_{i-1})$
34. Define the pace bowler actions

35. $P_i^{t+1} = \delta \cdot \left(1 - \frac{t}{max_{iter}}\right) \cdot \left(\begin{matrix} P_i^t \\ -P_b \end{matrix}\right) \cdot L(D) + P_b \cdot \left(1 - \frac{t}{max_{iter}}\right) + \left(\begin{matrix} P_m^t \\ -P_b \end{matrix}\right) \cdot f\left(\frac{Q}{f_j}\right) \cdot f_j$
36. $Lévy = \mathrm{P} * \frac{\omega \cdot \sigma}{|Z|^{\frac{1}{\lambda}}}$
37. Explore the skills of the pace bowlers
38. $W_{i,j}^{t+1} = \begin{cases} E_{i,j}^t \cdot exp\left(\frac{i^2}{|2\beta - 1| \cdot T_{max}}\right) \\ , Q_1 < EN \\ E_{i,j}^t + exp\left(-\frac{2i}{T_{max}}\right) \cdot O \cdot A \\ , Q_1 \geq EN \end{cases}$
39. Define the spin bowlers action
40. $S_i^{t+1} = \delta \cdot \left(1 - \frac{t}{max_{iter}}\right) \cdot \left(\begin{matrix} S_i^t \\ -S_b \end{matrix}\right) \cdot L(D) + S_b \cdot \left(1 - \frac{t}{max_{iter}}\right) + \left(\begin{matrix} S_m^t \\ -S_b \end{matrix}\right) \cdot f\left(\frac{Q}{f_j}\right) \cdot f_j$
41. $S_i^{t+1} = \begin{cases} \omega \cdot S_i^t \cdot exp\left(\frac{-t}{R \cdot max_{iter}}\right), Q < W \\ \omega \cdot S_i^t + Q \cdot M(1, d), Q > W \end{cases}$
42. Define the batsman approach
43. $Y_i^{t+1} = Q \cdot exp\left(\frac{Y_e^{t2} - y_i^{t2}}{i^2}\right)$
44. Assess the quality of the bowlers
45. $R_{i,j}^{t+1} = \begin{cases} R_b^{t+1} + B \cdot \left(R_{i,j}^t - R_b^{t+1}\right), \\ i < H/2 \\ R_{i,j}^t + B \cdot \left(R_{i,j}^t - R_{i,j,k}^{t+1}\right), \\ i \geq H/2 \end{cases}$
46. Outline the coach analysis
47. $D_{i,j}^t = \frac{1}{3} \cdot \left(D_b^{t+1} + D_b^{t+1} + D_{i,j}^t\right)$
48. $G_{i,j}^{t+1} = \delta \cdot G_{i,j}^t + \gamma \cdot G_{i,j}^t$
49. $\mathrm{t} = \mathrm{t} + 1$
50. End

4 Results and Discussion

Optimal operation of nuclear power plant inspired (ONP) algorithm is substantiated in 7 benchmark functions [6]. Results are displayed in Table 1.

$$f_1 = \sum_{i=1}^{D} x_i^2$$

$$f_2 = -20\, exp\left\{-0.2\sqrt{\frac{1}{D}\sum_{i=1}^{D} x_i^2}\right\}$$

$$-exp\left\{\frac{1}{D}\sum_{i=1}^{D} cos(2\pi x_i)\right\} + 20 + e$$

$$f_3 = \sum_{i=1}^{D-1}\left\{\begin{matrix} 100\begin{pmatrix} x_{i+1} \\ -x_i^2 \end{pmatrix}^2 \\ +(x_i - 1)^2 \end{matrix}\right\}$$

$$f_4 = \frac{1}{4000}\sum_{i=1}^{D} x_i^2 - \prod_{i=1}^{D} cos\left(\frac{x_i}{\sqrt{i}}\right) + 1,$$

$$f_5 = -\sum_{i=1}^{4} c_i exp\left[-\sum_{j=1}^{3}\frac{a_{ij}}{(x_j - p_{ij})^2}\right]$$

$$f_6 = 4x_1^2 - 2.1x_1^4 + \frac{1}{3}x_1^6 + x_1x_2 - 4x_2^2 + 4x_2^2$$

$$f_7 = \left\{1 + (x_1 + x_2 + 1)^2\begin{pmatrix} 19 - 14x_1 + x_1^2 \\ -14x_2 + 6x_1x_2 + 3x_2^2 \end{pmatrix}\right\} \times \left\{\begin{matrix} 30 + (2x_1 - 3x_2)^2 \\ (18 - 32x_1 + 12x_1^2 + 48x_2 - 36x_1x_2 + 27x_2^2) \end{matrix}\right\}$$

Optimal operation of nuclear power plant inspired (ONP) is corroborated in IEEE 118 bus system. Table 2, Fig. 1 displays examination analysis. AIL (MW)—Active power loss,VGT (PU)—Voltage eccentricity.

Table 1 ONP examination in 7 benchmark functions

F	HN [6]	EN [6]	QTGO
F1	0.8874	0.0026	0.0070
F2	0.0918	0.0125	0.0152
F3	22.9814	16.5622	16.476
F4	1.1092 e−2	3.459 e−4	3.73 e−4
F5	−3.8629	−3.8629	−3.8629
F6	−1.0318	−1.0316	−1.0310
F7	3.0280	3.0130	3.0133

Table 2 Examination of results—IEEE 118 bus system

Technique	AIL(MW)	VGT(PU)
GAO [2]	128.137	1.12945
ETO [1]	123.511	1.21790
ONP	120.595	1.10892

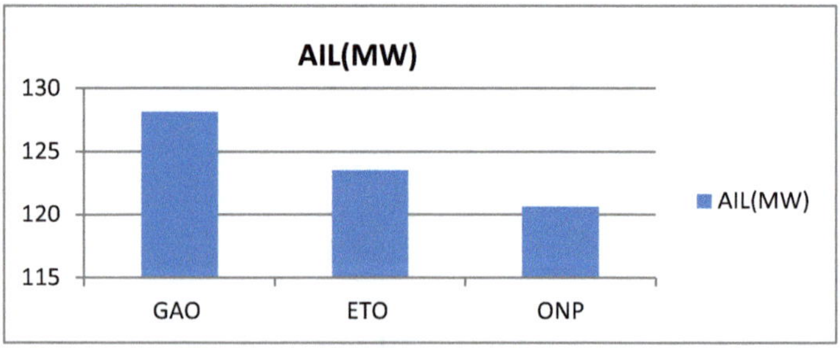

Fig. 1 AIL (MW) examination—IEEE 118 bus system

Table 3 Analysis of results -IEEE 300 bus system

Method	AIL(MW)	VGT(PU)
IBE [4]	410.31	1.1276
MN [3]	390.60	1.1079
ONP	384.13	1.0032

Optimal operation of nuclear power plant inspired (ONP) is corroborated in IEEE 300 bus system. Table 3 and Fig. 2 displays the examination analysis.

Computation time of Optimal operation of nuclear power plant inspired (ONP) algorithm is presented Table 4.

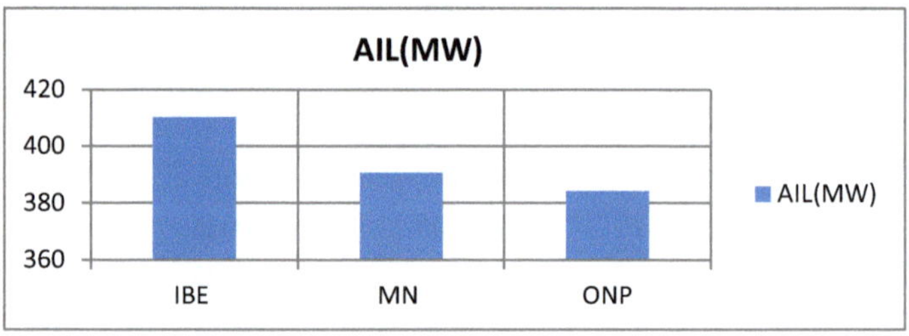

Fig. 2 AIL (MW) examination—IEEE 300 bus system

Table 4 Computation time of ONP

Technique	118 BUS T(S)	300 BUS T(S)
ONP	30.29	50.18

5 Conclusion

Optimal operation of nuclear power plant inspired (ONP) algorithm competently solved the problem. In the exploration section design, production of heat through the nuclear fission process is considered. Heat has to be engendered for the conversion of water into steam. This process is considered in the design. In the exploitation section design controlling the nuclear fission process in the power plant has been considered. Heat generation through fission process is vital one, but it is very important that fission process should not move to the state of uncontrolled condition. It is much more important to control the fission process throughout the operation of the plant. The movement of the Bedouin won't possess any permanent destination. Pace bowlers are contingent on the speediness of the ball delivery which bowled in the direction of the batsman. Pace bowler explores the skill of the batsman and tries to bowl rendering to the weakness of the batsman. Batsman facing the spin ball will be in challenging situation, ever since ball progression is in complicated, variable passage after the bounce. Special coaching camp will be organized by the coach to augment the bowling capabilities of pace and spin blowers. *Spheniscus humboldti* are conjoint beings which doohickey bountiful accounts for predominant pursuit and riddling in varied temperature conditions. During the movement *Spheniscus humboldti* throng together. Optimal operation of nuclear power plant inspired (ONP) algorithm is corroborated in 7 benchmark functions and IEEE 118, 300 bus systems.

References

1. Shaheen, M. A. M., Ullah, Z., Hasanien, H. M., Tostado-Véliz, M., Ji, H., Qais, M. H., Alghuwainem, S., & Jurado, F. (2023). Enhanced transient search optimization algorithm-based optimal reactive power dispatch including electric vehicles. *Energy (Oxford, England), 277*(1), 1–20.
2. Varan, M., Erduman, A., & Menevşeoğlu, F. (2023). A Grey wolf optimization algorithm-based optimal reactive power dispatch with wind-integrated power systems. *Energies, 16*(13), 1–10.
3. Lima, G. B., Neto, A., Alves, D. A., Minussi, C. R., Da Silva Amorim, E., & Silva, L. C. P. (2023). Technique for reactive loss reduction and loading margin enhancement using the curves of losses versus voltage magnitude. *Energies, 16*(16), 1–15.
4. Roy, R., Das, T., & Mandal, K. K. (2021). Optimal reactive power dispatch using a novel optimization algorithm. *Journal of Electrical Systems and Information Technology, 8*(1), 1–15.
5. Kanagasabai, L. (2022). Novel empress SARANI optimization algorithm for active power loss reduction and voltage stability enhancement. *Heliyon, 10*(22), 1–64.
6. Yapıcı, H., & Çetinkaya, N. (2017). An improved particle swarm optimization algorithm using eagle strategy for power loss minimization. *Mathematical Problems in Engineering, 2017*(1), 1–11. https://doi.org/10.1155/2017/1063045
7. Mullin, J. R., & Kotval, Z. (2024). The next generation of nuclear power plants and the role of the local planner. *Planing Practice Research, 39*(3), 547–557.
8. Susiati, H., Sunardi, N. A., Taufik, M., & Kholiq, M. A. (2024). A review on environmental management for nuclear power plant sites. *International Journal of Nuclear Energy Science and Technology, 17*(1), 49–65.

9. Zhang, Q., Wu, H., Ma, C., Wang, Y., & Zheng, X. (2024). Environmental risk assessment and management of nuclear power plants based on big data analysis. *Intelligent Decision Technologies, 18*(2), 1259–1269.
10. Felton, P. J., McCaig, S., & King, M. A. (2024). Cricket fast bowling: The relationship between range of motion and key performance and injury technique characteristics. *Journal of Sports Sciences, 41*(2), 112–120.
11. Fuss, F. K., Doljin, B., & Ferdinands, R. E. D. (2024). Identification of spin bowling deliveries with an advanced smart cricket ball. *Sensors (Basel, Switzerland), 24*(22), 1–20.
12. Lindsay, C., Crowther, R., Clark, B., Middleton, K., Keegan, R., & Spratford, W. (2024). Bowler and coach experiential knowledge of new ball swing bowling in elite cricket. *Journal of Sports Sciences, 42*(2), 146–159.
13. Muñoz, G., Ulloa, M., Alegría, R., Quezada, B., Bennett, B., Enciso, N., Atavales, J., Johow, M., Aguayo, C., Araya, H., & Neira, V. (2024). Stranding and mass mortality in humboldt penguins (*Spheniscus humboldti*), associated to HPAIV H5N1 outbreak in Chile. *Preventive Veterinary Medicine, 227*(1), 1–12.

Role of Electric Vehicles with Its Challenges

Nakul Soni and Akhil Nigam

Abstract Since the utilization of fossil fuels has increased the problem of carbon di oxide gas emission and has arisen alternative of transportation. The cost of such fuels is increasing day by day hence Electric Vehicle technology has been introduced. The problem of gas emission and transportation has been resolved but apart from this there are many challenges that have inbuilt in such technology. The charging problem of EV technology is one of the most important issues. To analyze and optimize proper structure of EV enables to ensure proper charging as well as its maintenance cost. During charging, the capacity of battery may be affected due to insecure heat management. Hence reducing gas emission and improving fuel economy, EV technology can be developed under suitable framework by using smart intelligent techniques. This book chapter covers role of electric vehicles with its challenges for accomplishment of its charging capacity by using emerging trends.

Keywords Electric vehicle technology · Battery · Artificial intelligence · Neural network

1 Introduction

The perceptions towards energy conservation & protecting environment have been growing over the years globally. Earth's ecosystems should be balanced for ecological balance, with four components contributing to environmental aspects like social, human, environmental and economical [1, 2]. Specifically, modes of transportation contribute to the increases in carbon footprint worldwide. Environmental suitability is at stake because human induced climate change contributes to air pollution, toxic acid rain, coastal and ocean acidification, and melting glaciers and polar ice. Human activities that produce greenhouse emissions have impacted the environment in a many ways. In the automotive industry, internal combustion engines (ICEs) have

N. Soni · A. Nigam (✉)
Chandigarh University, Mohali, Punjab, India
e-mail: akhilnigam03@gmail.com

M. A. Mellal et al. (eds.), *Computational Intelligence for a Greener Future: Innovations in Renewable Energy Systems*, Studies in Computational Intelligence 1240,
https://doi.org/10.1007/978-3-032-06732-6_15

been powering vehicles for centuries, which rely on petroleum and produce harmful emissions. Meanwhile, Electric Vehicles (EVs) have arisen, and they possess vast advantages as compared to vehicles powered by ICEs. EVs offers instant acceleration, advanced energy efficiency, and potential to overcome greenhouse gas emissions & air pollution [3–5]. Multiple types of electric vehicles have been designed according to different configuration and applications that they are available in hybrid vehicle, fuel cell and plug-in type vehicles. Fuel cell based electric vehicles are also driven by electric motors but initially use fossil fuel for starting purpose. As compared to conventional energy systems, electric vehicles provide offer enhanced energy efficiency due to property of regenerative braking system. During this operation, kinetic energy is then transformed into required electric energy which is further stored in EV batteries. Unlike continuously burning of conventional fuels, EV consumes no any energy when it is not working [6]. EVs do not require a conventional gearbox system as well as electric motors have capability for providing efficient driving torque across wide range of speed. This has reduced the need of transmission system which tend to reduce vehicle weight and energy loss further improves EV's efficiency. Electric vehicles also have potential for developing environmental impacts across residential and commercial areas which may be in negative minds. This is specially for oil-based industries such as in middle east countries where government-led adaption of EVs which may disrupt economic balance [7]. One of the biggest hurdles is to widespread adaption of EV system. In addition to this, an overall awareness is necessary among people for accepting this emerging trend into transportation system. The acceptance of battery driven vehicles is majorly concern among users. Therefore, continuous research work through practical implications have developed a new revolution into transportation market.

The use of lithium-ion batteries into EV system may pose some environmental concerns. Although technical advancements have led to larger energy density and reduction of cost in recent times. Traditional materials are inadequate for completing electric vehicle requirements which have demanded higher capacity for supporting longer distance range. Instead, enhanced technologies and rare costly materials like cobalt is important which may further contribute to higher cost (Fig. 1).

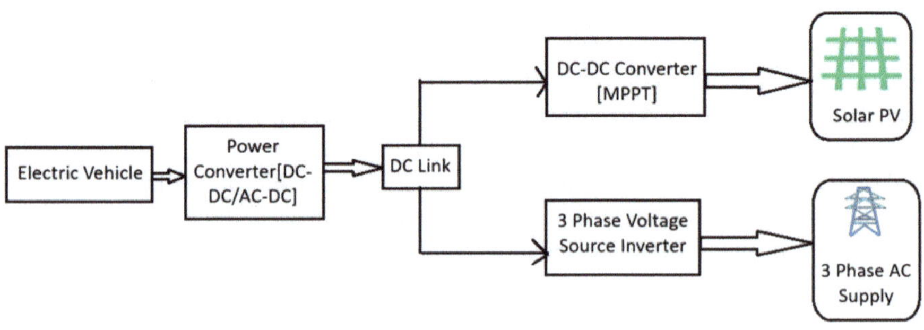

Fig. 1 Representation of EV Block diagram

1.1 Economic Challenges

The EV itself comes with a lot of economic challenges. While developing it causes EV face various issues competing with its cost as compared to conventional vehicles which operate on ICEs. To promote EVs, it requires help from many parties, especially the government [8]. Even though gasoline and diesel are more expensive than electricity, the problem of electric vehicle systems has caused the price of EVs to be higher due to high initial investment [9]. Since the systems of EV is not matured enough and most of the configuration has been done manually in order to reduce overall cost of electric vehicles [10]. Hence such an advancement in emerging techniques for configuration of electric vehicles have been changed into transportation market.

1.2 Role of Battery

The battery cost is always a major concern over using EV systems. In some recent years, prices have lowered while energy density of EV has enhanced. An extensive utilization of expensive materials and advancement of manufacturing framework have continued the cost of EV systems. Latest technologies are still unable to develop high performance batteries as compared to conventional batteries [11]. Li-ion based batteries giving excitation to EVs, also have economic challenges through themselves. However, a large level production of EV system may also create a lot of challenges. Conventional materials for making batteries are leading towards inadequacy for EV systems. Then after accepting advanced technologies in EV systems have also raise concerns over using of cobalt material for achieving larger driving range. Therefore due to such enhancement in EV systems it is introducing high cost of EV purchasing [12]. Safety is one of priority concern among humans and environment when industries use lithium-ion batteries. This type of material is majorly used in electronic devices like mobiles, laptops etc. in which it may lead to overheating or explode condition. In many cases we have found hazardous accidents among human while using such electronic devices for longer times. Electric vehicle batteries mostly rely on efficient material like lithium-ion under normal working conditions. In this world, Chile is currently one of the largest investing countries into servicing of lithium-ion production followed by China and Argentina. In addition to lithium-ion material, EV system may also require other material like copper, aluminum, nickel, cobalt and sometimes occasionally non-metallic materials. Out of these, cobalt is most collectively material among miners and women in areas like Republic of Congo. However, cobalt is one of most toxic elements which may rise health and environmental issues [13].

1.3 Integration with Artificial Intelligence

In this 21st Century, energy conservation and protection environmental aspects have become major concerns with advancement in sustainable development. Such concerns have enhanced EV technology with integration of AI technique based systems with employment of traffic safety. AI technology in electric vehicles (EVs) makes them more attractive by adding ADAS and autonomous driving capabilities. The main advantages are better charging, energy management, battery management system, estimative maintenance, and smart charging framework. Researchers are now more interested in autonomous vehicles, using unmanned test platforms with different components like driving motors & sensors to monitor autonomous EV system. AI has important environmental implementations, specifically in the case of energy and carbon footprint during training large models. This process also involves large water resources and presents cybersecurity threats, such as data breaches, Lags, traffic. These have been alarming cases of electric vehicles (EVs) losing control and causing deaths. Most AI algorithms, deep learning (DL), neural networks (NNs), and reinforcement learning (RL), are very intensive. While cloud-based AI can make use of heavy GPUs and distributed computing, the on-board vehicle processors do not have enough resources to get result in power consumption costs and less energy efficiency during real-time AI processing.

2 How to Overcome These Challenges?

It is widely identified that electric vehicles provide personal and societal benefits as compared to combustion engine systems. However, in recent studies, it has identified there are several challenges which may occur in utilization of electric vehicles. The majorly challenges are high cost, limitability of driven range, inadequate EV infrastructure and perceptions of users. Although electric vehicle driving ranges are gradually enhancing especially for long distances.

The replacing EVs in place of combustion engines may significantly improve energy efficiency and quality of environmental aspects. These EVs may considered climate friendly because they drive on batteries and do not produce unwanted gas emissions. Previously combustion engines introduce carbon emissions during burning petroleum which release toxic pollutants into air which suffer both environment and humans. Hence it is important to know that greenhouse gas may still emit during power generation process used to charge electric vehicles [12]. An emission of greenhouse gas into the atmosphere may cause atmospheric or climatic issues. In an EV, the batteries require proper charging and once the batteries may deploy so these require again charging at specified charging station. These stations intake electricity for charging EVs through area-specific grids, where it may often huge CO_2 emissions into environment then it confronts to pollution into the environment.

Fast DC charging provides suitable response for reducing charging times and it is able to extend driving range of EVs. So numerous driving ranges result into energy consumption and it may provide extensive infrastructure planning for EV battery management system. The battery is one the main limitations in electric vehicles (EVs).

Presently battery configurations for EV systems undergo lower energy density which can limit driving ranges. To enhance electric vehicle efficiency, numerous battery technologies have been configured over time. Hence such advancement in these configurations may lead towards higher capacity of EV battery as compared to combustion engine system.

The current electric grid is under continuous observation to assess its capacity to achieve increasing demand. Enhanced smart charging systems allow customers to regulate automatically charging state during OFF times with improving more efficiency. In addition to this, vehicle to grid (V2G) system enables EV to supply energy back to grid which enhances energy stability. So, investors are encouraged to allocate suitable resources towards exploring renewable energy sources in EV system. The increasing demand for minerals for the battery of electric vehicles has led to increasing mining activities. These pursuits have sparked confrontations amidst local communities & mining companies. Land areas preferred for extraction of raw material often hinder the establishment and development of large and small business-like agriculture, forestry and fisheries etc.

To handle such conflicts, a dynamic and suitable involvement of flexible & inclusive method is required in government system. So different pathways can be considered for enhancement of optimal result. Even if there is a mineral resource discovered, among every 1000 primary resources only one or two of them can host a mine, and among started mining operations some of them can be prematurely closed. The renewable energy sector plays an important role in reduction of carbon gas emissions across whole EV sector chain. However, it requires extensive land area for infrastructure for different energy sources like solar photovoltaic cells, wind turbine, hydro power plant etc. because an energy source is also required for EV specially in remote areas. In addition with this, energy sources employed for EV may produce carbon gas emissions. During operation, carbon materials mix and it can substantially increase emissions specially in the production of battery. Consequently, graphical location for the establishment of EV manufacturing system specially proximity towards non-conventional energy sources may become an important aspect for energy production.

3 Future Trends in Electric Vehicle Technology

The future of transportation is now developing towards into EV market through the understanding behavior of its operation over the past decades. EV markets have enhanced by employing intelligent optimization techniques that users can get a reliable solution in the transportation field. Even in the filed of electricity production,

EVs have developed sustainable approach that can be preferred any time. In future work, EV can be developed in hybrid topologies with renewable energy sources like solar, wind, geothermal, hydro etc. so at different time interval no any obstacle can be introduced during energy production. The rising demand towards EV markets have created a research environment among authors and researchers. By preferring lithium-ion batteries can get more efficiency in charging of EVs. By fixing charging problems in future battery may be portable that users can take anywhere. Hence by providing improvements in durability of battery, energy density, reliability and charging/discharging states have created more usage of EV system [11].

3.1 Artificial Intelligence in Electric Vehicle

As Artificial Intelligence techniques are rapidly employing through out the word in each sector with its providing of optimal solution. So in the filed of EV these techniques have been accepted. Some proposed techniques can provide optimal battery charging, smart infrastructure and best energy efficient model. An approach like Artificial Neural Network (ANN) has been used to enhance suitable temperature management and to maintain EV operation under temperature limit. Regarding this many research have been developed that can create suitable relation between EV parameter and thermal behavior of batteries. An Advanced Driver Assistance Systems (ADAS) achieves EV's efficiency through the optimization of battery usage. This proposed system has represented compelling applications of AI techniques in electric vehicles as such techniques provide optimal solution towards Ev charging. In addition with this, ADAS also helps in reducing EV cost by lowering insurance premiums so that it can easy reachable to the customers. AI-powered driver assistance also diminishes low impact incidents, eliminating investment in repairs and maintenance.

3.2 Public Policies

Many government policies have been developed in the field of electric vehicles. Our government have addressed to EV field for attaining clean environment. Such widespread acceptance of EV has also introduced challenges like adoption such emerging trend and shifting from conventional fuel to intelligent fuel. India has reduced dependency upon oil import from other countries for creating sustainable and cleaner future in EV system. Game theory models address prospect solution to such an issue. Numerous factors can influence significantly towards strategies of subsidy programs like demanding of bargaining power between conventional vehicles and electric vehicles (Table 1).

Table 1 Existing literature survey of electric vehicles

Year	Ref	Findings
2017	[16]	The passenger vehicle markets are directed to achieve saturation level till year 2030 with 12 M active EVs. With this increment the expected number of EVs to reach nearly 1.43 M till year 2040
2017	[17]	Functioning EV by using batteries for producing electricity form national or utility grid results in reducing carbon gas emissions. Basically EVs produce approximately 18% less as compared to conventional engines
2022	[18]	Overview of global scientific contribution towards EV from year 1955 to 2021 has been presented
2021	[19]	An advancement in EV market related to battery technology, charging techniques have been reached towards future scope in the field of transportation
2022	[20]	The primary objectives in the field of EV are to identify by analysis of market of batteries and to monitor technological aspects energy density and charging rate
2017	[21]	The valuable data of EV topologies electric equipment, energy sources for charging, environmental impacts has been discussed
2019	[22]	The complications and possibilities of EVs in mass market implementation have been discussed
2025	[23]	Recent advancements in EVs market with focus on transformative impact of AI techniques on battery management system have been analyzed

4 Conclusion

This chapter has concluded about expansion or growth of EV represents towards best transportation service throughout the world leads to provide reliable transportation for the people. Innovation in charging technology with emerging approaches gives direction to improve battery management system. For achieving net zero gas emissions, environmentally infrastructure needs to built up. In recent years many authors and industries have done extensive research towards charging techniques, reducing emissions and maintenance cost.

The implementation of EV markets faces many challenges like awareness of people, distance range, speed, charging infrastructure & insight barriers. So, some strategies have been adopted like charging speed, best infrastructure, best electrical machines and equipment, emerging battery technology etc. which raise awareness of benefits of accepting EV market.

Through overcoming such challenges, it has become essential to accelerate towards reliable transportation and to mitigate environmental impacts. EV markets still require sustainable research and innovation towards each key factor of EV technology with its manufacturing, charging methods, modernization of electric grid. In addition to this economic concern, overcoming environmental impacts with EV production have been addressed to ensure reliable and sustainable transition of transportation system.

References

1. Faizal, M., Chelvan, R.K., Amirah, A. (2017) Energy, economic & environmental impact of wind power in Malaysia. *International Journal of Advanced Scientific Research and Management, 2*, 81–87
2. Bauer, P. (2019) Future of EV charging. In: *2019 international Aegean conference on electrical machines and power electronics (ACEMP) & 2019 international conference on optimization of electrical and electronic equipment (OPTIM)* (pp. 1–9). IEEE.
3. Marshall, B. M., Kelly, J. C., Lee, T. K., Keoleian, G. A., & Filipi, Z. (2013). Environmental assessment of plug-in hybrid electric vehicles using naturalistic drive cycles & vehicle travel patterns: A Michigan case study. *Energy Policy, 58*, 358–370.
4. Jaunky, V. C. (2011). The CO_2 emissions-income nexus: Evidence from rich countries. *Energy Policy, 39*(3), 1228–1240.
5. Adorsky, P. (2014). The effect of urbanization & industrialization on energy use in emerging economies: Implications for sustainable development. *American Journal of Economics and Sociology, 73*(2), 392–409.
6. Kihm, A., & Trommer, S. (2014). The new car market for electric vehicles & the potential for fuel substitution. *Energy Policy, 73*, 147–157.
7. Peng, T., Ou, X., Yan, X. (2018). Development & application of an electric vehicle's life-cycle energy consumption and greenhouse gas emissions analysis model. *Chemical Engineering Research and Design, 131*, pp. 699–708
8. Timilsina, L., Badr, P. R., Hoang, P. H., Ozkan, G., Papari, B., & Edrington, C. S. (2023). Battery degradation in electric & hybrid electric vehicles: A survey study. *IEEE Access, 11*, 42431–42462.
9. Cao, J., Harrold, D., Fan, Z., Morstyn, T., Healey, D., & Li, K. (2020). 'Deep reinforcement learning-based energy storage arbitrage with accurate lithiumion battery degradation model. *IEEE Transactions on Smart Grid, 11*(5), 4513–4521.
10. Sapinski, B., Snamina, J. J. (2017). Automotive vehicle engine mount based on an MR squeeze-mode damper: modeling and simulation. *Journal of Theoretical and Applied Mechanics* 377–388
11. Hidrue, M. K., Parsons, G. R., Kempton, W., & Gardner, M. P. (2011). Willingness to pay for electric vehicles and their attributes. *Resource and Energy Economics, 33*(3), 686–705.
12. Smith, J., Singh, R., Hinterberger, M., Mochizuki, M. (2018). Battery thermal management system for electric vehicle using heat pipes. *International Journal of Thermal Sciences, 134*(6), 517–529.
13. Guha, A., Patra, A., Vaisakh, K. V. (2017) Remaining useful life estimation of lithium-ion batteries based on the internal resistance growth model. In: *Proc. Indian Control Conf. (ICC)* (pp. 33–38).
14. Outlook, G. E. V. (2018). *Towards cross-modal electrification.* International Energy Agency.
15. Laayouj, N., & Jamouli, H. (2016). Lithium-ion battery degradation assessment & remaining useful life estimation in hybrid electric vehicle. *Renewable Energy and Sustainable Development, 2*(1), 37–44.
16. Azmi, M., & Tokai, A. (2017). Electric vehicle and end-of-life vehicle estimation in Malaysia 2040. *Environment Systems and Decisions, 37*, 451–464.
17. Onn, C. C., Mohd, N. S., Yuen, C. W., Loo, S. C., Koting, S., Rashid, A. F. A., Karim, M. R., & Yusoff, S. (2018). Greenhouse gas emissions associated with electric vehicle charging: The impact of electricity generation mix in a developing country. *Transportation Research Part D: Transport and Environment, 64*, 15–22.
18. Novas, N., Salvador, R. M. G., Portillo, F., Robalo, I., Alcayde, A., Fernández-Ros, M., Gázquez, J. A. (2022) Global perspectives on & research challenges for electric vehicles. *Vehicles, 4*(4), 1246–1276.
19. Sanguesa, J. A., Torres-Sanz, V., Garrido, P., Martinez, F. J., Marquez-Barja, J. M. (2021) A review on electric vehicles: Technologies and challenges. *Smart Cities, 4*(1), 372–404.

20. Roy, H., Nath Roy, B., Kamaruzzaman, Md., Islam, Md. S., Abdel-Khalik, A. S., Hamad, M. S., Ahmed, S. (2022) Global advancements and current challenges of electric vehicle batteries and their prospects: A comprehensive review. *Sustainability, 14*(24), 1–30.
21. Un-Noor, F., Sanjeevikumar, P., Mihet-Pop, L., Mollah, M. N., Hossain, E. (2017) A comprehensive study of key electric vehicle (EV) components, technologies, challenges, impacts, and future direction of development. *Energies, 10*(8), 1–84.
22. Faizal, M., Feng, S. Y., Zureel, M. F., Sinidol, B. E., Wong, D., Jian, G. K. (2019) A review on challenges & opportunities of electric vehicles (EVS). *Journal of Mechanical Engineering Research and Developments, 42*(4), 130–137.
23. Cavus, M., Dissanayake, D., & Bell, M. (2025). Next generation of electric vehicles: AI-driven approaches for predictive maintenance and battery management. *Energies, 18*(5), 1–41.

Storage-Integrated Resilient Frequency Regulation in an EV-Enabled Renewable Microgrid

Subhranshu Sekhar Pati and Umamani Subudhi

Abstract This research presents the design and modelling of an autonomous microgrid with advanced frequency control, integrating renewable energy sources and Electric Vehicles (EVs), including their associated battery storage, as essential system components. Frequency control is critical in microgrid operations to maintain power system stability, prevent cascading failures, and ensure a reliable power supply despite fluctuations in generation and demand of renewable sources. The study focuses on modeling a local voltage-based microgrid that incorporates solar photovoltaic (PV) and wind turbine generator (WTG) systems, a fleet of EVs, and storage solutions to address the intermittency of renewable energy sources. To achieve coordinated control, an enhanced and advanced modified fractional-order controller, termed the 1 + MFOPID controller, is introduced. This controller incorporates a constant value of 1 and is optimized using an advanced optimizer with a newly designed cost function. The superiority of the proposed control strategy is validated in ensuring precise frequency regulation and delivering ancillary services, thereby enhancing the stability and efficiency of microgrid operations. Further, the effectiveness of the proposed model is evaluated under various operational scenarios, including sharp step load variations and dynamic load profiles. Comparative analysis of overshooting and settling time in frequency demonstrates the superior performance of the 1 + MFOPID controller over conventional control mechanisms such as the fractional-order proportional-integral-derivative (FOPID) and classical PID controllers. Furthermore, the impact of battery storage, including EVs, on frequency regulation is examined, and its influence on system performance is thoroughly discussed. The results underline the adaptability and reliability of the suggested management scheme in maintaining system stability while accommodating the dynamic interaction of renewable energy sources and EVs.

S. S. Pati (✉) · U. Subudhi
Department of Electrical Engineering, IIIT Bhubaneswar, Bhubaneswar, Odisha 751003, India
e-mail: c120007@iiit-bh.ac.in

U. Subudhi
e-mail: umamani@iiit-bh.ac.in

M. A. Mellal et al. (eds.), *Computational Intelligence for a Greener Future: Innovations in Renewable Energy Systems*, Studies in Computational Intelligence 1240,
https://doi.org/10.1007/978-3-032-06732-6_16

Keywords Microgrid · Renewable energy · 1 + MFOPID controller · Electric vehicle · Battery

1 Introduction

Renewable energy sources (RESs) present a promising alternative for meeting the growing demand for electrical energy in the modern era [1]. These sources are environmentally sustainable and contribute to a cleaner energy system. However, their inherent intermittency poses significant challenges. The variability and unpredictability associated with RESs introduce technical complexities that can impact the security, stability, and reliability of power grid operations [2].

In recent years, autonomous microgrid(MG) systems have been developed as sustainable energy solutions for electrifying remote areas. These systems integrate renewable energy, such as wind and solar power, for electricity generation. As independent energy sources, MGs offer a feasible and reliable power supply, particularly in isolated regions [3]. However, ensuring the safe, steady, and uninterrupted operation of MGs requires maintaining system frequency at its nominal value. Frequency stability dependence upon the power generation and consumption at a particular time. Furthermore, MGs are more susceptible to frequency variations due to their inherently higher imbalance rates between generation and consumption. The primary factors contributing to these fluctuations include: the uncertainty and intermittency of RESs and the stochastic nature of load demand in remote areas [5, 6].

To address the aforementioned challenges, energy storage systems (ESSs) have emerged as effective solutions for enhancing the stability and reliability of MGs [7, 8]. The gaining popularity of EV in microgrid can now be used as ESS for better frequency stability [9]. Various ESS technologies, including battery-energy-storage-systems (BESS), flywheel-energy-storage-systems (FESS), redox-flow-batteries (RFB), and superconducting-magnetic-energy-storage (SMES) systems, have been explored in the literature for modern power system applications [10]. However, in MGs integrating both renewable energy and ESSs, system parameters and operating conditions exhibit rapid variations. Consequently, the design of a robust frequency controller capable of accommodating a wide range of operating conditions is essential. Further, it ensure the frequency fluctuation and tie-line power variation restrains at a specific limit with balance of power generation and scheduled load demand [11].

MGs with its inter-related units communicates the control signals through communication mediums, leading to time delays between the control center, generation units, and sensing elements. These delays, coupled with the intermittency of RESs, degrade system performance and reliability, potentially causing instability or blackouts. Various control algorithms for load frequency control (LFC) have been proposed to address these challenges. While existing studies examine time-delay effects in microgrids with electric vehicles and isolated or interconnected configurations, limited research focuses on multi-area interconnected microgrids with time delays. This study investigates such a complex system, incorporating practical nonlinearities to enhance frequency control.

Proportional-Integral-Derivative (PID) controllers and their variants are simple to design, real-time applicability, and the availability for optimizing its parameter, thus used in most of the practical scenario including power system application [12]. Recently, modified fractional calculus based controller such as 1 + MFOPID controller has gained increased research attention, demonstrating superior performance across various control applications [13]. Nevertheless, optimal tuning of controller parameters becomes complex in nonlinear, multi-dimensional, or interconnected coupled systems, such as power networks. Consequently, researchers have shifted from purely analytical methods to numerical optimization approaches. FOPID-type controllers are now commonly tuned using stochastic optimization algorithms, including swarm-based, evolutionary-based, physics-based, and human-based methods [14].

After reviewing the literature, global optimization algorithms were found to meet the control system design requirements [6–8, 15, 16]. Thus, this chapter employs the harris hawk-based optimization (HHO) algorithm that has been enhanced for 1 + MFOPID controller tuning. The HHO, inspired by harris hawks, mathematically models that mimics the cooperative hunting strategy of Harris hawks. Hence, this study introduces improvised version of HHO (iHHO) to tune the controller parameters.

The effectiveness of an optimization is inherently dependent on the formulation of the fitness or objective function. A review of the literature indicates that researchers frequently retune controllers across various states or cases to assess their efficacy and working performance [15]. This repeated tuning is necessary because fitness functions are typically defined for specific conditions. To address this limitation, this paper introduces a novel objective function that incorporates information from multiple scenarios. By utilizing the proposed objective function, the controller requires only a single tuning while ensuring improved performance across all scenarios, effectively accounting for the nonlinearities present in practical power systems [16]. The presentation of this chapter is as follows: Sect. 2 provides a comprehensive summery of the microgrid system. Section 3 presents the controller design along with the objective function formulation. Section 4 introduces the proposed improvements in the optimization approach. Section 5 discusses the results, providing a comprehensive analysis and interpretation. Finally, Sect. 6 concludes the paper with key findings and potential future directions.

2 System Study

In this study, an interconnected microgrid is analyzed as a physical system, as illustrated in Fig. 1. The purpose of the study is to demonstrate the microgrid's robustness as a resilient system capable of preventing power system blackouts and maintaining a stabilized frequency deviation profile despite uncertainties in load variations [1,

2]. To achieve this objective, the proposed system integrates two renewable energy sources—photovoltaic (PV) and wind power—along with three energy storage units, specifically flywheel energy storage. Additionally, a distributed controller arrangement is implemented within the control center to manage microgrid data and facilitate communication of state information. A sensory network is incorporated to receive output signals, such as frequency deviations, and subsequently transmit commands to actuators. These actuators adjust generation levels accordingly to minimize the imbalance between load demand and generation through set point modifications, which are regulated by the controller with optimized tuning parameters. The study considers both simple load variations and dynamic load profiles, both of which influence the area control error, potentially leading to significant frequency deviations. Such deviations can push the system into instability. However, appropriate countermeasures are implemented to prevent system failure. Through various control actions, the frequency deviation is effectively minimized, ultimately maintaining stability by reducing deviations to zero or a negligible value. The detailed modeling of the proposed system is summarized below. Wind Turbine Generators (WTGs) are increasingly adopted due to their established technological maturity and growing advantages. The pitch control coordinated wind turbine involves linkage dynamics arising from an additional control system, comprising a transducer, set-point controller, feedback system, and hydraulic pitch actuator. The transfer function of the Hydraulic Pitch Actuator (HPA), characterized by gain K_{P1} and time constant T_{P1}, is provided in Eq. (1) as referenced in [3].

$$\frac{\Delta H(s)}{X(s)} = \frac{K_{p1}(1 + sT_{p1})}{(1 + sT_{p1})(1 + s)} \tag{1}$$

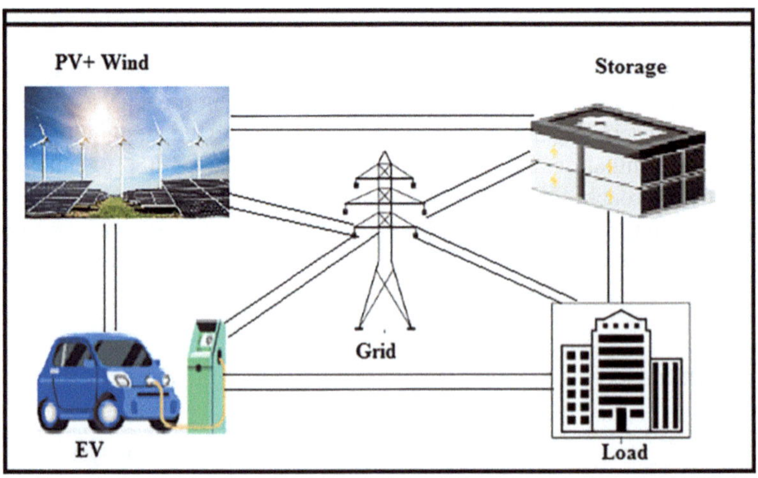

Fig. 1 Generalized microgrid model considered for investigation

The gain determines the actuator's responsiveness to control input changes, while the time constant defines the response duration to a step input, affecting overall system speed. The transfer function of the Data Pitch Response (DPR), given in Eq. (2), defines the relationship between the control input and pitch angle adjustment, considering system dynamics like sensors, controllers, and actuators [4].

$$\frac{\Delta H(s)}{X_1(s)} = \frac{K_{p2}}{(1+s)} \tag{2}$$

Solar photovoltaic energy promotes clean energy by lowering fossil fuel emissions and improving energy independence. The PV plant's transfer function, based on Black's expression with tachometric feedback, is presented in Eq. (3) [5].

$$T/F_{solar\,PV} = \frac{1}{1+\pi(1+k)s+100\pi s^2} \tag{3}$$

A first-order transfer function model of simple EV used in Frequency Control in an autonomous microgrid be represented in the following equation, capturing the dynamic response of EVs participating in frequency regulation [7]. The general form of the transfer function is discussed in which K_{EVi} is the gain representing the EV's participation factor in frequency regulation and T_{EVi} is the time constant representing the response delay of EVs in adjusting power output [8]. In general, the gain quantifies the proportional contribution of EVs to frequency support, often influenced by factors such as the available power capacity, state of charge (SoC) constraints, and control policy design. A higher gain implies greater sensitivity and responsiveness of the EV fleet to frequency deviations [8]. The time constant, models the inherent latency in EVs' response, which arises from communication delays, inverter dynamics, and coordination time among distributed EVs. It determines how quickly the system can approach the desired corrective action, with smaller values indicating faster responses [9].

$$T/F_{\mathrm{EVi}} = \frac{K_{EVi}}{1+sT_{EVi}} \tag{4}$$

Energy storage play a vital role in maintaining power system stability due to their fast response capabilities. This chapter evaluates the performance of various ESDs, including ultracapacitors (UC), SMES, and batteries, each integrated into a test system for analysis. Batteries play a crucial role in frequency control for modern power grids, with lithium-ion, lead-acid, sodium-sulfur, and flow batteries being commonly used. Lithium-ion batteries are preferred for their high energy density, efficiency, and fast response, while lead-acid batteries offer a cost-effective solution for short-term regulation. Sodium-sulfur batteries, with high energy capacity, suit longer-duration applications, whereas flow batteries provide scalability and long cycle life. Each type contributes to grid stability and facilitates renewable energy integration [10]. The battery considered in the system is represented by a first order transfer function as presented in Eq. (5), where $K_{Battery}$ and $T_{Battery}$ is the gain and time constant of the battery, respectively.

$$T.F_{Battery} = \frac{K_{Battery}}{1 + sT_{Battery}} \tag{5}$$

A SMES unit comprises a superconducting coil with negligible resistance, cooled by helium or liquid nitrogen to maintain superconductivity. It includes a step-down transformer and a power conversion unit, which facilitate charging and discharging. SMES stores energy during off-peak periods and releases it during generation loss, offering superior performance over other storage devices. Its transfer function is given in Eq. (6), where K_{SMES} and T_{SMES} refers gain and time constant of the SMES [6]. Ultracapacitors or supercapacitors, store significant power between two plates and are used for peak power demands, often alongside batteries. Despite high costs, they offer easy maintenance, long lifespan, and high power density. The transfer function is provided in Eq. (7) in which K_{UC} is gain constant and T_{UC} is time constant of the UC [10, 11].

$$T.F_{SMES} = \frac{K_{SMES}}{1 + sT_{SMES}} \tag{6}$$

$$T.F_{UC} = \frac{K_{UC}}{1 + sT_{UC}} \tag{7}$$

3 Formulation of Controller and Objective Function

The 1 + MFOPID (One Plus Modified Fractional-Order PID) controller is an advanced control strategy designed to enhance frequency regulation in microgrid systems. Microgrids often experience frequency fluctuations due to variations in load demand, renewable energy intermittency, and communication delays [12].

Traditional controllers, such as PID and FOPID, have limitations in handling these challenges effectively. The 1 + MFOPID controller addresses these issues by introducing a modification function that improves adaptability, robustness, and dynamic response. The formation of the 1 + MFOPID controller is based on the FOPID controller, which generalizes the conventional PID structure by incorporating fractional-order integral and derivative terms [13]. The 1 + MFOPID controller further enhances this framework by introducing a corrective term that dynamically adjusts the control action. This modification enables better compensation for system uncertainties and external disturbances, making it particularly effective in frequency regulation applications.

The arrangement of proposed controller, that is an extension of the FOPID controller, is illustrated in Fig. 2. This arrangement consists of key control parameters, including the fractional-integrator order (λ), integral-gain (K_I), proportional-gain (Kp), derivative-gain (Kd), fractional derivative order (μ), filter gain (n), and an additional optimized constant, represented as "1". In the conventional FOPID controller, feedforward control is applied to all gain parameters [13]. In contrast, the proposed 1 + MFOPID controller adopts a hybrid approach, wherein feedforward control is utilized specifically for the integral gain (K_I) and fractional integrator order (λ), as feedback control is incorporated for the remain parameters. This modification enhances the controller's adaptability, stability, and overall performance in dynamic systems. The transfer function of the discussed controller and the overall system is represented as:

$$\frac{Y(s)}{R(s)} = \frac{K_i G_p(s)}{S^{\lambda} + S^{\lambda} K_p + G_p(s) K_I + K_d S^{\lambda} S^{\mu}} \tag{8}$$

One of the key benefits of the 1 + MFOPID controller in microgrid frequency control is its improved robustness against system variations. In a microgrid, frequency deviations arise due to the integration of distributed energy resources, communication delays, and fluctuating loads [14]. The 1 + MFOPID controller effectively mitigates these disturbances by offering a more flexible and adaptive control response compared to traditional controllers. Additionally, the 1 + MFOPID controller provides better transient performance by reducing overshoot and settling time. This ensures that frequency deviations are quickly corrected, preventing instability in the microgrid. Furthermore, the fractional-order nature of the controller allows for fine-tuned control actions, making it well-suited for microgrid applications where conventional integer-order controllers may struggle to achieve the desired performance. One of the significant advantage of the discussed controller is its capability to handle time delays. Communication delays are common in microgrid operations,

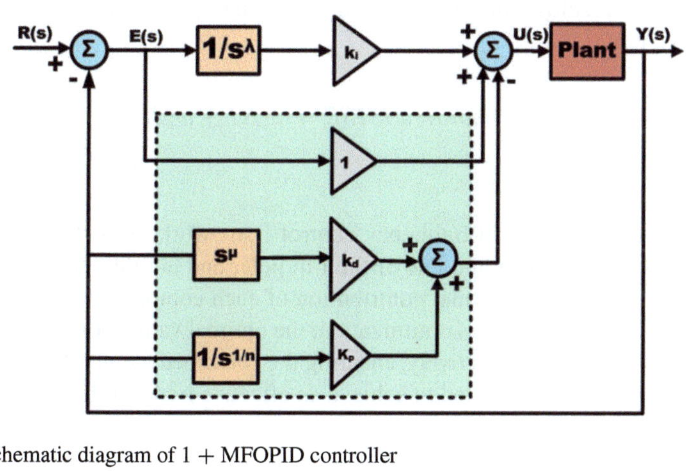

Fig. 2 Schematic diagram of 1 + MFOPID controller

especially in multi-area configurations where information exchange occurs between different regions [14]. The 1 + MFOPID controller effectively compensates for these delays, ensuring reliable and stable frequency regulation even in the presence of network-induced latencies.

In frequency management, cost functions optimize control strategies for stability and performance. In frequency control, selecting the right function impacts response time and control effort. While IAE and ISE are simple, ITAE, ISAE, and ITSE offer better long-term stability [3–5]. Integral of the Squared Time-weighted Squared Error (ISTSE) is preferred for its balance of fast error correction and damping, minimizing overshoot and oscillations. In contrast, ITAE slows responses, and ISAE may cause oscillations [9]. Equation (9) defines ISTSE, incorporating frequency deviation, tie-line power change, and time t, with simulation time $t_{simul.}$ starting from 0.

$$J = ISTSE = \int_{0}^{t_{simul.}} (\Delta F_i + \Delta P_{tie\,ij})^2 \times t^2 \times dt; \;\; i, j = 1, 2, \ldots.., n \, and \; i \neq j \quad (9)$$

Traditional cost functions in microgrid frequency control prioritize time-domain performance, often overlooking frequency-domain aspects and potentially compromising stability. To address this issue, an improved cost function based on ISTSE (iISTSE), as defined in Eq. (10), is projected. This function minimizes the ISTSE while maximizing the sum of the damping ratios (DRs) of the system's poles. Consequently, it ensures stability by preventing the presence of unstable poles in the right half of the s-plane, where damping ratios are −1, thereby improving microgrid performance. Such design inherently avoids unstable poles (with negative damping), ensuring the system remains robust under diverse operating conditions. This is particularly critical in microgrid environments, where frequent disturbances and nonlinearities demand both rapid response and high resilience. Moreover, the incorporation of damping ratios adds a physically interpretable dimension to the optimization process, allowing control designers to balance performance metrics like overshoot and settling time effectively, resulting in improved reliability and operational efficiency of the microgrid system.

$$J_{improved} = improved\ ISTSE = \omega_1 \Delta J + \omega_2 \Delta \frac{1}{\sum_{i=1}^{N} \zeta_i} \quad (10)$$

In the proposed microgrid frequency control framework, N is the total number of system's poles, ξ_i signify the DR of the i-th pole, and ω refers to the weighting parameters that govern the virtual contribution of each component in the improved cost function. Through rigorous optimization, the optimal values for ω_1 and ω_2 were identified as 0.2 and 0.8, respectively, ensuring the combined weight does not exceed unity. These values were selected to achieve an effective trade-off between time and

Table 1 Optimal value obtained based on the defined cost function

Control mechanism	Optimal value obtained based on the defined cost function				
	ISE	ITAE	ISTE	ISTSE	iISTSE
PID	0.0584	0.0086	0.0767	0.0047	0.0042
FOPID	0.0516	0.0084	0.0669	0.0037	0.003
1 + MFOPID	0.0467	0.0071	0.0560	0.0022	0.0014

frequency domain accuracy and stability respectively. Accordingly, the improved cost function is presented as:

Minimize: Objective Function Represented in Eq. (10)

Subject to constraints, min. value < Kp, Ki, Kd, α,μ < max. value.

The proposed cost function is vital for optimizing microgrid frequency control, acting as a benchmark to assess how well a solution satisfies system constraints and performance goals. It guides the optimization process by quantifying deviations from desired outcomes. Optimization technique minimize this function by searching both global and local search spaces, ensuring robust and efficient control strategies. The global search identifies the best possible solution across a wide range set by the constraints maximum and minimum value, while the local search refines it by exploring neighbouring values once the global search identify the approximate optimal range. This dual approach ensures an optimal balance between global exploration and local exploitation, leading to improved system stability and performance. As discussed, Table 1 presents a comparative performance analysis of the improved ISTSE against various cost functions, considering different controllers within the microgrid framework. The results indicate that the improved cost function consistently yields superior performance across all controller types. Notably, when employed in conjunction with the 1 + MFOPID controller arrangement, it delivers the most favorable outcomes, highlighting the synergistic effectiveness of this specific arrangement in enhancing the microgrid system performance.

4 Improved HHO Optimization Algorithm

Proper tuning of controller parameters using advanced optimization techniques is essential for enhancing stability, minimizing frequency fluctuation, and ensuring a rapid and improved reaction to disturbances, especially in microgrids employing various storage with nonlinear interactions [15]. The Beta Chaotic Harris Hawks Optimization (iHHO) combined with Opposition-Based Learning (OBL) has recently emerged as a sophisticated hybrid algorithm, offering significant improvements over the classical HHO algorithm [16]. Motivated by Harris hawks' hunting behavior, HHO alternates between exploration and four exploitation phases. Enhancements like chaotic maps, Beta distribution, and Opposition-Based Learning (OBL) improve convergence speed, prevent local optima entrapment, and refine search efficiency.

These features give the Improved HHO algorithm a superior balance between exploration and exploitation, leading to higher convergence accuracy in complex problems [15, 16].

IHHO integrates chaotic function in its mechanism which is further supported by the OBL concept to avoid local minima, outperforming traditional methods in complex, nonlinear search spaces [15]. So, the dynamic behavior adjustment improves robustness and solution quality. Compared to conventional algorithms like PSO, SCA optimization, IHHO demonstrates faster convergence with reduced computational overhead [16].

In the context of microgrid applications, integrating the Beta-Chaotic HHO-OBL technique for tuning the 1 + MFOPID controller offers substantial advantages. The suggested controller, with its additional fractional calculus-based flexibility, enhances the dynamic performance of microgrid systems by improving robustness, adaptability, and disturbance rejection. Optimizing this controller using Beta-Chaotic HHO-OBL ensures precise parameter tuning, leading to improved voltage and frequency stability, reduced oscillations, and superior disturbance mitigation [16]. The advanced optimization approach prevents local optima entrapment, allowing faster convergence and improved dynamic response of the microgrid. This is particularly beneficial in handling the uncertainties associated with renewable energy integration, ensuring that the controller adapts effectively to fluctuations in generation and load demand.

Additionally, the hybrid optimization strategy enhances the controller's ability to manage nonlinearities, optimizing energy storage utilization and ensuring efficient power distribution. By achieving an optimal balance between exploration and exploitation, the tuned 1 + MFOPID controller contributes to a more resilient, stable, and efficient microgrid operation, ultimately improving power quality, reliability, and economic performance. The steps involving the optimizer described as follows:

Step 1: Initialization the parameter.
Step 2: The best and worst solution is identified from the initialize population's size.
Step 3: Utilizing the update equation, the optimum solution is updated.
Step 4: Check the solution that is better than previous reported solution.
Yes : Ultimate solution is replace by the precedding value.
No : No change.
Step 5: Evaluting termination procedure.
Yes : Declared the optimum value of the controller and stops.
No : Again proceed to step 2 and reiterates.

5 Result and Analysis

This study examines an interconnected microgrid model that consist of conventional and sustainable power sources, along with an energy storage system, to maintain load-demand balance and ensure overall system stability. Given the simultaneous operation of multiple energy sources, efficient control is essential, requiring precise tuning of controller parameters. To achieve optimal performance, an advanced optimization technique is employed for its superior capability in handling complex dynamic interactions. A proposed controller tuned by iHHO optimizer is compared against PSO, SCA and standard HHO optimizer, as illustrated in Fig. 3, to highlight its improved performance in regulating microgrid stability. The controller values are controlled within predefined ranges: K_p, K_I, K_d (0–2) and fractional orders constants (less than 1).

The coefficients for the PID, FOPID and the 1 + MFOPID controllers, derived using the iHHO algorithm, are presented in Table 2. To evaluate the effectiveness of these controllers, simulations were conducted under nominal system parameter conditions, incorporating a time delay of 2.28 s. The system response was examined under a step load perturbation of 2% to assess its dynamic performance. Figure 4 illustrates the frequency deviation responses for three different control strategies: the conventional PID controller, the FOPID controller, and the modified Fractional-Order PID controller, denoted as 1 + MFOPID, which was designed using the improved Harris Hawks Optimization technique. The results demonstrate that all controllers maintain the frequency deviation within acceptable operational limits, ensuring the secure, consistent, and stable performance of the interrelated microgrid system, even under the influence of renewable energy and storages (Table 3).

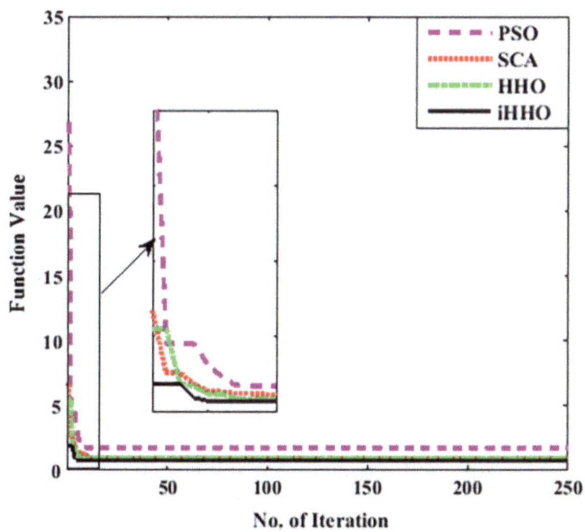

Fig. 3 Effectiveness of the iHHO optimization

Table 2 Optimal controller parameters

Controller	Plant	Optimal controller Parameters				
		K_p	K_I	K_d	λ	μ
1 + MFOPID	Solar PV	1.552	1.750	2.541	0.65	0.18
	WTG	2.547	1.895	1.872	0.78	0.33
	EV	2.655	1.585	1.241	0.87	0.67
FOPID	Solar PV	2.855	1.428	1.366	0.49	0.82
	WTG	1.264	2.028	1.981	0.71	0.28
	EV	1.853	2.822	2.014	0.52	0.34
PID	Solar PV	2.255	1.750	2.255	–	–
	WTG	1.058	2.979	1.270	–	–
	EV	2.364	0.956	1.741	–	–

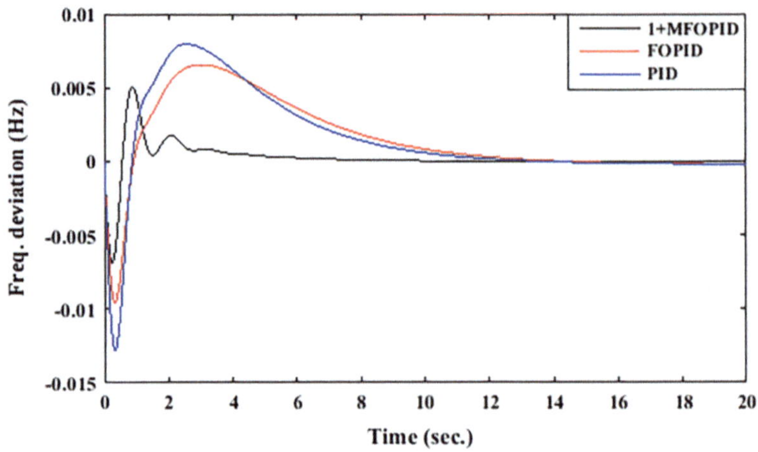

Fig. 4 Graphical representation of frequency deviation

Table 3 Mathematical data for the frequency deviation with step load

Controller	Overshoot (10^{-3})	Settling time (2%)	iISTSE (10^{-3})	ISTSE (10^{-3})	ITAE (10^{-3})	ISAE (10^{-3})
PID	8.19	12.54	9.27	12.65	15.55	14.50
FOPID	6.81	11.84	8.52	11.87	13.61	14.11
1 + MFOPID	4.94	8.61	7.36	10.21	11.80	10.97

A comprehensive analysis of these indices provides deeper insights into the controllers' ability to regulate system dynamics efficiently. Based on the comparative assessment, the 1 + MFOPID controller demonstrates superior performance compared to both the FOPID and PID controllers under nominal operating conditions. The results indicate that the 1 + MFOPID controller outperforms both the PID and FOPID controllers across all performance metrics. Specifically, the overshoot/undershoot is significantly reduced to 4.94×10^{-3} in the 1 + MFOPID controller compared to 6.81×10^{-3} in FOPID and 8.19×10^{-3} in PID, demonstrating better transient response and system stability. Similarly, the settling time for the 1 + MFOPID controller (8.61 s) is the lowest among the three controllers, reflecting its superior capability in achieving faster system stabilization. Furthermore, the cost function based performance indices (iISTSE, ISTSE, ITAE, and ISAE) consistently show lower values for the 1 + MFOPID controller, signifying improved control accuracy and reduced steady-state error. Notably, the ITAE and ISAE values for 1 + MFOPID are 10.21×10^{-3} and 10.97×10^{-3}, respectively, which are lower compared to the corresponding values in the FOPID and PID controllers.

This enhanced performance highlights its effectiveness in improving system stability and response characteristics. Such comparative analysis of the control strategies reveals that the 1 + MFOPID controller exhibits superior performance relative to both the FOPID and the traditional PID controllers. This superiority is attributed to its enhanced tuning flexibility and improved robustness against system uncertainties and intermittent nature of RES. The 1 + MFOPID controller achieves more effective mitigation of frequency deviations, leading to improved system stability and reliability. These findings underscore the effectiveness of advanced optimization based controller designs in enhancing the resilience of time-delayed microgrids.

Furthermore, a continuous load patterns are applied to the system, following dynamic load variation, as illustrated in Fig. 5. The load variation is introduced in steps, fluctuating within ±2% of the rated values at different time intervals, with a periodicity of 200 s. This dynamic load variation is implemented to simulate real-world operating conditions where sudden changes in demand occur within an interconnected microgrid system or showcasing the intermittent behaviour of renewable energy for a short period. The main purpose of this analysis is to examine the impact of these dynamic load fluctuations on system frequency and to evaluate the effectiveness of the proposed control mechanism in mitigating frequency deviations. The results indicate that all controllers effectively mitigate frequency deviations and stabilize the system after each load perturbation as shown in Fig. 6. However, a closer examination reveals that the 1 + MFOPID controller exhibits superior performance compared to the FOPID and PID controllers. The zoomed-in section highlights a significant difference in transient behavior, where the 1 + MFOPID controller achieves lower peak deviations and faster settling times, thereby enhancing system stability. In contrast, the PID and FOPID controllers exhibit higher oscillatory responses with prolonged settling periods.

These findings demonstrate the effectiveness of the proposed 1 + MFOPID control approach in maintaining frequency stability in microgrid systems under dynamically varying loads. By minimizing frequency deviations more efficiently, the 1 + MFOPID controller enhances the resilience and reliability of the microgrid, particularly in the presence of time delays, sudden load change. This underscores the

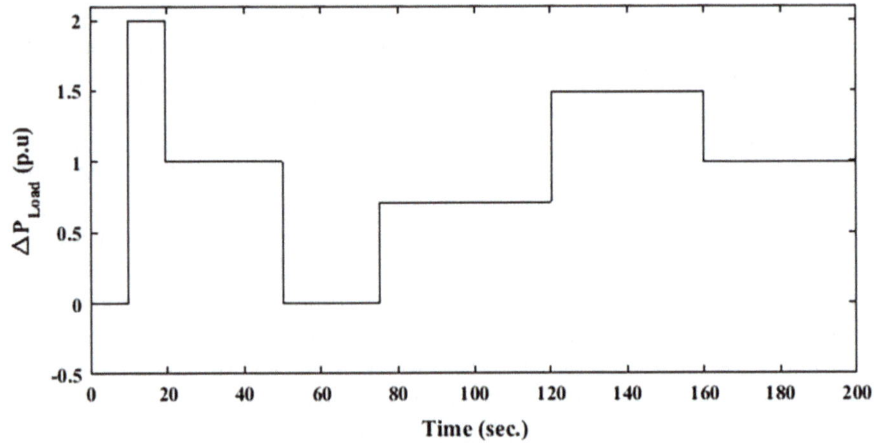

Fig. 5 Dynamic load profile applied to the considered microgrid

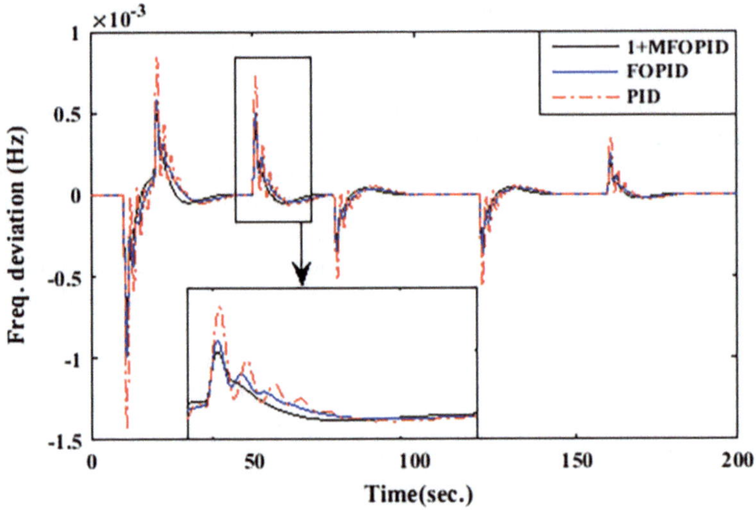

Fig. 6 Graphical representation of frequency deviation under dynamic load profile

necessity of advanced control strategies to ensure stable microgrid operation under real-world dynamic conditions.

The synergetic effect of all storages and battery on the microgrid is further evaluated for the effective frequency management. From the Fig. 7, it is evident that the presence of multiple storage units leads to improved frequency stability compared to the scenario where only the battery storage is utilized. The system with all storage units exhibits lower peak deviations and faster settling time, demonstrating enhanced

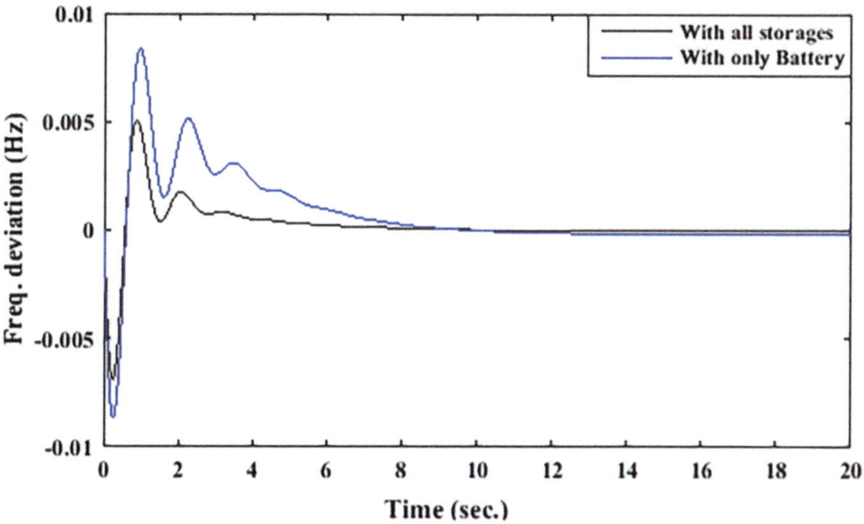

Fig. 7 Graphical representation of effect of storage on frequency change

damping of oscillations. In contrast, the case with only battery storage experiences higher frequency deviations and prolonged oscillatory behaviour before stabilizing.

The results highlight the significance of integrating diverse energy storage systems in microgrid operations for efficient Frequency Control. The superior performance of the system with all storage units including EVs suggests that a hybrid energy storage approach provides better frequency regulation by effectively mitigating disturbances and reducing transient effects. This observation is crucial for ensuring stable and resilient microgrid operation, particularly under dynamic load conditions and external disturbances such as renewable energy fluctuations.

6 Conclusion

This study presents an in-depth discussion of an autonomous microgrid equipped with a hybrid energy system, integrating various renewable and conventional energy sources. A novel 1 + MFOPID controller is proposed, whose parameters are finely tuned using the improved Harris Hawk Optimization in conjunction with an improved Integral Squared Time-Squared Error cost function. The proposed control strategy is evaluated against traditional PID and FOPID controllers. Simulation results demonstrate that the 1 + MFOPID controller significantly outperforms the conventional methods by effectively reducing overshoot, undershoot, and oscillations. This leads to improved damping characteristics and enhanced stability, particularly under step-loading disturbances. In dynamic loading scenarios, the analysis further validates the robustness of the system, ensuring that it remains stable without drifting toward instability. Additionally, the integration of storage systems and EVs in discharging mode exhibits a synergistic effect, where battery systems provide sustained frequency support. The coordinated operation of EVs with other storage devices effectively mitigates transient disturbances, minimizes steady-state deviations, and enhances the overall resilience of the autonomous microgrid. Future research can explore the scalability of the proposed approach in larger microgrid networks, including the distributed generators and other operational challenges like cyber attack.

References

1. Bevrani, H. (2009). *Robust power system frequency control*. Springer.
2. Elgerd, O. I. (2000). *Electric energy systems theory: An introduction*. McGraw-Hill.
3. Hirsch, A., Parag, Y., & Guerrero, J. (2018). Microgrids: A review of technologies, key drivers, and outstanding issues. *Renewable and Sustainable Energy Reviews, 90*, 402–411.
4. Tudu, A. K., Naguru, N., Dey, S. H. N., & Paul, S. (2024). Load frequency control of an isolated microgrid using optimized model predictive control by GA. *Electrical Engineering, 106*(4), 4171–4183.
5. Khalil, A. E., Boghdady, T. A., Alham, M. H., & Ibrahim, D. K. (2024). A novel cascade-loop controller for load frequency control of an isolated microgrid via dandelion optimizer. *Ain Shams Engineering Journal, 15*(3), Article 102526.
6. Pati, S. S., & Subudhi, U. (2024). Stability analysis of a multi-area renewable system and frequency control with improved chaotic Harris Hawk optimization algorithm. *Arabian Journal for Science and Engineering, 49*(5), 6531–6550.
7. Alizadeh, M., Tightiz, L., & Nasab, M. A. (2024). Providing an intelligent frequency control method in a microgrid network in the presence of electric vehicles. *World Electric Vehicle Journal, 15*(7), 276.
8. Debbarma, S., & Dutta, A. (2017). Utilizing electric vehicles for LFC in restructured power systems using fractional order controller. *IEEE Transactions on Smart Grid, 8*(6), 2554–2564.
9. Mohamed, A., Wazeer, E. M., El Masry, S. M., Ghany, A. A., & Mosa, M. A. (2024). A novel scheme of load frequency control for a multi-microgrid power system utilizing electric vehicles and supercapacitors. *Journal of Energy Storage, 89*, Article 111799.
10. Yousef, M. Y., Mosa, M. A., El Masry, S. M., Ghany, A. A., & Ali, A. A. (2024). Decentralized utilization of distributed energy storage resources for simultaneous frequency regulation in a microgrid. *Journal of Energy Storage, 86*, Article 111357.
11. Aftab, M. A., Latif, A., Al-Durra, A., & Hussain, S. S. (2024). An optimized cascaded controller for frequency regulation of energy storage integrated microgrid considering communication delays. *Journal of Energy Storage, 90*, Article 111856.
12. Pati, S.S., Subudhi, U. (2021). Frequency regulation of solar-wind integrated multi-area system with SMES and SSSC. In *IEEE International Power and Renewable Energy Conference (IPRECON 2021)* (pp. 1–4).
13. Daraz, A., Malik, S. A., Basit, A., Aslam, S., & Zhang, G. (2023). Modified FOPID controller for frequency regulation of a hybrid interconnected system of conventional and renewable energy sources. *Fractal and Fractional, 7*(1), 89.
14. Tepljakov, A., Alagoz, B. B., Yeroglu, C., Gonzalez, E. A., Hosseinnia, S. H., Petlenkov, E., & Cech, M. (2021). Towards industrialization of FOPID controllers: A survey on milestones of fractional-order control and pathways for future developments. *IEEE Access, 9*, 21016–21042.
15. Alabool, H. M., Alarabiat, D., Abualigah, L., & Heidari, A. A. (2021). Harris Hawks optimization: A comprehensive review of recent variants and applications. *Neural Computing and Applications, 33*, 8939–8980.

16. Hussien, A. G., Abualigah, L., Abu Zitar, R., Hashim, F. A., Amin, M., Saber, A., & Gandomi, A. H. (2022). Recent advances in Harris Hawks optimization: A comparative study and applications. *Electronics, 11*(12), 1919.

Solar Powered Single Switch Controlled DC–DC Converter for Battery Charging Applications

Vaishnav Tripathi, Akshat Shukla, and Gunabalan Ramachandiran

Abstract In this chapter, a new approach to address the challenge of efficient storage of renewable energy in batteries is solved by presenting the design and execution of a modified half-bridge converter capable of operating in both step-up and step-down modes to regulate the desired output voltage. The main aim is to design and simulate a single switch-controlled modified half bridge converter used for battery charging applications for an extensive range of solar input variations. The presented converter employs a single switch control methodology, where the duty cycle of one switch is adjusted while the other is kept constant. By dynamically varying the duty cycle of a single switch, the converter efficiently converts input power to match the requirements for charging batteries with a specified nominal voltage. Simulation with solar input and a battery load for both buck and boost test cases are performed on Simulink. A lithium-ion battery of nominal voltage 24 V and rated capacity 20.8 Ah is used to simulate the battery load. For the experimental setup, an Arduino microcontroller is used for driving both IGBTs, through the TLP250 driver IC. The converter circuit PCB is designed, fabricated and tested for both buck and boost modes of operation (12–46 V) for the desired output of 30 V in simulation and 20 V input voltage to 30 V output voltage in the solar input.

Keywords Battery charging · DC–DC converter · Electric vehicle · Solar PV

V. Tripathi · A. Shukla · G. Ramachandiran (✉)
School of Electrical Engineering, Vellore Institute of Technology-Chennai, Chennai, Tamil Nadu 600127, India
e-mail: gunabalan.r@vit.ac.in

V. Tripathi
e-mail: vaishnavtripathi15@gmail.com

A. Shukla
e-mail: akshatkuhu1@gmail.com

M. A. Mellal et al. (eds.), *Computational Intelligence for a Greener Future: Innovations in Renewable Energy Systems*, Studies in Computational Intelligence 1240,
https://doi.org/10.1007/978-3-032-06732-6_17

1 Introduction

Solar and wind power have observed substantial growth in recent years because of advancements in technology, declining costs, and increasing environmental awareness. Energy storage systems, such as batteries, serve as essential components in renewable energy installations, enabling the capture and storage of extra energy during high generation for further use.

Renewable energy sources offer several advantages over conventional fossil fuels, including sustainability, reduced environmental impact, and energy independence. With the increasing adoption of solar and wind energy sources in modern power systems, efficient storage of energy has become a critical concern.

Power converters play a crucial role in modern battery charging applications, enabling efficient energy transfer between various power sources and batteries. Recent advancements in power converter technology have significantly impacted the way batteries are charged, particularly in the context of renewable energy integration and electric vehicles.

One notable area of application is the use of power converters in solar PV systems for battery charging. These converters facilitate the conversion of solar energy into a form suitable for charging batteries, optimizing the utilization of renewable energy sources and enabling energy storage for off-grid or grid-tied systems. The ability to efficiently convert variable solar power into stable charging currents is essential for maximizing the effectiveness of solar battery systems.

Additionally, power converters are vital components in electric vehicle (EV) charging infrastructure. DC–DC converters are preferred to balance the voltage levels between charging stations and EV batteries, ensuring safe and efficient charging processes.

A novel controller for buck-boost converters was presented for tight voltage regulation and good transient convergence [1]. High efficiency as well as smooth transitions enables steady-state operation over the whole conversion ratio range. An experimental verification of the controller's functionality was conducted with a 2–15 V input for the desired output of 3.3 V. An immittance-network based resonant dc converter that can function well across a broad range of voltage variations was discussed [2]. Two phase-shifted half bridges on the inverter side allow for output current control over a broad input voltage range with lower circulation currents. For efficiency improvement over a wide choice of battery voltages, a dual-stage isolated converter was proposed and analyzed for use in electric car charging applications [3]. The proposed two stage DC–DC converter circuit comprises CLLC resonant assembly and a buck controller. A prototype converter module with a 10-kW rating, an input voltage of 800 V, and an output of 250–500 V using silicon-carbide and gallium-nitride power semiconductor switches was used to report the experimental results. A non-isolated converter with soft switching under various output voltage ranges was implemented, and promising results were found [4]. The coupling coefficient was adjusted for efficiency improvement and soft switching. A new parallel input and series output buck-boost converter (IOBBC) was discussed, which displays reduced ripples in source current and output voltage and provides numerous output voltages, improved efficiency, and high voltage gain. IOBBC operates by combining two distinct buck-boost converter topologies in a parallel-series configuration [5]. During a single switching cycle, the two converters run in an interleaving condition with a 180-degree phase displacement. Unlike the traditional single switch BBC, the suggested converter provides non-inverting output voltage, and a larger range of output voltage is possible for changes in duty cycle. Buck-boost converters play a pivotal role in voltage regulation, capable of stepping up or stepping down input voltage to match desired output levels. As such, comparing existing buck-boost converter topologies is essential for understanding their strengths and weaknesses in different scenarios. Table 1 showcases a detailed breakdown of the converter's attributes in relation to established buck-boost converter topologies:

The on-board, off-board DC–DC converters under isolated and non-isolated categories were reviewed in detail for light electric vehicle applications [6]. Various bidirectional DC–DC boost converters such as isolated, non-isolated, switched capacitor (SC) and non-switched capacitor were reviewed, and their performances were compared [7]. The limitations of a hard switching three port converter were overcome by a bidirectional soft switching multiport DC–DC converter with minimum number of switches [8]. Multi input bidirectional power converters are needed to utilize the combinations of super capacitor-battery and supercapacitor-fuel cell effectively in EVs [9]. The features are independent bidirectional control, buck mode or boost mode operation. The converter was tested in real time with an operating frequency of 20 kHz for 5 kW load. The hardware in loop (HIL) was used for validation and performance analysis. To maximize the output of a solar based wide input DC–DC converter, adaptive genetic algorithm with hybrid MPPT controller was discussed [10]. The simulation and experimental results were presented for various irradiation levels. The power MOSFET switches and TLP350 driver ICs were used. A modified grey wolf algorithm was introduced for a single switch high power rating solar based DC–DC converter [11]. A comparison table was presented to understand the importance of the single switch converter. A single switch converter with lossless passive elements and a coupled inductor were employed to reduce the stress on the switching devices [12]. The switching frequency was 60 kHz, and the efficiency of the converter was 98.6%.

Table 1 Comparison with different topologies and performance

Converter topology	No. of switching devices	Switching technique	No. of switches under control	Max. efficiency (%)
[1]	4	CPM	4	–
[2]	4	Phase shift control	4	89.5
[3]	6	PWM	6	–
[4]	4	PWM	4	95
[5]	3	PWM	3	91

This chapter presents a new approach to address this challenge by presenting the design and implementation of a modified half-bridge converter operated in both buck and boost modes to a desired output voltage under variable dc input conditions, specifically tailored for battery charging applications. Thus, it enables the storage of extra energy during high generation for later use when energy demand exceeds supply.

2 Block Diagram

Figure 1 illustrates the various components of the single switch-controlled DC–DC converter. The solar dc input is provided through the PV array, which gets regulated into the desired output voltage by the converter. The proposed converter employs a single switch control methodology, where the duty cycle of one switch is adjusted while the other is kept constant. By dynamically varying the duty cycle of a single switch, the converter efficiently converts input power to match the requirements for charging batteries with a specified nominal voltage. Moreover, the versatility of the converter in operating as both buck and boost modes enables flexibility in adapting to varying input and output voltage requirements.

The converter circuit structure is depicted in Fig. 2, where the input and output are denoted as V_{in} and V_o,, respectively. This DC–DC converter is composed of one inductor (L), one capacitor (C) and four switches (S_1, S_2, D_1, D_2)—two controlled and two uncontrolled, which conduct alternatively during a complete switching period (T_s). In a closed-loop control system, the controller continuously monitors the system output and adjusts the input to maintain the desired output. This is done by comparing the actual output with the desired output (reference voltage, V_{ref}), and using the difference (error) to adjust the system input. The load is connected to the V_O side while the V_{in} side is connected to the solar DC source. The equivalent converter circuit diagram for each mode is presented with the respective loop equations.

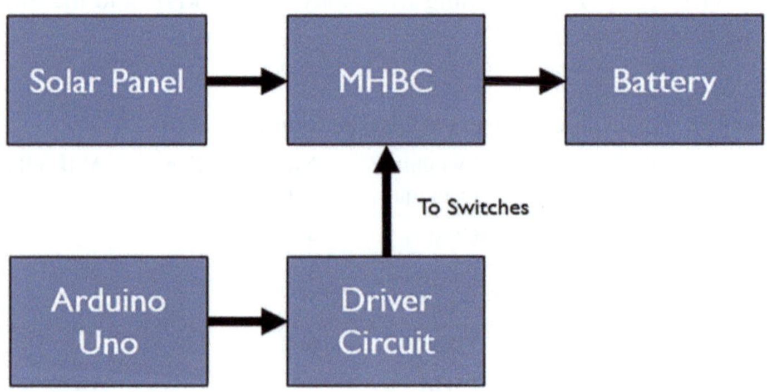

Fig. 1 Block diagram

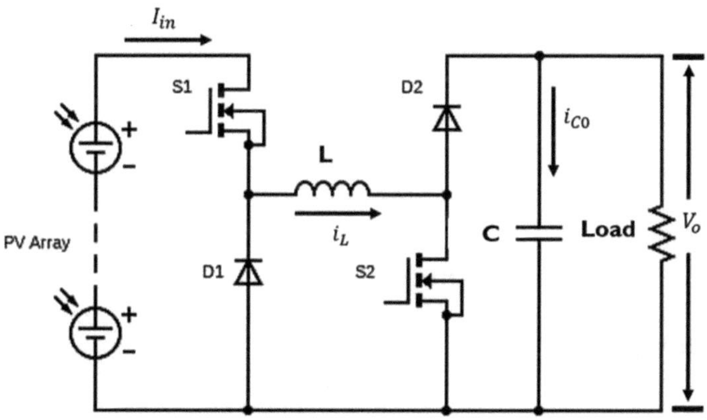

Fig. 2 Single switch-controlled converter circuit diagram [13]

3 Converter Design

The inductor and capacitor design plays a major role in the choice of converter topology. The elements are designed for continuous conduction operation. The design equations derived and presented in [13] are reproduced here for better understanding.

$$\mathbf{L} > \frac{V_O(1 - D_B)^2}{I_O f_{sw}} \tag{1}$$

For $V_O = 48$ V, $f_{sw} = 20$ kHz, and $I_O = 6$ A, and varying D_B from 0.1 to 0.8,

L > 324 μH (at $D_B = 0.1$)

Taking **L = 3 mH**

$$\Delta I_L < \frac{2V_o(1 - D_B)}{f_{sw}}$$

$$\Delta I_L < \frac{2 * 48 * 0.9}{20000}$$

$$\Delta I_L < 4.32\text{mA}$$

$$C_o = \frac{\Delta I_L}{8 f_{sw} \Delta V_O} \tag{2}$$

For $\Delta I_L = 3.2$ mA and $\Delta V_O = 0.001$ V,

$C_o = 20$ MF

The input voltage range of the proposed converter is 30 V to 80 V and the desired output voltage is 48 V. Keeping the duty cycle of one switch (S_1) constant at 55%, the minimum and maximum duty cycle of switch S_2 is calculated as follows:

CASE 1: $V_{in} = 12$ V

$$D_B = 1 - \left(\frac{D_{AX} V_{in}}{V_O}\right)$$

$$D_B = 1 - \left(\frac{0.55 * 12}{30}\right)$$

$$\text{DB} = 0.78(\text{or})78\%$$

CASE 2: $V_{in} = 46$ V

$$D_B = 1 - \left(\frac{D_{AX} V_{in}}{V_O}\right)$$

$$D_B = 1 - \left(\frac{0.55 * 46}{30}\right)$$

$$\text{DB} = 0.157(\text{or})15.7\%$$

4 Simulation Results and Discussions

Table 2 gives a detailed description of all key components with their part number and specifications. Table 3 gives the rating of the converter.

Table 2 List of components used in MHBC testing setup

Components	Part no	Specifications
Solar panel	Eldora 230	30 V, 250 W
Battery	Micronix Li-Ion	24 V, 20.8 Ah
IGBT	KGT25N120NDA	1200 V, 25 A
Diodes	10A10	1000 V, 10 A
Inductor	–	3 mH
Capacitor	–	20 μF
Driver IC	TLP250	–
Microcontroller	Arduino UNO	–

Table 3 Converter ratings

Circuit parameters	Symbol	Value/type
Switching frequency	f_{sw}	20 kHz
Input voltage	V_{in}	12–46 V
Load Voltage	V_o	30 V

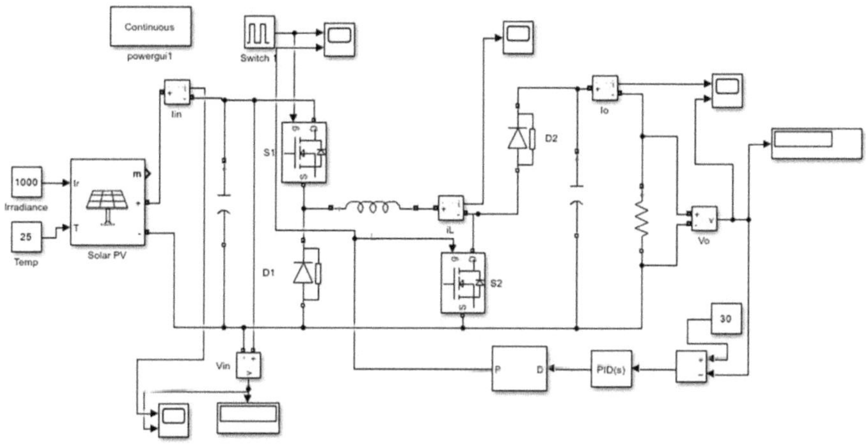

Fig. 3 Closed loop simulation diagram

4.1 Buck Mode ($V_{in} = 46$ V)

Figure 3 shows the closed loop simulation diagram of the converter in the closed loop operation. The PI controller parameters are obtained by random tuning method. Waveforms under closed loop operation in buck mode are shown on Fig. 4. In Fig. 4a, the current input varies from 0.2 A to 0.35 A and the output voltage is almost constant. Figure 4b depicts the switch pulse waveforms, with duty ratio 55% and 15.7% for switch 1 and switch 2 respectively. Figure 4c shows the inductor current waveform with an average value of 0.4 A. The inductor current is continuous, and the ripple is less. Figure 4d shows the output current and voltage. The current varies from 0.349 A to 0.351 A and voltage between 29.9 V and 30.1 V. The switching pulses are in accordance with the duty cycle calculation value. The pulse width for switch 2 is automatically adjusted by the PWM generator in the closed loop.

4.2 Boost Mode ($V_{in} = 12$ V)

The waveforms under closed loop operation in boost mode are presented in Fig. 5. In Fig. 5a, the input current varies from 0 to 4 A. Average input voltage is 12 V.

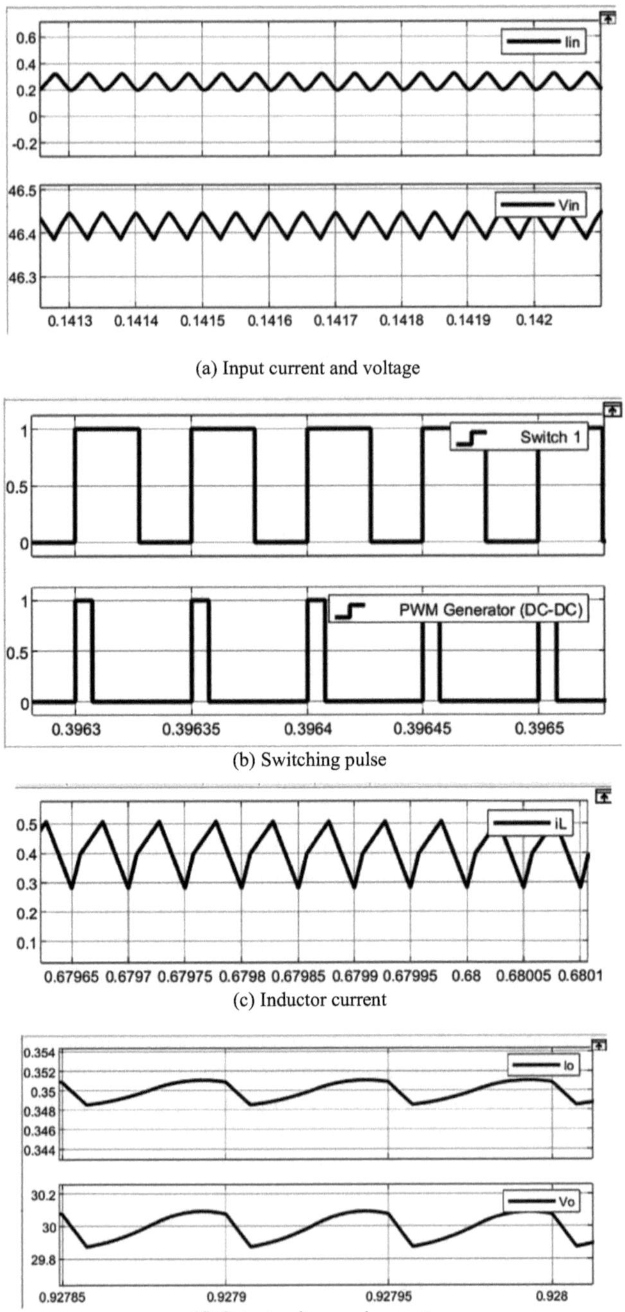

(a) Input current and voltage

(b) Switching pulse

(c) Inductor current

(d) Output voltage and current

Fig.4 Closed loop waveforms under buck mode of operation

Figure 5b depicts the switching pulse waveforms, with duty ratio 55% and 78% for switch 1 and switch 2 respectively. The duty cycle of switch 1 is fixed in both buck and boost modes. Figure 5c presents the inductor current with an average value of 2 A and the current is not saturated. Figure 5d illustrates the measured output current and voltage. The measured current is 0.24 A, and the voltage is 31 V. It is inferred that the switching pulses are in accordance with the duty cycle calculation. The pulse width for switch 2 is automatically adjusted by the PWM generator in the closed loop operation by the PID controller for the fixed duty cycle of switch 1.

5 Experimental Results and Discussions

The complete hardware has been designed by a three-step process. Figure 6 illustrates the block diagram of the hardware design. Firstly, the component selection is based on the converter specifications and the requirements. It is followed by the assembly of the entire setup. This includes the design of the driver circuit and the PCB design of the power circuit. Lastly, the assembled hardware is tested. In this converter, two IGBTs are used as power switches. IGBTs are voltage-controlled devices and require a certain amount of voltage to fully turn ON. This is accomplished by the driver's circuit. It consists of a TLP250 IC and its biasing circuit. Figure 7 shows the driver circuit diagram, and it provides the necessary isolation between the control circuit and the power circuit.

Figure 8 shows the experimental setup of the converter. It consists of three regulated power supplies, for driving each driver IC and providing the dc input, two driver circuits for each IGBT, PCB, digital storage oscilloscope (DSO) to observe the waveforms, multimeter to measure the voltages and a resistive load.

5.1 RPS Input

The circuit is tested with a dc input voltage directly obtained from a dc power supply. Figure 9a shows the gate pulse waveform observed on the DSO. The duty cycle of S_1 is 55% and S_2 is 25%. The peak-peak control voltage of S_1 is 13.6 V and S_2 is 14 V. Figure 9b shows the test results for buck mode with a resistive load. The performance parameters are measured by multimeters. An input voltage of 31.8 V is applied to get the output voltage of 18.2 V. The measured input and output currents are 0.4 A and 0.36 A respectively.

Figure 10a shows the gate pulse waveforms observed on the DSO under boost mode. The duty cycle of S_1 is 55% and S_2 is 65%. The peak-peak control voltage of S_1 is 11.8 V and S_2 is 11.8 V. Figure 10b shows the test results for boost mode. An input voltage of 14 V is applied to get the desired output voltage of 29.6 V. It is observed that the converter is operated under boost mode. The measured input current is 0.4 A and output current is 0.36 A.

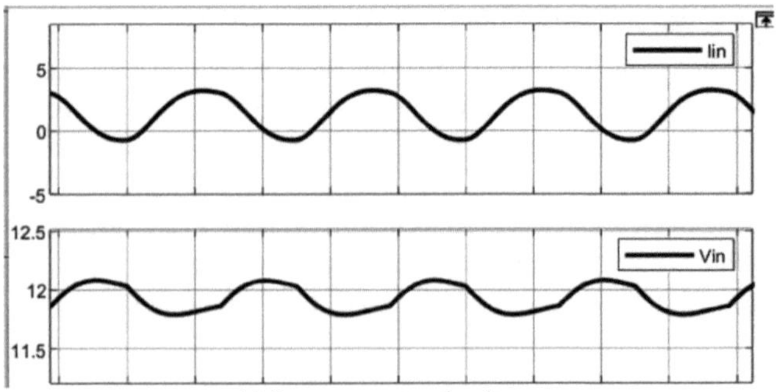

(a) Input current and voltage

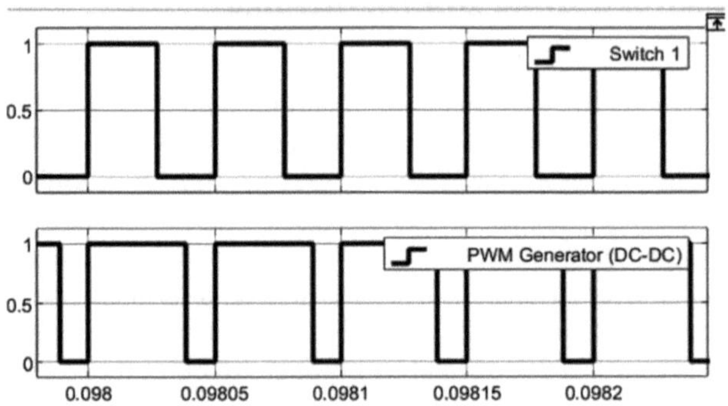

(b) Switching pulse

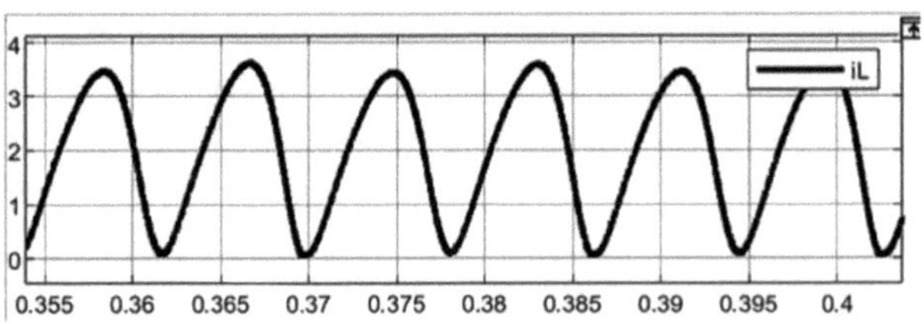

(c) Inductor current

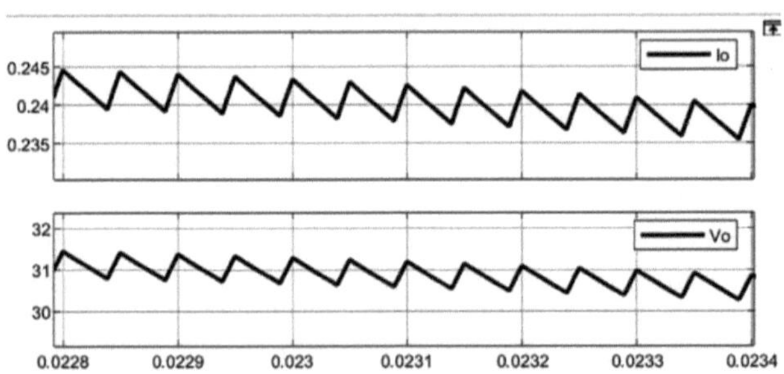

(d) Output voltage and current

Fig. 5 Closed loop waveforms under boost mode of operation

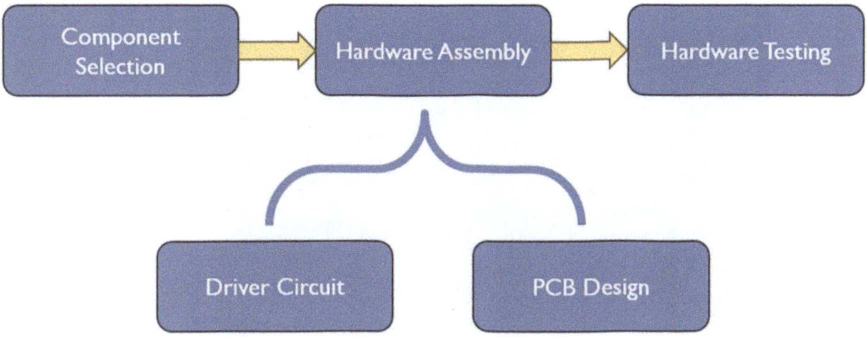

Fig. 6 Block diagram of the hardware design

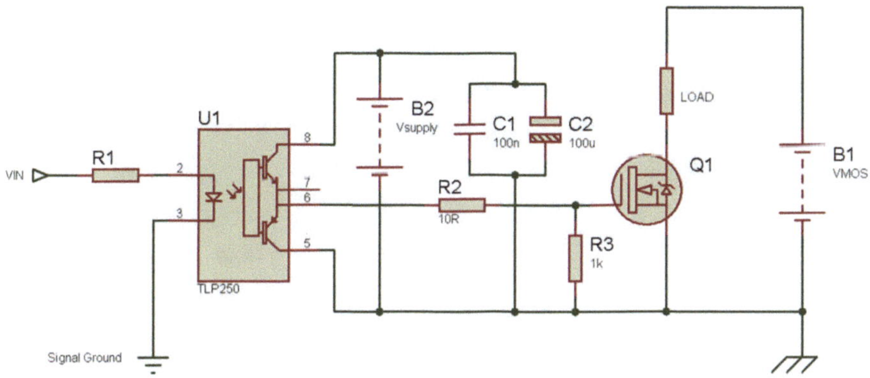

Fig. 7 TLP250 driver circuit

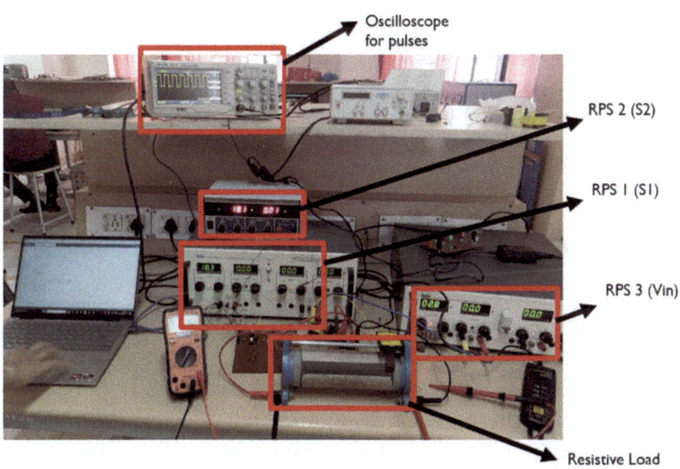

Fig. 8 Experimental setup

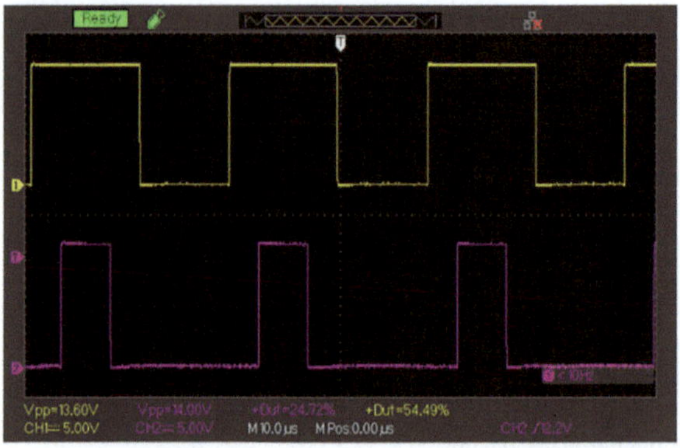

$D_A = 55\%$

$D_B = 25\ \%$

(a) Switching pulse waveforms for switch 1 and switch 2 under buck mode

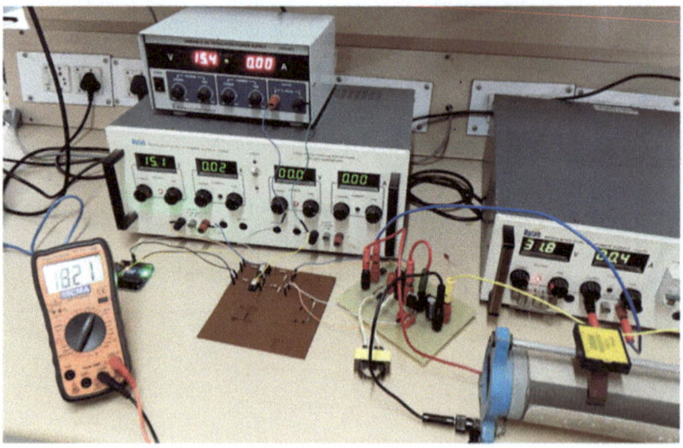

$V_{in} = 31.8\ V$ $V_o = 18.2\ V$

$I_{in} = 0.4$ A $I_o = 0.36$ A

(b) Output voltage and current measured by a multimeter

Fig. 9 Experimental test results for buck mode

5.2 *Solar Input*

Figure 11 shows the experimental setup of the converter with solar input. The setup consists of two regulated power supplies, for driving each driver IC, solar panel for providing the DC input, two driver circuits for each IGBT, PCB, digital storage oscilloscope to observe the waveforms, multimeter to measure the voltages and a resistive load. A current probe is used to measure the source and load currents. The test was conducted in electrical machines lab and the solar panel was fixed nearer to the window for solar irradiation as shown in Fig. 11.

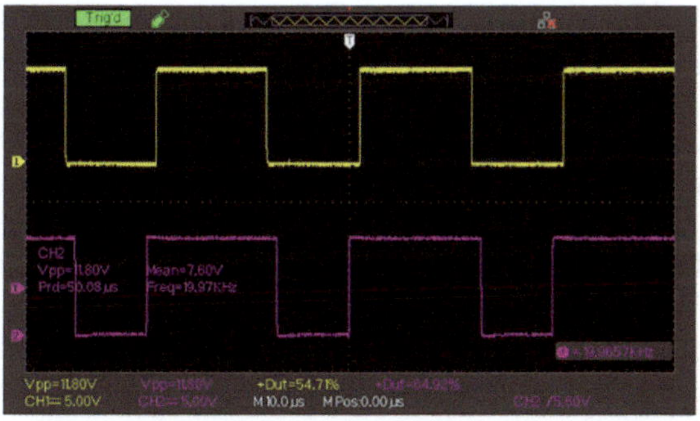

D_A = 55%

D_B = 65 %

(a) Switching pulse waveforms for switch 1 and switch 2 under boost mode

$V_{in} = 14\ V$ $\quad$ $V_o = 29.6\ V$

I_{in}= 1.3 A $\quad$ I_o= 0.2 A

(b) Output voltage and current measured by a multimeter

Fig. 10 Experimental test results for boost mode

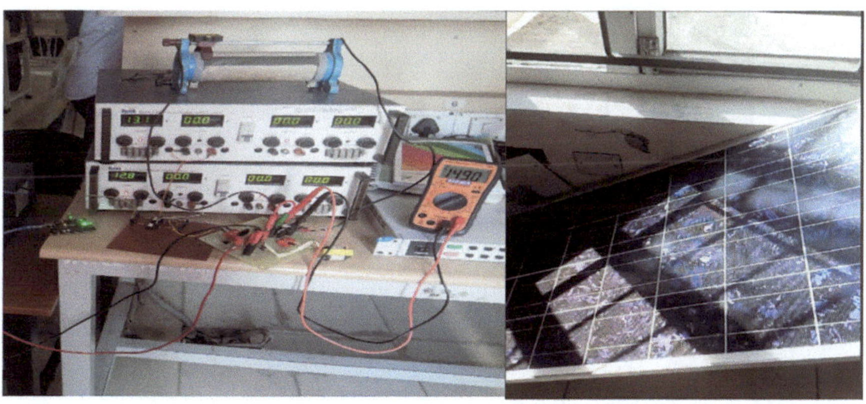

Fig. 11 Experimental setup with solar input

Figure 12a shows the gate pulse waveform observed on the DSO. The duty cycle of S_1 is 55% and S_2 is 25%. Figure 12b shows the test results for buck mode. The solar input voltage of 20 V is stepped down to 14.9 V. As the voltage rating of the solar panel is 20 V, the expected output voltage in the buck mode is less than 20 V. The working condition of the converter is validated in buck mode using solar input source.

Figure 13a shows the gate pulse waveform observed on the DSO. The duty cycle of S_1 is 55% and S_2 is 65%. The peak-peak voltage of S_1 is 11.8 V and S_2 is 11.8 V. Figure 13b shows the test results for boost mode. The solar input voltage of 20 V is stepped up to 33.6 V which is measured in multimeter.

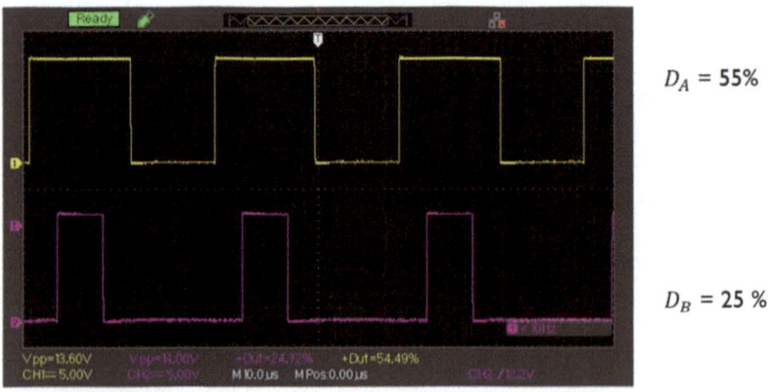

(a) Switching pulse waveforms under buck mode with solar input

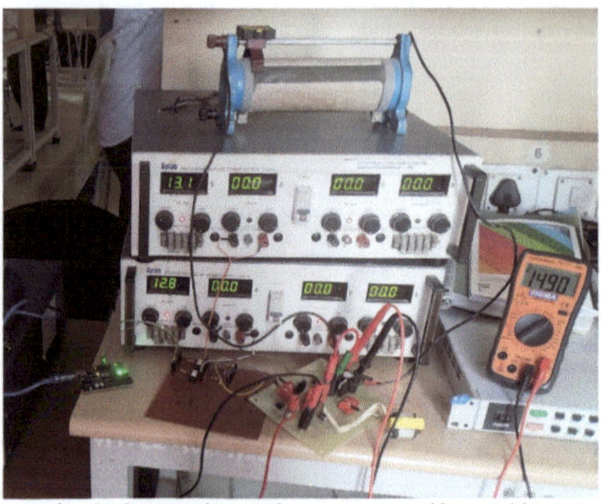

(b) Output voltage and current measured by a multimeter

Fig. 12 Experimental test results for buck mode with solar input

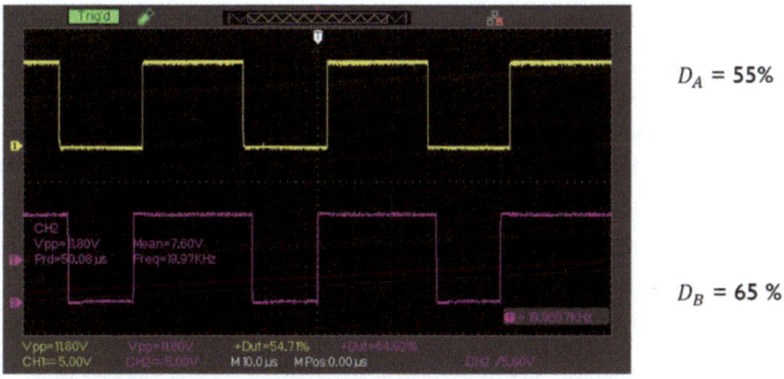

(a) Switching pulse waveform under boost mode with solar input

(b) Output voltage and current with solar input

Fig. 13 Experimental test results for boost mode with solar input

6 Conclusion

In this chapter, a solution has been developed to tackle the challenge of efficiently storing renewable energy in batteries. A modified half-bridge converter was designed to operate in both buck and boost modes, effectively regulating the desired output voltage for a variable dc input supply. Simulations were conducted using solar input and a battery load for both buck and boost scenarios. The results were validated against theoretical calculations, showcasing the effectiveness and efficiency of the proposed converter for such critical applications.

In conclusion, this research successfully demonstrates the practical viability of the modified half-bridge converter in efficiently charging batteries with renewable energy sources. The converter's dual-mode capability ensures optimal energy management, enabling reliable storage and utilization of surplus power, thus contributing significantly towards sustainable energy practices.

The future scope of the project includes optimizing efficiency, exploring multi-output functionalities, integrating with other renewable energy systems, developing smart charging algorithms, incorporating wireless charging capabilities, and enhancing control and monitoring features. These areas offer opportunities to improve converter performance, expand their applicability across different domains, and contribute to advancements in battery charging technologies for diverse applications including portable electronics, renewable energy systems, electric vehicles, and industrial equipment.

The advanced computational intelligence techniques will help to optimize the duty cycle of both switches. The PI controller will be replaced by advanced controllers in future. The combination of genetic algorithms, grey wolf algorithms and a few adaptive controllers will enhance the performance of the single switch DC–DC converter for solar photovoltaic applications.

References

1. Urkin, T., & Peretz, M. M. (2020). Digital CPM controller for a non-inverting Buck-Boost converter with unified hardware for steady-state and optimized transient conditions. *IEEE Transactions on Power Electronics, 35*(8), 8794–8804.
2. Mukherjee S, Sepahvand A, Maksimovic D. (2018). High-frequency LC3L resonant DC–DC converter for automotive LED driver applications. In *IEEE Applied Power Electronics Conference and Exposition (APEC), IEEE, San Antonio, TX* (pp 797–802).
3. Zanatta, N., Caldognetto, T., Biadene, D., Spiazzi, G., & Mattavelli, P. (2023). A Two-stage DC–dc isolated converter for battery-charging applications. *IEEE Open Journal of Power Electronics, 4*, 343–356. https://doi.org/10.1109/OJPEL.2023.3271227
4. Zhang, Y., Cheng, X.-F., & Yin, C. (2019). A soft-switching non-inverting Buck-Boost converter with efficiency and performance improvement. *IEEE Transactions on Power Electronics, 34*(12), 11526–11530.
5. Rana, N., & Banerjee, S. (2020). Development of an improved input-parallel output-series Buck-Boost converter and its closed-loop control. *IEEE Transactions on Industrial Electronics, 67*(8), 6428–6438.
6. Akhtar, M. F., Rohani, S., Raihan, S., Abd Rahim, N., Akhtar, M. N. (2023). Recent developments in DC–DC converter topologies for light electric vehicle charging: a critical review. *Appl. Sci. 13*(3), 1676.
7. Singh, A. K., Mishra, A. K., Gupta, K. K., & Siwakoti, Y. P. (2022). High voltage gain bidirectional DC–DC converters for supercapacitor assisted electric vehicles: a review. *CPSS Transactions on Power Electronics and Applications, 7*(4), 386–398. https://doi.org/10.24295/CPSSTPEA.2022.00035
8. Faraji, R., Ding, L., Rahimi, T., Farzanehfard, H., Hafezi, H., & Maghsoudi, M. (2021). Efficient multi-port bidirectional converter with soft-switching capability for electric vehicle applications. *IEEE Access, 9*, 107079–107094. https://doi.org/10.1109/ACCESS.2021.3097750
9. Hintz, A., Prasanna, U. R., & Rajashekara, K. (2015). Novel modular multiple-input bidirectional DC–DC power converter (MIPC) for HEV/FCV application. *IEEE Transactions on Industrial Electronics, 62*(5), 3163–3172. https://doi.org/10.1109/TIE.2014.2371778
10. Kumar, S. S., & Balakrishna, K. (2024). A new wide input voltage DC–DC converter for solar PV systems with hybrid MPPT controller. *Science and Reports, 14*, 10639. https://doi.org/10.1038/s41598-024-61367-x

11. Touti, E., Rafikiran, S., Aoudia, M., et al. (2024). A new single switch universal supply voltage DC–DC converter for PV systems with MGWM-AFLC MPPT controller. *Scientific Reports, 14*, 12103. https://doi.org/10.1038/s41598-024-62171-3
12. Kamarajam, A. S., Karuppasamypandiyan, M., Karthik Kumar, K. (2024). Enhancing electric vehicle battery charging efficiency through advanced DC to DC conversion in solar PV-integrated PRHGHECIC for improved energy storage. *Journal of Energy Storage, 99*(Part A), 113247. https://doi.org/10.1016/j.est.2024.113247
13. Sureshkumar, A., & Gunabalan, R. (2022). Design of robust guaranteed margin stability region PI controller for automotive LED lighting with parameter uncertainty. *IEEE Access, 10*, 15657–15670.

Central Grid Frequency Control in Power System: A Comprehensive Review

D. Murugesan and K. Jagatheesan

Abstract The growing integration of renewable energy in modern electric utility system has made load frequency control (LFC) increasingly important. This study explores various power system models, refinement techniques, and control methods to strengthen the efficiency and resilience of LFC. It focuses on Proportional Integral Derivative (PID)-based LFC in microgrids, examining seven optimization techniques across four cost functions, with random step load perturbations to assess response reliability. Preliminary results show that the new Cohort Intelligent Optimization (CIO) algorithm outperforms other methods in achieving optimal results while maintaining efficient computational costs. The CIO-ITAE (Integral Time Absolute Error) optimized LFC also exhibits improved resilience in addressing microgrid challenges. This review helps researchers bridge gaps in current advancements, applications, and future developments of microgrids in LFC.

Keywords Power system optimization and control · Intelligent control · Computer optimization algorithm · Grid frequency control · Machine learning · Integral time absolute error

1 Introduction

Renewable energy has become a key global priority as governments tackle rising energy constraints and environmental concerns. Its lower carbon emissions, cost-effective operation, and broad operational range position it as a leading alternative. However, the variability of renewable energy, influenced by weather conditions, can lead to mismatches between production and demand, finally in power imbalances

D. Murugesan (✉)
Paavai Engineering College, Namakkal, Tamilnadu, India
e-mail: benignsofts@gmail.com

K. Jagatheesan
M. Kumarasamy College of Engineering, Karur, Tamilnadu, India
e-mail: jaga.ksr@gmail.com

M. A. Mellal et al. (eds.), *Computational Intelligence for a Greener Future: Innovations in Renewable Energy Systems*, Studies in Computational Intelligence 1240,
https://doi.org/10.1007/978-3-032-06732-6_18

and frequency inconsistency. These fluctuations can threaten voltage profile and, in severe cases, enable protective relays, leading to power outages. Consequently, integrating renewable energy into load frequency control (LFC) systems presents a significant challenge, as LFC aims to correct load-generation discrepancies and diminish frequency and tie-line wattage fluctuations across control regions.

In a practical integrated power scheme, the overall generation capacity includes various sources namely thermal, tidal, nuclear, hydro, biomass, wind, fuel cell and solar energy. Typically, nuclear power as one of base load facilities operates close to its maximum capacity, while gas power generation is often used to meet variable load demands during peak hours. As a result, LFC mainly focuses on thermal or hydro generating units. The primary goal of LFC include: sustain the frequency at very close to its optimal value, Optimize the generation of each unit to maximize efficiency and ensure that tie-line power flow remains within predefined tolerances [1, 2]. The assistance and innovation of the current work are outlined below:

a. A complete review of non-renewable and renewable power systems, computational algorithms, and various control techniques for grid frequency regulation.
b. Tuning of PID regulator gains and investigation of its dynamic performance in microgrid energy systems using four cost functions.
c. Graphical comparison of the dynamic performance of PID controllers based on GA, DE, ACO, PSO, TLBO, ABC, and CIO algorithms against an optimal controller for a microgrid power system.
d. Evaluation of the robustness of the algorithm utilized regulator with optimized values and its capability to manage a wide range of loading pattern variations.

This article is organized into six sections: Sect. 1 reviews conventional and renewable energy power scheme, examining the suitability of control areas for one-area, two-area, triple-area, four-area, five-area, and six-area models. Section 2 investigates LFC classification using various machine learning algorithms. Section 3 discusses different control techniques to categorize LFC. In Sect. 4, the study explores LFC within the proposed microgrid power system using seven soft computing techniques and a PID controller. Finally, Sect. 5 provides the conclusions of the review, focusing on the dynamic performance of recently developed microgrid systems.

1.1 Brief Review of Power Systems

Conventionally, the power generated and transmitted to consumers has been principally provided by thermal, nuclear and hydroelectric power systems that form the prevailing overall power organisation. However, with increasing focus on the reduction of fossil fuel which is detrimental to the surroundings, renewable sources like solar and wind energy assume an important part in the power scheme. Similarly, the effective implementation of LFC in power scheme is a significant topic to be addressed with regards to both conventional as well as non-conventional resources.

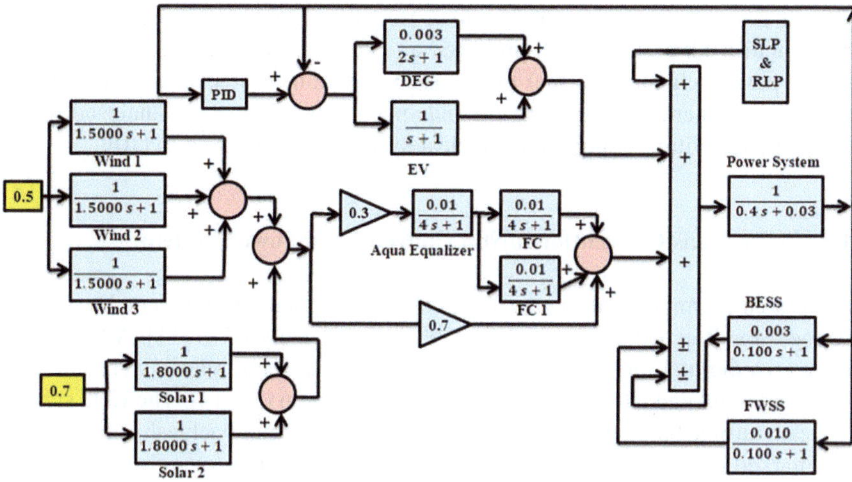

Fig. 1 Schematic diagram of μG system for frequency regulation

1.1.1 Renewable/Non Renewable Power Systems

Each power scheme has a crucial control task of managing the actual power production to meet consumer requirements and overcoming deficits, known as LFC. The greater part of the investigative studies conducted till now has focused on designing different linearised system models of renewable/non-renewable of one area with multiple sources or interconnected multi-area power organisations. The conventional electrical power scheme structures tended to be uncomplicated and inflexible. The essential sources of electrical power generation used to be the hydro plant, thermal plant, or both, despite the increments in power requests that eventually led to the rebuilding of power architectures. So, the addition of adaptable transmission arrangements and other secondary equipments came into the picture. Furthermore, new autonomous players entered into the electrical power organisation market promoting the idea of power deregulation. Subsequently, the LFC control issue with the non-renewable and renewable power frameworks alone became the target of researchers in the recent years. Figure 1.Micro grid power scheme is considered for frequency regulation in this article [3]. The following Sects. 1.1.1–1.1.6 present a few related studies that addressed LFCs in non-renewable and renewable power frameworks.

1.1.2 Single Area Conventional/Non-Conventional Energy Scheme

The LFC issue for one area thermal energy scheme with constant time and delayed system [4] performances were introduced in [5]. Aftab and colleagues (Aftab et al., [6]) have explored LFC for a Hybrid one area microgrid network employing a PI(1 + ID) regulator. Additionally, a study introducing an alkaline electrolyser-based one area power scheme to enhance load frequency control was outlined in [7]. Seyedmahdi Izadkhast et al. [8] proposed an electric vehicle application involving one area power scheme. Thermal-hydro-gas based multi-source single-area systems with various generation rate pressure and speed regulations were analysed in [9]. LFC challenge within a single-area nuclear system was discussed in [10].

1.1.3 Two Areas Conventional/Non Conventional Power System

Double area thermal-thermal power schemes, incorporating Governor Dead Band or Generation Rate Constraint, were established and analysed in [11]. Additionally, double area thermal-thermal power plants, with and without reheat, were interconnected and regulated using various control techniques in [12–14]. The LFC scheme of a steam turbine based power scheme considering governor backlash nonlinearities were marked out in [15]. A frequency power system model for a thermal-thermal turbine system with governor dead-band in double area power scheme was delineated in [16]. In [17], a frequency control design for a two-area network plan seeing the commitment of a hybrid system with thermal was described. LFC model for two areas hydro–hydropower systems considering fuzzy logic control was proposed in [18] along with generation rate constraint and governor dead band in [19]. Double area solar-thermal power scheme [20] and solar photovoltaic systems with diesel generators were controlled through an LFC scheme in [21]. Effect of solar and wind power in two areas thermal-thermal power networks along with GDB and GRC considerations was presented in [22]. Two area thermal hydropower plant was presented in [21, 23]. In [24–27], other works explored superconducting magnetic energy storehouse in multiple areas and cooperative energy flow regulator in tie line in [28].

1.1.4 Three Areas Conventional/Non Conventional Power System

Several studies [29–31] have dealt with three equal/unequal area thermal power systems considering one or more constraints together, such as the GDB, delay and GRC in the examination of the LFC characteristics of a dynamic system. Researchers [32] reported that the PD-PID cascade controller reduced positive and negative side oscillation values and also obtained a tracking of nominal value less than 45 s. Researchers [33, 34] devised an load frequency control model for a triple area incorporated network scheme, where thermal-thermal power plants were designated for the one and double areas, as the third area featured a hydropower plant. In addition, achievement of the proposed LFC was tested by sequential and random use of disturbances in all the three areas.

A study [33] showed that an ANFIS based system improved transient LFC response over an integral LFC controller for a dynamic power system. Another study [34]designed an LFC for variable load requests of 0%,1% step load, 1% step load with GRC, and obtained settling time of zero, 22 and 50 s respectively. The study showed that the addition of suitable constraints to the system can enhance the overall system dynamics. In [35] LFC of three area consisting of a reheated thermal area, non-reheated turbine of the thermal area, and hydro area were presented. In [36]three areas hydro nuclear thermal power system controlled by polar fuzzy logic control provided a improved settling time (less than 10 s) as compared to the PI and Fuzzy logic mechanism. In [37] triple-area thermal power scheme for a Plug-in Electric Vehicle (PEV) was considered for analysis. The authors reported that the proposed power system with PEV provided better transient response in terms of settling time less than or equal to 2 s as well as improve the transient result were devised for three areas with four-unit hydrothermal power plants in [38]. Additionally, in [39], LFC based on FOPID controllers for multi-source network scheme, has achieved a settling time of under 4 s and mitigated oscillations.

1.1.5 Four Areas Conventional/Non Conventional Power Systems

Researchers [40–42] have examined different Load Frequency Control (LFC) challenges in four-area interconnected power scheme, as typically large-scale power scheme are segmented into many control areas to ensure frequency stability. An LFC system, incorporating thermal-hydro and thermal-wind units interconnected via a tie-line, was presented in [43]. An LFC model for four equal areas interconnected energy network having triple reheat turbine-based thermal area and one non-reheat turbine thermal area was considered in [25]. Conventional regulator based automatic generation control for a four areas (thermal-thermal-hydro-hydro) power scheme was considered for LFC model in [52]. In [55] an interconnected four equivalent thermal areas system was introduced with transient parameters and was compared with one area, two areas, and three areas systems. The authors observed that the model with four areas demonstrated enhanced dynamic performance, showing reduced settling time and minimized overshoots and undershoots.

1.1.6 Five and Six Areas Power System

Some researchers [44, 45] studied load frequency control of 200 MW rated capacity-based generator for five equal area thermal power systems with reheater with/without GRC. In [46], a ring arrangement of five area power system was presented for LFC. Similarly, a PID based LFC controller for automatic gain control was introduced for six areas thermal electric network with non-reheat, single reheat, and two reheat turbine in [45].

2 Optimization Techniques for LFC

Soft computing methods are employed to find optimal outcome and remove ambiguity and vagueness in discovery the best answer of a process. These techniques are designed to obtain practical and sensitive solutions at minimal costs. Hence, these techniques form the premise of the varied and impressive artificial intelligence strategies. Numerous methods draw inspiration from swarm logic observed in nature, giving rise to swarm-based algorithms such as those inspired by particles, ants, bees, and others. On the other hand, bio-inspired soft computation approaches are designed to solve problems through algorithms that mimic and/or simulate biological processes. Socio inspired algorithms are yet another class of techniques which find optimal solutions by mimicking social behaviours of a group of participants. Figure 2 illustrates an overview of computer algorithms and their publications related to frequency regulation.

2.1 Genetic Algorithm

The GA is a well known search optimization technique based on natural option i.e., the survival of the fittest. Firstly, an initial population (solution set) is generated, comprising of various 'chromosomes' where every chromosome (features set) forms an individual (solution) to the issue whose realisation is examined by fitness values. Fundamentally, GA comprises of three operators namely the selection, crossover and

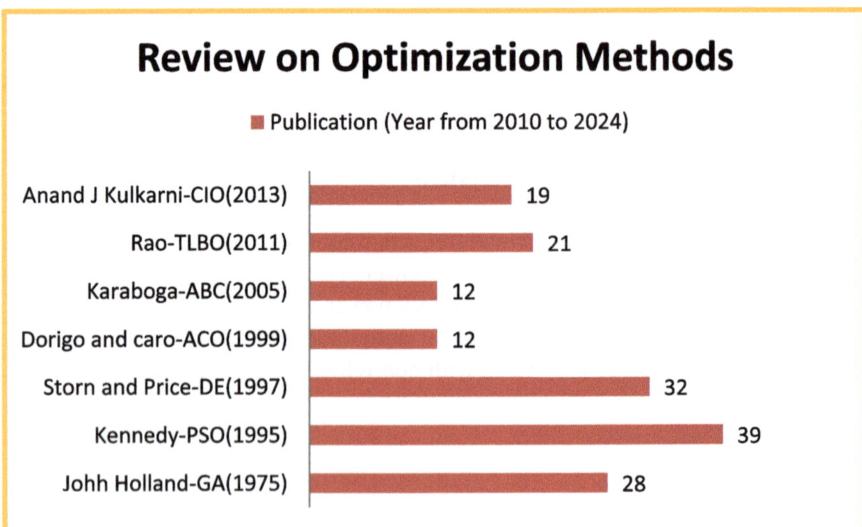

Fig. 2 Review of computer algorithms and its publication for frequency regulation

mutation. The use of these three operations permit the generation of new population which is preferable over the previous generations due to the iterative selection of fitter 'parents' to produce better 'offspring's in the next generation. This iterative method repeats until reaching the best solution for the problem. Due to its simplicity and adaptability, GA can be effectively utilized across a wide spectrum of problems, including load frequency control, economic dispatch, deregulated power systems, and physiological control tasks such as regulating blood glucose levels, among others.

Fatemeh Danesh far and Hassan Bevrani [47] introduced PI regulators for gain tuning using genetic algorithm in a three region power system. A marine micro grid LFC control using GA was proposed in [48]. Another study [49] proposed a GA based versatile acknowledgment strategy for an AC multi-micro grid plant framework. This study used GA for versatile recognition of the optimal micro grid framework based on plant boundary changes due to frequency variations. In [50], a PID-based LFC regulator for a single-area network with multiple sources, optimized using GA, was examined. Meanwhile, Debashish et al. [51] suggested a GA-based LFC approach for an interconnected island microgrid framework incorporating solar and wind power. The authors used Fuzzy PID and Fuzzy FOPID regulators on the network under various load conditions. Murugesan et al. [52] have addressed the training of PID regulators for load frequency control applications by employing various genetic algorithm variants.

2.2 *Differential Evolution*

DE is more are less similar to GA in working principle and its operation begins with mutation, followed by crossover and selection. In summary, Differential Evolution (DE) utilizes a population size denoted by Np, with a minimum requirement of four for creating mutant vectors. DE operates with three vectors: target, donor, and trial. In each generation, target vectors are selected, and mutation generates donor vectors by summing weighted differences between two randomly chosen target vectors. The crossover process then combines donor and target vectors to create trial vectors. If trial parameters result in improved fitness compared to target parameters, trial vectors replace target vectors in the next iteration. The merits of DE are effectiveness, quickness, and local searching behaviour. Researchers have explored DE for designing LFC of two areas interconnected power systems [6]. DE optimised fractional order LFC controllers for various interconnected power systems are designed in [53].

2.3 *Particle Swarm Optimization*

It is a swarm intelligence-based metaheuristic that searches for the optimal solution by simulating particles moving through a search space. Each particle represents a potential solution, with its position and velocity updated iteratively. A fitness function evaluates each particle, and the one with the best score is selected as the global best solution. PSO was utilised for designing two areas integrated network LFC in [54]. In [51], PSO optimised fuzzy regulator was developed for a multi-area interconnected power scheme, and its effect on the dynamic performance was analysed. Multi-area LFCs were designed using PSO in [55] and employing FOPID PSO among other optimization methods was analysed in [56]. The utilization of PSO for LFC in deregulated power systems was investigated in [57].

2.4 Ant Colony Optimization

Ant Colony Optimization (ACO) [12, 58, 59] mimics how real ants find food: artificial "ants" explore a solution space and deposit virtual pheromones. Over iterations, paths with stronger pheromone trails—representing better solutions—are more likely to be chosen. The algorithm consists of three phases: initializing parameters and pheromone levels, constructing solutions via probabilistic walks guided by pheromone and heuristic information, and updating pheromone trails (including evaporation and reinforcement).Jagatheesan et al. [45] designed single and double reheat turbines for single and multi-area electrical networks using PID regulators optimized with the ACO algorithm. ACO was applied to zero carbon multisource system economics and robust control by Ebrahim et al. [60]. In [61], ACO was utilised to optimise the PID regulator gain parameters considering non-linearities and boiler dynamics for a one area thermal power scheme. Murugesan et al. [62, 63] analysed performance of a PID regulator in an integrated thermal network scheme based on the ACO technique.

2.5 Teaching–Learning-Based Optimization

The Teaching–Learning-Based Optimization (TLBO) algorithm emulates a classroom scenario with two phase's teacher and learner where a population of candidate solutions (students) iteratively improves. It requires only population size and iteration count no additional algorithm specific parameters making it simple, memory-efficient, and effective for finding global optima in [4, 37, 64, 65]. Several recent studies have leveraged TLBO to optimize LFC and AGC in various power systems. Reference [66] deployed TLBO to develop an LFC strategy for a hydrothermal setup inclusive of plug-in electric vehicles. Gorripotu et al. [67] applied TLBO to tune a fractional-order PID (FOPID) regulator for AGC in an interconnected hydrothermal system, demonstrating improved frequency and tie-line performance. Srinivasarathnam et al. [57] optimized PID gains for LFC in a renewable multi-microgrid using TLBO. Following that, study [38] presented a fuzzy-PID LFC controller refined through TLBO, and Sahoo et al. [68] proposed a TLBO-tuned LFC for a multi-area configuration with an HVDC link.

2.6 Cohort Intelligence Optimization

Another soft computing technique known as Cohort Intelligence Optimization (CIO) was developed by Kulkarni et al. [69]. This technique is based on the self-improving behaviour of an individual solution called cohort in a gathering of cohorts. The mutual learnings among the cohorts with a common objective prompts progressively better behaviour of the individual cohort members. Every cohort member notices and gains from others in the group and attempts to improve its individual characteristics, bringing about a superior general conduct of the entire group. The cohort performance reaches saturation when there is no further development in the individual performances after multiple iterations. In CI, the solution quality often depends on the number of cohorts employed.

Different applications of CI are discussed in literature [70] such as healthcare, supply chain management, sea cargo mix issue and cross-border shoppers' problem among many others. DC motor speed control using CI was presented in [71]. In [72], the travelling salesman problem and knapsack problem were addressed using CI. Fractional order PID controller was tuned using CI in [73] and its applications with variations in CI were examined in [74]. Patankar et al. [75] have reported that the roulette wheel selection rule consumed the least process time as compared to other variations of CI. CI-assisted image processing is addressed in [76]. CI-assisted mesh smoothing of complex objects such as a pentagonal, hexagonal, and hexagonal prism with holes was investigated in [77]. Inter and intra-group learning performance improvement using multi-cohort intelligence was tested in [78]. LFC in isolated and integrated power scheme using CI was examined in [79]. Pareto dominance using multi objective CI was addressed in [80]. Flexible robotic design using CI was interpreted in [81]. Design of mechanical engineering components such as helical spring, hollow shaft, and belt pulley drive using CI was analysed in [82]. Power electronics buck converter design using CI was discussed in [83].

3 Control Techniques

This section addresses the categorization and evaluation of LFC (Load Frequency Control) across various methodologies, which vary according to the scale of power system models. These methodologies encompass traditional control techniques as well as more contemporary advanced control designs. Over the years, many LFC controllers have been designed and implemented successfully and gradually newer control techniques have evolved incorporating hybrid and soft computing approaches. Figure 3. Depicts the roadmap of different control methods for frequency regulation. The mathematical model of the controller is presented in Eq. (1), while the mathematical formulation of the optimization problem is detailed in Eqs. (2)–(5).

Transfer function PID Controller

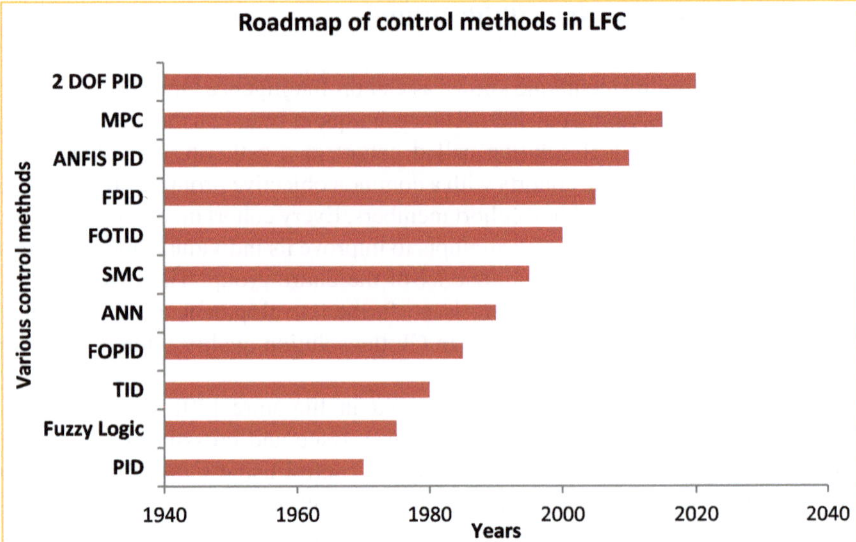

Fig. 3 Roadmap of various control methods for frequency regulation

$$U(t) = K_P e(t) + K_I \int e(t)dt + K_d \frac{d}{dt} e(t) \tag{1}$$

Objective function as below

$$ISE = \int_0^t (ACE^2)dt \tag{2}$$

$$ITSE = \int_0^t t.(ACE^2)dt \tag{3}$$

$$IAE = \int_0^t |ACE|dt \tag{4}$$

$$ITAE = \int_0^t t.|ACE|dt \tag{5}$$

Here, U(t) represents the control signal, e(t) is the error signal, Kp, K_I and Kd denote the proportional, integral and derivative gains, respectively. ACE refers to the Area Control Error, while t denotes the simulation time in seconds. The optimization problem can be formulated as follows: the objective is to minimize the defined objective functions, subject to the constraint that the controller parameters are bounded as follows: $0 \leq K_P, K_I, K_d \leq 1$

3.1 Conventional Controller

Khodabakhshian et al. [84] addressed Nichols chart supported PID regulator for AGC of a hydropower system. Attya et al. [85] addressed the LFC of a hydro-pumped storage scheme. Dahiya et al. [86] proposed an optimum regulator for AGC using various optimization techniques in a multi-area electrical network. Dhundhara et al. [28] addressed the reliable frequency operation in micro grid systems using conventional PID controllers. Additionally, researchers have employed sliding mode control [87] and crone controller [88] as further control methods.

3.2 Fuzzy Logic Controller

Ozbay et al. [89] presented the ramp-type test signal that takes a shorter settling time than step input by using the adaptive fuzzy controller in small hydropower plants. Dhanalakshmi et al. [90] proposed a multi-stage fuzzy logic control in LFC. Researchers [91, 92] have implemented fuzzy logic in power systems for self regulating generation control. Anilannamraju et al. [93] designed multistage PID controller using fuzzy logic based LFC in a micro grid system.

3.3 ANN Control

Researchers showcased the utilization of Adaptive Neuro-Fuzzy Inference System in adapting input/output datasets through the back propagation gradient descent method for managing LFC in distributed generation systems [94]. The application of an ANFIS controller in an isolated microgrid power network to address LFC challenges was elucidated in [95], revealing its functional equivalence to FIS (Fuzzy Inference System).

3.4 Fractional Order Controller

Shah et al. explored fractional order control for higher order control systems [96], undamped control scheme [97] and a detailed review of FOPID in [98, 99]. Sondhi et al. [100] put forward the concept of utilizing a fractional-order system specifically tailored for LFC within the context of a single-area network. Researchers [101] have focused on training TID regulators for LFC applications and utilizing green energy sources.

3.5 Hybrid Controller

A hybrid conjunctional of fuzzy and fractional order systems was suggested by researchers in [49, 51]. Ismail et al. [102] addressed fuzzy and fractional order controller-based PID controllers for multi-area smart grid systems. Babaei et al. [103] proposed a recurrent adaptive neuro-fuzzy system based on LFC problem in a microgrid power scheme.

4 Result and Discussion

To validate the PID control strategy, a renewable energy-based microgrid, as shown in Fig. 1, is analyzed under various case studies using the MATLAB/SIMULINK R2014A environment on a computer with an AMD PRO A4-3350B APU processor. This study examines seven different algorithmic techniques.

4.1 Case 1: Evaluation of Various Algorithmic Techniques Under Conditions Where All Cost Functions Are Available

This case study presents a performance evaluation of GA, DE, ACO, PSO, TLBO, and CIO techniques across various cost functions: ITAE, IAE, ISE, and ITSE. The optimized regulator parameter values are shown in Table 1. The dynamic responses, along with decision indicators namely peak frequency overshoot, and frequency settling time, are compared in Figs. 4 and 5, clearly demonstrating the superiority of the CIO technique. The developed frequency control model effectively achieves a controlled frequency signal with minimal deviation. Figures 4 and 5 clearly illustrate the bar chart analysis, demonstrating that the proposed CIO-PID technique achieves the smallest deviation in overshoot and the shortest settling time compared to other computational algorithms.

Table 1 Optimized gain values for various computer algorithm using different cost function

Algorithm	Cost function	Controller gain values		
		Kp	Ki	Kd
GA	ITAE	0.42026	0.99995	0.50434
	IAE	0.94282	0.99983	0.13368
	ISE	0.61392	0.99999	0.28148
	ITSE	0.64596	0.99994	0.44634
DE	ITAE	0.73658	0.99816	0.50861
	IAE	0.51385	1	0.35875
	ISE	0.8909	0.99994	0.9691
	ITSE	1	1	0.59118
ACO	ITAE	0.42	1	0.52
	IAE	0.98	1	0.17
	ISE	1	1	0.98
	ITSE	0.98	1	0.33
PSO	ITAE	0.0155	1	0.4668
	IAE	0	1	0.6646
	ISE	0.0160	1	1
	ITSE	0	1	0.2490
TLBO	ITAE	0.57146	1	0.31048
	IAE	0.42002	0.99998	0.50546
	ISE	1	1	0.5552
	ITSE	1	1	0.33475
ABC	ITAE	0.8750	0.34857	0.04192
	IAE	0.0949	0.37542	0.546
	ISE	4.7953	2.2241	0.83251
	ITSE	0.40982	0.12338	0.44302
CIO	ITAE	0.44413	0.9693	0.61743
	IAE	0.60866	0.97671	0.56043
	ISE	0.9531	0.9993	0.7664
	ITSE	0.9858	0.9857	0.5124

4.2 Case 2: Assessment of Computational Algorithms Based on Error Values and Execution Time

Figure 6 presents a bar chart illustrating the error values computed for the various computational algorithms. In contrast, Fig. 7 shows the processing time in seconds for each proposed algorithm. In Fig. 7, it is observed that the execution time of the CIO-PID technique is 83.44% faster than GA-PID, 65.21% faster than DE-PID, 69.23% faster than ACO-PID, 12.5% faster than PSO-PID, 71.42% faster than TLBO-PID,

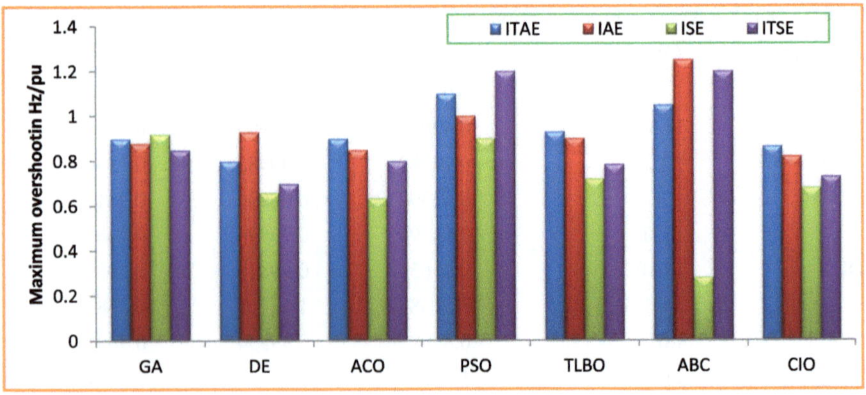

Fig. 4 Bar chart comparing different computer algorithms and their maximum overshoot

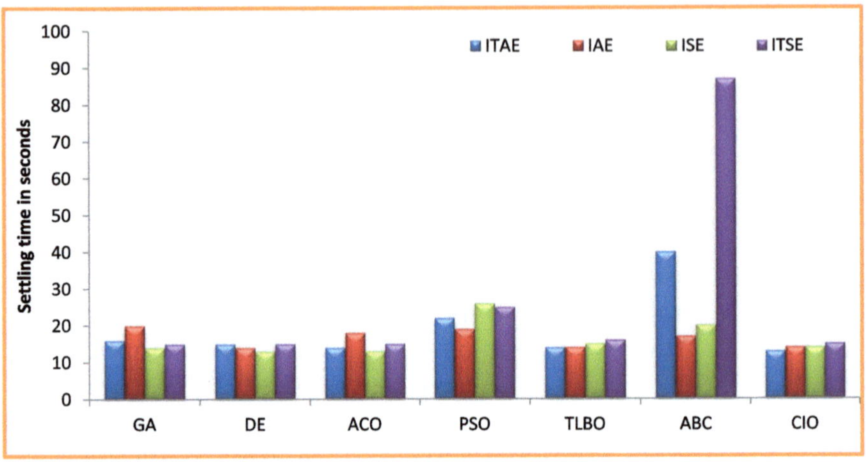

Fig. 5 Bar chart comparing different computer algorithms and their settling times

and 45.45% faster than ABC-PID techniques. Overall, the assessment indicates that the tuned CIO-PID controller parameters effectively manage the errors and execution time of the proposed computational algorithms.

4.3 *Case 3: Analysis of the Robustness of Various Computer Algorithms Under Uncertain System Loading*

In this scenario, step and multi-step random load changes are applied as disturbances to the LFC. Figure 9 illustrates the load steps, while Fig. 8 shows the frequency

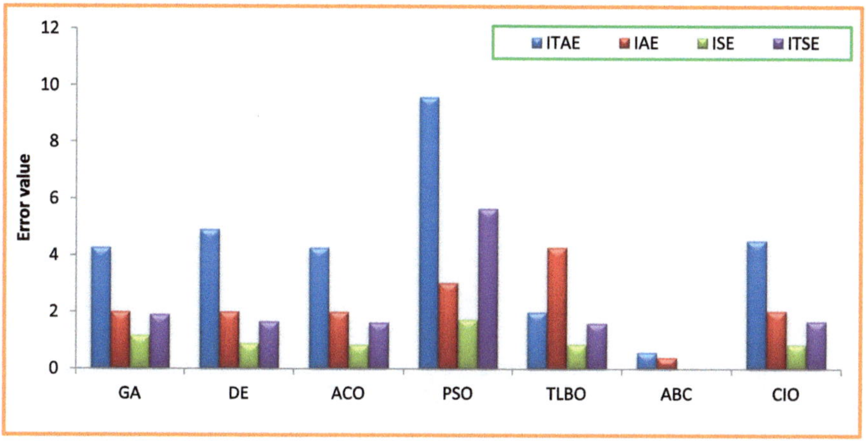

Fig. 6 Bar chart comparing different computer algorithms based on their error values

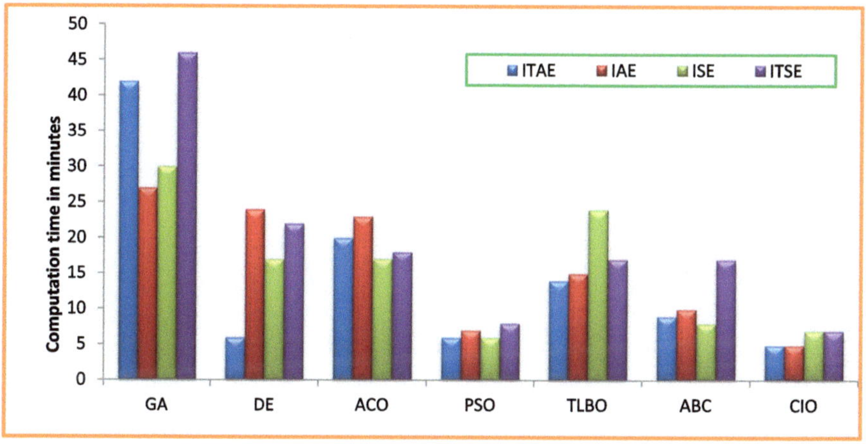

Fig. 7 Bar chart comparing different computer algorithms and their execution times

inconsistency responses of the CIA, ABC, TLBO, PSO, ACO, DE, and GA-based PID controllers.

As shown in Fig. 8, the frequency changes is reduced, and the LFC control can more effectively mitigate the impact of the load imbalance using the CIO-PID controller method compared to other methods. The performance of the regulator was evaluated by considering a large random step load as a severe condition at t = 60 s during the simulation. As indicated in Fig. 10, the LFC's ability to eliminate the load disturbance is enhanced by using the CIO algorithm over other techniques, particularly with a significant reduction in the settling time of the frequency

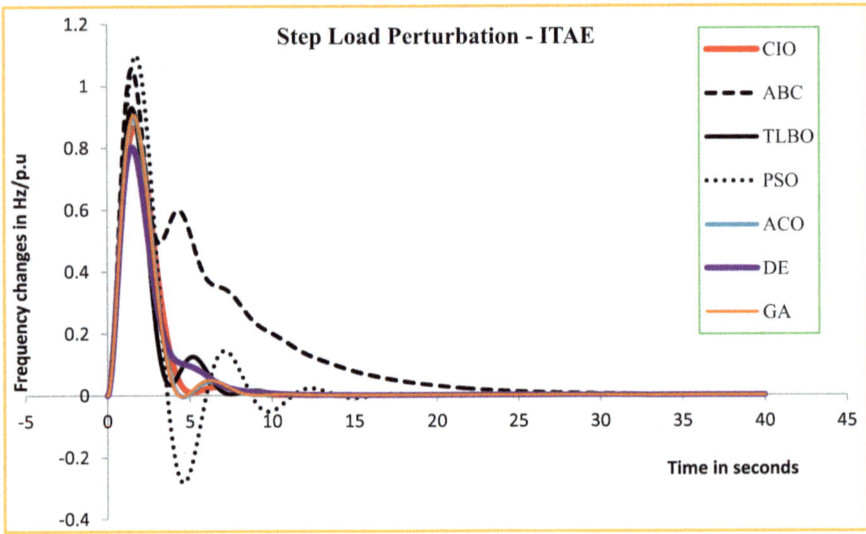

Fig. 8 Analysis of the robustness of different computer algorithms using step load perturbation

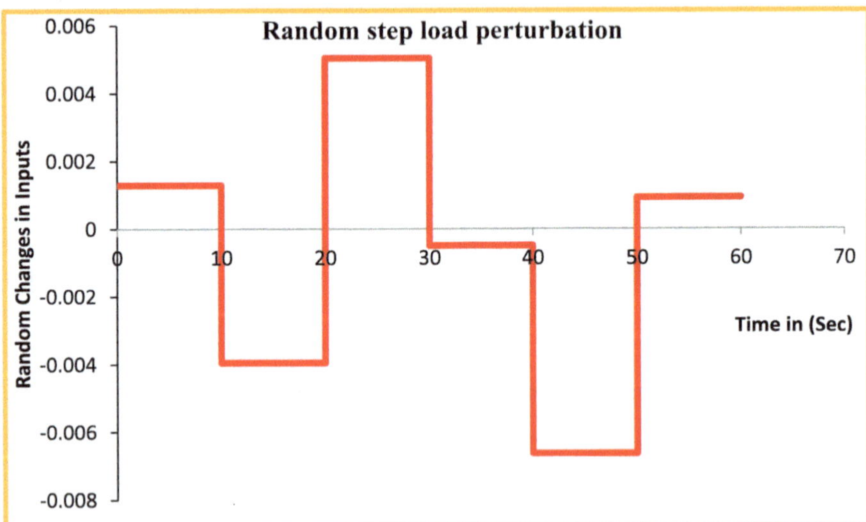

Fig. 9 Graphical representation of random step load perturbation input

response. Overall, the assessment suggests that the tuned CIO-PID controller parameters demonstrate strong robustness in handling the uncertainties of the proposed hybrid microgrid network.

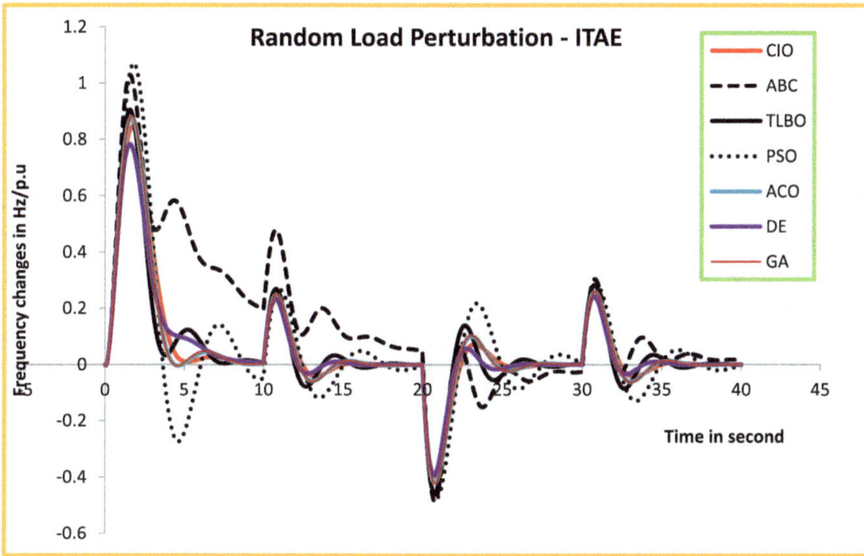

Fig. 10 Analysis of the robustness of different computer algorithms under random load perturbation

5 Conclusion

This study implemented single PID controllers for LFC in a microgrid power network comprising multiple solar systems, multiple wind systems, a diesel generator, electric vehicles, a fuel cell, and energy storage. Seven optimization techniques were applied to determine the optimal gain parameters across four objective error functions. The Cohort Intelligence Optimization (CIO) method demonstrated superior performance in terms of computation time, achieving optimal solutions within 5 min for the ITAE and IAE cost functions, and 7 min for the ITSE and ISE cost functions, outperforming other techniques. Additionally, CIO significantly improved settling time compared to alternative methods, with enhancements of 16% (GA), 2% (DE), 64% (PSO), 7% (ACO), 5% (TLBO), and 193% (ABC). The results were compared under transient conditions, showing that the CIO approach offered superior performance in computational efficiency, settling time, minimal error values, and consistency. Cohort intelligence exhibited faster convergence and was simpler to implement than existing methods. Robustness testing confirmed the effectiveness of the CIO-ITAE combination. Moreover, the industrial objectives of PID regulator tuning include minimizing frequency deviations, maintaining system stability, responding quickly to load variations, and improving energy efficiency. Future research is encouraged to explore hybrid optimization methods to further improve microgrid network performance, particularly with the increasing role of green energy sources and the integration of pumped storage systems for effective energy storage.

References

1. Elgerd, O. I. (1970). *Electric energy system theory: An introduction.* Tata Mc-Graw Hill Publishing Company limited.
2. Kundur, P. (1994). *Power system stability and control.* Tata Mc-Graw Hill Publishing Company limited.
3. Mahto, T., Malik, H., Mukherjee, V., Alotaibi, M. A., & Almutairi, A. (2021). Renewable generation based hybrid power system control using fractional order-fuzzy controller. *Energy Reports, 7*, 641–653.
4. Saxena, S., & Hote, Y. V. (2018). PI controller based load frequency control approach for single-area power system having communication delay. *IFAC-PapersOnLine, 51*(4), 622–626.
5. Jiang, L., Yao, W., Wu, Q. H., Wen, J. Y., & Cheng, S. J. (2011). Delay-dependent stability for load frequency control with constant and time-varying delays. *IEEE Transactions on Power systems, 27*(2), 932–941.
6. Aftab, M. A., Hussain, S. S., Latif, A., Das, D. C., & Ustun, T. S. (2021). IEC 61850 communication based dual stage load frequency controller for isolated hybrid microgrid. *International Journal of Electrical Power & Energy Systems, 130*, Article 106909.
7. Kiaee, M., Cruden, A., Infield, D., & Chladek, P. (2014). Utilisation of alkaline electrolysers to improve power system frequency stability with a high penetration of wind power. *IET Renewable Power Generation, 8*(5), 529–536.
8. Izadkhast, S., Garcia-Gonzalez, P., & Frías, P. (2014). An aggregate model of plug-in electric vehicles for primary frequency control. *IEEE Transactions on Power Systems, 30*(3), 1475–1482.
9. Parmar, K. S., Majhi, S., & Kothari, D. P. (2012). Load frequency control of a realistic power system with multi-source power generation. *International Journal of Electrical Power & Energy Systems, 42*(1), 426–433.
10. Murugesan, D., Jagatheesan, K., Kulkarni, A. J., Shah, P. (2023). A socio inspired technique in nuclear power plant for load frequency control by using cohort intelligence optimization-based PID controller. In A. Khosla, & M. Kolhe (Eds.), *Renewable energy optimization, planning and control. Studies in infrastructure and control.* Springer. https://doi.org/10.1007/978-981-19-8963-6_1
11. Nguyen, G. N., Jagatheesan, K., Ashour, A. S., Anand, B., & Dey, N. (2018). Ant colony optimization based load frequency control of multi-area interconnected thermal power system with governor dead-band nonlinearity. In *Smart trends in systems, security and sustainability* (pp. 157–167). Springer.
12. Dorigo, M., & MB, S. T. (2006). Ant colony optimization-artificial ants as a computational intelligence technique. Universit Libre de Bruxelles. Tech. Rep. TR/IRIDIA/2006–023.
13. Naidu, K., Mokhlis, H., & Bakar, A. A. (2014). Multiobjective optimization using weighted sum artificial bee colony algorithm for load frequency control. *International Journal of Electrical Power & Energy Systems, 55*, 657–667.
14. Roy, A., Dutta, S., & Roy, P. K. (2015). Load frequency control of interconnected power system using teaching learning based optimization. *International Journal of Energy Optimization and Engineering (IJEOE), 4*(1), 102–117.
15. Tsay, T. S. (2011). Load–frequency control of interconnected power system with governor backlash nonlinearities. *International Journal of Electrical Power & Energy Systems, 33*(9), 1542–1549.
16. Mohanty, B., Panda, S., & Hota, P. K. (2014). Differential evolution algorithm based automatic generation control for interconnected power systems with non-linearity. *Alexandria Engineering Journal, 53*(3), 537–552.
17. Pandey, S. K., Mohanty, S. R., Kishor, N., & Catalão, J. P. (2014). Frequency regulation in hybrid power systems using particle swarm optimization and linear matrix inequalities based robust controller design. *International Journal of Electrical Power & Energy Systems, 63*, 887–900.

18. Anand, B., & Jeyakumar, A. E. (2011). Load frequency control of interconnected hydro-hydro system with fuzzy logic controller. In *2011 International conference on process automation, control and computing* (pp. 1–4). IEEE.
19. Jagatheesan, K., Anand, B., Dey, N., Ashour, A. S., & Balas, V. E. (2018). Load frequency control of hydro-hydro system with fuzzy logic controller considering non-linearity. In *Recent developments and the new direction in soft-computing foundations and applications* (pp. 307–318). Springer.
20. El-Ela, A., Adel, A., El-Sehiemy, R. A., Shaheen, A. M., & Diab, A. E. G. (2021). Enhanced coyote optimizer-based cascaded load frequency controllers in multi-area power systems with renewable. *Neural Computing and Applications, 33*(14), 8459–8477.
21. Jeddi, S. A., Abbasi, S. H., & Shabaninia, F. (2012). Load frequency control of two area interconnected power system (Diesel Generator and Solar PV) with PI and FGSPI controller. In *The 16th CSI international symposium on artificial intelligence and signal processing (AISP 2012)* (pp. 526–531). IEEE.
22. Jood, P., Aggarwal, S. K., & Chopra, V. (2019). Performance assessment of a neuro-fuzzy load frequency controller in the presence of system non-linearities and renewable penetration. *Computers & Electrical Engineering, 74*, 362–378.
23. Chen, G., Li, Z., Zhang, Z., & Li, S. (2019). An improved ACO algorithm optimized fuzzy PID controller for load frequency control in multi area interconnected power systems. *IEEE Access, 8*, 6429–6447.
24. Mohanty, B., Panda, S., & Hota, P. K. (2014). Controller parameters tuning of differential evolution algorithm and its application to load frequency control of multi-source power system. *International Journal of Electrical Power & Energy Systems, 54*, 77–85.
25. Kumar, A., & Suhag, S. (2017). Multiverse optimized fuzzy-PID controller with a derivative filter for load frequency control of multisource hydrothermal power system. *Turkish Journal of Electrical Engineering and Computer Sciences, 25*(5), 4187–4199.
26. Gupta, D. K., Jha, A. V., Appasani, B., Srinivasulu, A., Bizon, N., & Thounthong, P. (2021). Load frequency control using hybrid intelligent optimization technique for multi-source power systems. *Energies, 14*(6), 1581.
27. Murugesan, D., Jagatheesan, K., & Boopathi, D. (2021). Meta-heuristic Strategy Planned Controller for Frequency Supervision of Integrated Thermal Plant with Renewable Source. In *2021 IEEE 3rd PhD colloquium on ethically driven innovation and technology for society (PhD EDITS)* (pp. 1–2). IEEE.
28. Dhundhara, S., & Verma, Y. P. (2020). Application of micro pump hydro energy storage for reliable operation of microgrid system. *IET Renewable Power Generation, 14*(8), 1368–1378.
29. Sahu, B. K., Pati, S., & Panda, S. (2014). Hybrid differential evolution particle swarm optimisation optimised fuzzy proportional–integral derivative controller for automatic generation control of interconnected power system. *IET Generation, Transmission & Distribution, 8*(11), 1789–1800.
30. Jagatheesan, K., Anand, B., & Omar, M. (2015). Design of proportional-integral-derivative controller using ant colony optimization technique in multi-area automatic generation control. *International Journal on Electrical Engineering and Informatics, 7*(4), 541.
31. Jagatheesan, K., & Anand, B. (2018). AGC of multi-area hydro-thermal power systems with GRC non-linearity and classical controller. *Journal of Global Information Management (JGIM), 26*(3), 11–24.
32. Dash, P., Saikia, L. C., & Sinha, N. (2015). Automatic generation control of multi area thermal system using Bat algorithm optimized PD–PID cascade controller. *International Journal of Electrical Power & Energy Systems, 68*, 364–372.
33. Golpira, H., & Bevrani, H. (2011). Application of GA optimization for automatic generation control design in an interconnected power system. *Energy Conversion and Management, 52*(5), 2247–2255.
34. Kaliannan, J., Baskaran, A., & Dey, N. (2015). Automatic generation control of thermal-thermal-hydro power systems with PID controller using ant colony optimization. *International Journal of Service Science, Management, Engineering, and Technology (IJSSMET), 6*(2), 18–34.

35. Dong, L., Zhang, Y., & Gao, Z. (2012). A robust decentralized load frequency controller for interconnected power systems. *ISA Transactions, 51*(3), 410–419.
36. Chaturvedi, D. K., Umrao, R., & Malik, O. P. (2015). Adaptive polar fuzzy logic based load frequency controller. *International Journal of Electrical Power & Energy Systems, 66*, 154–159.
37. Dixit, M., & Roy, R. (2015). Impact of PEVs on automatic generation control using TLBO algorithm. In *2015 IEEE international conference on signal processing, informatics, communication and energy systems (SPICES)* (pp. 1–5). IEEE.
38. Gheisarnejad, M., & Khooban, M. H. (2019). Design an optimal fuzzy fractional proportional integral derivative controller with derivative filter for load frequency control in power systems. *Transactions of the Institute of Measurement and Control, 41*(9), 2563–2581.
39. Kanimozhi, R. (2022). Load frequency control of three area power system using optimal tuning of fractional order proportional integral derivative controller with multi objective grey wolf optimization. *Journal of Engineering Research.*
40. Padhan, D. G., & Majhi, S. (2013). A new control scheme for PID load frequency controller of single-area and multi-area power systems. *ISA Transactions, 52*(2), 242–251.
41. Jagatheesan, K., Anand, B., Dey, N., Ashour, A. S., & Balas, V. E. (2016). Conventional controller based AGC of multi-area Hydro-Thermal power systems. In *2016 international conference on computer communication and informatics (ICCCI)* (pp. 1–6). IEEE.
42. Jagatheesan, K., Anand, B., Dey, K. N., Ashour, A. S., & Satapathy, S. C. (2018). Performance evaluation of objective functions in automatic generation control of thermal power system using ant colony optimization technique-designed proportional–integral–derivative controller. *Electrical Engineering, 100*(2), 895–911.
43. Sahoo, A. K., Chitra, V., & Kanagapriya, T. (2012) Load frequency control for a distributed grid system involving wind and hydro power generation. In *2012 2nd international conference on power, control and embedded systems* (pp. 1–6). IEEE.
44. Saikia, L. C., Nanda, J., & Mishra, S. (2011). Performance comparison of several classical controllers in AGC for multi-area interconnected thermal system. *International Journal of Electrical Power & Energy Systems, 33*(3), 394–401.
45. Jagatheesan, K., Anand, B., Dey, N., Omar, M., & Balas, V. E. (2017). AGC of multi-area interconnected power systems by considering different cost functions and Ant Colony Optimization technique based PID controller. *Intelligent Decision Technologies, 11*(1), 29–38.
46. Dhillon, S. S., Lather, J. S., & Marwaha, S. (2015). Multi area load frequency control using particle swarm optimization and fuzzy rules. *Procedia Computer Science, 57*, 460–472.
47. Daneshfar, F., & Bevrani, H. (2012). Multiobjective design of load frequency control using genetic algorithms. *International Journal of Electrical Power & Energy Systems, 42*(1), 257–263.
48. Fayek, H. H., & Mohammadi-Ivatloo, B. (2020). Tidal supplementary control schemes-based load frequency regulation of a fully sustainable marine microgrid. *Inventions, 5*(4), 53.
49. Mishra, S., Patel, S., Prusty, R. C., & Panda, S. (2020). MVO optimized hybrid FOFPID-LQG controller for load frequency control of an AC micro-grid system. *World Journal of Engineering.*
50. Murugesan, D., Jagatheesan, K., Kulkarni, A. J., Shah, P., & Sekhar, R. (2023). Comparison of biologically inspired algorithm with socio-inspired technique on load frequency control of multi-source single-area power system. In *Applied genetic algorithm and its variants: case studies and new developments* (pp. 185–208). Springer Nature.
51. Mishra, D., Sahu, P. C., Prusty, R. C., & Panda, S. (2021). A fuzzy adaptive fractional order-PID controller for frequency control of an islanded microgrid under stochastic wind/solar uncertainties. *International Journal of Ambient Energy*, 1–10.
52. Murugesan, D., Jagatheesan, K., Kulkarni, A. J., & Anand, B. (2024). Variants of the genetic algorithm on load frequency control application. In *Handbook of Formal Optimization* (pp. 1–25). Springer Nature.
53. Delassi, A., Arif, S., & Mokrani, L. (2018). Load frequency control problem in interconnected power systems using robust fractional PIλD controller. *Ain Shams Engineering Journal, 9*(1), 77–88.

54. Fayek, H. H. (2020). 5G poor and rich novel control scheme based load frequency regulation of a two-area system with 100% renewables in Africa. *Fractal and Fractional, 5*(1), 2.
55. Saxena, S., & Hote, Y. V. (2016). Decentralized PID load frequency control for perturbed multi-area power systems. *International Journal of Electrical Power & Energy Systems, 81*, 405–415.
56. Bingul, Z., & Karahan, O. (2018). Comparison of PID and FOPID controllers tuned by PSO and ABC algorithms for unstable and integrating systems with time delay. *Optimal Control Applications and Methods, 39*(4), 1431–1450.
57. Mohan, A. M., Meskin, N., Mehrjerdi, H., Yue, Z., Zhang, S., Xiao, W., Annamraju, A., Nandiraju, S., Bhukya, L., Ramesh, M., & Yadav, A. K. (2019). Load frequency control of multi-microgrid system considering renewable energy sources using grey wolf optimization. *Journal: Smart Science*, (3), 198–217.
58. Dorigo, M., & Di Caro, G. (1999). Ant colony optimization: a new meta-heuristic. In *Proceedings of the 1999 congress on evolutionary computation-CEC99* (Cat. No. 99TH8406) (Vol. 2, pp. 1470–1477). IEEE.
59. Murugesan, D., Jagatheesan, K., Shah, P., & Sekhar, R. (2024). PID controller trained using ant colony algorithm for load frequency control problems on three equal-area interconnected thermal power with renewable energy sources. In *Applications of ant colony optimization and its variants: case studies and new developments* (pp. 123–146). Springer Nature.
60. Ebrahim, M. A., Becherif, M., & Abdelaziz, A. Y. (2021). PID-/FOPID-based frequency control of zero-carbon multisources-based interconnected power systems underderegulated scenarios. *International Transactions on Electrical Energy Systems, 31*(2), Article e12712.
61. Jagatheesan, K., Anand, B., Dey, N., & Ashour, A. S. (2018). Optimal PID parameters based ant colony optimization algorithm of power system with non-linearity and boiler dynamics. *WSEAS Transactions on System and Control. 13*, 436–451.
62. Dhasagounder, M., Kaliannan, J., Shah, P., & Sekhar, R. (2024). Comparing the performance of various soft computing approaches in load frequency control for interconnected power systems combining thermal-hydro and thermal-diesel sources. *DYNA, 99*(5).
63. Dhasagounder, M., Kaliannan, J., Shah, P., & Sekhar, R. (2024). An application of artificial bee colony and cohort intelligence in the automatic generation control of thermal power systems. In *Intelligent methods in electrical power systems* (pp. 23–41). Springer Nature.
64. Rao, R. V., Savsani, V. J., & Vakharia, D. P. (2012). Teaching–learning-based optimization: An optimization method for continuous non-linear large scale problems. *Information Sciences, 183*(1), 1–15.
65. Murugesan, D., Jagatheesan, K., Shah, P., & Sekhar, R. (2023). Optimization of load frequency control gain parameters for stochastic microgrid power system. *Journal of Robotics and Control (JRC), 4*(5), 726–742.
66. Prabu, R. G. (2018). Effect of plug-in electric vehicles on load frequency control. In *2018 8th IEEE India international conference on power electronics (IICPE)* (pp. 1–5). IEEE.
67. Gorripotu, T. S., Samalla, H., Rao, J. M., Azar, A. T., & Pelusi, D. (2019). TLBO algorithm optimized fractional-order PID controller for AGC of interconnected power system. In *Soft computing in data analytics* (pp. 847–855). Springer.
68. Sahoo, P. K., Mohapatra, S., Gupta, D. K., & Panda, S. (2020). Multi verse optimized fractional order PDPI controller for load frequency control. *IETE Journal of Research*, 1–14.
69. Kulkarni, A. J., Durugkar, I. P., & Kumar, M. (2013). Cohort intelligence: A self supervised learning behavior. In *2013 IEEE international conference on systems, man, and cybernetics* (pp. 1396–1400). IEEE.
70. Kulkarni, A. J., Baki, M. F., & Chaouch, B. A. (2016). Application of the cohort-intelligence optimization method to three selected combinatorial optimization problems. *European Journal of Operational Research, 250*(2), 427–447.
71. Bhambhani, A., & Shah, P. (2016). PID parameter optimization using cohort intelligence technique for DC motor control system. In *2016 international conference on automatic control and dynamic optimization techniques (ICACDOT)* (pp. 465–468). IEEE.

72. Kulkarni, A. J., Krishnasamy, G., & Abraham, A. (2017). Solution to 0–1 knapsack problem using cohort intelligence algorithm. In *Cohort intelligence: a socio-inspired optimization method* (pp. 55–74). Springer.
73. Topno, P. N., & Chanana, S. (2018). Differential evolution algorithm based tilt integral derivative control for LFC problem of an interconnected hydro-thermal power system. *Journal of Vibration and Control, 24*(17), 3952–3973.
74. Shah, P., & Kulkarni, A. J. (2019). Application of variations of cohort intelligence in designing fractional PID controller for various systems. In *Socio-cultural inspired metaheuristics* (pp. 175–192). Springer.
75. Patankar, N. S., & Kulkarni, A. J. (2018). Variations of cohort intelligence. *Soft Computing, 22*(6), 1731–1747.
76. Sarmah, D. K., & Kulkarni, A. J. (2018). Image steganography capacity improvement using cohort intelligence and modified multi-random start local search methods. *Arabian Journal for Science and Engineering, 43*(8), 3927–3950.
77. Sapre, M. S., Kulkarni, A. J., Chettiar, L., Deshpande, I., & Piprikar, B. (2021). Mesh smoothing of complex geometry using variations of cohort intelligence algorithm. *Evolutionary Intelligence, 14*(2), 227–242.
78. Shastri, A. S., & Kulkarni, A. J. (2018). Multi-cohort intelligence algorithm: An intra-and inter-group learning behaviour based socio-inspired optimisation methodology. *International Journal of Parallel, Emergent and Distributed Systems, 33*(6), 675–715.
79. Murugesan, D., Shah, P., Jagatheesan, K., Sekhar, R., & Kulkarni, A. J. (2022). Cohort intelligence optimization based controller design of isolated and interconnected thermal power system for automatic generation control. In *2022 Second international conference on computer science, engineering and applications (ICCSEA)* (pp. 1–6). IEEE.
80. Patil, M. V., & Kulkarni, A. J. (2020). Pareto dominance based multiobjective cohort intelligence algorithm. *Information Sciences, 538*, 69–118.
81. Sekhar, R., & Shah, P. (2020). Predictive modeling of a flexible robotic arm using cohort intelligence socio-inspired optimization. In *2020 1st International conference on information technology, advanced mechanical and electrical engineering (ICITAMEE)* (pp. 193–198). IEEE.
82. Marde, K., Kulkarni, A. J., & Singh, P. K. (2019). Optimum design of four mechanical elements using cohort intelligence algorithm. In *Socio-cultural Inspired Metaheuristics* (pp. 1–25). Springer.
83. Warrier, P., & Shah, P. (2021). Optimal fractional PID controller for buck converter using cohort intelligent algorithm. *Applied System Innovation, 4*(3), 50.
84. Khodabakhshian, A., & Hooshmand, R. (2010). A new PID controller design for automatic generation control of hydro power systems. *International Journal of Electrical Power & Energy Systems, 32*(5), 375–382.
85. Attya, A. B. T., & Hartkopf, T. (2015). Utilising stored wind energy by hydro-pumped storage to provide frequency support at high levels of wind energy penetration. *IET Generation, Transmission & Distribution, 9*(12), 1485–1497.
86. Dahiya, P., Mukhija, P., Saxena, A. R., & Arya, Y. (2016). Comparative performance investigation of optimal controller for AGC of electric power generating systems. *Automatika: Časopiszaautomatiku, Mjerenje, Elektroniku, Računarstvo i Komunikacije, 57*(4), 902–921.
87. Qian, D., Zhao, D., Yi, J., & Liu, X. (2013). Neural sliding-mode load frequency controller design of power systems. *Neural Computing and Applications, 22*, 279–286.
88. Shah, P., Sekhar, R., & Shah, M. (2022) Design and implementation of CRONE controller for Automatic Voltage Regulator (AVR) System. In *2022 IEEE Bombay section signature conference (IBSSC)* (pp. 1–6). IEEE.
89. Özbay, E., & Gençoğlu, M. T. (2010) Load frequency control for small hydro power plants using adaptive fuzzy controller. In *2010 IEEE international conference on systems, man and cybernetics* (pp. 4217–4223). IEEE.
90. Alomoush, M. I. (2010). Load frequency control and automatic generation control using fractional-order controllers. *Electrical Engineering, 91*(7), 357–368.

91. Rajesh, K. S., & Dash, S. S. (2019). Load frequency control of autonomous power system using adaptive fuzzy based PID controller optimized on improved sine cosine algorithm. *Journal of Ambient Intelligence and Humanized Computing, 10*(6), 2361–2373.
92. Arya, Y. (2019). A new optimized fuzzy FOPI-FOPD controller for automatic generation control of electric power systems. *Journal of the Franklin Institute, 356*(11), 5611–5629.
93. Annamraju, A., & Nandiraju, S. (2020). A novel fuzzy tuned multistage PID approach for frequency dynamics control in an islanded microgrid. *International Transactions on Electrical Energy Systems, 30*(12), Article e12674.
94. Debnath, M. K., Agrawal, R., Tripathy, S. R., & Choudhury, S. (2020). Artificial neural network tuned PID controller for LFC investigation including distributed generation. *International Journal of Numerical Modelling: Electronic Networks, Devices and Fields, 33*(5), Article e2740.
95. Eshetu, W., Sharma, P., & Sharma, C. (2018) ANFIS based load frequency control in an isolated micro grid. In *2018 IEEE international conference on industrial technology (ICIT)* (pp. 1165–1170). IEEE.
96. Shah, P., & Agashe, S. (2013). Design and optimization of fractional PID controller for higher order control system. In *International conference of IEEE ICART* (pp. 588–592).
97. Shah, P., Agashe, S.D., & Singh, A.P. (2013) Design of fractional order controller for undamped control system. In *2013 Nirma University international conference on engineering (NUiCONE)* (pp. 1–5). IEEE.
98. Shah, P., & Agashe, S. (2016). Review of fractional PID controller. *Mechatronics, 38*, 29–41.
99. Murugesan, D., Jagatheesan, K., Shah, P., & Sekhar, R. (2023). Fractional order PIλDμ controller for microgrid power system using cohort intelligence optimization. *Results in Control and Optimization, 100218*.
100. Sondhi, S., & Hote, Y. V. (2014). Fractional order PID controller for load frequency control. *Energy Conversion and Management, 85*, 343–353.
101. Murugesan, D., Jagatheesan, K., Shah, P., & Sekhar, R. (2024). Application of a TID controller for assessing frequency stability in a thermal power system with renewable energy sources. In *Emerging materials, technologies, and solutions for energy harvesting* (pp. 158–178). IGI Global.
102. Ismail, M. M., & Bendary, A. F. (2018). Load frequency control for multi area smart grid based on advanced control techniques. *Alexandria Engineering Journal, 57*(4), 4021–4032.
103. Babaei, M., Abazari, A., & Muyeen, S. M. (2020). Coordination between demand response programming and learning-based FOPID controller for alleviation of frequency excursion of hybrid microgrid. *Energies, 13*(2), 442.

Comprehensive Analysis of Demand Response: Benefits and Challenges in Modern Power System

Mariam Naseem, Muhammad Arshad Shehzad Hassan, Umar Farooq, Asim Khan, Muhammad Zeeshan Khan, and Majid Hussain

Abstract Attentiveness towards the demand response (DR) is of great concern as it is the most probable technique to reduce the electricity system's cost and its consistency. DR can balance supply and demand more defensibly and economically, specifically when the energy is produced by unpredictable renewable resources such as solar and wind. Consumers can be encouraged if they provide DR through timebased and dynamic appraising. While executing the DR in the electricity system, the key issues and concerns are in both loads. Explicitly inflexible and flexible loads are served by the energy scheduling issues of a load-serving entity, which are also discussed. This chapter provides a comprehensive and state-of-the-art review of the operational analysis of the DR. Several comparisons of different pricing models have been made, leading to a complete conclusion of the authors' implementation benefits of DR.

Keywords Demand response · Distributed generation · Renewable energy resources · Smart grid · Price based demand response · Incentive based demand response

M. Naseem · M. A. S. Hassan (✉) · A. Khan
Department of Electrical Engineering, The University of Faisalabad, Faisalabad, Pakistan
e-mail: arxhad@yahoo.com

M. A. S. Hassan · M. Z. Khan
German Institute of Engineering, Chongqing College of Mobile Communication, Chongqing, China

U. Farooq
Shenzhen International Graduate School (SIGS), Tsinghua University, Shenzhen, China

M. Hussain
Department of Computer Sciences, The University of Faisalabad, Faisalabad, Pakistan

M. A. Mellal et al. (eds.), *Computational Intelligence for a Greener Future: Innovations in Renewable Energy Systems*, Studies in Computational Intelligence 1240,
https://doi.org/10.1007/978-3-032-06732-6_19

1 Introduction

Renewable energy resources (RERs) are ifntensively aggregated in a power system. At the same time, irregularity in their penetration levels is an immense challenge for policy fabricators and controllers magnificently. The RERs can be integrated into low- and high-voltage grids for electricity production. Enhancement in distributed generation (DG), electric vehicles (EVs), and electric battery storage may cause saturation in distributed energy resources (DERs). Moreover, that may disturb the functioning of distribution grids [1]. Although a demand response (DR) is a very economical method to manage peak demand, reliability, and risk, it lessens carbon emissions and energy costs, and the industrial and residential sectors have different aspects. The main goal of the industrial sector is maximizing profits [2]. As energy consumption cost is an essential fragment of the generation cost, DR can decrease the energy cost. Therefore, the industrial sector is more willing to adopt DR than domestic users [3].

The DR can also be expressed within the frame of a broader progress stream in a demand-side management (DSM) system. The DSM can be represented as a sequence of actions to be taken to improve the overall energy consumption on the user end. Specific methods comprise improving energy efficacy by using improved technologies and materials by end-user, manipulating smart energy dues along with incentives on specific usage behaviors, and employing compound control systems, including DERs [4]. With inadequate transmission aptitude, it will be expensive to attempt to meet the essential reliable supply of electric power to standby flexible power generation [5]. The governing authorities are intensively considering load flexibility in DR to enrich system synchronization. The consumer's capability to be flexible, respond quickly, and accept economic warnings referred to DR. Actual charges of numerous power utilities can be simulated by ample pricing signals. The price signals, which result in the adjustment of load demand, play a key role in influencing society in an optimistic financial way by entirely simulating the power consumption process [6]. Only a few months are managed during rush hour time span for peak demand, and there is a higher level of investment due to electricity generation, transmission, and electricity distribution systems. This is the result of an increase in the cost of electricity supply. In contrast, in developing conditions, such as the disruption or traffic line, the power system's assets out of the unit have many questions, passing the remaining assets out of the demand and supply of traffic or causing many expenses [4, 7–9].

The minimization of possible system collapse, remaining within technical limitations of intermediate and miniature tenure, may result in the modifications of loads because of system capability restrictions by the signaling of DR. Whereas, for an extensive duration, DR is beneficial for minimizing the investment required for power network and generation by lowering both generation and diminishing long-lasting grid congestion [10]. The DR has been intensively increasing over the last few decades due to saturation in air conditioning in homes and commercial buildings. It may result in needle peaks in the demand curve and a drop in load factor in the demand contour of the system. This enforces utilities to manage the loads and resist cost escalation by controlling the peak loads as a dedicated source [11]. During the early 1980s, subsequently passing "The Public Utilities Regulatory Policy Act (PURPA)" United States Act, load-management programs were elevated by adopting and promoting different strategies to cut off the demand peaks. It convoluted both prices based controlling and direct controlling of the large industries [12]. Large industrial clients are responsible for DR flexibility to stabilize system operations. The worth of DR is acknowledged by the European Commission [13]. The national governing authorities desire to elevate demand-side resources in combination with DR to contribute to the supply in retail and wholesale markets without any discrimination to ensure equilibrium in network operators.

Additionally, while promoting the major European regulations used for DR to contribute besides supply in an extensive market, it is important to accumulate the directions regarding general rules defined for the core energy market. The directive "ENTSO-E 2013 Demand Connection Code", besides "the ACER 2012 Framework Guidelines on Electricity Balancing" [1]. Hereafter, techniques opted for executing DR are intensively attaining devotion through governing establishments and associations. The capacity to develop DR in the European Union (EU) electricity marketplace is under practice now which is believed to raise in the future [14]. However, it will raise for domestic customers specifically because existing institutional schedules may accommodate increasing generator and commercial consumers.

The smart meters are installed along with the infrastructure of information and communication technology (ICT). These are used to facilitate the exemplar modification in the power system's functioning method, transforming conventionally passive customers to be part of dynamic market groups [14, 15]. Diversified rates stimulate a collection of incentives for consumers to transform consumption profiles, consequently providing an advantage to the system in attaining consistent goals. To persuade consumers to be responsible for load flexibility, price-reliant DR refers to financial encouragement or punishments [16]. Various options are available to manipulate and execute electricity tariffs [17]. Different pricing techniques might anticipate diverse forms of load flexibility because of indirect encouragements or incentives that may flourish due to the tariff design. So far, to stimulate large industrial consumers, time-based pricing has been considered frequently, while the technique for residential consumers has been overlooked. Literature about time-oriented pricing emphasizes DR to aid the aim of electricity resources, network purposes [18], and balancing [19]. Attention towards designing network and gird controls is attaining momentum specifically, in the power systems with extra saturation stages of renewable energy sources, which provides mutual benefits aimed at both networks and generation resolutions [20]. However, the literature does not interpret the corresponding consequences of the time-based pricing of the ultimate electrical energy bill of energy consumed by the user.

Gradual increase in DR is an asset in building the power system's flexibility and reducing the carbon emission of power sources [21]. Power obliges amendments within fixed facts of limitation progressively, which has usually been executed by increasing generation and optimizing transmission and distribution to achieve the expected requisite. Furthermore, this opportunity will be analyzed to determine if electric heating and transport addition meet the peak power demand and if power structures integrate a higher penetration of low flexible generation.

Hence, in both the real-time experiences and theoretical structure, this chapter plays a significant role in understanding reviews on DR. Additionally, existing challenges are mentioned. Also, recommendations are described to overcome them by an end-to-end modification to currently implemented rules and techniques learned from supplementary strategy frameworks as a proactive measure to implement DR in manual or automatic mode. Manual control of specific product responses generally refers to the same day in a different time segment. Semi-automated DR involves a pre-programmed DR strategy to be adopted by one centralized control system. Fully automated DR does not include human interference but has begun receiving external communication signals in homes, buildings, and facilities [22, 23].

DR schemes usually have a better existence in commercial and industrial regions than residential regions, keeping in mind that buildings like industrial units are extensive consumers of energy [24]. On the other hand, the DR perspective is particularly favorable for the essentially intact residential sector [25]. Furthermore, the residential sector is characterized by a wide variety of user behaviors and habits, thus representing a big challenge but, at the same time, a great potential for DR programs.

Previously, most of the reviewers had emphasized the advantages and challenges of existing DR schemes that can attain particular objectives. In this chapter, the primary emphasis is on a review analysis of different parameters of DR schemes by which we can motivate domestic, commercial, and industrial customers to participate and effectively utilize energy [26]. We can perform DSM by integrating RERs and introducing a pricing model for real-time electricity usage and real-time metering [27]. For this, instead of just focusing on price-based DR and incentive-based DR, we should encourage customers to focus on a hybrid price/incentive-based DR scheme in which users can avail themselves of a dynamic package of services of their choice. We should also make awareness schemes for consumers so they can be aware of the benefits of the DR and integration of RERs, by which user can play vital role in reducing the carbon dioxide emission gasses and can reduce the pollution. DR may also help financially to the customer by reducing the billing amounts and may also be provided by incentives on proficiently utilization of the energy. Whereas, inadequate attention has been given to DR schemes for commercial buildings, load scheduling methods, and the modern communication tools for DR implementation for commercial and industrial user. This aims to subsidize filling this research gap and mention the existing issues and challenges for future directions.

The content of this chapter is arranged in the corresponding sections. In Sect. 2, a theoretical explanation of DR is discussed. Section 3 provides mandatory energy consumption features aimed at incentivizing DR and time-based pricing and incentive-based pricing techniques for DR. Section 4 discusses different methods, their benefits, and their limitations. Section 5 summaries the main barriers to stimulating DR. In Sect. 6, future recommendations are provided as strategy endorsements. In Sect. 7, DR is monitored by conclusions.

2 Definition of Demand Response

As per the literature, many DR descriptions are provided. On the other hand, a common subject is that DR redirects energy demand that is reactive with intent (flexible) toward financial indications Table 1. Elaborate frequently mentioned DR descriptions specified in lite strategy documents and literature. A vital change in DSM and DR is realizing a curve model that can incorporate DR, energy efficiency, and electricity storage, as determined by DSM guidelines and strategies [28].

The power system facing overcrowding problems requires location-specific flexibility in demand. While DR is achieved by sustaining system stability through market provisions, the accumulated size, direction, and timing significance differ from that of the site-like power system.

2.1 Various Levels of Automation of Demand Response

Programmed solutions consent to more control over the consumers' discontinuation so that DR can be managed more appropriately. Real-time pricing schemes are programmed to charge EVs stations by considering the prevention of the voltage profiles [33]. In discussing communications between the DR performers, procedures have been suggested for the exchange of information and the balancing of the demand of consumers by optimizing the system [34, 35].

The uncertainty of usage motivates the insertion of direct digital controllers in buildings for cooling and heating services. Such controllers consent to reasonable levels of energetic effectiveness, which efficiently integrate DR [36]. Moreover, they

Table 1 Definitions of the DR, which are mentioned frequently

Sr no	Definition	Citations
1.	*DR comprises all intended electricity usage pattern variations by the consumers which are anticipated to modify the timing, range of instantaneous demand, and the entire electricity consumption*	[29]
2.	*Any modifications in load pattern by the customers from the usual load shapes in response to variations in electricity billing rates, encouragement payments intended to regulate electricity consumption, or feedback to the approved customer's offer, including through aggregation*	[30]
3.	*DR states that as an extensive series of arrangements that have to be accomplished by the consumer, the electricity pricing meter displays the specific conditions of the electricity system "(such as peak period network congestion or high prices)."*	[31]
4.	*The extensive description of DR comprises variations in electricity usage by the customers in feedback to the rate and application of supplementary energy-efficient equipment*	[32]

support dynamic operation on loads, fulfilling economic and technical criteria and restraints [37]. Resolving the challenges concerning automated control in any electric network involves combining DR with hierarchy-based schemes to control usage. Such schemes can decrease the redundant circuit requirement, handle better flexible loads, and fulfill customers' power and comfort needs through the dynamic operations over the network that leads to betterment on the lowest system voltage and improved distribution of loads [38, 39]. Most of the programmed load management prototypes motivate consumers to manage their energy usage within the permitted limit of consumption distribution according to the DR rate schemes suggested to reduce bills in the case of price-based programs [40]. There are various levels of automation for DR that can be implemented.

- Manual DR

DR is executed by the supervising staff, who physically turn off the electric appliances upon receiving the intimation of forthcoming DR events.

- Partially Automated DR

In semiautomatic DR, instead of executing all of the services physically by the labor, it reduces the load of labor, and the centralized, programmable control system performs the DR schemes, which increases the reliability of DR.

- Entirely Automated DR

Facilitates remotely produced event signals to control loads directly or to initiate pre-programmed DR approaches at the site. Fully automated DR can function deprived of human involvement. It is suggested that service providers remain informed of the method and can "opt out" of a DR event if needed.

2.2 Traditional Billing Scheme

The challenges facing different electrical systems are low efficiency, high reliability of energy loss, high emissions, and high probability of market forces exercise. The traditional flat tariff of electricity market prices and retail and wholesale price separation lead to inefficient use of resources because consumers are not motivated to adjust their use to provisioning costs [41–44]. Electricity bills comprise several factors, like unanimous electricity contract acquisitions, unconventional penalties, and grid charges [45]. For the upcoming days, the customers commit the consumption of electricity that loads on top of their longer-term contract commitments. The customer has to follow this demand shape to avoid paying penalties [45]. Energy pricing is a significant technique through which consumers can be incentivized by sustaining voluntary options. The ultimate electricity bill remains subject to particular cost mechanisms, and other (policy) rules related to the rates are endorsed under the billing tariff. This Fig. 1 defines one after the other cost provision:

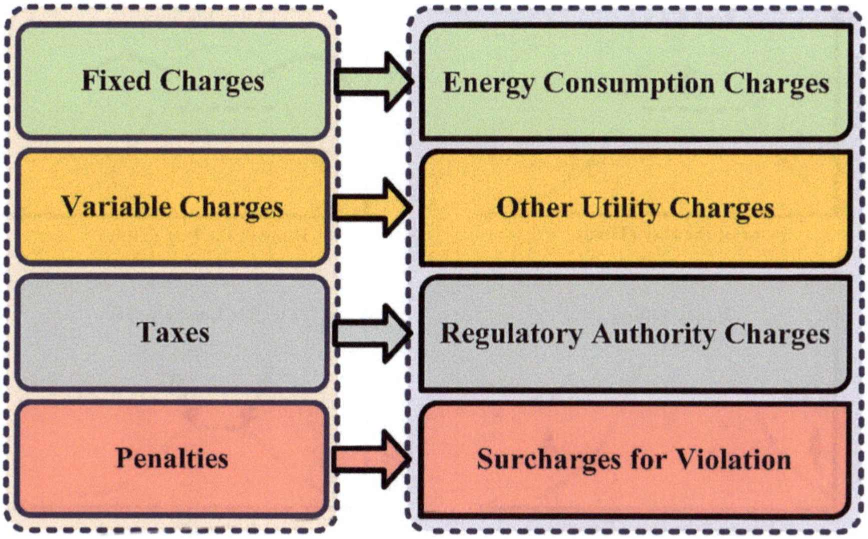

Fig. 1 Billing scheme

2.3 Demand Response Effects on the Power System

Techniques adopted in DSM are intended to modify conventional energy demand using multiple methods. Figure 2 demonstrates the graphical view of different load shapes. Techniques used to lessen consumption in peak hours can be considered peak clipping forms of DR. Decreasing the usage during peak hours to balance the consumption in off-peak hours is typically called load shifting. During extraordinary renewable energy production, DR, intended to enhance the consumption level, can be considered as load shifting, valley filling, or flexible load patterns.

2.3.1 Demand Response Advantages to Power System

The significant objectives of DR are to benefit society economically, be environmentally friendly, and increase reliability. In economics and market trends, DR decreases the overall cost of an electricity supply while preserving sufficient backup limits and reducing price instability by adapting short-term DR to power market situations [46]. Environment-friendly DR serves socially by reducing power consumption, enhancing energy efficiency, describing assurance to climate-friendly generation, and decreasing emission of greenhouse gasses. At the same time, network-driven DR aims to sustain system reliability by reducing consumption for a short duration and subsiding the requirement to improve generation or transmission capability in the power system emergency programs to reduce the consumption levels, mainly through the direct load control (DLC) program and emergency load reduction programs. As

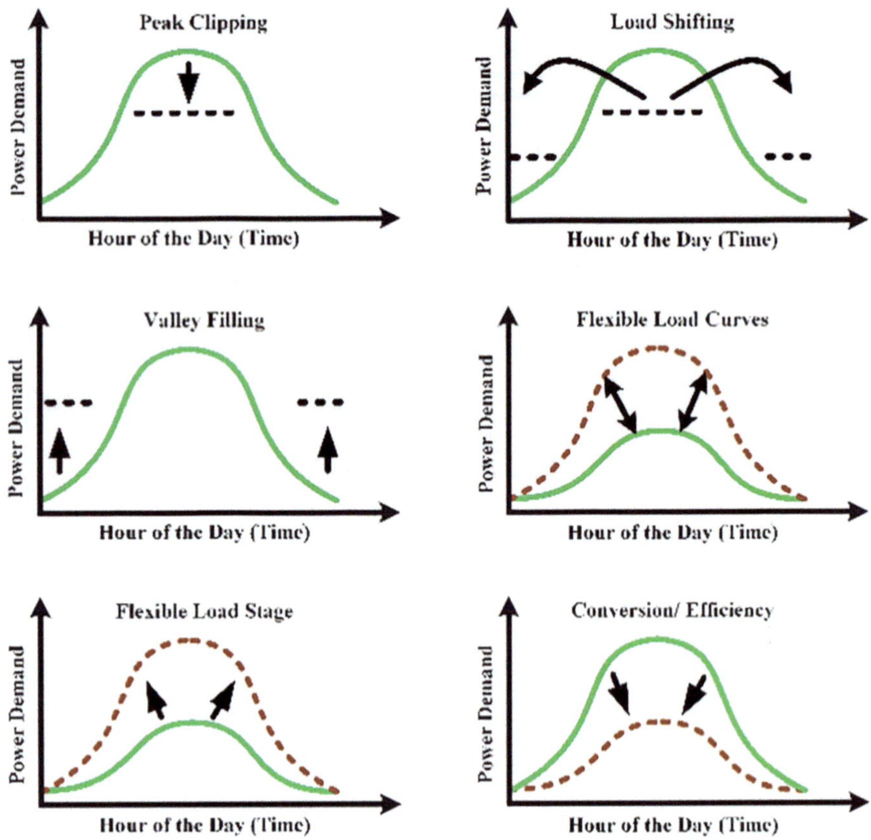

Fig. 2 Shape of load at different types of DR [1]

a result, the power system is reduced; in a sense, the system operator does not have to shed some loads, and in affirmation, power system reliability is improved.

DR benefits are classified according to the duration of the treatment. In the short term, by keeping power flow within technical constraints, DR can reduce the network peak and threat of system failure. However, in the long term, by increasing energy efficiency, DR can reduce the network investment requirements and power generation [10]. DSM could bring a variety of prospective advantages in terms of (1) reducing the investment required for the new network; (2) increasing the number of DG, which can be associated with the current distribution network structure [47]; (3) the regulator voltage limited transfer of power, (4) to relieve congestion, distribution substations, (5) streamline the interrupt management and improving its quality and security of supply, the critical load, and (6) provide the appropriate carbon reduction [48].

2.4 Consumer Response

First, a consumer can decrease their power consumption in critical peak hours when extra charges are to be paid, deprived of any change in usage pattern compared to other periods, as shown in Fig. 3. This opportunity includes a temporary loss of ease. This response can be attained by temporarily changing the settings of control devices of electric heaters and other electric equipment [49]. Furthermore, consumers may avoid extra electricity costs by transferring their peak hour's load to the leaner hours. For example, they can shift some activities (e.g., washing, vehicle charging, water pumps) to leaner hours. Domestic consumers will pay no cost in the above-mentioned cases. However, it is difficult for industrial consumers to reschedule activities to avoid extra energy costs during peak hours. Additionally, the consumer response is to generate electricity onsite, or consumers have DG [50]. Consumers who generate their electricity attain a very negligible change in their power consumption pattern, whereas from a utility perspective, a significant change can be seen in the power consumption pattern, and demand will decrease.

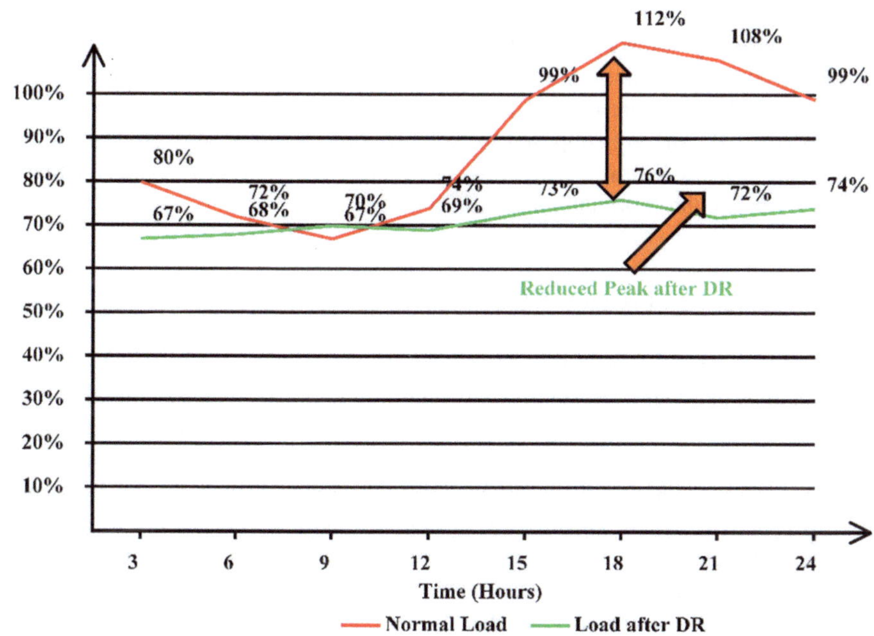

Fig. 3 Reduced consumption after DR

2.5 *Applications of Demand Response*

Reducing power consumption can provide multiple remunerations. A primary goal is to reduce the market price of electricity units due to decreased demand, comply with primary structure investments that include the latest generation, transmission, and distribution technology, and decrease greenhouse emissions. Further benefits include system safety, reliability, frequency control, integration of RES, etc. The DR is an approach targeted to meet these objectives resourcefully.

2.5.1 Integration of Renewable Energy Resources

DR is implemented extensively to aid in the improved and more effective integration of solar, wind, and other renewable resources. In RERs, the ambiguity, inconsistency, and instability of energy resources can be balanced by the DR in a quick and cost-effective method [51]. On the way to achieving a low potential carbon energy method, the portion of the production of electricity from RERs in the European power supply system has increased expressively during the past few years. Continuing concerns regarding environment variation and the main objective of numerous nations to become autonomous from importing energy will probably promote further escalation in RERs share [52]. Wind and solar energy are considered the primary sources of RERs and are supposed to be the premier sources of energy in Europe [53]. Because it is a natural source, electricity production from such sources fluctuates intensely during the short and long term [54, 55].

2.5.2 Voltage and Frequency Control

To attain the state of equilibrium of the power system and balance out the change in demand as per change in supply, two stability factors must be sustained: voltage and frequency. DR can be integrated into the power system to maintain the voltage and frequency stability points. The uncertainty of RERs, as well as DR to the power system, is accumulating a novel dimension to the issues of voltage and frequency control [56]. The synchronization of various DG entities is necessary to meet the aggregate electricity demand. Numerous control schemes, like droop control, average current sharing control, and master–slave control, have been implemented extensively worldwide to control load sharing by parallel connected inverters in DG networks. Amongst such techniques, the droop control method has been extensively recognized in the systematic community due to the deficiency of serious communication tools between parallel linked inverters to synchronize the DG entities contained by a microgrid [57]. In a microgrid, load-sharing accuracy will drop when line impedance is used. In this way, the droop control technique is used to analyze the line impedance's effect on the load distribution's precision [58].

The operational approach of the microgrid can be converted between grid-connected and islanded approaches. The main problem for parallel connected DG in the microgrid is avoiding circulating current. Among the solutions, droop control is the best technique, and it is extensively implemented in uninterruptable power supplies and conventional synchronous generators. Its main benefit is that each DG can operate independently without critical communication. P-f and Q–V droop control are typically implemented when a microgrid's correspondent impedance connecting DG to the grid is inductive. Using P-f droop control, the active power can be distributed accurately based on the frequency of the microgrid. Hence, the precision of reactive power allocation using Q–V droop control is generally significantly affected by DG's corresponding line impedance and impedance. To boost reactive power-sharing accuracy, droop control must be optimized. In a microgrid, this method is valid only with a diminutive distribution line where the line impedance is insignificant; for the power system founded on rotating generation, frequency and active power are closely connected. An increase in load indicates that load torque increase is deprived of the equivalent increase in the primary torque, which indicates that the rotating speed exhibits a direct decrease in the frequency. The primary method, deriving the droop control, reflects the complex shift by the transmission line.

$$S = P + jQ$$

Here, S represents complex power (VA), P denotes real power (W) representing energy transfer, Q signifies reactive power (VAR) representing energy oscillation, and j is the imaginary unit. This formulation enables unified analysis of both power components using complex mathematics.

$$S = VI^* \tag{1}$$

This expression involves voltage phasor V and current conjugate I^*, derived from instantaneous power analysis. The complex conjugate ensures proper phase accounting between voltage and current, establishing the foundation for AC power calculations.

$$\mathrm{I}^* = \left(\frac{\mathrm{V1} - \mathrm{V2}}{Z}\right)$$

where V1 and V2 are voltage phasors at buses 1 and 2, and Z represents line impedance. This relationship defines current flow between network nodes based on voltage differences and impedance characteristics.

Putting the value of I* in Eq. (1)

$$\mathrm{S} = \mathrm{V}_1\left(\frac{\mathrm{V1} - \mathrm{V2}}{Z}\right)^*$$

$$S = V_1\angle 0\left(\frac{V^1\angle 0 - V_2\angle - \delta}{Z\angle\theta}\right)^*$$

Now, convert the polar form $V1\angle 0$, $V2\angle - \delta$, $Z\angle\theta$ to the exponential form $V1e^{j\theta} = V1$, $V2e^{-j\delta}$, $Ze^{j\theta}$.

$$S = V_1\left(\frac{V_1 - V_2e^{-j\delta}}{Ze^{j\theta}}\right)^* \tag{2}$$

$$S = V_1\left(\frac{V_1 - V_2e^{-j\delta}}{Ze^{j\theta}}\right)^* = V_1\left(\frac{V_1 - V_2e^{j\delta}}{Ze^{-j\theta}}\right)$$

The power angle δ represents the phase difference between V_1 and V_2, while θ denotes the impedance angle. Exponential form facilitates separate analysis of magnitude and phase components.

$$S = \frac{V_1^2 e^{j\theta}}{Z} - \frac{V_1 V_2 e^{j(\theta+\delta)}}{Z} \tag{3}$$

As Euler's identity defines

$$e^{j\theta} = \cos\theta + j\sin\theta$$

By using Euler's formula and separating the total power into the real and imaginary components,

$$S = P + jQ$$

$$P = \frac{V_1^2 cos\theta}{Z} - \frac{V_1 V_2 \cos(\theta+\delta)}{Z} \tag{4 Real Part}$$

$$Q = \frac{V_1^2 sin\theta}{Z} - \frac{V_1 V_2 \sin(\theta+\delta)}{Z} \tag{5 Imaginary Part}$$

The real component represents net energy transfer, while the imaginary component represents reactive energy circulation between inductive and capacitive elements. We can also change the polar form to the rectangular form.

$$Z\angle\theta = Ze^{j\theta} = R + jX$$

Also, $Z = \sqrt{R^2 + X^2}$, $R = Z\cos\theta$, $X = Z\sin\theta$.

Where R represents resistance (energy dissipation) and X represents reactance (energy storage). This transformation enables analysis based on physical circuit parameters. For real Power,

$$\mathrm{P} = \frac{\mathrm{V}_1}{Z}[V_1 cos\theta - V_2\cos(\theta + \delta)] \tag{6}$$

As, cos(θ + δ) = cosθcosδ–sinθsinδ.
Therefore,

$$P = \frac{V_1}{Z^2}[V_1 Z\cos\theta - V_2 Z\cos\theta\cos\delta + V_2 Z\sin\theta\sin\delta]$$

$$P = \frac{V_1}{Z^2}[Z\cos\theta(V_1 - V_2\cos\delta) + Z\sin\theta(V_2\sin\delta)]$$

$$P = \frac{V_1}{R^2 + X^2}[R(V_1 - V_2\cos\delta) + X(V_2\sin\delta)] \tag{7}$$

This derivation demonstrates real power dependence on both resistance and reactance components, with contributions from voltage magnitude differences and phase angles. For reactive Power,

$$\mathrm{Q} = \frac{\mathrm{V}_1}{Z}[V_1 sin\theta - V_2\sin(\theta + \delta)] \tag{8}$$

As, sin(θ + δ) = cosθsinδ + sinθcosδ.
Therefore,

$$Q = \frac{V_1}{Z}[V_1\sin\theta - V_2(\cos\theta\sin\delta + \sin\theta\cos\delta)]$$

$$Q = \frac{V_1}{Z.Z}[V_1 Z\sin\theta - V_2 Z(\cos\theta\sin\delta - \sin\theta\cos\delta)]$$

$$Q = \frac{V_1}{Z^2}[V_1 Z\sin\theta - V_2 Z\sin\theta\cos\delta - V_2 Z\cos\theta\sin\delta]$$

$$Q = \frac{\mathrm{V}_1}{Z^2}[Z\sin\theta(V_1 - V_2\cos\delta) - Z\sin\theta(V_2\sin\delta)]$$

$$Q = \frac{V_1}{R^2 + X^2}[X(V_1 - V_2\cos\delta) - R(V_2\sin\delta)] \tag{9}$$

Reactive power is primarily influenced by reactance and voltage magnitude differences, establishing the basis for voltage control in power systems. Typical transmission lines are modeled with inductance much greater than reactance. So, resistance is commonly neglected.

$R \approx 0$

$$So,\ P = \frac{V_1}{X^2}[X(V_2\sin\delta)] = \frac{V_1}{X}[V_2\sin\delta] \tag{10}$$

$$Q = \frac{V_1}{X^2}[X(V_1 - V_2\cos\delta)] = \frac{V_1}{X}[V_1 - V_2\cos\delta] \tag{11}$$

This approximation applies to high-voltage transmission systems where reactance dominates resistance, simplifying analysis while maintaining essential system behavior characteristics. Where, $\cos\delta = 1$ and $\sin\delta = \delta$.

$$\mathrm{P} = \frac{V_1}{X}[V_2\delta]$$

$$\mathrm{Q} = \frac{\mathrm{V}_1}{X}[V_1 - V_2]$$

Hence,

$$\delta = \frac{PX}{V_1V_2} \tag{12}$$

$$[V_1 - V_2] = \frac{QX}{V_1} \tag{13}$$

The derived power flow equations demonstrate that power angle δ depends primarily on active power P, while voltage differences depend primarily on reactive power Q. This mathematical framework provides the theoretical foundation for understanding, analysing, and controlling electrical power systems, from traditional networks to modern smart grids incorporating renewable energy sources [59].

2.5.3 Demand Response Services

DR originates facilities that flatten the demand shape [60, 61], implementing the structure in the duration of high usage. Rigid stability is attained in the meantime [62]. By introducing new performers in power systems such as EVs and DG, DR proposes schemes for power systems that are directed to attain stability [63, 64]. DR pursues to propose services by improving the demand shape, which interprets into moderately dropping the electricity using in valley time slots or slots of peak hours; the peak lessening approach anticipates to smear DR in high usage hours, and the peak relocation technique induces the customer to shift their usage to time slots somewhere it is typically not that high. Table 2 demonstrates the utmost illustrative network facilities founded on the DR technique.

2.5.4 Consumers

Classification of consumers that contribute to the DR technique, which is conventionally based on standards that enhance from the regular level of energy usage up to

Table 2 Services presented by DR techniques

Remuneration	Related features	Citations
Voltage profiles	An integrated controlling programmed algorithm is designated to deal with auxiliary power network services by controlling the energy resources. The controlling schemes comprise voltage contained by adequate parameters attaining an enhancement in the voltage profile	[65]
Electric vehicle recharge	Charging EVs by regulating loads can be disturbed by the high rates of energy instants	[63]
	A controlling strategy for the home energy system is established to regulate the charging and discharging of EVs	[63]
	In the power system, the writer proposed a structure to control the charging of EVs and numerous controlling algorithms unified into a multi-mediator network to explore the effect of the diffusion of EVs	[66]
System reliability	A controllable load is used as a standby to ensure the network's uncontrollable load is supplied	[67]
Distributed generation support	In directive to improve the consistency of the power system, DR works as a backup to the stochastic system for generating energy from RERs	[1]
	The researcher emphasizes the strategy of demand algorithm, which can be opted to access several forms of energy by generating equipment, founded on the inquiry and a reminiscence of DR contrivances	[68]
Photovoltaic energy feature	A methodology is suggested to adopt the optimum DR in the domestic zone to reduce voltage peak curves, specifically in the extraordinary photo voltaic generation	[69]
Network stability	The intelligent system persists in the demand administration, including undertakings to modify the voltage profile completely	[70]
Peak Reduction	DR techniques are enhanced by the differential development procedure and implemented during peak hours	[71]
Voltage stability	DR is used to observe voltage safety in emergency circumstances. Simulated neural systems suggested a rapid inspection to attain voltage stability and achieve valuation	[72]

social and economic parameters. In this regard, the arrangement has been prepared to agree to the consumer's profile, separating them into domestic consumers whose usage observes the energy needs at homes; for commercial consumers, the electrical energy usage be determined by the sale and purchase of services in addition to industrial consumers the consumption of power energy usually reliant on the transformation of supplies into usable goods. The reconsideration indicates that the habit of variation in the use of home appliances is encouraged [73–75]; however, possession contented circumstances at home [76]. In existing circumstances, few loads like air conditioned are interfered with [77, 78], whereas refrigeration is seen in scenarios of the commercial user [79, 80]. Table 3 elaborates on the features of improvements concerning the mentioned consumer types.

Table 3 DR schemes for various kinds of consumers

Consumer	Emphasized Features	References
Residential	A demonstrating method in DR is suggested using an algorithm in which contributing consumers propose their discontinuation	[81]
	Modeled the Markov chains to describe the prospect of the contribution of residential customers in the DR	[82]
	Smart appliances at home can transform photovoltaic methods in reaction to vibrant deviations in energy rates	[83]
	Enormous data strategies are utilized to assemble and process data from various domestic clients to remember them for DR schemes	[84]
	A technique of artificial systems is recommended in residential energy organizations considering the occasions of DR	[85]
	Intelligent systems are publicized as power energy market service providers that offer multiple advantages for residential customers and the atmosphere	[86]
	The adequacy of DR dues was evaluated for residential consumers, including the actual use time that provides advantages to the system	[87]
	A methodology is projected with the goal that the administration framework retains the utilization of the customer energy under a well-defined time limit	[88]
	An uncertain subtractive combination technique is implemented to the contingency situation specified by the categorization of behaviors adopted by residential consumers towards the DR	[89]
	It expresses an amplification technique for the social assistance of the customers when changing the ongoing pricing strategy	[90]
Commercial	Commercial buildings act as a worthy contender to deal with flexible demand because of their capacity and constancy in loads, which are utilized to consider the relationship of components that influence the DR expectation	[91]
	The forecast of energy utilization is attained by constructing modeling methodologies using an estimated structure of DR	
	The writer considers the execution of DR actions by a simulation prototype entitled Energy Plus in a particular building	[92]
	The writer explores a joint approach of onsite power generation while decreasing its actual demand by the DR programs for commercial area buildings in Thailand	[61]
	An agreeable plan of DR is suggested aimed at commercial buildings with a foreseen pricing strategy	[23]
Industrial	The researcher implemented a DR scheme to reduce the load in manufacturing enterprises. Moreover, they encourage electricity usage other than peak domestic periods and dues that double per kWh	[93]

2.5.5 Distributed Generation

By introducing renewable sources, DG plays a key role in decentralizing energy production by allowing users to contribute to their community by generating electricity on their own. With the evolution of renewable energy, the prospect of decreasing energy costs has risen through strategies and plans such as the DR. This set these sources irrespective of their essential stochastic [94]. A significant cause of destabilizing issues in voltage for numerous nodes of the system is an increase in the diffusion of DG sources founded on renewable energy sources like wind and photo-voltaic sources. Meanwhile, the occurrence of electric vehicle loads also plays a role in unbalancing the network [95].

For a stable DG network, the flow control is essentially null in a neutral line. During the unbalanced state of the network, the neutral line current will not be zero. Some researchers suggested practicing the current flow from the neutral line as an indication signal to control energy storage entities and their capability to add energy to the network, decreasing the non-stabilized consequence in voltage [96, 97]. To deal with the need to reduce voltage unbalancing problems, a few solutions are recommended to use DR with dynamic load controllers, like on load tap changers (OLTCs) transformers. In order to apply DR, optimization complications associated with consumers' choices are resolved, such as the minute disruption in the comfort of the user [98].

It is significant to identify those variations in the demand that can infrequently lead to destabilizing complications in power and voltage, generally the commutation procedures of inductive loads that include conventional measures like capacitor banks [99]. A few researchers recommend that DG essentials take responsibility for justifying destabilizing complications [100, 101]. Consequently, hybrid micro-grids with wind and photo-voltaic energy sources have been deliberated on the use of "maximum power point tracking (MPPT)" controllers to maximize the storage of energy in altering load setups [102, 103]. This way out sustains the balance in power and a voltage with less than 0.1 s response times, which assure a dynamic system procedure. Table 4 Illustrates the conventional association between DG and DR schemes.

3 Classifications of Demand Response

There are numerous means that utility adopts for DR. Several techniques are presented and are further in development to industrialize the procedure of DR. Through these techniques, the need for load shedding can be detected and communicated to the customer. These techniques can be categorized into three main classes: incentive-based technique (IBT), price-based technique (PBT), and hybrid price-based/incentive-based technique (HP/IBT). Price-based DR applies universally to all consumers under the same utility, but incentive-based DR comprises customer choice and varies from user to user. At the same time, the HP/IBT is a combination

Table 4 DR schemes and their association with DG

Type of DG	Differentiating factors	References
Numerous categories	To manage the DG, a grid is added in the cycle with DR programs	[104]
Wind energy with ice-based energy storage	An energy prototype in domestic buildings that integrates the three-phase electricity wire through a transformed thermal ice storage system is introduced. This prototype can be modified per the system's needs and react dynamically to fluctuate the energy generation	[105]
Numerous categories	A microgrid was recommended to set the essentials of the DG in the cycle using price-based DR schemes	[61, 104]
Numerous categories	Customers can effortlessly adjust to the energy rates by implementing DR schemes in reasonable production so the DG can come in on a considerable scale	[106]
Numerous categories	DR supports the incorporation of renewable energies and is advantageous for electricity consumers, proposing economic incentives to rearrange their electricity usage patterns	[107]
Energy generation via photovoltaic	The generation of solar energy by using photovoltaic energy is encouraged in domestic zones by implementing the processes that ideally detect the DR during over-voltage peaks	[69]
Energy generation via photovoltaic	The best energy system model is a compulsory undertaking of the aggregator, which observes the load circumstances like the effect of the atmosphere while operating renewable sources	[108]
Numerous categories	An intelligent optimization set of rules is recommended composed of vector machines to resolve the incorporation of DG and DR in the power network	[27, 109]
Wind energy	A diversified energy price is advised to encourage consumers to captivate extra wind energy all night. The financial remunerations for consumers and the electric service companies are evaluated, and the wind instability as a corresponding load	[110, 111]
Wind energy	Smart load schemes are deliberated to aid the energy system in high saturations of renewable energy	[112]
Energy generation via photovoltaic	Ideal software design is anticipated for domestic utilization in smart homes with solar panels and battery storage systems	[113]
Wind energy storage	The demand dislocation can occur due to the assimilation of wind energy storage	[114]
Wind energy storage	The increasing contribution of renewable energies focuses on the significance of DR. Customers have to acclimate their usage to sustain the stability of the network and minimize the cost of electricity	[115]

(continued)

Table 4 (continued)

Type of DG	Differentiating factors	References
Photovoltaic and wind energy	DR server decides and distributes the amount of energy based on the data accumulated by all customers. Then, the respective customers can decide on their own to participate, and the system is adjustable regarding wind and solar energy penetration charges	[116]
Photovoltaic energy	Billing rates were designed for a case study using solar panels with battery storage. Subsequently, the result matched the present billing scheme	[117]
Numerous categories	The DR mechanism is proposed to dampen the massive appearance of extremely random sources of DG	[118]

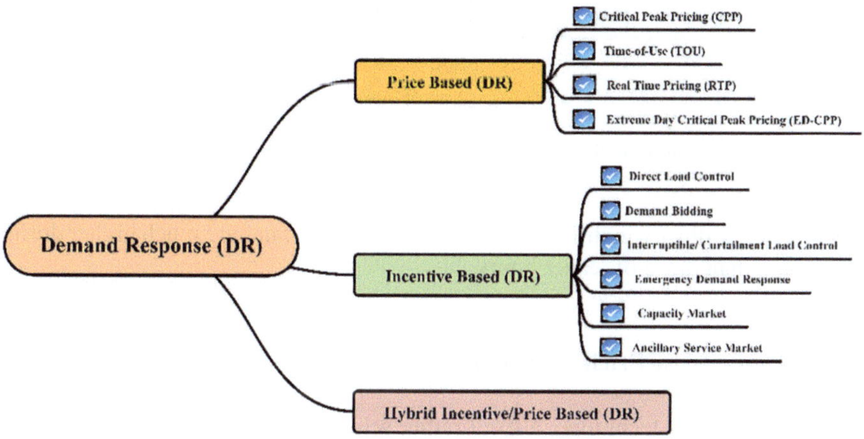

Fig. 4 Categories of DR techniques

of the other two techniques. The categorical elaboration of DR schemes is shown in Fig. 4.

3.1 Price Based Demand Response Techniques

Price-based programs rely on dynamic prices by which energy tariffs fluctuate; the variation in rate ensures the actual rate of energy. Eventually, the primary objective of all mentioned techniques is to uniform the consumption curve by subscribing high rates in peak hours and reducing the rates during leaner hours. These charges are comprised of the "time-of-use (TOU) scheme, critical peak pricing (CPP) scheme, extreme day CPP (ED-CPP) scheme, and real-time pricing (RTP) scheme. Common forms in the price based technique are the TOU scheme, in which the cost per unit

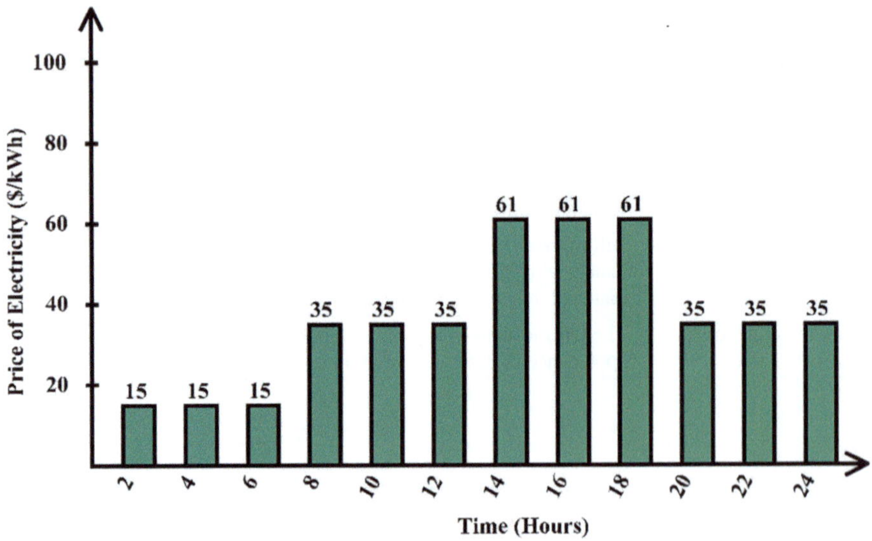

Fig. 5 Time-of-use on-peak, mid-peak and off-peak

energy usage will vary in different time slots [119]. The rates are much higher during peak hours than those of leaner hours.

3.1.1 Time-of-Use Pricing Technique

In the TOU pricing technique, a fixed rating scheme is charged on already defined prices for that particular duration, as elaborated in Fig. 5. These price patterns generally remain the same throughout the season. New assorted prices are suggested annually to meet the service-providing corporations' operational charges and long-term investment. The suggested rates are sanctioned through the rate directive authorities, whose principal objective is to defend the customers from unavoidable circumstances.

3.1.2 Critical Peak Pricing Technique

The CPP technique is an event-controlled technique in which utilities opt for extra high energy rates to regulate energy consumption in peak hours, as shown in Fig. 6. This pricing method is formulated to assist energy conservation inside and outside the home. CPP technique can be executed for that particular hour when the system is badly overburdened in extreme cold and extremely warm weather or an extreme increase in oil prices. In such conditions, customers may obtain incentives by lowering their consumption or shifting electricity usage to off-peak hours to secure the system's consistency. As system restrains occur rarely so, CPP does not take place

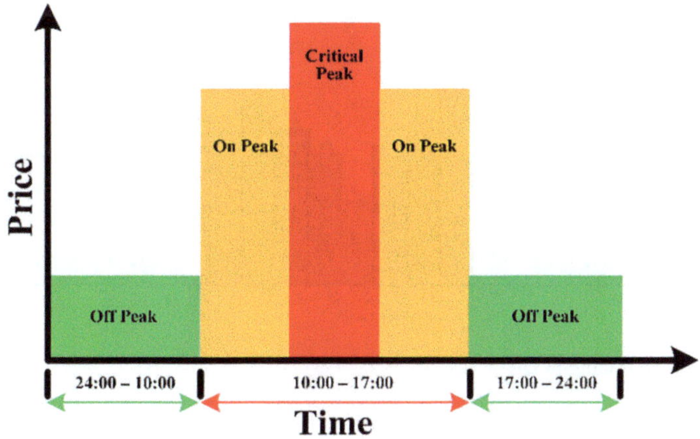

Fig. 6 Critical peak pricing in DR

regularly. Furthermore, CPP rates are much higher than TOU rates. The remuneration of the CPP technique is: a) simple to follow, b) effectual in shifting peak electricity consumption, and c) extra incentives. The CPP is a floating time slot that may or may not affect that particular day. Figure 6 reflects the results of CPP technique in DR against pricing.

3.1.3 Real-Time Pricing Technique

In the RTP technique, information about the original cost of electricity is provided at some anticipated time, or we can say in the RTP tariffs, as shown in Fig. 7. Customers paying the cost of power, which vary for small time intervals, it usually is defined for hours and is cited one day or less in advance to reflect contemporaneous marginal supply costs. RTP techniques need to be installed in the system, such as smart meters, by which information can be transformed regarding energy rates and provide more information to the customer about the usual consumption [120]. Numerous techniques are introduced to send and receive this information, like text messages, e-mails, telephone calls, etc. The RTP technique permits the consumer to utilize that information to decide energy consumption by giving exact information about energy rates determined by total demand. During the leaner hours, regulating individual energy consumption can help save customers money, taking advantage of low rates.

3.1.4 Extreme Day Acute Peak Pricing Technique

Depending on the customers' entire consumption, two pricing levels are offered in extreme day acute peak pricing technique (EDAPPT), as shown in Fig. 8, when

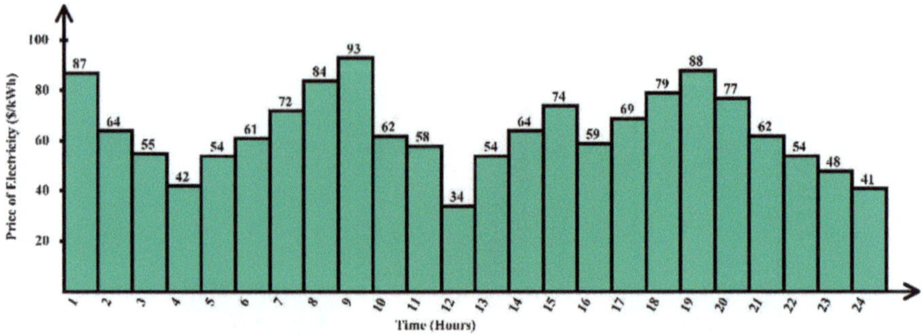

Fig. 7 Real-time pricing in DR

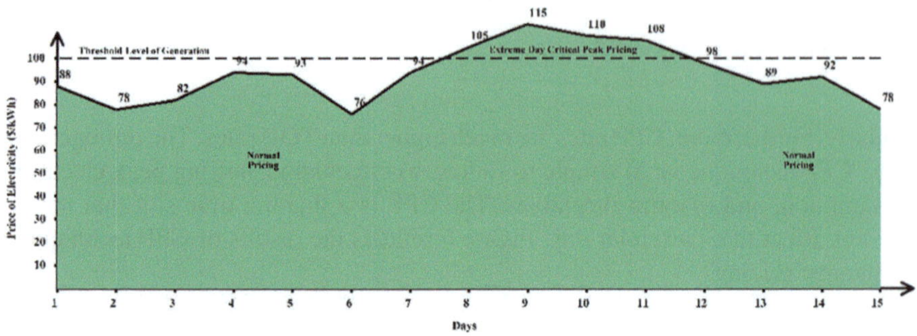

Fig. 8 Extreme day acute peak pricing in DR

consumption outstrips the threshold level of generation. The energy rate will increase to a higher level when the usage reaches the threshold. By implementing this technique, the redundant investments for generating, transmitting, and distributing more power can be reduced. This technique has been widely adopted by many of the services providing companies.

3.2 Incentive Based Pricing Techniques

In the incentive-based technique, enrolling consumers and bearing the system installation cost of control equipment and the requirement to pay incentive overheads to consumers is expensive. Figuring out appropriate electrical consumers is a taking and mandatory part of incentive-based DR [16]. Specifically, to overcome the upcoming uncertainties, an intelligent network is implemented to forecast the unidentified rates and power demands. Afterward, reinforcement culture is implemented to get the optimal incentive charges for different consumers, bearing in mind the profits for both utility and consumers [121].

- In the direct load control technique (DLCT), some consumers or devices have a registered scheme that allows the utility to turn off or, when necessary, to cycle (usually in peak demand or event) [122, 123]. Consumers receive certain payments on behalf of permitting the service provider company to regulate the operation of some of the appliances (commonly air condition or heaters) and to remotely shut appliances or turn them on/off, deprived of interfering with the other electrical appliances over definite hours of the day and in extreme weather [29]. Small and intelligent remote control is mounted, free of cost, on the consumer's appliances, which are to be controlled directly by the utility. Direct load control is typically implemented when the power system is critically overloaded, leaving the customer with no choice.
- In interruptible/curtailable control (ICC), the consumer will get encouragement payments for off concessions or cash credits when they decrease the load on request to a pre-defined value in uncertain conditions. In the form of fines, penalties are usually imposed if the consumers cannot provide the anticipated limitations [124].
- In demand bidding control (DBC), consumers can bid offers to the service provider to limit some portion of their loads through periods with extensive market rates [125]. These programs are usually available to large consumers (greater than 1 MW). Contingent liabilities or peak demand periods, consumers might limit their consumption to reach a particular part of the bid price redemption price.
- In the emergency DR technique, consumers can be rewarded for voluntarily decreasing the loads in emergency peak hours to escalate the grid consistency. Under this plan, consumers have paid many incentives to reduce their usage in serious coincidences. These programs can help improve the consistency of the power system.
- In the capacity market technique (CMT), consumers get payments or concessions to decrease the loads and reduce the capacity requirements in critical peak demand hours.
- In the ancillary services market technique (ASMT), consumers get credits as a subsidiary by the service provider, as they require consumers to adjust their loads quickly.

3.3 Hybrid Price/ Incentive Based Demand Response

Price-based DR and incentive-based DR techniques can probably be combined in several ways. It was figured out that price-based schemes can be effective only if an electricity contractor has more consumers than its electric supply ability and can purchase electricity from various electricity-providing companies/markets. It is indicated that the approach to gaining huge benefits is possible by the combinations of DR programs in several arrangements in targeted marketplaces. The foremost focus of this effort is to enumerate the benefits of DR. To obtain this exploration, a hybrid structure for the market with various pricing methods is anticipated [126]. A dynamic incentive scheme in a dual tariff system is proposed based on user elasticity and energy procurement cost analysis. A dynamic incentive strategy aims to give the opportunity to the people who want to participate while guaranteeing profit [127]. The research explores the effect of incentive payments over different smart grid technologies in several utility companies in North America. It presents how numerous incentives impact the accomplishment and scalability of smart grid DR schemes [128]. Discriminating features of DR schemes are illustrated in Table 5.

Table 5 Discriminating aspects of DR schemes

Remuneration	Discriminating Aspect of the Reviewed Papers	References
Price-based	A dynamic pricing structure for electricity in the intelligent network is recommended, considering probable DR and customer behavior	[129]
	Smart measuring bidirectional devices are used	[130]
	Between the micro network and the price-based technique of DR, the synchronization scheme is anticipated to regulate loads	[104]
	The RTP technique endorses the choice of electricity cost based on the jeopardy consumers are willing to undertake	[131]
	Multi-mediator demonstrating and optimization procedures under the DR techniques for existent prices	[66]
Incentive-based	The DR technique is presented from the financial point of view based on optimum incentives	[132]
	The authors encourage the contribution of smart homes and the assistance of a controller capable to handle the electrical restructuring and installation of the shape of demand by varying the operation of flexible loads	[133]
	An uncertain dynamic incentive-based structure is suggested for residential customers that integrate the program efficiently	[134]
	In network development, demand uncertainty is demonstrated, and the essential mechanism of load discontinuation is used to search for minimum charges	[135]
	The vibrant modification of the accessible rates is examined to decrease the demand and maximize its performance within the days	[136]
	The authors approximate encouraging customers to contribute in a spontaneous DR scheme with the persistence of reimbursing discrepancies between proposal and demand	[137]
	A programming strategy based on priorities for the loads to be detached is detailed	[134]
Hybrid	Consumers modify their usage in reaction to the pricing schemes and a dispatchable power that converts into high-demand hours. This uncertainty is assessed in contribution	[138, 139]
	It describes the choices based on incentives and prices to recognize which is most auspicious to the immense access of electric vehicles during the hours of their recharge	[66]

4 Discussions

Any technique that encourages the consumer to shift the peak demand for energy is considered as DR. The contribution of the consumer is feedback that depends on multiple factors, which can be incentives, different tariff schemes, awareness, and an intense sense of responsibility. As we know, DR is not directly an energy efficiency measure, but it can complement energy efficiency. DR is a technique used to shift energy consumption instead of reducing it. It does not change the area under the curve but only varies the nature of the demand curve. As shown in Fig. 9, DR can be categorized into price-based or incentive-based DR. In a price-based scheme, the consumer can change their consumption patterns in response to variations in the electricity rates that are exposed to the end user. Meanwhile, controlling devices are installed at the consumer end in an incentive-based scheme. Using these devices, certain electric appliances can be turned ON or OFF by the LSEs, aggregators, or system operators at certain times [140].

The current research on price-dependent DR methods has also identified numerous concerns. It is determined that problems of peak demand during the evening were not resolved by the TOU method and were directed to an increase in power demand for substations during the peak hours. The CPP technique is an easy method to follow by the participants as they can visualize their inducement rewards, which results in effective peak shifting. The deficiency of quantitative consideration of customers' performance and end-user action adaptive capabilities are significant barriers to projecting and installing effective DR. Every single DR technique has advantages and barriers to different features of flexible demand. All classifications of DR are presented for the demand price elasticity model to design the finest scheme to achieve the maximum advantage from DR. The primary objective is not only to decrease costs and increase reliability but also to escalate consumer acceptance of DR packages by preventive price instability, as in Table 6. Evaluation of various DR techniques is elaborated.

Initially, our analysis distinguishes a scope of variables that impact consumer's commitment to DR; these include:

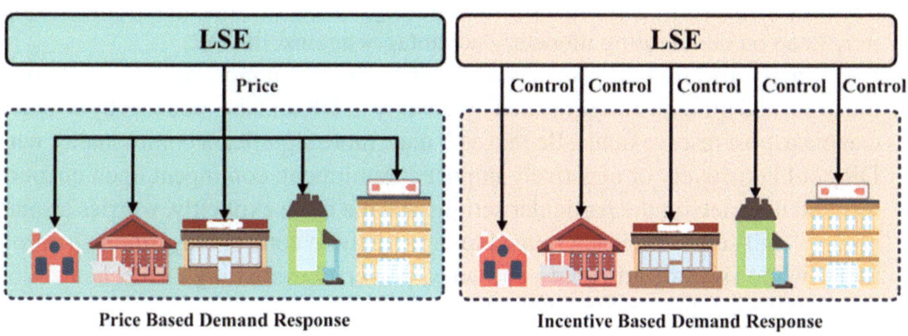

Fig. 9 Application of DR [140]

Table 6 Comparison of different DR techniques

DR method	Benefits	Disadvantages	Pricing display at the consumer end	Smart metering	User end demand control
TOU	Easy to implement Well aware consumer	Integration of RERs is limited	Not required	Mandatory	Optional
CPP	Easy to implement Efficient in shifting peak energy consumption	It does not work to reduce the cost of electricity It does not reduce carbon emission gasses	Mandatory	Mandatory	Optional
RTP	Supportive to the integration of RERs Reflect demand–supply relationship	Accurate metering is needed Accurate communication is mandatory	Mandatory	Mandatory	Optional
EDCPP	Figure out uncertain condition	It does not work to reduce the cost of electricity	Mandatory	Mandatory	Mandatory
DLC	Reliable control More efficient	No choice at the consumer end	Mandatory	Mandatory	Not required
Incentive based DR	Visualize incentive Customer's choice	Comparatively difficult to implement as it varies form one user to another	Mandatory	Mandatory	Mandatory

- **Financial and different inspirations for enrolment and response**: Financial and ecological advantages were the most ordinarily recognized inspirations for enrolment, and where these were thought about, money-related advantages were commonly found to have the most elevated significance. After some time, clients may keep on deliberating monetary advantages against the DR.
- **Familiarity and trust**: Users may question the apparent motivations of DR coordinators, and this can be aggravated by novelty to DR and the manners by which it can make cost reserve funds. Be that as it may, more significant commonality with DR could positively or negatively impact commitment, contingent upon encounters and notoriety in the particular setting. All the more explicitly, worries around security and self-sufficiency related to direct burden control may diminish enrolment, while specialized is-sues and hazy, powerful evaluating or mechanization timetables may disintegrate trust after some time.
- **Perceived hazard and control**: Enrolment might be discouraged by apparent hazards and encouraged by apparent control by clients. DLC, significant expense

levels, and less predictable appraising schedules might be related to the impression of hazard and decreased control. Some types of empowering innovation might decrease the apparent danger of dynamic evaluation, and clients may consider mechanization to permit more noteworthy control than direct burden control. The experience could diminish worries about direct burden control, yet concerns may likewise increment if clients experience specialized issues. The danger of higher power bills may support response; however, it decreases enrolment after some time.

- **Complexity and exertion**: Users may weigh anticipated randomness and exertion against anticipated advantages from DR when concluding whether to enlist. Responses might be restricted if more dynamic estimating builds the trouble of changing demand designs. However, clients consider various degrees of multifaceted nature and exertion adequate. Empowering innovations such as computerization or direct burden control may diminish the multifaceted nature and exertion related to reacting.
- **Interaction with client schedules and exercises**: Manual and mechanized DR could be encouraged if it is less disruptive to existing schedules. Be that as it may, various gatherings of clients seem to encounter various degrees of disturbance or be pretty much willing or ready to change their schedules to react.
- **User attributes**: Evidence on the convenience of socio-segment information, for example, salary and family size to anticipate adaptability, is combined. Possibly gradually adaptable clients may incorporate those investing additional time at home, families without wards, and clients with information on and access to empowering innovations and elective approaches to acquire vitality administrations.

5 Summary of Main Barriers to Implementing Demand Response

As we have discussed previously, DR is incredibly beneficial, as it helps participants save on bills and enhances reliability. Meanwhile, there are a lot of barriers and challenges that exist in the deployment of DR [141]. One of the possible barriers of DR is the customer's discrepancy of acquaintance with the market utilities and lack of interest. Another discrepancy in the consumer's mind is regarding the utility and belief in the effectiveness of electricity consumption [86]. Another limitation is aiding fundamental infrastructure or primary technology investments in control and automation technology, smart meters, display units, communication equipment, etc. For the DR services, this may limit utility, service providers, and participants from contributing effectually [142]. Many uncertain barriers may exist, a few of them as shown in Fig. 10. For example, if all of the basic infrastructures are installed but the service provider lacks in properly transforming information due to network issues on the communication side, then it may violate the reliability of the DR. Table 7 presents an analysis of reviews.

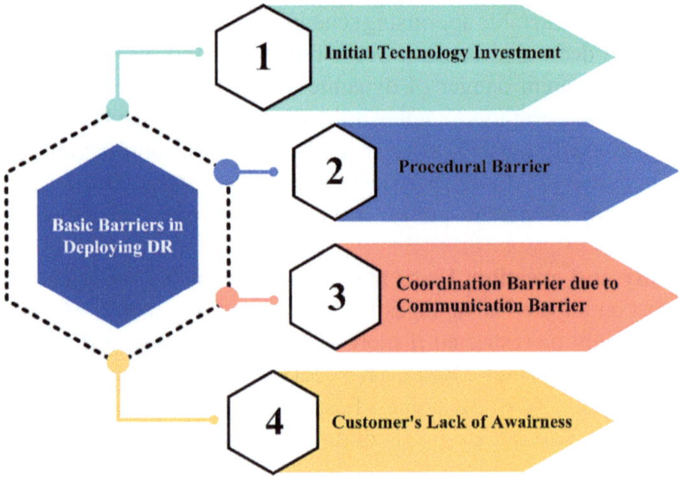

Fig. 10 General barriers to deploying DR

6 Future Recommendations and Strategy Endorsements

It is observed that most of the pricing models are designed for the flexible load, specifically for residential customers. However, by convincing inflexible consumers and introducing pricing models for commercial and industrial users, we can utilize energy more effectively, as commercial users utilize a considerable part of the energy. So, to convince the inflexible load to participate in DR, the utility has to be aware of the complementary benefit by introducing a reliable model that will be cost-effective in terms of reduced utility bills. The industry can play an immense role in attaining DR benefits socially and economically. Manifold options can be introduced to the inflexible load to encourage them to participate in DR. Imbalance in power supply and demand due to infrastructure schemes and commercial development may result in undesirable economic influences and social disturbance. These influences are intensified by uncertain extreme cold or hot weather and failure of energy infrastructure. Electrical service-provider companies gradually want significant industrial and commercial consumer demand cutbacks to control robust supply–demand imbalance. Customers have used thermal storage schemes, onsite electricity generation, shifting of manufacturing procedures, and curtailment for the short term as a technique to control and manage their demand for peak demand intervals. Utility announcements may be sent to appeal for a decrease in load for a certain duration when demand is limited to a specified percentage of accessible supply [143].

Although the efforts made, most commercial customers cannot respond or register to DR schemes. Emerging approaches to DR schemes that appeal to primarily commercial users are suggested for forthcoming research.

Table 7 An analysis of reviews in the field of industrial DR

Citation	Goals and objectives
[152]	Emerging tools interrelated to DR, such as smart metering, communication systems, and energy control, are cited, whereas few DR techniques are described, containing some actual industrial case studies
[153]	An influencing summary of strategy and regulation variations, electricity market improvements, and practical innovations on industrial DR contribution
[41]	A review of the main features of DR, including schemes, methods, concerns, and upcoming extensions
	Exploration of the currently designed simulations and complications monitored by the contemporary methodologies and solutions to resolve these concerns
	Familiarizing the techniques and pricelists of the power utility takes to incentivize industrial consumers to reorganize the pattern of their energy consumption
[24]	A summary of the obstacles that hold back an extensive consumption of industrial DR schemes
	Inspecting and familiarizing with various kinds of high prospective industries for DR schemes
	An analysis and exploration of the participation of auxiliary services and its influence on industries
[154, 155]	A review and study of DR modeling methods and their influence on operations and consequences of domestic energy management schemes
[156]	Exploration of DR to propose and program the operational techniques of RERs
	Numerous valuation approaches in accordance to find the influences of DR on electricity rates
	A complete pros and con evaluation of DR
[157]	Valuation of potential business schemes for DR and energy efficiency suppliers in various electricity markets. The survey analyses three features: "DSM transaction characteristics, renewable energy correlation, and load control characteristics."
[158]	An appraisal of industrial DR within the present applications and a potential operational scheme in smart grid locations
	A conversation on the cataloging and standing of industrial DR schemes in various U.S. electricity arcades
[159]	Classification of several DR curricula and strategies as per their control strategies, the inspirations offered in order to decrease the power usage and decision variables of the DR
	Classification of several prototypes for optimization to get optimum control founded on the objective of the optimization process

6.1 Installation of Flexible Equipment

Implementing intelligent hardware in buildings, like light-controlling units, dimmable stabilizers, and intelligent HVAC systems, allows energy supervisors to manage and adjust energy consumption and demand during peak intervals while guaranteeing consumer comfort. Adapting the installation of flexible equipment for

DR optimization problems more accurately is suggested. As per the review, typically, DR optimization is done for domestic users, and in rare cases, it is done for commercial or industrial users. Additional research efforts on the DR scheme for commercial and industrial customers is suggested.

6.2 *Commercial Building Contribution in Demand Response Schemes Sustained by Smart Lighting Super Vision*

Due to the significance of DR schemes in microgrids or smart grids, various efforts are being made to transform the pattern of usage by consumers and also to upsurge the use of RERs. Most energy usage happens in domestic, commercial, and office buildings. Most of the buildings are provided with mechanisms that can be used to participate in DR schemes. From such a perspective, SCADA is a system that plays an important role in facilitating the building technician to monitor and control usage and generation. There are many real-time optimized implementations of SCADA systems in various buildings, which allow numerous monitoring and controlling techniques to manage the usage and generation of the building to decide on the provision and contribute to DR actions [144]. Numerous physical controller modules manage the utilization of the building's air conditioning and lighting systems, which are constructed on an optimized prototype. The SCADA system is reflected as an accumulate model, deliberated as a managing entity for DR and its performance while surveying the DR proceedings. The consequences illustrate that with a suitable small reduction in air conditioning and lighting systems, electricity consumers can dynamically contribute to saving power energy using DR schemes and correspondingly for internal effectual electricity consumption.

6.3 *Integrating IoT Protocols in Demand Response*

The upcoming tenure of ICT will be freestanding the outdated desktop domain. IoT foresees the prospect of interconnected physical and digital entities using suitable data and communication equipment to facilitate customers and utilities [145]. The IoT prototype consists of various objects surrounding us, such as smart grids, building automation, social domains, health care, and other intelligent entities that will be computerized on the system in one form or another [146–148]. This model possibly includes many associated devices and will significantly intensify the scope and size of the internet, as displayed in the Fig. 11. These latest novel way-outs can exchange and use information with the most extensively used internet protocols. IoT has been extensively integrated into various applications for the armed forces, atmospherically monitoring, intelligent transport, and medicinal schemes [149, 150]. The smart grid depends on the communication infrastructure for real-time control

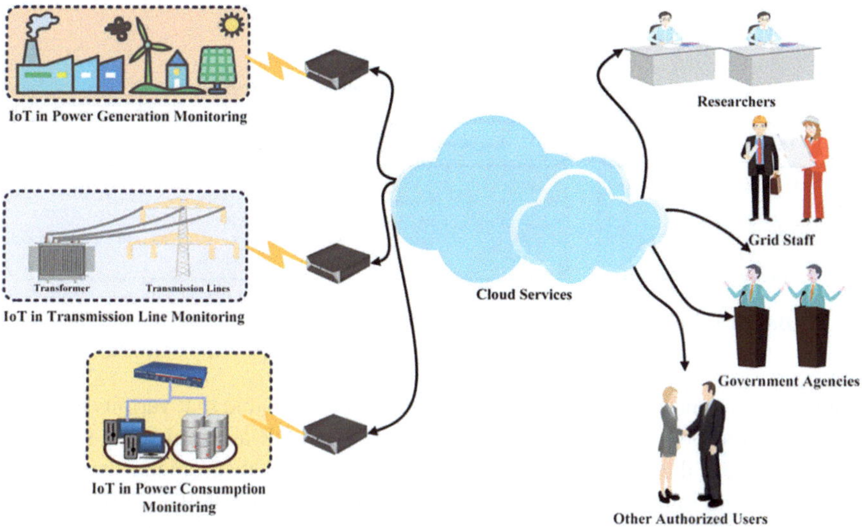

Fig. 11 IoT illustration in power network [151]

and monitoring. Such kind of grid may also have the features of IoT, like intellectual processing, consistent transmission, and frivolous frameworks. IoT can and will significantly influence the smart grid and DR in upcoming years, providing controlling applications and operative solutions to DR issues [151].

6.4 *Insufficient Research on Environmental Effects and Basic Barriers to Demand Response*

The major problem we are facing so far is a polluted environment. Very inadequate research has been done on the ecological effects of DR schemes and considering pollution in the impartial function of the DR optimization problem. Exploring the ecological consequences of DR programs and making an allowance for pollution in the objectives of the DR optimization problem is suggested for imminent research. Industrial DR schemes can play a vibrant role in reducing environmental pollution. Figure 12 shows the fundamental barriers to implementing the industrial DR.

7 Conclusions

DR is valuable in figuring out and resolving numerous power system issues. Different comparisons are performed in this chapter to show the importance of DR. This review paper also anticipates bringing issues to light regarding the alternatives and the related

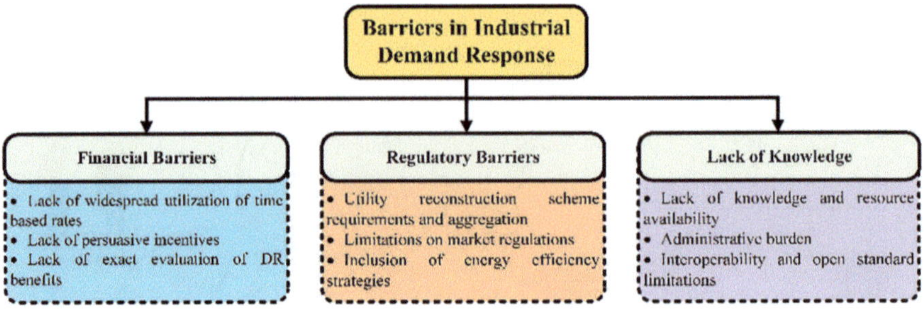

Fig. 12 Basic barriers to implementing industrial DR [24]

upsides and downsides with every choice. It could be possible that various load side entities (LSEs) provide services in specific areas, and DR assembler can select the LSE that offers the best DR prices. Hence, it will lead to problems with multi-aggregators and multi-users. So, it concluded that LSEs must define an optimal DR pricing model acceptable to more DR users and maximize their profits. DR can also play a vital role if commercial users participate; meanwhile, domestic users in DR consume a significant amount of electricity. It is also concluded that DR is an aim to be achieved and a potentially effective and defensible technique used in a power system with intensively increasing needs for flexibility. With the increasing significance of DR and smart grids, the future strategy endorsements and recommendations mentioned above are of enormous worth to individuals, establishments, and governments regarding more efficient, effective, and cleaner energy systems. While these might be potential advantages, we should not anticipate they should be complementary, as achieving them simultaneously is impossible. IoT plays a significant role in supporting advanced metering infrastructure (AMI) technologies. Therefore, it is proposed in this chapter that the DR schemes be enhanced to manage the schedules of home appliances with IoT.

References

1. Eid, C., Koliou, E., Valles, M., Reneses, J., & Hakvoort, R. (2016). Time-based pricing and electricity demand response: Existing barriers and next steps. *Utilities Policy, 40*, 15–25. https://doi.org/10.1016/j.jup.2016.04.001
2. Yuemin, D., & Hong, S. H. (2013). A model of demand response energy management system in industrial facilities. In *2013 IEEE International Conference on Smart Grid Communications (SmartGridComm)* (pp. 241–246). https://doi.org/10.1109/SmartGridComm.2013.6687964
3. Bartusch, C., Wallin, F., Odlare, M., Vassileva, I., & Wester, L. (2011). Introducing a demand-based electricity distribution tariff in the residential sector: Demand response and customer perception. *Energy Policy, 39*(9), 5008–5025. https://doi.org/10.1016/j.enpol.2011.06.013
4. Palensky, P., & Dietrich, D. (2011). Demand side management: demand response, intelligent energy systems, and smart loads. *IEEE Transactions on Industrial Informatics, 7*(3), 381–388. https://doi.org/10.1109/TII.2011.2158841

5. Sheng, Q., Chen, M., Li, Q., Wang, Y., & Hassan, M. A. S. (2018). Analysis for the influence of electric vehicle chargers with different SOC on grid harmonics. In K. Li, J. Zhang, M. Chen, Z. Yang, & Q. Niu (Eds.), *Advances in green energy systems and smart grid* (pp. 284–294). Springer.
6. Hassan, M. et al. (2020). Lyapunov based adaptive controller for power converters used in hybrid energy storage systems. *Sustainable Energy Technologies and Assessments, 42*, 100853. https://doi.org/10.1016/j.seta.2020.100853
7. Li, C., Yu, X., Yu, W., Chen, G., & Wang, J. (2017). Efficient computation for sparse load shifting in demand side management. *IEEE Transactions on Smart Grid, 8*(1), 250–261. https://doi.org/10.1109/TSG.2016.2521377
8. Werminski, S., Jarnut, M., Benysek, G., & Bojarski, J. (2017). Demand side management using DADR automation in the peak load reduction. *Renewable and Sustainable Energy Reviews, 67*, 998–1007. https://doi.org/10.1016/j.rser.2016.09.049
9. Zhang, S., Jiao, Y., & Chen, W. (2017). Demand-side management (DSM) in the context of China's on-going power sector reform. *Energy Policy,* 100, 1–8. https://doi.org/10.1016/j.enpol.2016.09.057
10. Batlle López, C., & Rodilla Rodríguez, P. (2009). Electricity demand response tools: current status and outstanding issues.
11. Cappers, P., Goldman, C., & Kathan, D. (2010). Demand response in U.S. electricity markets: Empirical evidence. *Energy, 35*(4), 1526–1535. https://doi.org/10.1016/j.energy.2009.06.029
12. Kofod, C., Togeby, M. (2004). Demand response offered by households with direct electric heating.
13. Europea, C. (2013). Incorporating demand side flexibility, in particular demand response, in eletricity markets. Commission Staff Working Document, SWD.
14. C. Communication from the, "Delivering the internal electricity market and making the most of public intervention." *Brussels, 5*, 16.
15. Pollitt, M. G. (2019). The European single market in electricity: An economic assessment. *Review of Industrial Organization, 55*(1), 63–87. https://doi.org/10.1007/s11151-019-09682-w
16. Wang, J., Bloyd, C. N., Hu, Z., & Tan, Z. (2010). Demand response in China. *Energy, 35*(4), 1592–1597. https://doi.org/10.1016/j.energy.2009.06.020
17. Reneses, J., & Rodríguez Ortega, M. P. (2014). Distribution pricing: theoretical principles and practical approaches. *IET Generation, Transmission & Distribution, 8*(10), 1645–1655.
18. Bartusch, C., & Alvehag, K. (2014). Further exploring the potential of residential demand response programs in electricity distribution. *Applied Energy, 125*, 39–59. https://doi.org/10.1016/j.apenergy.2014.03.054
19. Koliou, K., Eid, C., Hakvoort, R. A. (2013). Development of Demand Side Response in liberalized electricity markets: Policies for effective market design in Europe. In *2013 10th International Conference on the European Energy Market (EEM)* (pp. 1–8). https://doi.org/10.1109/EEM.2013.6607403
20. Conchado, A., Linares, P., Lago, O., & Santamaría, A. (2016). An estimation of the economic and environmental benefits of a demand-response electricity program for Spain. *Sustainable Production and Consumption, 8*, 108–119. https://doi.org/10.1016/j.spc.2016.09.004
21. Batista Abikarram, J., McConky, K., & Proano, R. (2019). Energy cost minimization for unrelated parallel machine scheduling under real time and demand charge pricing. *Journal of Cleaner Production, 208*, 232–242. https://doi.org/10.1016/j.jclepro.2018.10.048
22. Ding, Y. M., Hong, S. H., & Li, X. H. (2014). A demand response energy management scheme for industrial facilities in smart grid. *IEEE Transactions on Industrial Informatics, 10*(4), 2257–2269. https://doi.org/10.1109/TII.2014.2330995
23. Ma, K., Hu, G., & Spanos, C. J. (2015). A cooperative demand response scheme using punishment mechanism and application to industrial refrigerated warehouses. *IEEE Transactions on Industrial Informatics, 11*(6), 1520–1531. https://doi.org/10.1109/TII.2015.2431219
24. Shoreh, M. H., Siano, P., Shafie-khah, M., Loia, V., & Catalão, J. P. S. (2016). A survey of industrial applications of Demand Response. *Electric Power Systems Research, 141*, 31–49. https://doi.org/10.1016/j.epsr.2016.07.008

25. Wang, Y., Lin, H., Liu, Y., Sun, Q., & Wennersten, R. (2018). Management of household electricity consumption under price-based demand response scheme. *Journal of Cleaner Production, 204*, 926–938. https://doi.org/10.1016/j.jclepro.2018.09.019
26. Cui, H., & Zhou, K. (2018). Industrial power load scheduling considering demand response. *Journal of Cleaner Production, 204*, 447–460. https://doi.org/10.1016/j.jclepro.2018.08.270
27. Shehzad Hassan, M. A., Chen, M., Lin, H., Ahmed, M. H., Khan, M. Z., Chughtai, G. R. (2019). Optimization modeling for dynamic price based demand response in microgrids. *Journal of Cleaner Production, 222*, 231–241. https://doi.org/10.1016/j.jclepro.2019.03.082
28. Warren, P. (2014). A review of demand-side management policy in the UK. *Renewable and Sustainable Energy Reviews, 29*, 941–951. https://doi.org/10.1016/j.rser.2013.09.009
29. Albadi, M. H., & El-Saadany, E. F. (2008). A summary of demand response in electricity markets. *Electric Power Systems Research, 78*(11), 1989–1996. https://doi.org/10.1016/j.epsr.2008.04.002
30. van der Veen, R. A. C., & Hakvoort, R. A. (2016). The electricity balancing market: Exploring the design challenge. *Utilities Policy, 43*, 186–194. https://doi.org/10.1016/j.jup.2016.10.008
31. Torriti, J., Hassan, M. G., Leach, M. (2010). Demand response experience in Europe: Policies, programmes and implementation. *Energy, 35*(4), 1575–1583. https://doi.org/10.1016/j.energy.2009.05.021
32. Greening, L. A. (2010). Demand response resources: Who is responsible for implementation in a deregulated market? *Energy, 35*(4), 1518–1525. https://doi.org/10.1016/j.energy.2009.12.013
33. Chen, Q., et al. (2017). Dynamic price vector formation model-based automatic demand response strategy for PV-assisted EV charging stations. *IEEE Transactions on Smart Grid, 8*(6), 2903–2915. https://doi.org/10.1109/TSG.2017.2693121
34. Koh, J., Ray, S., & Hodges, J. (2017). Information mediator for demand response in electrical grids and buildings. In *2017 IEEE 11th International Conference on Semantic Computing (ICSC)* (pp. 73–76). https://doi.org/10.1109/ICSC.2017.26
35. Yaghmaee, M. H., Leon-Garcia, A., & Moghaddassian, M. (2018). On the performance of distributed and cloud-based demand response in smart grid. *IEEE Transactions on Smart Grid, 9*(5), 5403–5417. https://doi.org/10.1109/TSG.2017.2688486
36. Ahmed, M. H., Wang, M., Hassan, M. A. S., Mehmood, M. A. (2017). High performance three phase AC-DC PWM converter with decoupled controller using Matlab/Simulink. In *2017 IEEE Conference on Energy Internet and Energy System Integration (EI2)* (pp. 1–4). https://doi.org/10.1109/EI2.2017.8245707
37. Kiliccote, S., Piette, M. A., Watson, D. S., Hughes, G. (2006). Dynamic controls for energy efficiency and demand response: Framework concepts and a new construction study case in New York.
38. Mendaza, I. D. D. C., Szczesny, I. G., Pillai, J. R., Bak-Jensen, B. (2015). Flexible demand control to enhance the dynamic operation of low voltage networks. *IEEE Transactions on Smart Grid, 6*(2), 705–715. https://doi.org/10.1109/TSG.2014.2375894
39. Hassan, M. A. et al. Dynamic price-based demand response through linear regression for microgrids with renewable energy resources. *Energies*, 15(4). https://doi.org/10.3390/en15041385
40. Haider, H. T., See, O. H., & Elmenreich, W. (2016). A review of residential demand response of smart grid. *Renewable and Sustainable Energy Reviews, 59*, 166–178. https://doi.org/10.1016/j.rser.2016.01.016
41. Deng, R., Yang, Z., Chow, M. Y., & Chen, J. (2015). A survey on demand response in smart grids: Mathematical models and approaches. *IEEE Transactions on Industrial Informatics, 11*(3), 570–582. https://doi.org/10.1109/TII.2015.2414719
42. Good, N., Ellis, K. A., & Mancarella, P. (2017). Review and classification of barriers and enablers of demand response in the smart grid. *Renewable and Sustainable Energy Reviews, 72*, 57–72. https://doi.org/10.1016/j.rser.2017.01.043
43. Ipakchi, A., & Albuyeh, F. (2008). Grid of the future. *IEEE Power and Energy Magazine 7*, 52–62. *DOI*, 10, 1109.

44. Parvania, M., Fotuhi-Firuzabad, M., & Shahidehpour, M. (2014). ISO's optimal strategies for scheduling the hourly demand response in day-ahead markets. *IEEE Transactions on Power Systems, 29*(6), 2636–2645. https://doi.org/10.1109/TPWRS.2014.2316832
45. Ave, G. D., Harjunkoski, I., & Engell, S. (2018). Industrial demand side management formulation for simultaneous electricity load commitment and future load prediction. In M. R. Eden, M. G. Ierapetritou, & G. P. Towler (Eds.), *Computer aided chemical engineering* (Vol. 44, pp. 1237–1242). Elsevier.
46. Xia, H., et al. (2018). Distributed control method for economic dispatch in Islanded microgrids with renewable energy sources. *IEEE Access, 6*, 21802–21811. https://doi.org/10.1109/ACCESS.2018.2827366
47. Wang, Y. et al. (2018). Energy management of smart micro-grid with response loads and distributed generation considering demand response. *Journal of Cleaner Production, 197*, 1069–1083. https://doi.org/10.1016/j.jclepro.2018.06.271
48. Strbac, G. (2008). Demand side management: Benefits and challenges. *Energy Policy, 36*(12), 4419–4426. https://doi.org/10.1016/j.enpol.2008.09.030
49. Gayathri, N., Vineeth, V. V., & Radhika, N. (2015). A novel approach in demand side management for smart home. *Procedia Technology, 21*, 526–532. https://doi.org/10.1016/j.protcy.2015.10.043
50. Valero, S., Ortiz, M., Senabre, C., Alvarez, C., Franco, F. J. G., & Gabaldon, A. (2007). Methods for customer and demand response policies selection in new electricity markets. *IET Generation, Transmission & Distribution, 1*(1), 104–110.
51. Dietrich, K., Latorre, J. M., Olmos, L., & Ramos, A. (2012). Demand response in an isolated system with high wind integration. *IEEE Transactions on Power Systems, 27*(1), 20–29. https://doi.org/10.1109/TPWRS.2011.2159252
52. Skea, J. (2012). Roadmap 2050: A practical guide to a prosperous, low-carbon Europe. European Climate Foundation. Elsevier.
53. Jacobson, M. Z., & Delucchi, M. A. (2011). Providing all global energy with wind, water, and solar power, Part I: Technologies, energy resources, quantities and areas of infrastructure, and materials. *Energy Policy, 39*(3), 1154–1169. https://doi.org/10.1016/j.enpol.2010.11.040
54. Georgilakis, P. S. (2008). Technical challenges associated with the integration of wind power into power systems. *Renewable and Sustainable Energy Reviews, 12*(3), 852–863. https://doi.org/10.1016/j.rser.2006.10.007
55. Heide, D., Von Bremen, L., Greiner, M., Hoffmann, C., Speckmann, M., & Bofinger, S. (2010). Seasonal optimal mix of wind and solar power in a future, highly renewable Europe, *Renewable Energy, 35*(11), 2483–2489. https://doi.org/10.1016/j.renene.2010.03.012
56. Rabiee, A., Soroudi, A., Mohammadi-ivatloo, B., & Parniani, M. (2014). Corrective voltage control scheme considering demand response and stochastic wind power. *IEEE Transactions on Power Systems, 29*(6), 2965–2973. https://doi.org/10.1109/TPWRS.2014.2316018
57. Tayab, U. B., Roslan, M. A. B., Hwai, L. J., & Kashif, M. (2017). A review of droop control techniques for microgrid. *Renewable and Sustainable Energy Reviews, 76*, 717–727. https://doi.org/10.1016/j.rser.2017.03.028
58. Schoonenberg, W. C. H., & Farid, A. M. (2017). A dynamic model for the energy management of microgrid-enabled production systems. *Journal of Cleaner Production, 164*, 816–830. https://doi.org/10.1016/j.jclepro.2017.06.119
59. Shuai, Z., Mo, S., Wang, J., Shen, Z. J., Tian, W., & Feng, Y. (2016). Droop control method for load share and voltage regulation in high-voltage microgrids. *Journal of Modern Power Systems and Clean Energy, 4*(1), 76–86. https://doi.org/10.1007/s40565-015-0176-1
60. Ghosh, D. P., Schrader, D. E., Schulze, W. D., & Wicker, S. B. (2012). Economic analysis of privacy-aware Advanced Metering Infrastructure adoption. In *2012 IEEE PES Innovative Smart Grid Technologies (ISGT)* (pp. 1–4). https://doi.org/10.1109/ISGT.2012.6175740
61. Pasom, W., Therdyothin, A., Nathakaranakule, A., Prapanavarat, C., & Limmeechokchai, B. (2015). The potential of demand response measures of commercial buildings in Thailand. *Energy Procedia, 79*, 1015–1020. https://doi.org/10.1016/j.egypro.2015.11.602

62. Bayat, M., Sheshyekani, K., Hamzeh, M., & Rezazadeh, A. (2016). Coordination of distributed energy resources and demand response for voltage and frequency support of MV microgrids. *IEEE Transactions on Power Systems, 31*(2), 1506–1516. https://doi.org/10.1109/TPWRS.2015.2434938
63. Kumar, K. N., & Tseng, K. J. (2016). Impact of demand response management on chargeability of electric vehicles. *Energy, 111*, 190–196. https://doi.org/10.1016/j.energy.2016.05.120
64. Raghavan, S. S., & Khaligh, A. (2012). Impact of plug-in hybrid electric vehicle charging on a distribution network in a Smart Grid environment. In *2012 IEEE PES Innovative Smart Grid Technologies (ISGT)* (pp. 1–7). https://doi.org/10.1109/ISGT.2012.6175632
65. Christakou, K. (2016). A unified control strategy for active distribution networks via demand response and distributed energy storage systems. *Sustainable Energy, Grids and Networks, 6*, 1–6. https://doi.org/10.1016/j.segan.2016.01.001
66. Wang, Z., & Paranjape, R. (2014). An evaluation of electric vehicle penetration under demand response in a multi-agent based simulation. In *2014 IEEE electrical power and energy conference* (pp. 220–225). https://doi.org/10.1109/EPEC.2014.14
67. Syrri, A. L. A., & Mancarella, P. (2016). Reliability and risk assessment of post-contingency demand response in smart distribution networks. *Sustainable Energy, Grids and Networks, 7*, 1–12. https://doi.org/10.1016/j.segan.2016.04.002
68. Yingdan, F., & Xin, A. (2013). The review of load scheduling model research based on demand response method. In *2013 IEEE PES Asia-Pacific Power and Energy Engineering Conference (APPEEC)* (pp. 1–5). https://doi.org/10.1109/APPEEC.2013.6837244
69. Rahman, M. M., Arefi, A., Shafiullah, G. M., & Hettiwatte, S. (2016). Penetration maximisation of residential rooftop photovoltaic using demand response. In *2016 International Conference on Smart Green Technology in Electrical and Information Systems (ICSGTEIS)* (pp. 21–26). https://doi.org/10.1109/ICSGTEIS.2016.7885760
70. Hussain, I., Mohsin, S., Basit, A., Khan, Z. A., Qasim, U., & Javaid, N. (2015). A review on demand response: pricing, optimization, and appliance scheduling. *Procedia Computer Science, 52*, 843–850. https://doi.org/10.1016/j.procs.2015.05.141
71. Salami, A., & Farsi, M. M. (2015). A cooperative demand response program for smart grids. In *2015 Smart Grid Conference (SGC)* (pp. 99–104). https://doi.org/10.1109/SGC.2015.7857417
72. Gavgani, M. H., Abedi, M., Karimi, F., & Aghamohammadi, M. R. (2014). Demand response-based voltage security improvement using artificial neural networks and sensitivity analysis. In *2014 Smart Grid Conference (SGC)* (pp. 1–6). https://doi.org/10.1109/SGC.2014.7090863
73. Andreou, G. T., Symeonidis, A. L., Diou, C., Mitkas, P. A., & Labridis, D. P. (2012). A framework for the implementation of large scale demand response. In *2012 International Conference on Smart Grid Technology, Economics and Policies (SG-TEP)* (pp. 1–4). https://doi.org/10.1109/SG-TEP.2012.6642380
74. Hadis Pourasghar, K., & Javidi, M. H. (2013). An efficient home energy management system for automated residential demand response. In *2013 13th International Conference on Environment and Electrical Engineering (EEEIC)* (pp. 307–312). https://doi.org/10.1109/EEEIC-2.2013.6737927
75. Shipman, R., Gillott, M., & Naghiyev, E. (2013). SWITCH: Case studies in the demand side management of washing appliances. *Energy Procedia,* 42, 153–162. https://doi.org/10.1016/j.egypro.2013.11.015
76. Baoshi, L., & Qi, W. (2013). Home energy control algorithm research based on demand response programs and user comfort. In *Proceedings of 2013 2nd international conference on measurement, information and control* (Vol. 02, pp. 995–999). https://doi.org/10.1109/MIC.2013.6758127
77. Sheng, L., Cheng, C., Wuhuan, D., Yu, D., & Chunsheng, L. (2012). The research on technology of periodic stopping of central air conditioning based on modelling and simulation of demand response. In *2012 China international conference on electricity distribution* (pp. 1–4). https://doi.org/10.1109/CICED.2012.6508600

78. Shen, S., Zhao, Y., Pang, J., Yu, W., & Wu, K. (2012). Exploring the use value of air conditioners in residential peak demand response. In *2012 China international conference on electricity distribution* (pp. 1–5). https://doi.org/10.1109/CICED.2012.6508622
79. Oldewurtel, F., Sturzenegger, D., Andersson, G., Morari, M., & Smith, R. S. (2013). Towards a standardized building assessment for demand response. In *52nd IEEE conference on decision and control* (pp. 7083–7088). https://doi.org/10.1109/CDC.2013.6761012
80. Son, J., Hara, R., Kita, H., & Tanaka, E. (2014). Energy management considering demand response resource in commercial building with chiller system and energy storage systems. In *The 2nd IEEE Conference on Power Engineering and Renewable Energy (ICPERE) 2014* (pp. 96–101). https://doi.org/10.1109/ICPERE.2014.7067239
81. Sofana Reka, S., & Ramesh, V. (2016). A demand response modeling for residential consumers in smart grid environment using game theory based energy scheduling algorithm. *Ain Shams Engineering Journal, 7*(2), 835–845. https://doi.org/10.1016/j.asej.2015.12.004
82. Chaney, J., Hugh Owens, E., & Peacock, A. D. (2016). An evidence based approach to determining residential occupancy and its role in demand response management. *Energy and Buildings, 125*, 254–266. https://doi.org/10.1016/j.enbuild.2016.04.060
83. Klaassen, E. A. M., Kobus, C. B. A., Frunt, J., & Slootweg, J. G. (2016). "Responsiveness of residential electricity demand to dynamic tariffs: Experiences from a large field test in the Netherlands. *Applied Energy, 183*, 1065–1074. https://doi.org/10.1016/j.apenergy.2016.09.051
84. Vajjala, V. A. H. (2016). A novel solution to use Big Data technologies and improve demand response program in aggregated residential houses. In *2016 IEEE Conference on Technologies for Sustainability (SusTech)* (pp. 251–256). https://doi.org/10.1109/SusTech.2016.7897176
85. Ahmed, M. S., Mohamed, A., Shareef, H., Homod, R. Z., & Ali, J. A. (2016). Artificial neural network based controller for home energy management considering demand response events. In *2016 International conference on advances in electrical, electronic and systems engineering (ICAEES)* (pp. 506–509). https://doi.org/10.1109/ICAEES.2016.7888097
86. Torstensson, D., & Wallin, F. (2015). Potential and barriers for demand response at household customers. *Energy Procedia, 75*, 1189–1196. https://doi.org/10.1016/j.egypro.2015.07.570
87. Fell, M. J., Shipworth, D., Huebner, G. M., & Elwell, C. A. (2015). Public acceptability of domestic demand-side response in Great Britain: The role of automation and direct load control. *Energy Research & Social Science, 9*, 72–84. https://doi.org/10.1016/j.erss.2015.08.023
88. Ranjan, S., & Thomas, M. S. (2015). An efficient home energy management algorithm for demand response analysis in Indian scenario. In *2015 Annual IEEE India Conference (INDICON)* (pp. 1–6). https://doi.org/10.1109/INDICON.2015.7443662
89. Soares, T., Pereira, F., Morais, H., & Vale, Z. (2015). Cost allocation model for distribution networks considering high penetration of distributed energy resources. *Electric Power Systems Research, 124*, 120–132. https://doi.org/10.1016/j.epsr.2015.03.008
90. Sabounchi, M., Khazaei, H., & Sani, S. K. H. (2014). A new residential demand response management method based on a social welfare optimization framework. In *2014 2nd International Conference on Electrical, Electronics and System Engineering (ICEESE)* (pp. 146–151). https://doi.org/10.1109/ICEESE.2014.7154611
91. Wollsen, M. G., Kjærgaard, M. B., & Jørgensen, B. N. (2016). Influential factors for accurate load prediction in a demand response context. In *2016 IEEE Conference on Technologies for Sustainability (SusTech)* (pp. 9–13). https://doi.org/10.1109/SusTech.2016.7897135
92. Christantoni, D., Flynn, D., & Finn, D. P. (2015). Modelling of a Multi-purpose commercial building for demand response analysis. *Energy Procedia, 78*, 2166–2171. https://doi.org/10.1016/j.egypro.2015.11.308
93. Tonder, A. J. M. V., Kleingeld, M., & Marais, J. H. (2013). Investigating demand response potential in a mining group. In *2013 Proceedings of the 10th Industrial and commercial use of energy conference* (pp. 1–5).
94. Leithon, J., Lim, T. J., & Sun, S. (2018). Online demand response strategies for non-deferrable loads with renewable energy. *IEEE Transactions on Smart Grid, 9*(5), 5227–5235. https://doi.org/10.1109/TSG.2017.2684815

95. You, S., & Segerberg, H. (2014). Integration of 100% micro-distributed energy resources in the low voltage distribution network: A Danish case study. *Applied Thermal Engineering, 71*(2), 797–808. https://doi.org/10.1016/j.applthermaleng.2013.11.039
96. Chua, K. H., Wong, J., Lim, Y. S., Taylor, P., Morris, E., & Morris, S. (2011). Mitigation of voltage unbalance in low voltage distribution network with high level of photovoltaic system. *Energy Procedia, 12*, 495–501.. https://doi.org/10.1016/j.egypro.2011.10.066
97. Zhou, W., Li, Q., Wu, K., Zhang, L., Hassan, M. A. S., & Chen, M. (2022). Accelerated hierarchical optimization method for emergency energy management of microgrids with energy storage systems. *Energy Science & Engineering, 10*(3), 962–972.
98. Rahman, M. M., Arefi, A., Shafiullah, G. M., & Hettiwatte, S. (2018). A new approach to voltage management in unbalanced low voltage networks using demand response and OLTC considering consumer preference. *International Journal of Electrical Power & Energy Systems, 99*, 11–27. https://doi.org/10.1016/j.ijepes.2017.12.034
99. Montoya, S., Dominguez, R., Becerra, A., Arias, L., Chica, A., & Flórez, O. (2013). Automation and SCADA system for reactive compensation in a motor control center. In *2013 Workshop on Power Electronics and Power Quality Applications (PEPQA)* (pp. 1–7). https://doi.org/10.1109/PEPQA.2013.6614956
100. Han, J., Khushalani-Solanki, S., Solanki, J., & Schoene, J. (2012). Study of unified control of STATCOM to resolve the Power quality issues of a grid-connected three phase PV system. In *2012 IEEE PES innovative smart grid technologies (ISGT)* (pp. 1–7). https://doi.org/10.1109/ISGT.2012.6175773
101. Afzal, M. M. et al. A comparative study of supercapacitor-based STATCOM in a grid-connected photovoltaic system for regulating power quality issues. *Sustainability, 12*(17). https://doi.org/10.3390/su12176781
102. Ou, T.-C., & Hong, C.-M. (2014). Dynamic operation and control of microgrid hybrid power systems. *Energy, 66*, 314–323. https://doi.org/10.1016/j.energy.2014.01.042
103. Farooq, U. et al. (2019). A reliable approach to protect and control of wind solar hybrid DC microgrids. In *2019 IEEE 3rd conference on energy internet and energy system integration (EI2)* (pp. 348–353). https://doi.org/10.1109/EI247390.2019.9062101
104. Zhang, C., Xu, Y., Dong, Z. Y., & Wong, K. P. (2018). Robust coordination of distributed generation and price-based demand response in microgrids. *IEEE Transactions on Smart Grid, 9*(5), 4236–4247. https://doi.org/10.1109/TSG.2017.2653198
105. Luo, X., Lee, C. K., Ng, W. M., Yan, S., Chaudhuri, B., & Hui, S. Y. R. (2017). Use of adaptive thermal storage system as smart load for voltage control and demand response. *IEEE Transactions on Smart Grid, 8*(3), 1231–1241. https://doi.org/10.1109/TSG.2015.2513743
106. Aalami, H. A., Khodaei, J., & Fard, M. (2011). Economical and technical evaluation of implementation mandatory demand response programs on Iranian power system. In *2011 IEEE student conference on research and development* (pp. 271–276). https://doi.org/10.1109/SCOReD.2011.6148749
107. Kokos, I., & Lamprinos, I. (2016).*Demand response strategy for optimal formulation of flexibility services.*
108. Nguyen, D. T., Negnevitsky, M., & Groot, M. D. (2010). A comprehensive approach to Demand Response in restructured power systems. In *2010 20th Australasian Universities power engineering conference* (pp. 1–6).
109. Chen, H., Wang, Z., Yan, H., Zou, H., & Luo, B. (2015). Integrated planning of distribution systems with distributed generation and demand side response. *Energy Procedia, 75*, 981–986. https://doi.org/10.1016/j.egypro.2015.07.314
110. Cheng, Y., & An, S. (2012). Demand response mechanism and simulation research of cold storage air conditioner supporting consumption of the onshore wind power. In *2012 IEEE international conference on industrial engineering and engineering management* (pp. 401–405). https://doi.org/10.1109/IEEM.2012.6837770
111. Jia, X., Xia, Q., & Chen, Q. X. (2012). Piecewise price mechanism to induce demand response in Smart Grid. In *2012 China international conference on electricity distribution* (pp. 1–4). https://doi.org/10.1109/CICED.2012.6508670

112. Huang, S., Infield, D., Cruden, A., Frame, D., & Densley, D. (2013). Plug-in electric vehicles as demand response to absorb local wind generation in power distribution network. In *2013 World Electric Vehicle Symposium and Exhibition (EVS27)* (pp. 1–5). https://doi.org/10.1109/EVS.2013.6914713
113. Arora, M., & Chanana, S. (2014). Residential demand response from PV panel and energy storage device. In *2014 IEEE 6th India International Conference on Power Electronics (IICPE)* (pp. 1–6). https://doi.org/10.1109/IICPE.2014.7115731
114. Pazouki, S., & Haghifam, M. R. (2014). Comparison between demand response programs in multiple carrier energy infrastructures in presence of wind and energy storage technologies. In *2014 Smart Grid Conference (SGC)* (pp. 1–6). https://doi.org/10.1109/SGC.2014.7090862
115. Lindberg, C.-F., Zahedian, K., Solgi, M., Lindkvist, R. (2014). Potential and limitations for industrial demand side management. *Energy Procedia, 61*, 415–418. https://doi.org/10.1016/j.egypro.2014.11.1138
116. Matsumoto, J., & Zhong, W. (2013). New demand response architecture for stabilization of power quality in smart grid. In *2013 9th international conference on information, communications & signal processing* (pp. 1–5). https://doi.org/10.1109/ICICS.2013.6782849
117. Mackey, K., McCann, R., Rahman, K., & Winkelman, R. (2013). Evaluation of a battery energy storage system for coordination of demand response and renewable energy resources. In *2013 4th IEEE international symposium on Power Electronics for Distributed Generation Systems (PEDG)* (pp. 1–8). https://doi.org/10.1109/PEDG.2013.6785650
118. Cheng, Y., Dong, N., & Ren, Y. (2011). Investigation on electric load peak and valley characters and demand response evaluation based on spectral analysis. In *2011 International conference on advanced power system automation and protection* (Vol. 2, pp. 1258–1262). https://doi.org/10.1109/APAP.2011.6180571
119. Yan, X., Ozturk, Y., Hu, Z., & Song, Y. (2018). A review on price-driven residential demand response. *Renewable and Sustainable Energy Reviews, 96*, 411–419. https://doi.org/10.1016/j.rser.2018.08.003
120. Chughtai, G. R., Lee, J., Kabir, A., Abbasi, R., & Hassan, M. A. S. (2018). Ontology-based academic article recommendation. In *2018 15th International Computer Conference on Wavelet Active Media Technology and Information Processing (ICCWAMTIP)* (pp. 93–96). https://doi.org/10.1109/ICCWAMTIP.2018.8632601.
121. Lu, R., & Hong, S. H. (2019). Incentive-based demand response for smart grid with reinforcement learning and deep neural network. *Applied Energy, 236*, 937–949. https://doi.org/10.1016/j.apenergy.2018.12.061
122. Cui, Q., Wang, X., Wang, X., & Zhang, Y. (2016). Residential appliances direct load control in real-time using cooperative game. *IEEE Transactions on Power Systems, 31*(1), 226–233. https://doi.org/10.1109/TPWRS.2015.2391774
123. Zhang, F., de Dear, R., & Candido, C. (2016). Thermal comfort during temperature cycles induced by direct load control strategies of peak electricity demand management. *Building and Environment, 103*, 9–20. https://doi.org/10.1016/j.buildenv.2016.03.020
124. Krarti, M. (2018). *Chapter 4-utility rate structures and grid integration* (pp. 189–245). Butterworth-Heinemann.
125. Contreras, J., Asensio, M., de Quevedo, P. M., Muñoz-Delgado, G., & Montoya-Bueno, S. (2016). *Joint RES and distribution network expansion planning under a demand response framework*. Academic Press.
126. Yang, W., Yu, R., & Nambiar, M. (2014). Quantifying the benefits to consumers for demand response with a statistical elasticity model. *IET Generation, Transmission & Distribution, 8*(3), 503–515.
127. Cheng, T. et al. (2018). Adaptive robust method for dynamic economic emission dispatch incorporating renewable energy and energy storage. *Complexity, 2018*(1), 2517987.
128. Assad, U., et al. Smart grid, demand response and optimization: A critical review of computational methods. *Energies, 15*(6). https://doi.org/10.3390/en15062003

129. Sachdev, R. S., & Singh, O. (2016). Consumer‘s demand response to dynamic pricing of electricity in a smart grid. In *2016 International Conference on Control, Computing, Communication and Materials (ICCCCM)* (pp. 1–6). https://doi.org/10.1109/ICCCCM.2016.7918236
130. Minchala-Avila, L. I., Armijos, J., Pesántez, D., & Zhang, Y. (2016). Design and implementation of a smart meter with demand response capabilities. *Energy Procedia, 103*, 195–200. https://doi.org/10.1016/j.egypro.2016.11.272
131. VanderKley, T. S., Negash, A. I., & Kirschen, D. S. (2014). Analysis of dynamic retail electricity rates and domestic demand response programs. In *2014 IEEE Conference on Technologies for Sustainability (SusTech)* (pp. 172–177). https://doi.org/10.1109/SusTech.2014.7046239
132. Wijaya, T. K., Vasirani, M., Villumsen, J. C., & Aberer, K. (2015). An economic analysis of pervasive, incentive-based demand response. In *2015 IEEE International Conference on Smart Grid Communications (SmartGridComm)* (pp. 331–337). https://doi.org/10.1109/SmartGridComm.2015.7436322
133. Sarris, T., Messini, G., & Hatziargyriou, N. (2016). *Residential demand response with low cost smart load controllers.*
134. Gadham, K. R., & Ghose, T. (2016). Design of incentive price for voluntary Demand Response Programs using fuzzy system. In *2016 International Conference on Electrical Power and Energy Systems (ICEPES)* (pp. 363–367). https://doi.org/10.1109/ICEPES.2016.7915958
135. Schachter, J. A., Mancarella, P., Moriarty, J., & Shaw, R. (2016). Flexible investment under uncertainty in smart distribution networks with demand side response: Assessment framework and practical implementation. *Energy Policy, 97*, 439–449. https://doi.org/10.1016/j.enpol.2016.07.038
136. Khezeli, K., & Bitar, E. (2018). Risk-sensitive learning and pricing for demand response. *IEEE Transactions on Smart Grid, 9*(6), 6000–6007. https://doi.org/10.1109/TSG.2017.2700458
137. Jain, M. et al. (2015). Methodologies for effective demand response messaging. In *2015 IEEE International Conference on Smart Grid Communications (SmartGridComm)* (pp. 453–458). https://doi.org/10.1109/SmartGridComm.2015.7436342
138. Croce, D., et al. (2017). Overgrid: A fully distributed demand response architecture based on overlay networks. *IEEE Transactions on Automation Science and Engineering, 14*(2), 471–481. https://doi.org/10.1109/TASE.2016.2621890
139. Wang, B., Yang, X., Short, T., & Yang, S. (2017). Chance constrained unit commitment considering comprehensive modelling of demand response resources. *IET Renewable Power Generation, 11*(4), 490–500.
140. Wang, Y., Chen, Q., Kang, C., Zhang, M., Wang, K., & Zhao, Y. (2015). Load profiling and its application to demand response: A review. *Tsinghua Science and Technology, 20*, 117–129. https://doi.org/10.1109/TST.2015.7085625
141. Nolan, S., & O'Malley, M. (2015). Challenges and barriers to demand response deployment and evaluation. *Applied Energy, 152*, 1–10. https://doi.org/10.1016/j.apenergy.2015.04.083
142. Cappers, P. (2013). Market and policy barriers for demand response providing ancillary services in US markets.
143. Olivieri, S. J., Henze, G. P., Corbin, C. D., & Brandemuehl, M. J. (2014). Evaluation of commercial building demand response potential using optimal short-term curtailment of heating, ventilation, and air-conditioning loads. *Journal of Building Performance Simulation*, 7(2), 100–118. https://doi.org/10.1080/19401493.2013.783118
144. Khorram, M., Abrishambaf, O., Faria, P., & Vale, Z. (2018). Office building participation in demand response programs supported by intelligent lighting management. *Energy Informatics, 1*(1), 9. https://doi.org/10.1186/s42162-018-0008-4
145. Miorandi, D., Sicari, S., De Pellegrini, F., & Chlamtac, I. (2012). Internet of things: Vision, applications and research challenges. *Ad Hoc Networks, 10*(7), 1497–1516. https://doi.org/10.1016/j.adhoc.2012.02.016

146. Leggieri, M., & Hausenblas, M. (2014). Interoperability of Two RESTful protocols: HTTP and CoAP. In W. E. Pautasso, & R. C. Alarcon (Eds.), *REST: Advanced research topics and practical applications* (pp. 27–49). Springer.
147. Shelby, Z. (2010). Embedded web services. *IEEE Wireless Communications, 17*(6), 52–57. https://doi.org/10.1109/MWC.2010.5675778
148. Wang, J., Zhu, Q., & Ma, Y. (2013). An agent-based hybrid service delivery for coordinating internet of things and 3rd party service providers. *Journal of Network and Computer Applications, 36*(6), 1684–1695. https://doi.org/10.1016/j.jnca.2013.04.014
149. Attwood, A., Lamb, D. J., & Abuelmaatti, O. (2014). Position-relative identities in the internet of things: An evolutionary GHT approach. *IEEE Internet of Things Journal, 1*(5), 497–507. https://doi.org/10.1109/JIOT.2014.2353194
150. Paschou, M., Sakkopoulos, E., Sourla, E., & Tsakalidis, A. (2013). Health internet of things: Metrics and methods for efficient data transfer. *Simulation Modelling Practice and Theory, 34*, 186–199. https://doi.org/10.1016/j.simpat.2012.08.002
151. Guan, Z., Li, J., Wu, L., Zhang, Y., Wu, J., & Du, X. (2017). Achieving efficient and secure data acquisition for cloud-supported internet of things in smart grid. *IEEE Internet of Things Journal, 4*(6), 1934–1944. https://doi.org/10.1109/JIOT.2017.2690522
152. Siano, P. (2014). Demand response and smart grids—A survey. *Renewable and Sustainable Energy Reviews, 30*, 461–478. https://doi.org/10.1016/j.rser.2013.10.022
153. Shen, B., Ghatikar, G., Lei, Z., Li, J., Wikler, G., & Martin, P. (2014). The role of regulatory reforms, market changes, and technology development to make demand response a viable resource in meeting energy challenges. *Applied Energy, 130*, 814–823. https://doi.org/10.1016/j.apenergy.2013.12.069
154. Beaudin, M., & Zareipour, H. (2015). Home energy management systems: A review of modelling and complexity. *Renewable and Sustainable Energy Reviews, 45*, 318–335. https://doi.org/10.1016/j.rser.2015.01.046
155. Alamgir, M., Ashraf, M. H., Zeeshan, M., Hassan, M. A. S., Shah, S. H., & Farooq, U. (2024). A hybrid approach to balance lithium-ion cells by implementing SoC using kalman filter. In *2024 International Conference on Engineering & Computing Technologies (ICECT)* (pp. 1–6). https://doi.org/10.1109/ICECT61618.2024.10581299
156. Aghaei, J., & Alizadeh, M.-I. (2013). Demand response in smart electricity grids equipped with renewable energy sources: A review. *Renewable and Sustainable Energy Reviews, 18*, 64–72. https://doi.org/10.1016/j.rser.2012.09.019
157. Behrangrad, M. (2015). A review of demand side management business models in the electricity market. *Renewable and Sustainable Energy Reviews, 47*, 270–283. https://doi.org/10.1016/j.rser.2015.03.033
158. Shariatzadeh, F., Mandal, P., & Srivastava, A. K. (2015). Demand response for sustainable energy systems: A review, application and implementation strategy. *Renewable and Sustainable Energy Reviews, 45*, 343–350. https://doi.org/10.1016/j.rser.2015.01.062
159. Vardakas, J. S., Zorba, N., & Verikoukis, C. V. (2015). A Survey on demand response programs in smart grids: Pricing methods and optimization algorithms. *IEEE Communications Surveys & Tutorials, 17*(1), 152–178. https://doi.org/10.1109/COMST.2014.2341586

Automated Diagnosis for Heating Systems in Buildings

Roxane Koitz-Hristov, Liliana Prikler, and Franz Wotawa

Abstract Estimations of the energy consumption of buildings range from 25 to 40%. Approaches like predictive control can reduce energy consumption by up to 27%. However, it is also well-known that due to a lack of maintenance activities, e.g., heating devices perform at a lowered efficiency level. Therefore, there is a need for fault diagnosis, especially in the building sector, to reduce energy consumption further. This work introduces different diagnosis approaches, ranging from data-centric to knowledge-based approaches. We introduce the foundations and explain the advantages and disadvantages of the various diagnosis methodologies, including the application of machine learning and model-based and qualitative reasoning. In addition, we discuss using different diagnosis methods on a building's heating system and compare the obtained outcome of the approaches. The work aims to provide a profound introduction to diagnosis and its methods, focusing on the application domain of building automation.

Keywords Fault detection and localization · Diagnosis methodologies · Smart buildings

1 Introduction

It is well known that buildings account for a significant amount of energy consumption. In Europe, the estimated energy consumption of buildings is 40% [1], where heating and cooling are major causes. Interestingly, about 75% of greenhouse gas emissions are emitted during the operation phase of a building [2]. Therefore, reducing the energy consumption of buildings is highly relevant to reaching the

The authors are listed in alphabetical order.

R. Koitz-Hristov · L. Prikler · F. Wotawa (✉)
Institute of Software Engineering and Artificial Intelligence, Graz University of Technology, Inffeldgasse 16b/2, Graz, Austria
e-mail: wotawa@tugraz.at

M. A. Mellal et al. (eds.), *Computational Intelligence for a Greener Future: Innovations in Renewable Energy Systems*, Studies in Computational Intelligence 1240,
https://doi.org/10.1007/978-3-032-06732-6_20

desired climate targets. However, Fault Detection and Diagnosis (FDD) for heating, ventilation, and air conditioning systems (HVAC) in buildings primarily still relies heavily on human effort and deserves automation. Indeed, studies indicate that 15–30% of a building's energy consumption is due to faults in HVAC systems or inefficient operations [3], affecting existing and new buildings. Even highly modern equipment such as heat pumps is very likely to operate at only 70–80% efficiency due to a lack of maintenance activities. By detecting, diagnosing, and repairing faults in heat pumps early on, energy loss can be reduced by approximately 40% [4]. In addition, predictive maintenance based on Artificial intelligence, particularly in the field of diagnosis, can significantly contribute to the circular economy. FDD and repair are crucial components of maintenance that significantly extend the lifetime of products and services. Early detection and resolution of faults are key to maintaining the performance and reliability of HVAC systems, resulting in longer vehicle life and reduced waste from premature disposal [5].

Hence, in this book chapter, we focus on diagnosis methods to be used in the context of heating systems installed in buildings. In particular, we discuss two different kinds of diagnosis approaches, i.e., a model-based and a data-based one. Whereas the former relies on a model of a system for fault detection and localization, the latter utilizes measurement data from systems as the basis for learning the correct and, in some cases, incorrect behavior. The learned model is used to distinguish behavior during operation. Both approaches have advantages and disadvantages. The model-based diagnosis requires a (formal) model that is usually developed by engineers considering physical constraints. Hence, it comes with additional costs and effort. However, the model—when constructed as indented comprising components and their interconnections—can be re-used and easily adapted whenever system changes occur or for being used for other systems. It is also worth mentioning that the model-based approach can be used for systems immediately after deployment because no real data is required. In contrast, the model coming from learning fits one particular system and can not necessarily be used for different systems, even when their structure is the same. Adaptations require re-learning and more costs and effort. Because of the need for data, the approach requires that the system to be diagnosed is already running for a particular time until enough observations are available for training. This challenge is also referred to as the Cold Start Problem when utilizing machine learning.

The purpose of this work is to (i) outline the foundations of diagnosis considering two methods but focusing on solely fault detection, (ii) compare the different approaches considering a qualitative and quantitative analysis, and (iii) provide means for deciding which method to use in a practical setup. Hence, we want to enable the reader to understand the basics and underlying assumptions to make a well-informed decision when bringing diagnosis into practice.

We organize this book chapter as follows: In Section 2, we introduce the foundations of model-based reasoning and machine learning-based diagnosis approaches. Furthermore, we compare the two methods qualitatively, considering underlying dependencies, completeness, correctness, adaptability, and explainability. In Section 3, we introduce the application of the two diagnosis approaches to the heating domain, where we carried out a study based on simple heating systems comprising a pump and a heat exchanger. Finally, we summarize the findings in the conclusions section.

2 Foundations

This section covers the theoretical groundwork for model-based and machine learning-based diagnosis methods. It provides the necessary background to understand their underlying principles, enabling a more informed comparison and integration in later parts of the chapter.

2.1 Model-based Reasoning

The idea of utilizing models for diagnosis and other tasks relying on reasoning dates back to the 80 s of the last century. Davis and colleagues [6] were the first who formulated the idea considering the structure and function of systems for diagnostic reasoning aiming at overcoming certain challenges of diagnosis expert systems. The idea was later formalized by Reiter [7], and De Kleer and Williams [8].

In Fig. 1 we outline the basic principles of model-based reasoning applied to diagnosis. On the right side, we have a physical system, e.g., a heat pump, from which we obtain observations. On the left side of the figure, we have a model of the heat pump. Using this model, we can derive a behavior and compare it with the observations. In case of discrepancies, we directly use the model to localize the fault. The latter is obtained by introducing explicit health states of components that cause a particular behavior. By changing the health states, the model can be used to search for the elimination of discrepancies. If there are no discrepancies anymore after assuming health states, the latter indicates the root cause behind the discrepancies.

To be self-contained, we describe the foundations behind model-based diagnosis in brief. In particular, we rely on definitions stated in more recent publications, i.e., [9] and [10].

We start by defining a **diagnosis system** as a pair (SD, $COMP$), where SD is a system description representing a model of the system, and $COMP$ is a set of system components. A diagnosis system comprises the model and all components for which we can set a health state. In model-based diagnosis, a model states the behavior of components and their interactions formally, e.g., using logic. Particularly, the component models at least state the correct behavior of a component. To distinguish correct from incorrect health states of components, we introduce a predicate $AB(C)$ stating that a component C from $COMP$ is abnormal. Hence, $\neg AB(C)$ states that the component is working as expected.

In the following example, we model the behavior of a radiator, stating that it works as expected if the input temperature (in) equals the output temperature (out) plus the

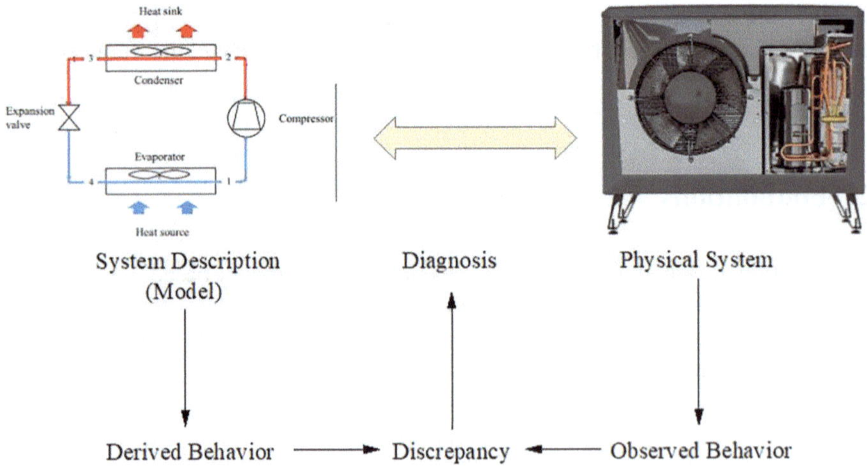

Fig. 1 The principle behind model-based reasoning applied to diagnosis

temperature difference (*diff*) that is supposed to heat a room. That is, if the radiator R is not abnormal ($\neg AB(R)$), and we know the input temperature is $T1$, the output temperature is $T2$, and the intended temperature difference is $T3$, then it must be the case that $T1 = T2 + T3$. Such a behavior can be formalized in logic as follows:

$$\neg AB(\mathrm{R}) \Rightarrow ((\mathrm{in}(\mathrm{R}, \mathrm{T1}) \wedge \mathrm{out}(\mathrm{R}, \mathrm{T2}) \wedge \mathrm{diff}(\mathrm{R}, \mathrm{T3})) \rightarrow \mathrm{T1} = \mathrm{T2} + \mathrm{T3})$$

This expresses that a correctly working radiator will have an input temperature equal to the sum of its output temperature and the difference it's supposed to produce.

Such a model of a radiator would be added to the model of a system comprising different radiators like $r1$, $r2$, etc. Hence, R in this context represents all potential radiators of a system and has to be replaced in the system model considering a specific radiator. It is worth noting that such rules are only describing the correct behavior. Only in the case $\neg AB(R)$ is true, the right part of the rule will be considered.

To identify whether a system behaves correctly or incorrectly, we need observations. Formally, we introduce a **diagnosis problem** as a tuple $(SD, COMP, OBS)$, where $(SD, COMP)$ is a diagnosis system and OBS is a set of observations.

For our example, we might measure the temperatures of a radiator as follows: $T1 = 55$, $T2 = 50$, and $T3 = 5$, which obviously is not in line with the stated behavior. Hence, we are interested in a diagnosis, i.e., an explanation of an observed behavior. Formally, a **diagnosis** for a given diagnosis problem is defined as a set $\Delta \subseteq COMP$ for which the logical sentence $SD \cup OBS \cup \{AB(C)|C \in \Delta\} \cup \{\neg AB(C)|C \in COMP \setminus \Delta\}$ is satisfiable and, therefore, not leading to any contradictions. In this definition, the health status of components is set such that a difference between the behavior of a model is not in discrepancy with the observed behavior. For our radiator example, we construct the following logical formula:

$$\{\neg AB(r) \Rightarrow ((in(r, T1) \wedge out(r, T2) \wedge diff(r, T3)) \rightarrow T1 = T2 + T3)\} \\ \cup \{in(r, 55), out(r, 50), diff(r, 3)\} \cup \{\neg AB(r)\}$$

This logical sentence combines.

- the system's model, which specifies the expected behavior of a correctly functioning radiator (i.e., $\neg$AB(R) $\Rightarrow \ldots$),
- the observations given as $in(r, 55)$, $out(r, 50)$, $diff(r, 3)$, and
- the assumption that the radiator r is functioning correctly, i.e., $\neg AB(r)$.

Substituting the observed values into the model results in the following logical implication:

$$\neg\mathrm{AB(r)} \Rightarrow ((\mathrm{in(r, 55)} \wedge \mathrm{out(r, 50)} \wedge \mathrm{diff(r, 3)}) \rightarrow 55 = 50 + 3)$$

Since $55 \neq 53$, the right-hand side of this implication is false. Therefore, if we assume $\neg AB(r)$ (i.e., the radiator is healthy), the entire implication becomes false, resulting in a logical contradiction. To resolve the contradiction and make the entire sentence satisfiable, we must conclude that $\neg AB(r)$ is false, i.e., $AB(r)$ is true, meaning the radiator is faulty. Hence, $\{r\}$ is a diagnosis and represents the minimal set of components that must be considered faulty to align the model with the observations.

It is worth noting that in practice, we are interested in only the best diagnosis. Hence, we focus on **minimal diagnoses** that are defined as diagnoses where none of its subsets is itself a diagnosis. A minimal diagnosis is also referred to as a parsimonious diagnosis. Note that sometimes, we might be interested in the most probable diagnosis. Knowing the fault probabilities of components, the likelihood of a diagnosis can be directly obtained (see [8]).

Diagnoses can be computed automatically from observations and models when considering the given definition. For algorithms we refer the interested reader to previous work, e.g., [11] or [12].

2.2 *Machine Learning for Diagnosis*

A wide variety of machine learning and data-driven techniques are used in diagnosis [13–16]. We distinguish between supervised and unsupervised techniques and various problems that these techniques can solve, such as classification, regression, clustering, or dimensionality reduction [17].

In ***supervised machine learning***, data is presented to an algorithm as a sequence of pairs $(\vec{x}, y)$, where $\vec{x}$ is a set of features and y a variable to be learned. We call a task **classification** if $y \in L$ for some finite, usually small set of labels L and **regression** if a parameter β is to be estimated, s.t. $y = f(\vec{x}, \beta) + \varepsilon)$ for some known function f. Figure 2a shown an example of a classification problem in which

f_1, f_2, and f_3 are decision functions, which attempt to separate points labeled a (blue dots) from points labeled b (red boxes). In Fig. 2b one can see a regression model $\hat{f}$ which approximates f with the areas around it indicating thresholds of acceptable deviation in behavior. In this example, the runtime behavior g is clearly faulty at $x = -3$ (triangle symbol) and possibly faulty, depending on the threshold used for classification at $x = 1$ (square symbol).

In ***unsupervised machine learning***, only the features $\vec{x}$ are presented to the algorithm. This makes classification and regression more difficult, if not impossible, but tasks such as clustering and dimensionality reduction do not require that y to be known. Clustering is similar to classification in that the aim is to find a partition of the features into a finite number of bins, typically based on a measure of proximity (see Fig. 2c). Dimensionality reduction instead aims to compress the features in a manner that preserves certain properties.

Fault detection quite naturally presents itself as a classification problem, particularly a binary classification problem with labels $L = \{\text{nonfaulty, faulty}\}$. In the most general setting, relations between individual features are not known and may even be absent. However, in real-world systems, there are often implicit correlations that the

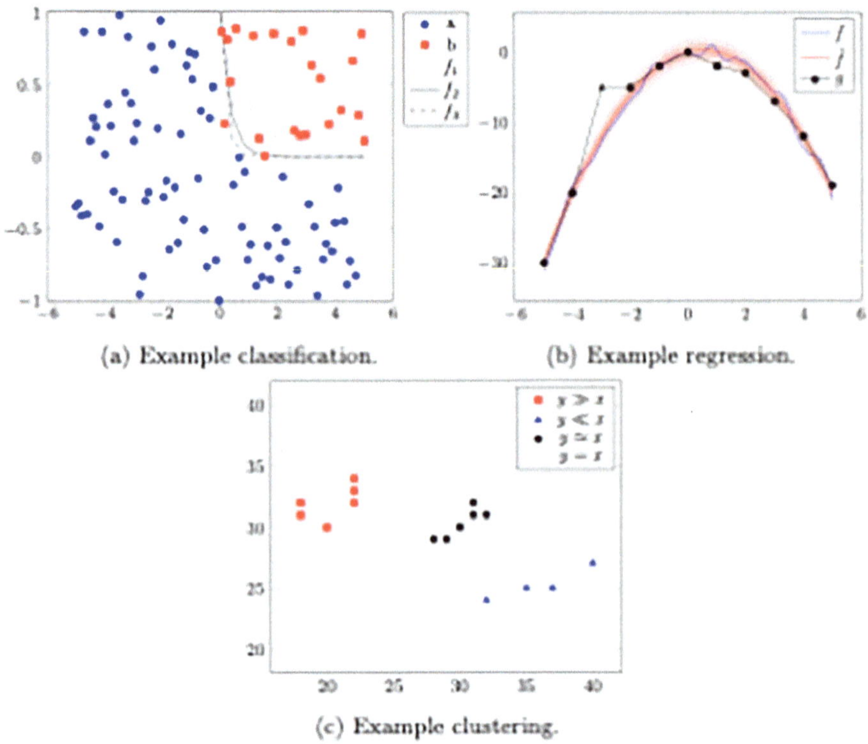

(a) Example classification.

(b) Example regression.

(c) Example clustering.

Fig. 2 Machine learning methods for fault detection and diagnosis

machine learning algorithm can exploit to draw decision boundaries between faulty and nonfaulty data.

If one feature in the system is known to depend on other features—e.g., the output temperature of a heater is proportional to the consumption of fuel or electricity—one may instead use a regression model to estimate the exact coefficients, such as energy efficiency. Supposing that this model matches the actual behavior of the system closely enough, one can then use the residual (the difference between the predicted value and the actual value of a feature) as an oracle to derive whether a system is faulty or nonfaulty. If the residual is low, that is, the system behaves as predicted by the regression model, it appears nonfaulty. If the residual is high, then the system is likely faulty.

A special case of classification is outlier detection, where a machine learning model is only trained on nonfaulty data or mostly nonfaulty data without labels. The task is to find particular points that do not appear as though they are drawn from the same distribution as the training data—and to label those as outliers. For example, a sudden spike in room temperature from 20 °C to 22 °C might be considered a (local) outlier. Likewise, if a model was trained on temperatures ranging from 20 °C to 25 °C, a temperature of 30 °C would be an outlier even if it persists over a longer period of time.

Fault isolation can be modeled as a multiclass classification problem, where each class identifies one possible fault in the system. Unsupervised machine learning techniques for clustering and dimensionality reduction may be used to distinguish particular failure modes. As an example, consider a mixing valve that is used to control the outlet temperature in a shower. Depending on the desired water temperature of the person showering and the actual water temperature, three clusters "too hot", "too cold", and "(roughly) as desired" may be observed, where different (internal) errors could skew the outlet temperature into being warmer or colder than expected.

2.3 Comparison

In this section, we compare the model-based and machine learning approach to fault diagnosis from a general perspective and with a focus on heating systems in the building domain. We focus on qualitative differences between two different methodologies among various dimensions.

3 Dependencies

Model-based diagnosis approaches rely on certain assumptions, such as the availability and reliability of the system description. System models can be difficult to obtain due to time constraints and the lack of in-depth domain knowledge. Even in cases where expert knowledge exists in the form of structured knowledge about devices held by designers, developers, and maintenance personnel, formalizing the expertise in the form of a suitable system description is challenging. While there is research on reusing system descriptions obtained from previous system-related tasks, such as the design phase [18], e.g., CAD systems [19] or fault analysis [20], e.g., fault tree analysis, model-based diagnosis is always associated with an initial modeling phase.

In the domain of HVAC systems in large commercial buildings, one can benefit from the fact that especially air handling units and heat pumps consist of standardized components and are embedded in BACnet[1] building management systems. This standardization aids in the creation of suitable system descriptions and can allow for the reuse of entire models or at least models of components.

Unlike model-based approaches that require explicit modeling of the system, machine learning can adapt to various systems by training on relevant data, removing the need for detailed domain knowledge to apply the technique. However, machine learning algorithms require large amounts of high-quality data to train effectively [21]. For HVAC systems, collecting sufficient and relevant fault data can be challenging, especially for rare or unprecedented fault types. While unknown anomalies can be detected, e.g. by using unsupervised learning techniques, determining the underlying root cause can be challenging [22]. Additionally, labeled datasets from real-world operations are rare and need to be created with high manual effort. In a new system (or building) usually not enough historical data has been accumulated to generate a machine learning model that can accurately diagnose faults [23]. Seasonal changes in outdoor temperature, humidity, and other environmental factors have a significant impact on building variables such as energy expenditure. Collecting a sufficient volume of data capturing seasonal changes and different external conditions is a time-consuming process. In general, one approach to mitigate this is transfer learning, where a model trained in one domain is adapted to another with minimal data [24]. However, in the building domain, this is problematic. Buildings vary significantly in terms of architecture, occupancy, usage, environmental conditions, or subsystem configurations. A model trained on one building might not perform well on another one. In addition, the building data is mostly largely imbalanced, as fault events do not occur as frequently compared to the normal operation mode or are not noticed by the building operator/building owner.

[1] BACnet (Building Automation and Control Network) is the worldwide standard for data communication in building automation and control systems.

4 Completeness, Resolution, and Correctness[2]

The completeness, resolution, and correctness of the model-based diagnosis approach largely depend on the quality of the system description [25]. A crucial aspect of modeling is the chosen level of abstraction, which involves a tradeoff between comprehensiveness, and thus completeness and resolution, and computational efficiency [26]. Complex systems can be challenging to model with accuracy, and any simplifying assumptions made during the modeling process can introduce inaccuracies in the resulting diagnoses. The success of modeling an HVAC system thus relies to a large portion on the level of detail that considers all relevant system behaviors while excluding unnecessary complexity. For instance, for the smallest replaceable units within a system, fault localization may be sufficient, making detailed fault identification unnecessary [27, 28]. Another obstacle in real world applications may be uncertainties due to imperfect sensor readings and operational data, introducing possible issues even if a valid model is used. In practice, such issues can often be mitigated using heuristics and data preprocessing [29] or by considering certain observations as uncertain within the model-based diagnosis framework [30].

Depending on the application domain and model structure, a given set of symptoms may lead to (i) multiple possible diagnoses and (ii) diagnoses involving multiple faults. (i) Since maintenance or repair tasks typically follow fault identification in practice, a way to discriminate or rank diagnoses is necessary. One approach is to gather additional observations, e.g., from an on-site inspection, to rule out some of the irrelevant diagnoses and thus reduce the set of possible solutions [31]. Because the diagnostic model captures the system's behavior, it can be directly utilized to identify useful probes and tests that help refine the diagnostic results. Additionally, probabilities or other ranking mechanisms can be integrated into the model to create a ranking of diagnoses [23]. Using probabilistic diagnostic models, however, comes with an increased complexity, and determining appropriate probabilities is a task in itself [33, 34]. (ii) While single faults, i.e., where only a single component malfunctions at a point in time, are common in practice, model-based diagnosis can identify all possible faults, including multiple-component failures, such as double or triple faults [7].

Despite challenges, model-based diagnosis is grounded in a solid theoretical background and has been proven to derive complete and correct results based on the behavior and structure represented in the models, the faults included (if any), and the given observations [19, 35].

Machine learning stands in contrast to model-based diagnosis as it captures empirical correlations from data rather than relying on fundamental physical and engineering principles and the deductions derived from them [19]. As a result, its completeness depends on how sufficient various faulty and nominal operation modes are represented in the training data, which is challenging in real-world applications

---

[2] *Completeness* indicates whether all possible valid diagnoses can be identified by the system, while *resolution* describes the level of detail or specificity of the diagnosis. *Correctness* refers to whether the diagnosis produced by the system is accurate and consistent with the available data.

such as HVAC systems. This is especially problematic for rare faults, as those may be misclassified or missed completely, given the lack of available data [36]. Depending on the machine learning approach used, the model typically performs either fault detection (e.g., using a binary classifier) or fault identification (e.g., using a multi-class classifier). In the case of fault identification, the classifier usually assigns each instance to one predefined class, although it often provides confidence or probability scores for all possible classes. By exploiting these scores, it is possible to generate a ranked list of potential diagnoses. Additionally, multi-label classification enables the detection of multiple faults simultaneously rather than limiting predictions to either "faulty" or "nonfaulty" [37]. Another factor influencing diagnosis is the quality of the features extracted from the raw data and the ability of the chosen machine learning approach to effectively discriminate between different classes of faults.

Furthermore, machine learning models lack guarantees on the correctness of the results as they are probabilistic and generally only uncover statistical patterns, correlations, and anomalies in the data. Noisy, imbalanced, or insufficient training data can lead to false positives or false negatives, which is why performance metrics for machine learning methods typically assess these aspects [38]. In addition, machine learning models may not recognize whether learned correlations conflict with real-world constraints and thus generate incorrect diagnoses [36].

5 Adaptability to Changes

Model-based system descriptions are static in the sense that they are usually generated for a system at a certain point in time. Yet, over time, an HVAC system may evolve or be updated. While model-based diagnosis can detect (new) "unknown" faults as deviations from the expected behavior, and those deviations contribute to the generation of possible solutions [34], adapting to entirely new conditions or operational contexts not covered by the initial model can be challenging and may require model revision or redevelopment. How to automate this in a structured, efficient, and integrity-preserving way has not yet been answered [26]. An advantage of the model-based approach is that diagnosing new systems is feasible without creating a model in its entirety from scratch, provided the system description is built using components whose behavior models are included in a model library as is possible in the HVAC domain [26].

The same issue arises with machine learning models, as they can become outdated in case the system evolves or undergoes modifications and the new conditions are not represented within the training data. In the case of fault detection, this might not be an issue; however, in the case of using classification to identify various faults, this becomes problematic. Thus, in practice, machine learning models need to be retrained or fine-tuned with new data that reflects the latest changes within a building [39].

6 Interpretability and Explainability[3]

Model-based diagnosis offers a high level of interpretability and explainability as its system model is explicit, and reasoning is transparent. The diagnostic process can be traced step-by-step, and diagnoses can be explained in terms of component behavior, connections, and causal relationships [40].

Complex machine learning models, including deep neural networks, are often referred to as "black boxes", which can make it difficult for humans to understand the reasoning behind their predictions or decisions. This research area is referred to as Explainable Artificial Intelligence (XAI). Understanding the cause and nature of a fault, however, is crucial for an effective fault resolution and is necessary in obtaining trust and acceptance of a diagnostic system. Especially since often machine learning models–in general, and in the building domain–support human decision-making rather than replacing it [41].

On the one hand, there are model-agnostic XAI methods that can explain the output of any model. For instance, LIME [42] creates a simpler model that approximates the original one but is easier to interpret, focusing on how changes in input affect predictions, while SHAP [43] explains a model's prediction by measuring the impact of each feature. Various Python libraries implement the most common model-agnostic XAI methods. On the other hand, there are model-specific approaches, such as attention mechanisms [44] in neural networks or path analysis in decision trees [45], that provide insights into which parts of the input data the model is focusing on to make its decisions. However, recent work [46, 47] shows that those previous methods fall short of delivering sound, irredundant minimal explanations even in the case of model-specific XAI, such as decision tree paths.

7 Temporal Aspects

HVAC systems typically produce multi-variant time series data from a combination of sensors, actuators, setpoints, and control signals, capturing the dynamic interactions between the system components. In online diagnosis, it is common to apply a temporal window that captures the system's evolution over time to detect gradual behavior changes.

The model-based diagnosis framework can be extended to incorporate temporal reasoning, including dynamic system behavior (e.g., the gradual increase in room temperature following a heating setpoint change) and temporal dependencies between variables (e.g., the delayed change in water flow rate after a valve position is adjusted). This enables the diagnostic system to reason not only about current inconsistencies but how deviations evolve over time, allowing it to detect faults that may

[3] In the context of machine learning *interpretability* denotes how transparent the internal workings of a machine learning model are, while *explainability* refers to how well the reasoning behind a model's predictions can be understood.

only manifest through temporal inconsistencies. However, incorporating temporal aspects within the model introduces additional complexity in both modeling and computation [48].

In contrast, machine learning techniques, such as Autoregressive Moving Average (ARMA) [49] or Long-Short-Term Memory (LSTM) encoder-decoder architectures [50], are specifically designed to handle sequential and time-series data. These models learn temporal dependencies and trends in the data, thus making them suitable for diagnosing faults that develop over time and/or manifest in temporal signatures. Yet, as with many data-driven approaches, their performance can be negatively affected by noise [51].

8 Case Study Heating System

In our use case, we consider a commercial building heating system consisting of multiple heat distribution lines connected to a pressurized distributor that supplies heat from the district heating transfer station to the heating system. Each heat distribution line provides heat to a different zone within the building.

For our case study, we focus on a single heat distribution line, as illustrated in Fig. 3. When the heating system is active, the central flow pipe distributes hot water from the district heating transfer station to the individual flow pipes of the heat distribution lines. The central return pipe then carries the cooled water from the heat distribution line back to the transfer station. The zone temperature is regulated by a motorized valve that adjusts the flow temperature. This valve's position is determined by the temperature setpoint, which depends on the outdoor temperature and the heating curve: As the outdoor temperature decreases, both the required flow temperature setpoint increases to maintain the desired zone temperature.

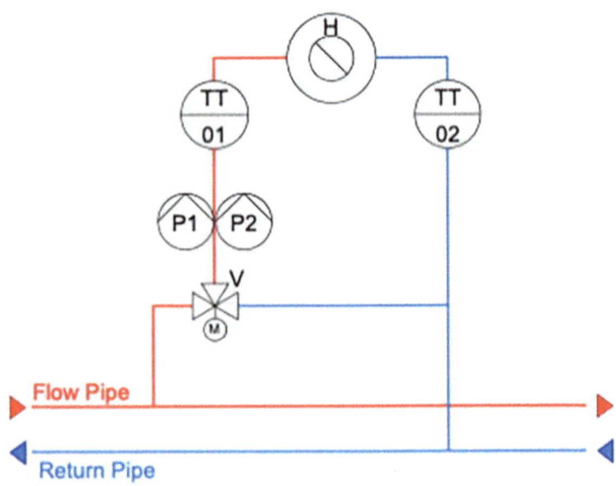

Fig. 3 Schematics of the heat distribution line

In addition to the flow and return pipe, the heat distribution line consists of the following components:

- **Heat Exchanger (H):** The heat exchanger converts the heat from the flow pipe into warm air, thereby heating the designated zone.
- **Temperature Sensors (T1,T2):** Sensor T1 measures the temperature of the water entering the heat exchanger and serves as a control input for valve positioning. Sensor T2 monitors the return flow temperature.
- **Pumps (P1,P2):** The pumps ensure consistent water flow from the valve to the heat exchanger, maintaining a stable pressure, operating either alternately or remaining inactive.
- **Valve (V):** The motorized valves regulate the mixture of hot water from the flow pipe and cooled water from the return pipe, with its position (0%–100%) determined by the desired temperature. At 0%, no hot water is injected, causing the flow temperature to match the return temperature. At 100%, the flow temperature equals the central flow pipe temperature.
- **Temperature Sensors (T1,T2):** Sensor T1 measures the temperature of the water entering the heat exchanger and serves as a control input for valve positioning. Sensor T2 monitors the return flow temperature.
- **Pumps (P1,P2):** The pumps ensure consistent water flow from the valve to the heat exchanger, maintaining a stable pressure, operating either alternately or remaining inactive.
- **Valve (V):** The motorized valves regulate the mixture of hot water from the flow pipe and cooled water from the return pipe, with its position (0%–100%) determined by the desired temperature. At 0%, no hot water is injected, causing the flow temperature to match the return temperature. At 100%, the flow temperature equals the central flow pipe temperature.
- **Outdoor Temperature Sensor:** In addition to the components depicted in Fig. 3 there is a sensor that provides the outdoor temperature for the control system.

Besides plausibility checks for sensor readings based on defined ranges, the following constraints must be satisfied under normal operating conditions:

- The measured flow temperature (T_f) must always be greater than the measured return temperature (T_r): $T_f > T_r$
- The measured flow temperature (T_f) must be within 2 K below and 1 K above the setpoint temperature (T_s): $T_s - 2 \leq T_f \leq T_s + 1$

8.1 Application of Fault Detection via Machine Learning

In order to apply machine learning, we first gather data from the running system using the available sensors as well as the control system, capturing the commands it sends to e.g. switch on a pump or change the position of the valve. This raw stream of data is first cleaned by sampling it at uniform intervals, where the nearest sensor

Table 1 Distribution of training data

Label	Count	Time frame
okay	3500	2023–01–2023–02
fault 1	3319	2023–04–2023–05
fault 2	4176	2023–11–2023–12

readings are used to interpolate the "actual" value and then labelled. Table 1 contains information on the types of data used in our empirical evaluation.

We then process the labeled data as follows:

1. We split each table separately into a training and test set.
2. We merge the training sets, using oversampling to balance them.
3. We fit outlier detection models to the "okay" training data
4. and binary classification models to the balanced training data.
5. We merge the test sets, using oversampling to balance them.
6. We test the models obtained in step 3 on the balanced test sets.
7. We report the performance metrics of the models.
8. We repeat this process ten times, constructing ten training and test splits at random and training and testing each model on each split once.

We repeat this process ten times, that is, constructing ten training and test splits at random, training and testing each model on each split once.

We consider four outlier detection models that are already implemented in scikit-learn [52] (version 1.4.2), using default parameters for each of them. We do not employ supervised machine learning methods, as they would require extensive faulty data, which is not easily procured in real-world settings; and we do not exploit clustering or dimensionality reduction as neither yields a "diagnosis" as a classification of data into faulty or non-faulty sets. The models we consider are (1) Local Outlier Factor, (2) Elliptic Envelope, (3) Isolation Forest, and (4) One-Class SVM. Figure 4 shows the classification results for each of the models. Most models appear to perform decently well, with f_1 scores of 0.8, but varying results on precision and recall. Only the Elliptic Envelope achieves both a very high precision (average 0.899) and recall (average 0.915), leading to an average f_1 score of 0.906.

Figure 5 shows the times spent on training and testing the models respectively. We see that Elliptic Envelope, the method that performs best according to classification metrics, takes the longest to train while being quite good at runtime. Local Outlier Factor and One-Class SVM are fast in training, but comparatively slow in testing.

8.2 *Application of Diagnosis via Model-based Reasoning*

For model-based reasoning, we formulate the constraints of the normal operating conditions into adequate logic formulae. We implement the following constraints in both Python and Clingo:

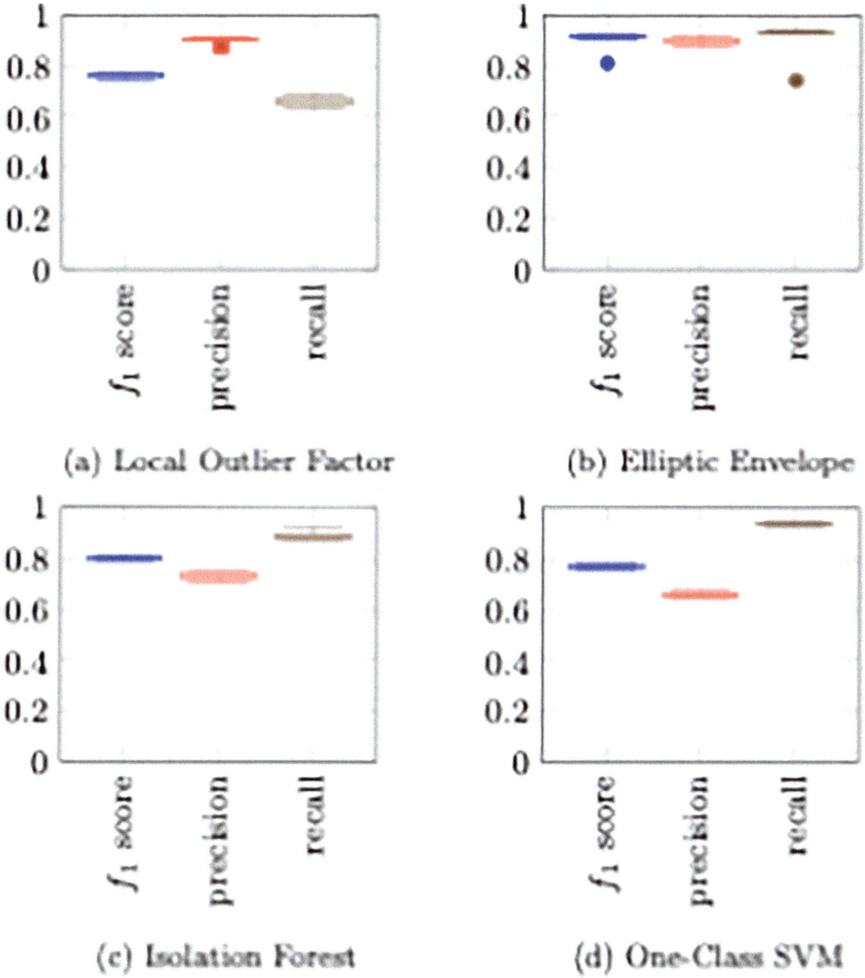

Fig. 4 Classification metrics of various outlier detection models. The box plots indicate the mean, 25% quartile, 75% quartile, lowest and highest value within 1.5 × *IQR* and outliers respectively. Local Outlier Factor produces high precision and low recall, Isolation Forest and One-Class SVM have high recall, but low precision. Elliptic Envelope has both high precision and high recall, with one outlier performing quite poorly

- The measured flow temperature must be greater than or equal to the measured return temperature.
- The measured flow temperature must be not more than 2 K below the setpoint temperature.
- The measured flow temperature must be not more than 1 K above the setpoint temperature.

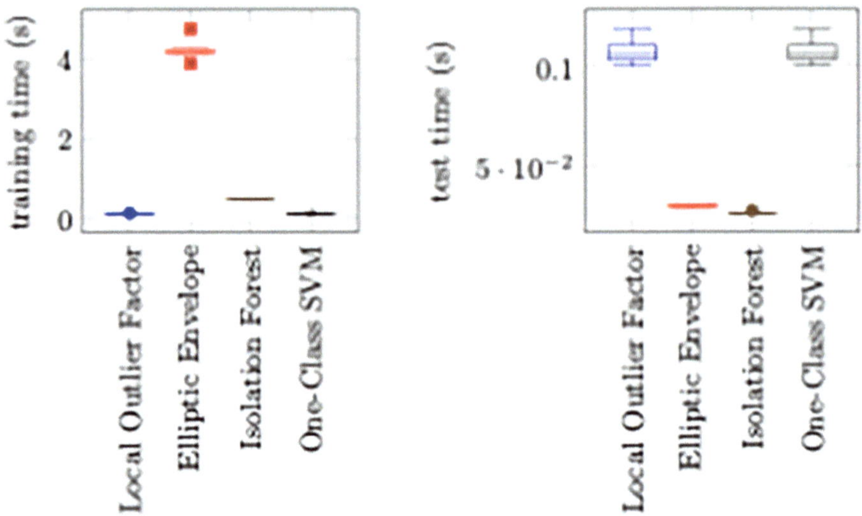

Fig. 5 Training and test times of machine-learning-based outlier detection. All methods take less than 5 s for training and less than 1 s for testing. Elliptic Envelope takes comparatively long to train, Local Outlier Factor and One-Class SVM comparatively long to test

These constraints are relaxations of the domain constraints listed before. Notably, all of them can be written in the form $a \leq b + c$, where c is a constant and a, b refer to some measured temperatures. In Python, we implement the entire system as a logical and of these three constraints. In Clingo, we instead detect each violation separately and mark an entry as *ok* only if all checks are satisfied. This structure allows for better interpretability, as the failing rule is highlighted in the answer set.

We reuse the same training and test splits as before, but export the tables to CSV and then translate this CSV into a logic program by means of a Python script. The translation to Clingo is lossy, as it transforms real numbers to integers and thus rounding errors may influence the result.[4] As a result, we expect worse classification results than for the pure Python implementation, which can make use of a floating point representation of the data.

Figure 6 shows the classification metrics for these two implementations. We can see, that the Python implementation is slightly, but barely noticeably better in terms of f_1 score, but considerably better in terms of recall (0.85 rather than 0.818). Figure 7 shows the runtime for both the Python implementation and Clingo. We do not report any training times, because neither implementation needs training. For Clingo, we use three statistics provided by Clingo itself: the solving time, the CPU time, and the total time. The total time includes the time spent converting the input, as we run the whole process as a pipeline. The solving time is the time spent on solving and is in our test runs always reported as "0.00 s". The difference between CPU time and

[4] We use commercial rounding rather than banker's rounding to prevent temperatures that are exactly 1 K apart being rounded towards the same integer.

total time is due to other processes (like the conversion process) using the CPU, as well as IO.

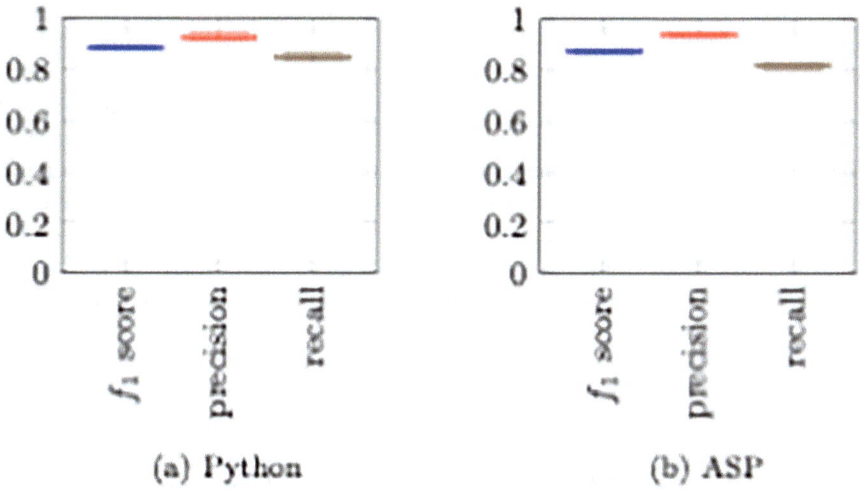

Fig. 6 Classification metrics of model-based fault detection. Both implementations show high precision and moderate recall

Fig. 7 Runtime of Clingo compared to a straight-forward Python implementation. While solving times are trivial, the total time spent (including input conversion) is significantly higher

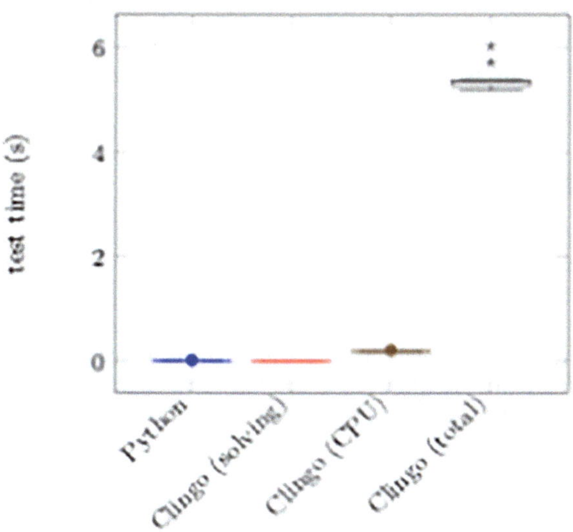

9 Conclusions

In this work, we present and compare two approaches for automated diagnosis: model-based reasoning and machine learning. Both methods are promising fault detection and localization methods to improve the efficiency and reliability of HVAC systems.

On the one hand, model-based diagnosis provides a transparent, explainable, and reusable framework grounded in formal logic. It is particularly well-suited in cases where the system behavior is standardized and well-understood. However, it requires an appropriate diagnostic model that is accurate enough to generate relevant diagnoses but includes a suitable level of abstraction to ensure solutions can be derived in a reasonable time. As the approach does not require historical data, it is ideal for newly developed systems. Our case study shows that, given a suitable model, the approach detects faults efficiently and reliably.

Machine learning, on the other hand, offers a data-driven alternative that can adapt to complex dynamic systems without the need to craft a model by hand. The empirical evaluation of our heating system demonstrates that outlier detection methods such as Elliptic Envelope can be very efficient at detecting faults with high precision and recall. Yet, the approach heavily depends on the availability and quality of data. Further, it suffers from generalization issues in the case of a new or evolving system.

In conclusion, the choice between these methods depends on the specific application context, data availability, system complexity, and operational requirements. In certain scenarios, a hybrid approach that integrates model-based and data-driven diagnosis methods, exploiting the strengths of each, can provide the most robust and scalable solution.

10 Competing Interests

The authors have no conflicts of interest to declare that are relevant to the content of this chapter.

Acknowledgements The work presented in this chapter has been supported by the FFG project Artificial Intelligence for Smart Diagnosis in Building Automation (ALFA) under grant 914932. We would like to express our gratitude to our industrial partner, DiLT Analytics FlexCo.

References

1. European Union. (2017). Commission welcomes agreement on energy performance of buildings
2. Deutsche Gesellschaft für Nachhaltiges Bauen (DGNB). Wegweiser klimapositiver Gebäudebestand. (2022).

3. Chen, Z., et al. (2023). A review of data-driven fault detection and diagnostics for building HVAC systems. *Applied Energy, 339*, Article 121030.
4. Bellanco, I., Fuentes, E., Vallès, M., & Salom, J. (2021). A review of the fault behavior of heat pumps and measurements, detection and diagnosis methods including virtual sensors. *Journal of Building Engineering, 39*, Article 102254.
5. Thuve, C. & Yuvaraj, H. P. (2023). Life cycle assessment of a combustion engine—mapping the environmental impacts and exploring circular economy. https://odr.chalmers.se/server/api/core/bitstreams/99284a9d-52ff-49b1-bef6-caffee5199c6/content
6. Davis, R. et al. (1982) Diagnosis based on structure and function. In *Proceedings AAAI* 137–142 (Pittsburgh).
7. Reiter, R. (1987). A Theory of Diagnosis from First Principles. *Artificial Intelligence, 32*, 57–95.
8. de Kleer, J., & Williams, B. C. (1987). Diagnosing Multiple Faults. *Artificial Intelligence, 32*, 97–130.
9. Wotawa, F. (2019). Reasoning from first principles for self-adaptive and autonomous systems. In *Predictive Maintenance in Dynamic Systems: Advanced Methods, Decision Support Tools an d Real-World Applications* Springer. https://doi.org/10.1007/978-3-030-05645-2
10. Wotawa, F. (2020). Using model-based reasoning for self-adaptive control of smart battery systems. In: Sayed Mouchaweh, M. (eds) *Artificial Intelligence Techniques for a Scalable Energy Transition.* Springer, Cham. https://doi.org/10.1007/978-3-030-42726-9_11
11. Nica, I., Pill, I., Quaritsch, T., Wotawa, F. (2013). The route to success—a performance comparison of diagnosis algorithms. In *Proceedings International Joint Conf Artificial Intelligence (IJCAI)*, pp 1039–1045
12. Kaufmann, D., Nica, I., & Wotawa, F. (2021). Intelligent agents diagnostics—enhancing cyber-physical systems with self-diagnostic capabilities. *Adv Intell Syst.* https://doi.org/10.1002/aisy.202000218
13. Abid, A., Khan, M. T., & Iqbal, J. (2021). A review on fault detection and diagnosis techniques: Basics and beyond. *Artificial Intelligence Review, 54*(5), 3639–3664. https://doi.org/10.1007/s10462-020-09934-2
14. Bi, J., Wang, H., Yan, E., Wang, C., Yan, K., Jiang, L., & Yang, B. (2024). AI in HVAC fault detection and diagnosis: A systematic review. *Energy Reviews, 3*(2), Article 100071. https://doi.org/10.1016/j.enrev.2024.100071
15. Dai, X., & Gao, Z. (2013). From model, signal to knowledge: A data-driven perspective of fault detection and diagnosis. *IEEE Transactions on Industrial Informatics, 9*(4), 2226–2238. https://doi.org/10.1109/TII.2013.2243743
16. Melo, A., Câmara, M. M., & Pinto, J. C. (2024). Data-driven process monitoring and fault diagnosis: A comprehensive survey. *Processes, 12*(2), 251. https://doi.org/10.3390/pr12020251
17. Mohri, M., Rostamizadeh, A., & Talwalkar, A. (2012). *Foundations of machine learning.* MIT Press, Cambridge.
18. Struss, P., Rehfus, B., Brignolo, R., Cascio, F., Console, L., Dague, P, et al. (2002). Model-based tools for the integration of design and diagnosis into a common process—a project report. DX 2:2–4
19. Dressler, O., Puppe, F. (1999). Knowledge-based diagnosis—survey and future directions. In German Conference on Knowledge-Based Systems. Springer, pp. 24–46.
20. Koitz, R., Wotawa, F. (2016). Integration of failure assessments into the diagnostic process. In: *Annual Conference of the PHM Society 8*(1)
21. Dash, B. M., Bouamama, B. O., Boukerdja, M., Pekpe, K. M. (2022). A comparison of model-based and machine learning techniques for fault diagnosis. In: *2022 23rd International Middle East Power Systems Conference (MEPCON)*, Cairo, Egypt, pp 1–7. https://doi.org/10.1109/MEPCON55441.2022.10021712
22. Mirnaghi, M. S., & Haghighat, F. (2020). Fault detection and diagnosis of large-scale HVAC systems in buildings using data-driven methods: A comprehensive review. *Energy Build, 229*, Article 110492.

23. Barata, R., Leite, M., Pacheco, R., Sampaio, M. O., Ascensão, J. T., Bizarro, P. (2021). Active learning for imbalanced data under cold start. In: *Proceedings of the Second ACM International Conference on AI in Finance*, pp 1–9.
24. Genkin, M., McArthur, J. J. (2022). Using reinforcement learning with transfer learning to overcome smart building cold start. In: 2022 *International Conference on Computational Science and Computational Intelligence (CSCI).*, IEEE, pp 432–439.
25. Console, L., & Torasso, P. (1991). A spectrum of logical definitions of model-based diagnosis. *Computational Intelligence, 7*(3), 133–141.
26. Travé-Massuyès, L., Milne, R. (1998). Gaps between research and industry related to model based and qualitative reasoning. In: *Proceedings of the European Workshop on Model Based Systems and Qualitative Reasoning*, pp 54–57.
27. Chittaro, L., & Ranon, R. (2004). Hierarchical model-based diagnosis based on structural abstraction. *Artificial Intelligence, 155*(1–2), 147–182.
28. Provan, G. (2001). Hierarchical model-based diagnosis. In: Proc DX 1, pp. 167–174.
29. Yellapu, V. S., Zhang, W., Vajpayee, V., & Xu, X. (2021). A multiscale data reconciliation approach for sensor fault detection. *Progress in Nuclear Energy, 135*, Article 103707. https://doi.org/10.1016/j.pnucene.2021.103707
30. Cazes, D., Kalech, M. (2020). Model-based diagnosis with uncertain observations. In *Proceedings of the AAAI Conference on Artificial Intelligence 34*(3):2591–2598
31. Koitz, R., Wotawa, F. (2015). From theory to practice: model-based diagnosis in industrial applications. In *Annual Conference of the PHM Society 7*(1)
32. De Kleer, J., & Kurien, J. (2003). Fundamentals of model-based diagnosis. *IFAC Proceedings Volumes, 36*(5), 25–36.
33. Flesch, I., Lucas, P., van der Weide, T.P. (2007). Conflict-based diagnosis: Adding uncertainty to model-based diagnosis. In *Proceedings of the 20th International Joint Conference on Artifical Intelligence*
34. De Kleer, J., Williams, B.C. (1989) Diagnosis with behavioral modes. In: *Proceedings of the 11th International Joint Conference on Artificial intelligence-Volume 2*
35. Darwiche, A. (2000). Model-based diagnosis under real-world constraints. *AI magazine, 21*(2), 57–57.
36. Tan, Y., Zhang, J., Tian, H., Jiang, D., Guo, L., Wang, G., & Lin, Y. (2021). Multi-label classification for simultaneous fault diagnosis of marine machinery: A comparative study. *Ocean Engineering, 239*, Article 109723. https://doi.org/10.1016/j.oceaneng.2021.109723
37. Zhang, M.-L., & Zhou, Z.-H. (2014). A review on multi-label learning algorithms. *IEEE Transactions On Knowledge and Data Engineering, 26*(8), 1819–1837. https://doi.org/10.1109/TKDE.2013.39
38. Mittal, P., Lallie, H. S., Titis, E. (2025). Impact of imbalanced datasets on ML algorithms for malware classification. *Information Security Journal: A Global Perspective*: 1–14. https://doi.org/10.1080/19393555.2025.2459736
39. Kawa, B., & Borkowski, P. (2023). Integration of machine learning solutions in the building automation system. *Energies, 16*(11), 4504.
40. Peischl, B., & Wotawa, F. (2003). Model-based diagnosis or reasoning from first principles. *IEEE Intelligent Systems, 18*(3), 32–37.
41. Dwivedi, R., et al. (2023). Explainable AI (XAI): Core ideas, techniques, and solutions. *ACM Computing Surveys, 55*(9), 1–33.
42. Ribeiro, M.T., Singh, S., Guestrin, C. (2016). Why should I trust you?: Explaining the predictions of any classifier. In: *Proceedings of the 22nd ACM SIGKDD International Conference on Knowledge Discovery and Data Mining*, pp. 1135–1144
43. Lundberg, S.M., Lee, S.-I. (2017). A unified approach to interpreting model predictions. Advances in neural information processing systems *30*
44. Vaswani, A. et al. (2017). Attention is all you need. Advances in neural information processing systems *30*
45. Quinlan, J. R. (1986). *Induction of decision trees. Machine learning, 1*, 81–106.

46. Marques-Silva, J., Ignatiev, A. (2022). Delivering trustworthy AI through formal XAI. In *Proceedings of the AAAI Conference on Artificial Intelligence 36*(11)
47. Darwiche, A., & Hirth, A. (2023). On the (complete) reasons behind decisions. *Journal of Logic, Language and Information, 32*(1), 63–88.
48. Brusoni, V., Console, L., Friedrich, G., Friedrich, H., Torta, G., & Williams, B. (1998). A spectrum of definitions for temporal model-based diagnosis. *Artificial Intelligence, 102*(1), 39–79.
49. Brockwell, P. J., & Davis, R. A. (1991). *Time series: Theory and methods.* Springer.
50. Qin, Y., Song, D., Chen, H., Cheng, W., Jiang, G., Cottrell, G. (2017). A dual-stage attention-based recurrent neural network for time series prediction. In *Proceedings of the 26th International Joint Conference on Artificial Intelligence*
51. Zhang, C., Song, D., Chen, Y., Feng, X., Lumezanu, C., Cheng, W., Chawla, N. V. (2019). A deep neural network for unsupervised anomaly detection and diagnosis in multivariate time series data. In: Proceedings of the *AAAI conference on artificial intelligence 33*(1):1409–1416
52. Pedregosa, F., Varoquaux, G., Gramfort, A., Michel, V., Thirion, B., Grisel, O., Blondel, M., Prettenhofer, P., Weiss, R., Dubourg, V., Vanderplas, J., Passos, A., Cournapeau, D., Brucher, M., Perrot, M., Duchesnay, E. (2011). Scikit-learn: machine learning in Python. *The Journal of Machine Learning Research 12*:2825–2830

Characterization of Sound Absorption Performance of Acoustic Metamaterials: Microstructure-Based Modeling and Machine Learning Technique

Van-Hai Trinh and Duong Vu

Abstract Acoustic metamaterials (AM) are considered a promising and innovative solution for sound absorption across desired frequency ranges, particularly at low frequencies. They offer potential advancements in meeting application-specific constraints, including spatial arrangements, environmental conditions, and additional functional requirements. Several AM structures have been developed based on the integration of porous materials with hybrid and Helmholtz resonators (i.e., micro-perforated plates, coiled-up/dead-end/extended-tube cavities). In the present work, we reconstruct neural network-based surrogates for estimating the sound absorption performance of acoustic metamaterials made from hybrid resonators. A neural network (NN) architecture is built and trained within the dataset involving input variables (i.e., microstructure and displacement factors) and the sound absorption considered as the output variable. Firstly, the structure–property correlations of acoustic metamaterials are reconstructed using theoretical calculations and numerical simulations. Then, surrogate models for input–output mappings, based on the truncated Karhunen–Loève decomposition combined with neural networks, are reconstructed for characterizing acoustic properties of the AM structure within a limited reference dataset. Finally, the established NN model is examined based on its efficiency and predictive performance with respect to a number of virtual and real material samples. The results obtained from designing acoustic metamaterials are present to demonstrate contributions of the proposed methodology.

Keywords Acoustic metamaterial · Helmholtz resonators · Local structure · Machine learning · Reduced modeling · Sound absorption

V.-H. Trinh
Institute of Vehicle and Energy Engineering, Le Quy Don Technical University, Ha Noi, Vietnam

D. Vu (✉)
Duy Tan University, Da Nang, Vietnam
e-mail: vuduong@duytan.edu.vn

M. A. Mellal et al. (eds.), *Computational Intelligence for a Greener Future: Innovations in Renewable Energy Systems*, Studies in Computational Intelligence 1240,
https://doi.org/10.1007/978-3-032-06732-6_21

1 Introduction

Recently, acoustic metamaterials or meta-absorbers are regarded as an innovative solution for sound absorption, leveraging the unique performance of acoustic metamaterials for different applications such as aeronautical, aerospace and automotive engineering [1–4]. Unlike conventional absorptive materials, which primarily rely on porous structures or fibrous/granular compositions, advanced meta-absorbers employ engineered resonant structures to manipulate acoustic wave interactions at subwavelength scales [3, 5]. These structures enable enhanced absorption at targeted frequencies, often achieving high efficiency with reduced material thickness. This characteristic makes meta-absorbers particularly attractive for applications where spatial constraints are a critical concern [6–8]. Through meticulous design of their internal structures—such as integrating Helmholtz resonators, membrane-based components, or hybrid periodic architectures—researchers have developed meta-absorbers capable of broadband absorption and tunable absorption performance [3, 9, 10]. This versatility allows them to address noise control challenges across a wide range of frequency bands, providing solutions for low-frequency noise mitigation (i.e., few hundreds hertz), which has traditionally been difficult to achieve with conventional materials [8, 11, 12].

Metamaterials are composed of artificially engineered sub-wavelength structures arranged in either a periodic or random pattern [1]. For acoustic applications, various metamaterial structures have been designed by combining porous materials with hybrid elements and Helmholtz resonators [10, 13–15], such as micro-perforated panels (MPPs), coiled-up cavities, dead-end tubes, and extended-tube cavities. MPPs consist of thin panels with numerous micro-scale perforations, backed by an air cavity. As Helmholtz resonators, these structures enhance their performance by dissipating sound energy through viscous losses within the perforations [16–18]. Coiled-up cavities, or labyrinthine structures, extend the path of sound waves within a compact volume, effectively lowering the frequency at which resonance occurs [14, 18, 19]. Dead-end and extended-tube cavities involve tubes that either terminate or extend to create resonant cavities for sound absorption [20, 21]. Hybrid acoustic metamaterial structures offer highly efficient sound absorption and can be extensively tailored to meet specific requirements [10, 13, 15, 17, 21]. A coupled resonator can achieve an sound absorption coefficient exceeding 0.99 at a central frequency of 511 Hz and a 140 Hz bandwidth with a sound absorption of 0.5 [15], see also Ref. [17].

Simulation and computation techniques are powerful tools in design and optimization of metamaterials for sound absorption, enabling the prediction and manipulation of wave behaviors to achieve desired acoustic properties. The finite element method (FEM) is extensively employed for modeling complex geometries and analyzing wave interactions within these materials [22, 23]. For instance, Roca et al. used FEM in their study on locally resonant acoustic metamaterials to analyze and optimize their performance [23]. Additionally, co-simulation optimization methods that integrate commercial finite element tools (i.e., COMSOL, ANSYS) have been developed to enhance the acoustic properties of metamaterials, as demonstrated by recent research

[16, 17, 24]. Furthermore, machine learning algorithms have been explored to optimize bandgap properties in acoustic metamaterials, offering efficient solutions to complex design challenges [25–28]. It can be said that the reference data for developing NN-based model can be obtained from numerical [25, 28], analytical [29] or experimental [30] estimations of the structure–property relationship of acoustic materials.

In this work, neural network-based models are developed to predict the functional behavior (i.e., sound absorption property) of acoustic metamaterials configured from Helmholtz resonators. On that basis, the constructed model is used to design and optimize this acoustic structure. The remainder of the chapter is presented as follows. In Sect. 2, the link between the structural characteristics and acoustic properties of metamaterials is determined through theoretical analysis and numerical simulations. Section 3 presents the mathematical foundation for generation of neural network based surrogate models via the truncated KL decomposition. Next, Sect. 4 studies the convergence characteristics of the reconstructed surrogate maps within a limited reference dataset. In the Sect. 5, the effectiveness and predictive capability of the trained NN model are assessed by applying for the design process of AM structures to demonstrate the advantages of the proposed approach.

2 Theoretical Framework

2.1 Microstructure Representation

Acoustic metamaterials based on Helmholtz resonators are engineered structures designed in this work. These metamaterials consist of periodic unit cells, each incorporating an array of different Helmholtz resonators to achieve specific acoustic properties. A Helmholtz resonator typically comprises a cavity and a narrow neck. Figure 1 illustrates an AM structure (a) made of a periodic pattern of Helmholtz resonators (b, c). As shown in the middle part (b), a single PUC consists of $n \times n$ resonators arranged in a square pattern, each contributing to the overall acoustic response. The detailed geometrical factors of a single resonator include the neck (with radius r_n and length l_n) and the cavity (with radius r_c and length l_c). These geometric factors directly influence the resonance frequency. The PUC periodicity (D) of defines the spacing between adjacent arrays of resonators.

2.2 Acoustic Property Estimations

As a series connection of a neck and a cavity, a single Helmholtz resonator can be formulated by the transfer matrix $\boldsymbol{T}_i$, calculated as the product of the component transfer matrices:

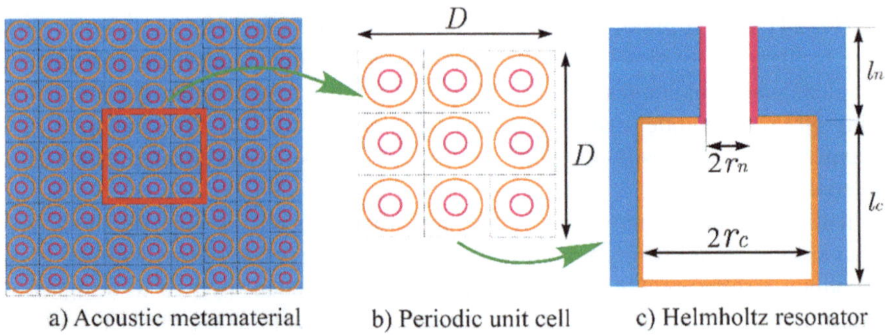

Fig. 1 Structure diagram of acoustic metamaterials configured as a periodic $n \times n$ pattern of Helmholtz resonators. Here is an example with $n = 3$

$$\boldsymbol{T}_i = \begin{bmatrix} T_{11} & T_{12} \\ T_{21} & T_{22} \end{bmatrix} = \boldsymbol{T}_n \boldsymbol{T}_c, \tag{1}$$

with

$$T_{n,c} = \begin{bmatrix} \cos(k_{n,c} l_{n,c}) & jZ_{n,c} \sin(k_{n,c} l_{n,c}) \\ \frac{j}{Z_{n,c}} \sin(k_{n,c} l_{n,c}) & \cos(k_{n,c} l_{n,c}) \end{bmatrix}. \tag{2}$$

In above equation, the components of the transfer matrices ($\boldsymbol{T}_n$, $\boldsymbol{T}_c$) are calculated through the corresponding wave numbers (K_n, K_c) and the characteristic impedances (Z_n, Z_c) [31]:

$$k_c = \omega / c_0; \;\; Z_c = \rho_0 c_0 \tag{3}$$

$$k_n = \omega \sqrt{\frac{\tilde{\rho}_e}{\tilde{K}_e}}; \;\; Z_n = \tilde{\rho}_e \tilde{K}_e \tag{4}$$

in which, ω denotes the angular frequency. c_0 and ρ_0 represents the sound speed in air and the air density, respectively. The parameters $\tilde{\rho}_e$ (equivalent density) and $\tilde{K}_e$ (bulk modulus) are used to model the viscous-inertial and thermal behaviors of the neck part. These equivalent properties can be characterized by the semi-phenomenological models (i.e., Johnson-Champoux-Allard (JCA) model) as [32]:

$$\tilde{\rho}_e(\omega) = \frac{\rho_0}{\phi} \left(\alpha_\infty + \frac{\sigma \phi}{j \omega \rho_0} F(\omega) \right); \tag{5}$$

and

$$\tilde{K}_e(\omega) = \frac{1}{\phi}\frac{\gamma P_0}{\gamma - (\gamma - 1)\left(1 + \frac{8\eta}{j\omega\rho_0 N_{pr} L_t^2} G(\omega)\right)^{-1}} \tag{6}$$

with

$$F(\omega) = \sqrt{1 + \frac{j\omega 4\eta\rho_0\alpha_\infty^2}{\phi^2\sigma^2 L_v^2}};\ G(\omega) = \sqrt{1 + \frac{j\omega\rho_0 N_{pr} L_t^2}{16\eta}} \tag{7}$$

where γ denotes the specific heat ratio, η denotes the dynamic viscosity of air, P_0 represents the atmospheric pressure, N_{pr} is the Prandtl number, and j is the imaginary unit, $-j^2 = 1$. Five fluid-acoustic properties of the above JCA model include the porosity ϕ, the thermal characteristic length L_t, the viscous characteristic length L_v, the static flow resistivity σ, and the high-frequency tortuosity α_∞. These transport properties are defined directly from the geometry factors of the resonator neck as [32]:

$$L_t = L_v = r_n, \tag{8}$$

$$\sigma = 8\eta / (\phi r_n^2), \tag{9}$$

$$\alpha_\infty = 1 + 2\varepsilon / l_n. \tag{10}$$

In Eq. (10), the correction length (ε) for the tortuosity factor is calculated by:

$$\varepsilon = 0.48\sqrt{s_n}\left(1 - 1.44\sqrt{\phi}\right); \tag{11}$$

where $s_n = \pi r_n^2$ and $\phi = r_n^2 / r_c^2$.

For the AM structure consisting of resonators arranged in parallel as above, applying the parallel circuit theory, the overall characteristic impedance Z_t can be calculated from the acosutic impedance of component resonators Z_{si} using the following formula [17]:

$$\frac{A_t}{Z_t} = \sum_{i=1}^{n\times n} \frac{A_i}{Z_{si}} \tag{12}$$

where $A_i = \pi r_c^2$ and $A_t = D^2$ are the cross-sectional area of the resonator cavity and the AM absorber. The characteristic impedance of a single resonator is defined from the transfer matrix $\boldsymbol{T}_i$ as:

$$Z_{si} = \frac{T_{21i}}{T_{11i}} \tag{13}$$

From the estimated total characteristic impedance, the sound absorption coefficient *α(ω) at normal incidence of the meta-absorbers* is estimated through,

$$\alpha(\omega) = 1 - |R(\omega)|^2, \tag{14}$$

with

$$R(\omega) = \frac{Z_t - Z_0}{Z_t + Z_0}. \tag{15}$$

where the $R(\omega)$ is the sound reflection coefficient and $Z_0 = Z_c$.

3 Mathematical Framework

3.1 Karhunen–Loève Expansion

Let $\Phi = \Phi(\omega, \boldsymbol{\mu})$ represent the forward mapping function that characterizes the response of interest. Consider $\boldsymbol{\mu}$ as the input vector defined by the parameters within the design space, while $\omega \in \mathrm{W}$ corresponds to the consider frequency range. In this section, we aim to develop a surrogate function for the aforementioned input–output response, written simply $\Phi = \Phi(\omega)$. First, the system response can be represented in the form of a KL expansion with the following mathematical formulation [33],

$$\Phi(\omega) = \underline{\Phi}(\omega) + \sum_{i=1}^{+\infty} \sqrt{\lambda_i}\eta_i\varphi_i(\omega) \tag{16}$$

here, $\underline{\Phi}(\omega)$ is the average dynamics response, while ($\{\lambda_i\}_{i\geq 1}$ and $\{\varphi_i\}_{i\geq 1}$ respectively) denotes the eigenvalues and the eigenfunctions of the covariance operator $C(\omega, \omega')$,

$$\int_{\mathrm{W}} C(\omega, \omega')\varphi_i(\omega')d\omega' = \lambda_i\varphi_i(\omega) \tag{17}$$

and $\{\eta_i\}_{i\geq 1}$ is the reduced variables estimated by the following integral equation:

$$\eta_i = \frac{1}{\sqrt{\lambda_i}} \int_{\mathrm{W}} \left\{\Phi(\omega) - \underline{\Phi}(\omega)\right\}\varphi_i(\omega)d\omega \tag{18}$$

In the above equation, the reduced variables are centered, unit variance, and pairwise uncorrelated.

The KL expansion in Eq. (16) can be written in the following form with a truncation order m (satisfying convergence $\lim_{p\to+\infty} \Phi_m(\omega) = \Phi(\omega)$):

$$\Phi(\omega) = \underline{\Phi}(\omega) + \sum_{i=1}^{m} \sqrt{\lambda_i}\eta_i\varphi_i(\omega) \tag{19}$$

The above truncated KL expansion is characterized via convergence properties related to the realization number N_s and the truncation order m by:

$$\varepsilon_N = \left\| \left[\tilde{C}(N_s) \right] \right\| \tag{20}$$

$$\varepsilon_m = 1 - \frac{\sum_{i=1}^{m} \lambda_i}{tr\left(\left[\tilde{C}\right]\right)} \tag{21}$$

where $\tilde{C}$ presents the covariance matrix of the discretization process.

3.2 Neural Network Architecture

The frequency-dependent sound absorption characteristics, as a function of input parameters $\Phi = \Phi(\omega, \boldsymbol{\mu})$, have been represented using a set of reduced variables $\{\eta_i(\boldsymbol{\mu})\}_{1\times m}$ with $\boldsymbol{\mu} = \{\mu_i\}_{1\times p}$ is the input vector of p design parameters. In this section, we will develop a neural network to analyze the absorption response based on a limited set of the calculated data. The n_L-layer NN architecture consits of an input layer, an output layer, and n_L-2 hidden layers in between, see Fig. 2. The numbers of neurons of the input and output layers are given as $n^{(1)} = p$ and $n^{(nL)} = m$, while that of neurons in the hidden layers $\left\{n^{(i)}\right\}_{i=2,3,..n_L-1}$ are designed to ensure the predictive capability of the NN structure formulated based on the reference dataset with a sample size of N_s, denoted by $S = \left[\left(\mu^{(i)}, \boldsymbol{\eta}^{(i)}\right) | \, i \in [1, \ldots, N_s]\right]$.

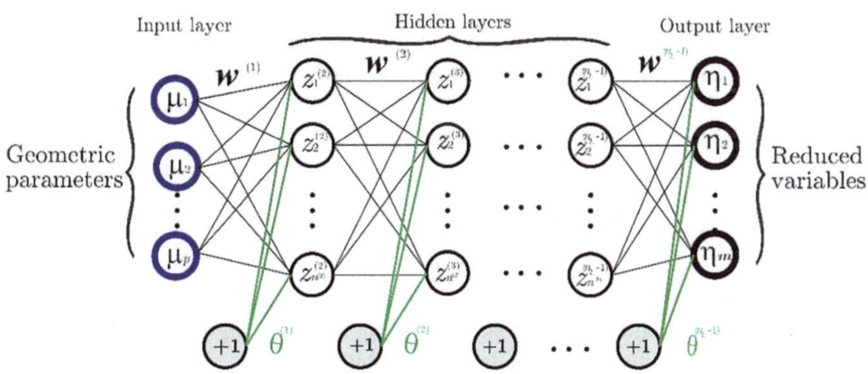

Fig. 2 Diagram of the generated NN architecture for characterization of the AM structure. Here, $n^{(l)}$ represents the number of neurons of the lth layer

The connection between two consecutive neuron layers is characterized by a weight matrix $\boldsymbol{w}^{(l)} = [w_{lk}^{(l)}]_{n^{(l)} \times n^{(l+1)}}$ and a bias vector $\theta(l) = [\theta_k^{(l)}]_{1\times n}^{(l+1)}$. Therefore, the output of the *jth* neuron in the layer (except for those in the output layer) is calculated from the information of the immediately preceding layer as,

$$z_j^{(l+1)} = \Gamma\left(\sum_{i=1}^{n^{(l)}} w_{ij}^{(l)} z_j^{(l)} + \theta_j^{(l)}\right), \tag{22}$$

where $\Gamma(.)$ represents the transfer function (i.e., $\Gamma(v) = 1/\{1 - \exp(-v)\}$). Thus, the NN prediction $\hat{\boldsymbol{\eta}}_{w,\theta}\left(\boldsymbol{\mu}^{(i)}\right)$ in the output layer can be calculated from the input data $\boldsymbol{\mu}^{(i)} = \left(\mu_1^{(i)}, \mu_2^{(i)}, ..., \mu_p^{(i)}\right)$, the weights $w = \left(w^{(1)}, \ldots, \boldsymbol{w}^{(nL-1)}\right)$, and the biases $\theta = \left(\theta^{(1)}, \ldots, \boldsymbol{\theta}^{(n_L-1)}\right)$. Clearly, design of the NN architecture (i.e., determination of the best weights $\boldsymbol{w}$ and the biases $\boldsymbol{\theta}$) is essentially solving the problem of minimizing the discrepancy between the network predictions and the corresponding target or reference value. More specifically, the L_k- norm loss function needs to be minimized:

$$L_k(\boldsymbol{w}, \boldsymbol{\theta}) = \frac{1}{N_n}\sum_{i=1}^{N_n} \left\| \hat{\eta}_{w,\theta}\left(\mu^{(i)}\right) - \eta^{(i)}\right\|^k, \; k = 1, 2 \tag{23}$$

A new learning iteration is carried out to reduce the discrepancy between the predicted outputs and the target or reference values $L_k(\boldsymbol{w},\theta)$. When training with the Levenberg–Marquardt, the collection $\boldsymbol{w}$, containing both weights and bias, is updated in the following form [34, 35],

$$\boldsymbol{w}^{(k+1)} = \boldsymbol{w}^{(k)} - \left[\boldsymbol{J}^T\boldsymbol{J} + \mu\boldsymbol{I}\right]^{-1}\boldsymbol{J}^T\boldsymbol{e} \tag{24}$$

where μ represents the learning rate, while $\boldsymbol{e}$ denotes the error vector. Symbols ($\boldsymbol{I}$,$\boldsymbol{J}$) represent the identity and Jacobian matrix, respectively.

The Neural Network MATLAB toolbox [36] is employed to construct the NN model are performed. It can be noted that the raw data obtained from numerical computations, including both inputs and outputs, are normalized before being used for training [37].

3.3 Optimization Technique

By assuming that the function $J(\boldsymbol{x})$ typically serves as the objective function in a maximization (or minimization) task, depending on the design parameters: $J(\boldsymbol{x}) : \mathbb{R}^P \mapsto \mathbb{R}$. This optimization problem can be stated as,

$$\min_{x \in S_x \subseteq R^P} J(\boldsymbol{x})$$

$$\text{subject to} \begin{cases} g_i(\boldsymbol{x}) = 0, \; i = 1, 2, \ldots m; \\ h_j(\boldsymbol{x}) = 0, \; j = 1, 2, \ldots n \end{cases} \tag{25}$$

where $\boldsymbol{x} \in S_x = \times_{i=1}^{P}\left[\, l_i \; u_i \,\right]$ is the design vector with $\left(\boldsymbol{L}_b = \{l_i\}_{1\times P}, \boldsymbol{U}_b = \{u_i\}_{1\times P}\right)$ representing upper and lower bound vectors, while $g_i(\boldsymbol{x})$ and $h_i(\boldsymbol{x})$ denote some equality and inequality constraints, respectively.

Originally proposed by Kennedy and Eberthart [38], Particle Swarm Optimization (PSO) is a metaheuristic optimization algorithm modeled on the collective behavior observed in bird flocks and fish schools. This technique is commonly used to solve engineering optimization problems where traditional gradient-based methods struggle due to nonlinearity, non-differentiability, or multimodality. The algorithm emulates a particle traversing the search space in pursuit of the global optimal solution. At iteration $k + 1$, the velocity and position of the particle i are updated from the previous state k as:

$$\boldsymbol{v}_i^{(k+1)} = \mu_0^{(k)} \boldsymbol{v}_i^{(k)} + \mu_1 r_1 \left(\boldsymbol{p}_i^{(k)} - \boldsymbol{x}_i^{(k)}\right) + \mu_2 r_2 \left(g_i - \boldsymbol{x}_i^{(k)}\right) \tag{26}$$

$$\boldsymbol{x}_i^{(k+1)} = \boldsymbol{x}_i^{(k)} + \boldsymbol{v}_i^{(k+1)} \tag{27}$$

where μ_0 is the particle inertia weight which controls the particle momentum [38, 39]. The a**cceleration coefficients** μ_1, μ_2 are related to control influence of personal and global best, while r_1, r_2 are random numbers in [0,1], ensuring stochastic behavior. $\boldsymbol{p}_i^{(k)}$ and g_i correspond the personal best position and the global best position, respectively. These two quantities guide the optimization process by balancing individual exploration and collective convergence. The update rule for the personal best position is given by,

$$\boldsymbol{p}_i^{(k+1)} = \begin{cases} \boldsymbol{x}_i^{(k+1)}, & \text{if } J(\boldsymbol{x}_i^{(k+1)}) > J\left(\boldsymbol{p}_i^{(k+1)}\right) \\ \boldsymbol{p}_i^{(k)}, & \text{otherwise} \end{cases} \tag{28}$$

For handling the constraints, a penalty function approach modifies the objective function as:

$$J'(\boldsymbol{x}) = J(\boldsymbol{x}) + \mu_\alpha \sum_{i=1}^{m} |g_i(\boldsymbol{x})| + \mu_\beta \sum_{i=1}^{n} \max(0, h_i(\boldsymbol{x})) \tag{29}$$

where μ_α and μ_β are penalty coefficients.

In the optimization problem of the AM structure, the details of the objective function and boundary conditions will be presented later, along with other relevant information such as variables and design space.

4 Assessment of Surrogate Models

4.1 Case Study and Validation

To provide data to illustrative numerical applications of proposed framework the presented above, the selected configurations of AM materials will be introduced. To ensure feasibility and simplicity in fabrication, the selected design parameters are the radius r_{ni} and the length l_{ni} of the neck. In addition, 2 configurations of the designed PUC are selected corresponding to the sizes 2×2 and 3×3 (see Fig. 1b for a case of 3×3-pattern PUC). To maintain notation consistency, the configuration of the AM with an $n \times n$-resonator array is denoted by AMn, and each individual Helmholtz resonator is represented as Ri with i from 1 to n^2.

The configuration AM2 (having structures given in Table 1) is selected to validate both the theoretical and numerical framework provided in Sect. 3. In order to simulate the acoustic properties of the resonator structure, an additional cavity layer is introduced right in front of the resonator neck (as shown in Fig. 3b, highlighted by two layers). The acoustic-thermoacoustic coupling model is subsequently solved, and the acoustic impedance is determined based on the acoustic pressure and velocity fields at the incident interface (highlighted layer, Fig. 3b).

The geometric of the PUC of this structure is present in Fig. 3ab, and the corresponding finite element model with 24,897 tetrahedron elements is shown in Fig. 3c. Using this mesh model and the numerical framework above mentioned, it can be obtain the solved fields of acoustic pressures in different frequencies as Fig. 4. It can be observed that around the resonance frequency of each resonator, the acoustic pressure is significantly higher than that of the other resonators. From the solution fields of the pressure and velocity, the numerical calculations of the system's acoustic response are defined. For each frequency, the computation time is approximately 2 s and 140 s for the single and multiple resonators, respectively. The numerical calculations were performed on a workstation with Intel(R) Xeon(R) CPU X5650, 2.67 GHz, 2.66 GHz (two processors), and 48 GB RAM.

Comparison results of sound absorption coefficient obtained from theoretical calculation and numerical simulation of single Helmholtz resonators and the structure AM2 are plotted in Figs. 5 and 6. Comparisons show a high similarity in the results of the two calculation methods. In terms of sound absorption characteristics, it can be stated that the absorption properties of a multi-element parallel structure are formed

Table 1 Geometric factors of the AM2. Noted that other factors are selected as: l_{ni} = 10 mm, r_{ci} = 24 mm, l_{ci} = 10 mm

Parameter	Unit	Resonator			
		R1	R2	R3	R4
Neck radius, r_{ni}	mm	1.0	1.5	2.0	2.5

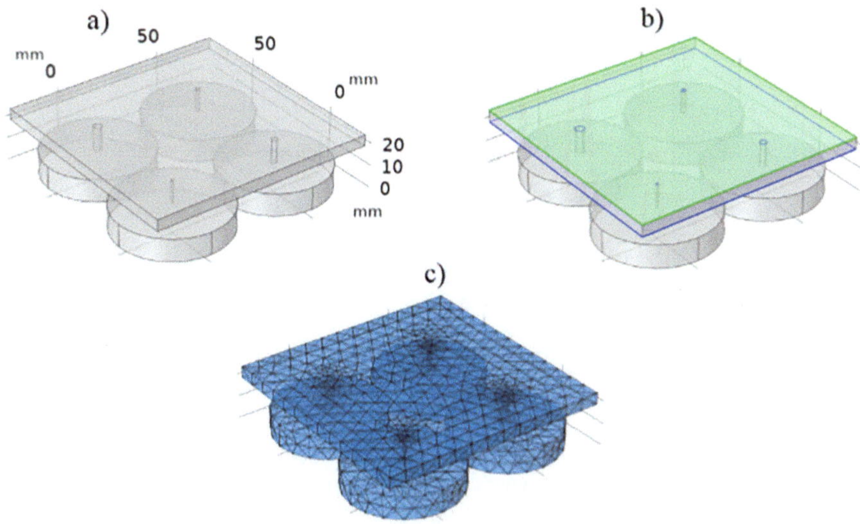

Fig. 3 Geometric (**a**, **b**) and finite element mesh (**c**) models of the PUC

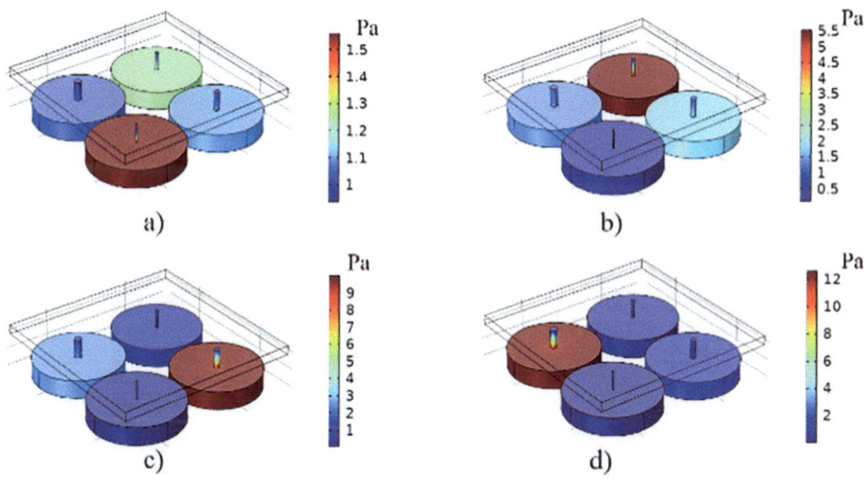

Fig. 4 Solved fields of acoustic pressures over the PUC domain in different frequencies: 100/200/300/400 Hz (corresponding to sub-figures **a** to **d**)

based on the resonance responses of its constituent resonators. The frequency positions and resonance amplitudes tend to change when combined; however, the number and order of resonance points remain relatively distinct. These characteristics are generally consistent with some previously reported results for hexagonal array [40] and square array [41]. This also demonstrates the potential to design and customize the acoustic properties of AM structures by utilizing the constituent elements.

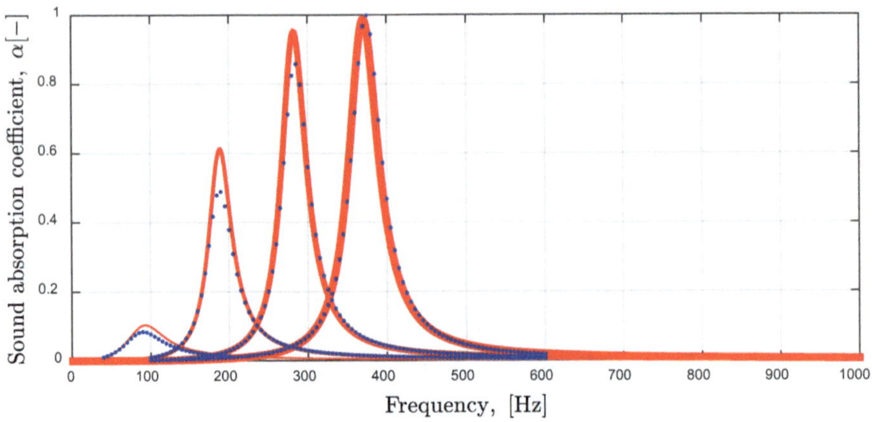

Fig. 5 Comparison of the sound absorption coefficients for four single resonators Ri, from theoretical calculations (solid lines) and numerical simulations (markers). Noted that the thicker lines the bigger neck radii, or responses left to right correspond to resonators R1 to R4

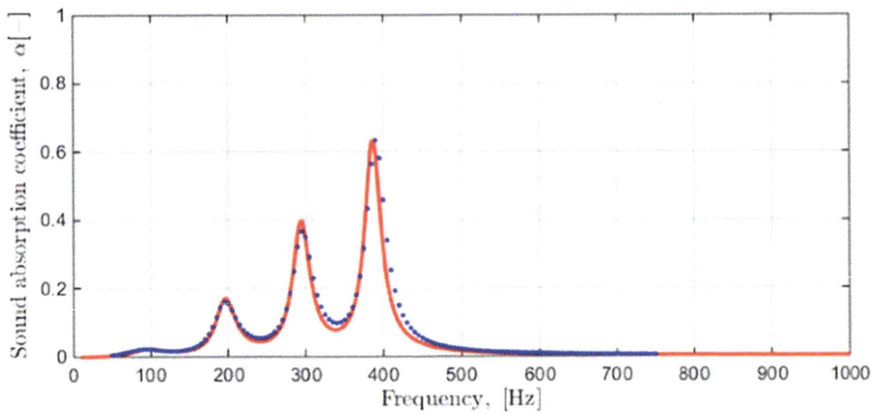

Fig. 6 Comparison of the sound absorption coefficients for the structure AM2, from theoretical calculations (solid line) and numerical simulations (markers)

4.2 *Surrogate Characteristics*

As mentioned above, herein we develop three KL expansion for three acoustic responses (i.e., the real part $\mathrm{Re}(Z_t)$ and the imaginary part $\mathrm{Im}(Z_t)$ of the characteristic impedance $Z_t(\omega, \boldsymbol{\mu})$, and sound absorption property $\alpha = \alpha(\omega, \boldsymbol{\mu})$) of a single Helmholtz resonator. These responses are functions of the resonator dimensions, including the radii (r_n, r_c) and the lengths (l_c, l_n), see Fig. 1c.

With sampling size $N_s = 1000$ resonators configured based on design parameters given in Table 2, the convergence characteristics of three truncated expansions are provided in Fig. 7 (a: convergence map in number of realizations N_s and b:

tolerance error maps with truncation order m). The results on $N_s \mapsto \varepsilon_N(N_s)$ show that a computational quantity N_s of 1000 ensures a stable convergence level for all three system responses. Conversely, the convergence level of the KL series with respect to the truncation order m exhibits a clear distinction between the impedance characteristic and the sound absorption coefficient. While both components of the impedance characteristic require expansion with low-order truncated KL series, the absorption coefficient necessitates a high-order expansion. Specifically, to achieve an error smaller than 1×10^{-4}, the required truncation orders m for the responses $Z_t(\omega)$ and $\alpha(\omega)$ are 4 and 40, respectively.

In terms of the used eigenfunctions, Fig. 8 plots four first eigenfunctions for representing the characterized acoustic responses. These functions also illustrate the greater complexity of the sound absorption response compared to the other two responses, especially in the resonance region. This is precisely the reason why higher-order expansion is needed for the sound absorption response. Figure 9 shows the KL-based predictions in comparison with their reference functions (for 4 configurations of Helmholtz resonators). The truncation order values $m = 40$ (Fig. 9a) and $m = 4$ (Fig. 9b-c) are used to ensure an error smaller than 10^{-4}.

After obtaining the KL expansion of the acoustic responses of the Helmholtz resonator elements, these KL paramters (i.e., reduced varaibles) can be used to study the response of the AM structure as follows. Neural network architectures will be

Table 2 Tunable ranges of the geometrical parameters of Helmholtz resonators

Parameter	Unit	Minimum value	Maximum value
Neck radius, r_{ni}	mm	0.5	3.5
Neck length, l_{ni}	mm	5	25
Cavity radius, r_{ci}	mm	10	20
Resonator length, $l_{ci} + l_{ni}$	mm	30	40

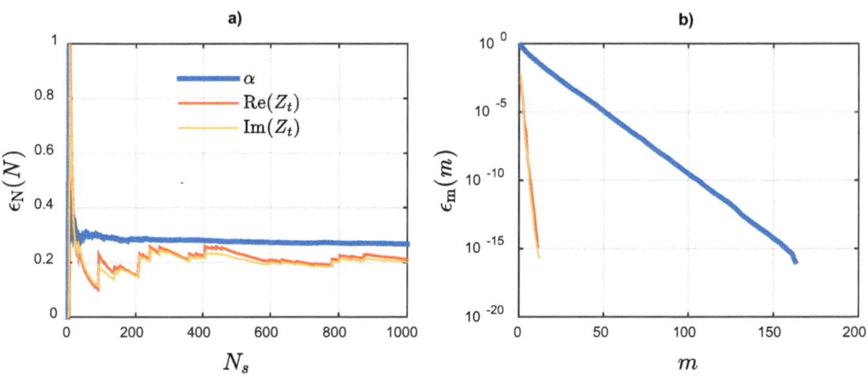

Fig. 7 Convergence maps of the KL truncated expansions for acoustic responses. The characteristic convergence maps in numbers of realizations N_s have been modularized

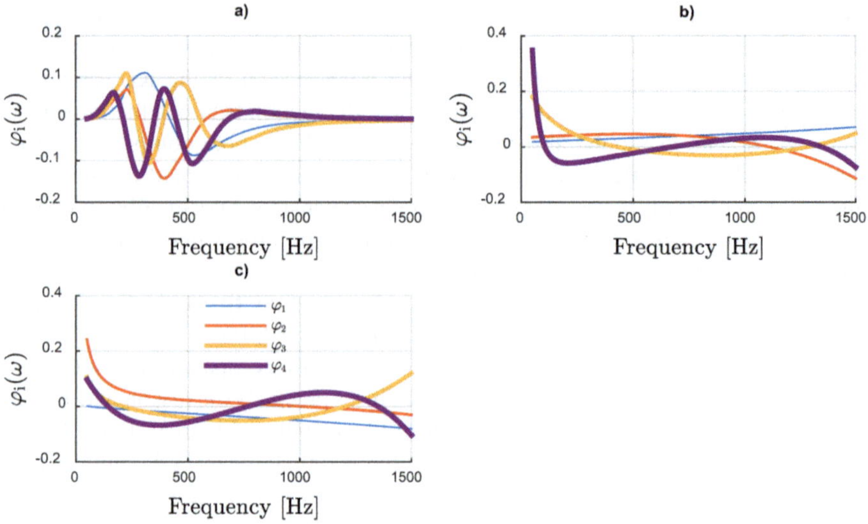

Fig. 8 Plots of a set of four first eigenfunctions for representing the acoustic responses: sound absorption (**a**) and characteristic impedance (**b**, **c**)

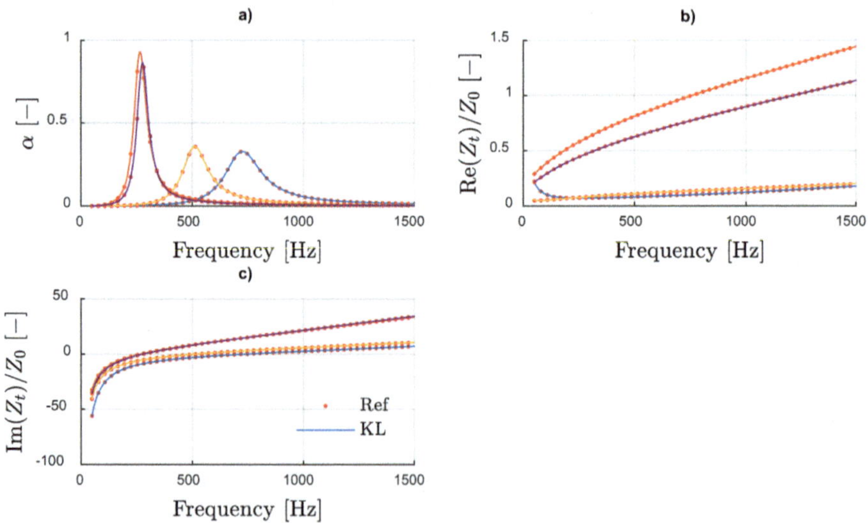

Fig. 9 Comparisons of four reference maps (dots) and the corresponding KL-based surrogates (solid lines)

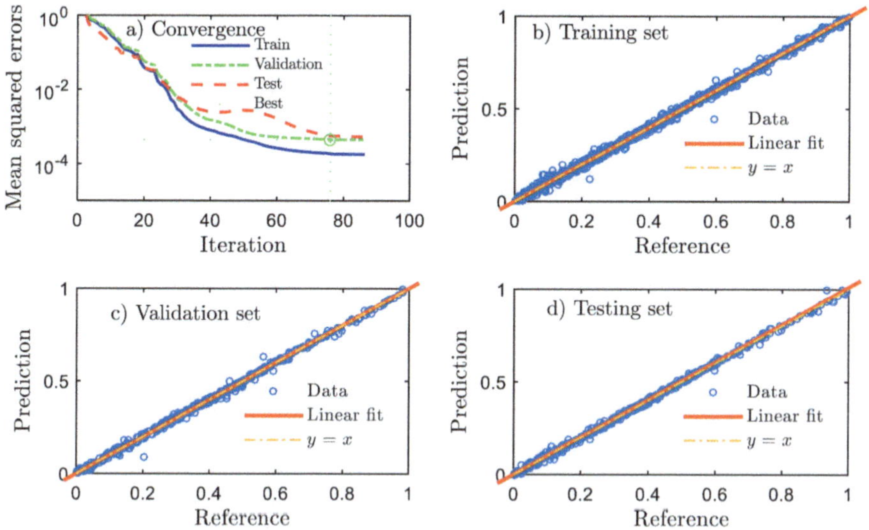

Fig. 10 Characteristics of the trained NN model is evaluated through two key aspects: **a** convergence behavior of the mean squared error during training, and **b–d** regression performance of the neural network across three distinct datasets—training, validation, and testing, respectively

constructed with input variables representing the geometric parameters and output variables corresponding to the reduced variables (i.e., 4 + 4 corresponding to the imaginary and real parts of the response Z_t). With 4 inputs and 8 outputs, the network is designed with 2 hidden layers—each containing 10 neurons—demonstrating good predictive capability for the characteristics shown in Fig. 10. Note that among the 1000 resonator configurations used for training the network, the data is split into 70% for training, 15% for validation, and 15% for testing. A mean squared error below 1×10^{-3} can be reached across all three datasets—demonstrating high predictive accuracy of the NN model.

To evaluate the predictive capability, the neural network model is applied to 10 resonator configurations with $l_n = 15$ mm, $r_c = 11$ mm, $l_c = 20$ mm, and r_n varying from 0.5 to 1.625 mm in steps of 0.125 mm. The results, presented in Fig. 11, demonstrate the high accuracy of the constructed network. Note that the results in Fig. 11c are obtained indirectly from the predicted characteristic impedance response shown in the two preceding sub-figures.

5 Applications for AM Design

In this section, we demonstrate the use of the previously initialized NN predictive model for optimizing the AM3 structure. A total of 20 design variables are selected, including 18 geometric parameters for the resonator neck (r_{ni}, l_{ni}) and 2 parameters

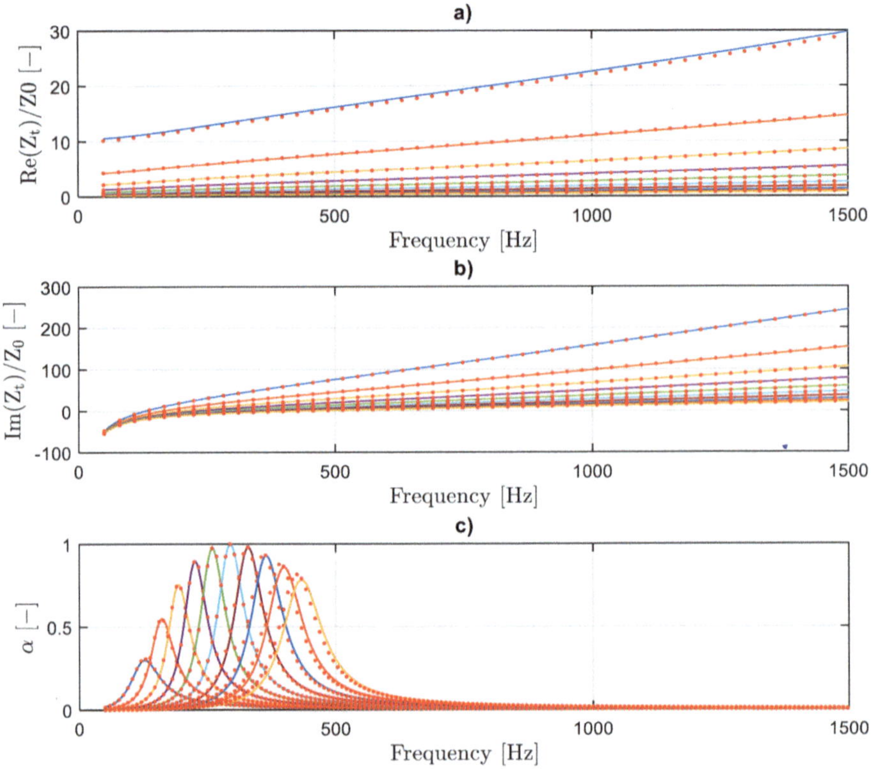

Fig. 11 Comparisons between the reference and NN-based prediction maps of acoustic responses. The results are presented for 10 Helmholtz configurations

for the resonator cavity (r_c, l_c). This means that the 9 resonators of the AM3 will have the different neck dimensions while the same cavity dimensions. The AM3 structure is designed for optimal performance in the following frequency ranges: low frequency (LF) [400–600] Hz and mid-frequency (MF) [600–1000] Hz. In the optimization process, the objective function is determined using the average absorption coefficient over the frequency range of interest, as follows

$$J(\boldsymbol{x}) := \alpha_A = \sum_{i=1}^{N} \alpha(\omega_i), \text{ with } \boldsymbol{x} = [\{l_{ni}\}_{1\times 9} \ \{r_{ni}\}_{1\times 9}\, r_c \ L] \tag{30}$$

here $h_1(\boldsymbol{x}) := D - 6x_{19}$ is added as an inequality constraint related to control the the total size of the pattern unit, $D = 90$ mm. It can be noted that the lower and upper vector are formed from the value in Table 1.

By utilizing the surrogate model based on the NN framework and the PSO optimization algorithm mentioned above, the optimized AM structure is found as follows.

Table 3 Dimensions of the resonator necks of the optimal AM3. The bold values indicate the coordinates of the population in the final iteration (left panels, Fig. 12)

Helmholtz resonators	LF		MF	
	Neck radius, r_{ni}	Neck length, l_{ni}	Neck radius, r_{ni}	Neck length, l_{ni}
R1	2.45	13.61	2.76	25.00
R2	2.16	5.29	3.10	10.32
R3	2.79	22.70	3.50	21.60
R4	2.92	6.63	1.98	5.00
R5	1.71	5.00	3.50	**5.06**
R6	2.09	10.56	2.48	**9.82**
R7	2.82	**14.43**	3.28	7.20
R8	2.26	**11.38**	2.89	10.51
R9	3.11	18.34	2.69	10.45

For descriptive purposes, the objective function results are obtained with 18 optimized variables, while the remaining 2 variables (neck length dimensions) vary within their defined range [5 25] mm. The objective function results for the low and mid-frequency ranges are 0.9625 and 0.8566, respectively. Table 3 presents the detailed optimal neck dimensions. The two remaining parameters of the cavity are given as, the cavity radius: $r_{ci} = 15.11$ mm (LF) and $r_{ci} = 11.78$ mm (MF), and the cavity length: $l_{ci} = L - l_{ni}$ with $L = 35.50$ mm (LF), and $L = 32.71$ mm (MF).

Figure 13 presents two aspects of the optimization process (left: the convergence properties of the objective function; right: the optimal absorption response). The convergence characteristics indicate that for iteration steps greater than 130, both populations are very close to the optimal solution. From the perspective of the acoustic characteristics of the AM3 structure, it can be observed that the overall response has sufficiently merged the individual responses of the resonators. For the low-frequency response of the AM3 structure, the number of resonance points is only 8. This can be explained by the fact that two resonators (R5 and R6) have very similar response characteristics, resonating at nearly 400 Hz.

6 Conclusions

The present study has developed and evaluated a neural network-based predictive model for sound absorption performance in acoustic metamaterials designed with hybrid resonators. By integrating microstructure-based modeling and surrogate modeling approaches, we have reconstructed structure–property correlations to enhance the understanding and optimization of AM configurations. The truncated KL expansion and neural networks enabled efficient approximation of acoustic properties with a limited dataset, offering a computationally feasible alternative to

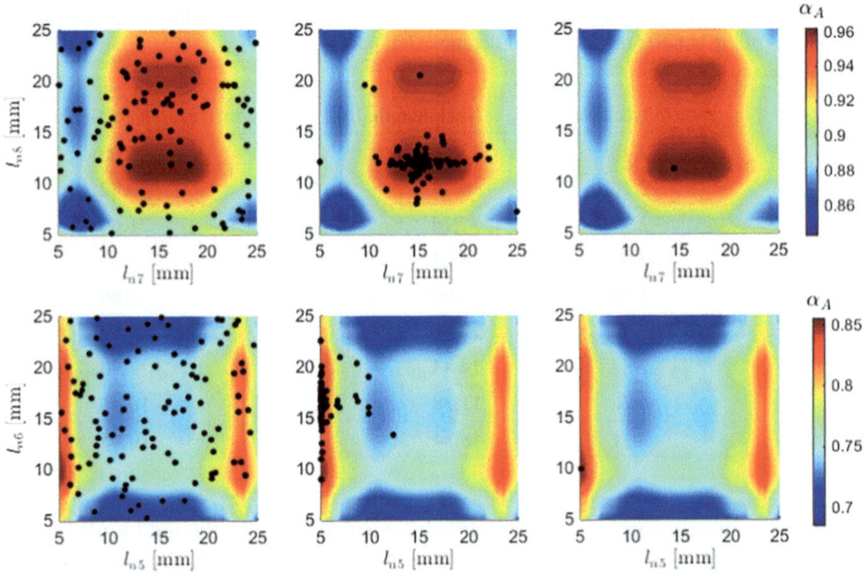

Fig. 12 Illustration of the movement of the swarm with 100 particles during the PSO optimization process: LF (top panels) and MF (bottom panels). Maps from left to right correspond to positions at the initial iteration, the 50th iteration, and the final iteration

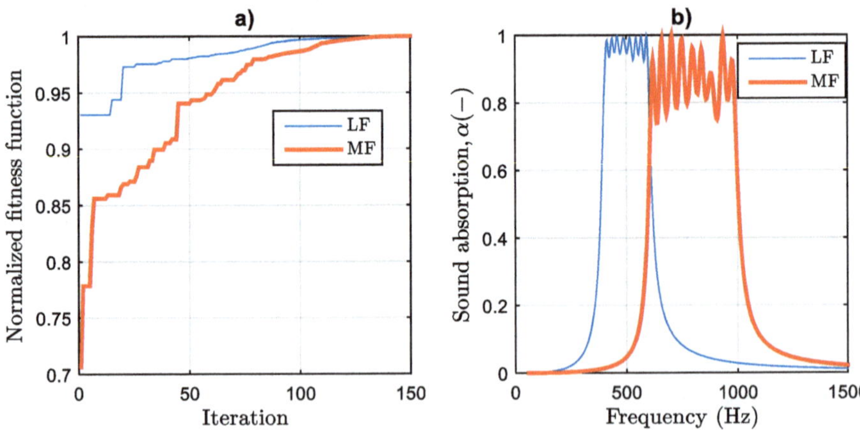

Fig. 13 **a** Convergence of the objective functions during the optimization process, and **b** the optimal sound absorption coefficient (i.e., the global best position)

extensive numerical simulations. The proposed methodology facilitates the systematic design and optimization of AM structures while addressing practical constraints such as spatial limitations and functional requirements. The results highlight the strong potential of machine learning-based approaches in accelerating development of acoustic metamaterials, offering valuable insights for future applications in noise control and sound absorption technologies.

Future research may focus on expanding the dataset, refining surrogate models, and exploring additional physical parameters to further enhance predictive accuracy and design flexibility.

References

1. Gao, N., et al. (2022). Acoustic metamaterials for noise reduction: A review. *Advanced Materials Technologies, 7*(6), 2100698.
2. Liao, G., et al. (2021). Acoustic metamaterials: A review of theories, structures, fabrication approaches, and applications. *Advanced Materials Technologies, 6*(5), 2000787.
3. Yang, M., & Sheng, P. (2017). Sound absorption structures: From porous media to acoustic metamaterials. *Annual Review of Materials Research, 47*(1), 83–114.
4. Kumar, S., & Lee, H. P. (2020). Recent advances in acoustic metamaterials for simultaneous sound attenuation and air ventilation performances. *Crystals, 10*(8), 686.
5. Zhang, J., Hu, B., & Wang, S. (2023). Review and perspective on acoustic metamaterials: From fundamentals to applications. *Applied Physics Letters, 123*(1).
6. Wu, X., et al. (2018). High-efficiency ventilated metamaterial absorber at low frequency. *Applied Physics Letters, 112*(10).
7. Zhu, Y., et al. (2019). Broadband ultra-thin acoustic metasurface absorber with coiled structure. *Applied Physics Express, 12*(11), Article 114002.
8. Abrahams, M. P., et al. (2024). Hybrid ultrathin metasurface for broadband sound absorption. *Applied Physics Letters, 124*(15).
9. Katiyar, N., et al. (2024). Low-frequency noise attenuation through hybrid perforated hemispherical shells and membrane-type acoustic metamaterial. *Journal of Applied Physics, 136*(1).
10. Krasikova, M., et al. (2024). Broadband noise-insulating periodic structures made of coupled Helmholtz resonators. *APL Materials, 12*(1).
11. Guan, Y. J., et al. (2024). An ultra-low-frequency sound absorber and its application in noise reduction in ducts. *APL Materials, 12*(1).
12. Donda, K., et al. (2019). Extreme low-frequency ultrathin acoustic absorbing metasurface. *Applied Physics Letters, 115*(17).
13. Mahesh, K., Ranjith, S. K., & Mini, R. (2024). Recent advancements in Helmholtz resonator based low-frequency acoustic absorbers: A critical review. *Archives of Computational Methods in Engineering, 31*(4), 2079–2107.
14. Almeida, G. N., et al. (2021). Low-frequency sound absorption of a metamaterial with symmetrical-coiled-up spaces. *Applied Acoustics, 172,* 107593.
15. Li, J., et al. (2016). A sound absorbing metasurface with coupled resonators. *Applied Physics Letters, 109*(9).
16. Singh, S. K., Prakash, O., & Bhattacharya, S. (2022). Hybrid fractal acoustic metamaterials for low-frequency sound absorber based on cross mixed micro-perforated panel mounted over the fractals structure cavity. *Scientific Reports, 12*(1), 20444.
17. Tang, Y., et al. (2017). Hybrid acoustic metamaterial as super absorber for broadband low-frequency sound. *Scientific Reports, 7*(1), 43340.
18. Wu, F., et al. (2019). Low-frequency sound absorption of hybrid absorber based on micro-perforated panel and coiled-up channels. *Applied Physics Letters, 114*(15).
19. Wang, Y., et al. (2018). A tunable sound-absorbing metamaterial based on coiled-up space. *Journal of Applied Physics, 123*(18).

20. Lopez, M., Dupont, T., & Panneton, R. (2024). Mass-spring model for acoustic metamaterials consisting of a compact linear periodic array of dead-end resonators. *The Journal of the Acoustical Society of America, 155*(1), 530–543.
21. Li, Y., et al. (2024). Low-frequency broadband sound absorption of the metastructure with extended tube resonators and porous materials. *Applied Acoustics, 217*, Article 109827.
22. Bashir, I., & Carley, M. (2020). Development of 3D boundary element method for the simulation of acoustic metamaterials/meta surfaces in mean flow for aerospace applications. *International Journal of Aeroacoustics, 19*(6–8), 324–346.
23. Roca, D., et al. (2019). Computational design of locally resonant acoustic metamaterials. *Computer Methods in Applied Mechanics and Engineering, 345*, 161–182.
24. Liu, B., & Chen, F. (2022). Optimization design of the acoustic metamaterial based on the co-simulation method. *AIP Advances,12*(7).
25. Bacigalupo, A., et al. (2020). Machine-learning techniques for the optimal design of acoustic metamaterials. *Journal of Optimization Theory and Applications, 187*, 630–653.
26. Yago, D., et al. (2024). Machine learning in solid mechanics: Application to acoustic metamaterial design. *International Journal for Numerical Methods in Engineering, 125*(14), Article e7476.
27. Chen, J., et al. (2024). Application of machine learning on the design of acoustic metamaterials and phonon crystals: A review. *Smart Materials and Structures.*
28. Liu, L., et al. (2022). Broadband acoustic absorbing metamaterial via deep learning approach. *Applied Physics Letters, 120*(25).
29. Casaburo, A., et al. (2021). Gaussian-based machine learning algorithm for the design and characterization of a porous meta-material for acoustic applications. *Applied Sciences, 12*(1), 333.
30. Trinh, V. H., et al. (2022). Learning acoustic responses from experiments: A multiscale-informed transfer learning approach. *Journal of the Acoustical Society of America, 151*(4), 2587–2601.
31. Allard, J., & Atalla, N. (2009). *Propagation of sound in porous media: Modelling sound absorbing materials* (2edn). John Wiley & Sons.
32. Atalla, N., & Sgard, F. (2007). Modeling of perforated plates and screens using rigid frame porous models. *Journal of Sound and Vibration, 303*(1–2), 195–208.
33. Lord, G. J., Powell, C. E., & Shardlow, T. (2014). *An introduction to computational stochastic PDEs* (Vol. 50). Cambridge University Press.
34. Marquardt, D. W. (1963). An algorithm for least-squares estimation of nonlinear parameters. *Journal of the Society for Industrial Applied Mathematics, 11*(2), 431–441.
35. Hagan, M. T., & Menhaj, M. B. (1994). Training feedforward networks with the Marquardt algorithm. *IEEE Transactions on Neural Networks, 5*(6), 989–993.
36. Demuth, H. B., & Beale, M. H. (1992). *Neural network toolbox user's guide.* Mathworks, Incorporated.
37. Kim, D. (1999). Normalization methods for input and output vectors in backpropagation neural networks. *International Journal of Computer Mathematics, 71*(2), 161–171.
38. Kennedy, J., & Eberhart, R. (1995). *Particle swarm optimization.* In *Proceedings of ICNN'95-International Conference on Neural Networks.* IEEE.
39. Shi, Y., & Eberhart, R. (1998). A modified particle swarm optimizer. In: *IEEE International Conference on Evolutionary Computation Proceedings.* IEEE world congress on computational intelligence (Cat. No. 98TH8360). IEEE.
40. Yang, F., et al. (2022). Optimal design of acoustic metamaterial of multiple parallel hexagonal Helmholtz resonators by combination of finite element simulation and cuckoo search algorithm. *Materials, 15*(18), 6450.
41. Guo, J., et al. (2021). An investigation on noise attenuation by acoustic liner constructed by Helmholtz resonators with extended necks. *Journal of the Acoustical Society of America, 149*(1), 70–81.

Clustering Applications in Water Distribution Networks—A Review

Shivangi Sharma, Nikita Palod, and Rajesh Gupta

Abstract Urban Water distribution networks (WDNs) are huge and complex infrastructures comprising of numerous components such as pipes, junctions, reservoirs, pumps, valves, etc. The primary function of any WDN is to convey treated water to the consumers at adequate pressure, in desired quantities. A computer model of the WDN is usually prepared to study its performance under various operating conditions. Computer model of the WDN may also be required for determining the best locations for placement of valves (PRVs and Isolation Valves), Sensors (Pressure, Water Quality, Acoustic, Structural health monitoring), Bulk meters, Booster stations (pressure or chlorination) etc. Further, model is useful to solve the problems related to leakage, contamination, aging of the network infrastructure, pipe bursts, pressure imbalances, poor operation & maintenance. It is desirable in many cases to club the pipes or nodes having similar properties to reduce the size of the problem. Clustering is a technique of data analysis and machine learning that divides the network into sub-networks or clusters by virtue of which the complex network becomes simpler. The aim of this study is to comprehensively review various clustering methods, algorithms, and their applications in the WDNs. The chapter introduces clustering and its significance in WDNs. It also explores several clustering methods and their diverse applications in WDNs. Furthermore, it highlights the different objectives for which clustering has been applied in the design and optimization of facility locations in WDNs.

Keywords Clustering · District metered areas · Facility locations · Water distribution networks

S. Sharma (✉) · N. Palod · R. Gupta
Department of Civil Engineering, Visvesvaraya National Institute of Technology, Nagpur, India
e-mail: shivangi0699@gmail.com

R. Gupta
e-mail: rajeshgupta@civ.vnit.ac.in

M. A. Mellal et al. (eds.), *Computational Intelligence for a Greener Future: Innovations in Renewable Energy Systems*, Studies in Computational Intelligence 1240,
https://doi.org/10.1007/978-3-032-06732-6_22

1 Introduction

Urban WDNs are intrinsic segments of a city's infrastructure that are accountable for transmitting potable and safe water from the source to the consumers at adequate pressure and in sufficient quantity. A conventional WDN is a complex, underground system that consists of diverse inter-connected components including pipes, pumps, valves, reservoirs, elevated storage tanks, and junctions. Pipes are primary conduits for water flow, while pumps take care of pressure required to overcome elevation changes and frictional losses. Valves regulate and isolate flow in various segments of the network, and reservoirs and tanks grant storage and pressure stabilization. WDNs can stretch from hundreds to thousands of kilometers and comprise of immense numbers of components.

Due to their intricate nature, operation and maintenance requirements, WDNs pose serious technical and administrative challenges. Typical issues include pipe bursts, leakages, pressure imbalances, contamination risks due to intrusions, and aging of critical infrastructure constituents. Identifying and reacting to such problems in real time is problematic due to the concealed nature of network and restricted accessibility. Since the urban population is increasing rapidly, WDNs are also changing frequently and expanding continuously to accommodate raised water demands which results in their increased complexity. This aggravates the already present issues, making it more difficult to monitor, analyze and control WDNs. This increasing intricacy necessitates utilization of advanced techniques and methodologies to simplify network structure for better and efficient management. One of the extensively acknowledged techniques is the creation of District Metered Areas (DMAs). DMA is a hydraulically severed zone of a WDN, designed to support monitoring and control of water inflow and outflow in the network. DMAs can be perceived as sub-networks that are obtained by dividing the larger network into smaller, manageable segments. These segments can be individually monitored and analyzed, favoring more effective leakage detection, pressure control and maintenance. DMA creation comprises of three steps involving clustering, boundary optimization and performance assessment. Clustering step divides the network into tentative sub-networks which are then physically isolated in the boundary optimization step by installing flow meters and boundary valves, and in the last step, the performance of the WDN is evaluated based on different performance indices after DMA creation.

So, clustering is the fundamental step of DMA creation which is crucial in deciding final configurations of the DMAs. However, the application of clustering is vast and diverse in the field of WDNs and has been applied for various other purposes other than DMA creation such as calibration in which pipes are clustered based on material/size/age, contamination control by placing isolation valves, water quality and pressure sensor placement to distribute sensors over the entire network by placing them cluster wise, etc. Clustering is a long-established and famous technique that is used for the simplification of the WDNs. The network gets subdivided into clusters or communities after simplification, which supports in close observation of the interactions happening amongst the network constituents.

Clustering is an unsupervised technique in machine learning and data mining. It is a process that groups the items into clusters based on a particular similarity metric. Grouping is such that the items in the same cluster are more identical to each other and are distinct than those in other clusters. For example, cluster of pipes based on the material of pipes used in calibration to calibrate pipe roughness coefficient, or cluster of nodes having HGL within certain range used to locate and fix the PRV and its setting. Clustering splits the network into sub-networks or clusters to make the intricate network structure and interaction between its components comprehensible. The purpose of clustering is to increase intra-cluster connections and reduce inter-cluster connections, thereby arranging data into meaningful structures. There are diverse types of clustering methods and algorithms that have different mechanisms and functions that are available in machine learning and data science.

In the upcoming sections, various clustering methods and algorithms are introduced, and the application of clustering in different areas of WDNs is reviewed. The aim is to review the different application areas of WDNs where clustering has been administered to solve the problems related to them and the different objectives for which clustering is utilized.

2 Clustering Methods and Algorithms

There are different types of clustering methods and algorithms which have miscellaneous applications in diverse fields including WDNs. In this section, different clustering methods and algorithms are introduced.

2.1 Hierarchical or Connectivity-Based Clustering Method

This clustering type involves construction of clusters based on the proximity distance or the closeness of an object with its neighbors. Identical objects with minimal distance are placed in the same cluster and the un-identical objects are placed farther away in the hierarchy. In this method, as mentioned by [67], the clusters are formed in either a top–bottom or bottom-up style. Hierarchical clustering is classified into two styles: the divisive (top-down) style and the agglomerative (bottom-up) style. In the divisive hierarchical style, all the data points are believed to belong to single large cluster and the immense cluster is divided into smaller sub-clusters. In contrast, each data point is presumed to be a cluster itself and is combined iteratively into smaller number of clusters in the agglomerative style. The clusters are represented as dendrograms, which are tree-like structures consisting of sequences of divisions or mergers. Cluster of the data points is acquired by breaking down the dendrogram at the required resemblance level.

2.2 Partition-Based or Centroid-Based Clustering Method

Partition-based clustering method involves discovering the clusters in the data set by considering the proximity of the data point with the selected central value. Central value can be the centroid of the data objects or the actual data object. The sensitive step which is involved in this method is predefining of total partitions required at the primary step which is to be decided by the user. It iteratively measures distance between data points and representative centroids using different distance metrics such as Euclidian distance, Manhattan distance, etc. Its simplicity, efficiency and practical implementation makes partition-based clustering the most popular and applied method amongst all the other types of clustering methods [56].

Some of the examples of algorithms in this category are K-means, K-medoids and Clustering LARge Applications (CLARA). K-means algorithm divides the dataset into 'k' clusters such that each data point resides in the cluster having closest mean. The points are allocated to the clusters by calculating the distance between mean and data points. Partitioning Around Medoid (PAM) or K-medoid algorithm is mostly identical to the K-means algorithm but has difference in the selected central value which is the actual data object. Both the algorithms require the initial value of 'k' which the user provides. The K-means algorithm is incapable of handling noisy data whereas PAM algorithm is better in handling noisy data because a medoid is less affected by the outliers or other boundary data points than the mean [67].

CLARA algorithm is a development of the PAM algorithm which is designed for administering larger datasets more efficiently. In CLARA, a sample is selected from the whole dataset and then PAM is employed to select medoids from the sample. The thought behind its working is that if the sample is selected in a random fashion, then its medoid will correctly represent the whole dataset [48].

2.3 Density-Based Clustering Method

This method, as the name suggests, utilizes the data points' density to generate clusters. Within a cluster, data points having higher density are clustered together and the data points with lower density remain out of the cluster. These data points which lie outside the clusters are recognized as outliers or noise and are efficiently handled by this method. This method helps in discovering clusters of various random patterns and sizes which are highly homogeneous due to similar density. Few examples of algorithms under this category are DBSCAN (Density-Based Spatial Clustering of Applications with Noise) and OPTICS.

DBSCAN (Density-Based Spatial Clustering of Applications with Noise) is commonly used DBSC. It groups the closely connected dense points while identifying the outliers. Regions having high concentrations of data points are clustered together and these clusters are surrounded by areas having low concentration of data points. In DBSCAN, the user is required to mention two parameters which are maximum radius of neighborhood and minimum points which should be included in the nearby area. The underlying concept of density-based cluster formation method is that density is decided by the number of points present in the neighborhood.

OPTICS (Ordering Points to Identify the Clustering Structure) is an advancement to DBSCAN, but it is not that sensitive to the input parameters. It is like DBSCAN but is designed to handle varying densities. In OPTICS, rearrangement of a point dataset in accordance with the densities in the proximity of each point takes place and then clustering is carried out using the re-arranged set as input [48].

2.4 Distribution-Based Clustering Method

The concept of probability distribution is taken into consideration for creation of clusters in this clustering type. The data points are grouped together based on the assumption that the data points may possess the same probability distribution. Those data points which are assumed to have a high possibility of being part of same probability distribution are clustered together. A principal or central point exists in every cluster and as distance of point extends from the principal point, possibility of data point being accommodated in that cluster is reduced. This clustering method performs best when it is provided with an artificial data or data set comprising data points most plausibly belonging to a distribution defined a prior, which becomes this method's major disadvantage as well because real dataset may not have a pre-defined probability distribution.

2.5 Fuzzy Clustering

Fuzzy clustering is a generalized adaptation of partition-based clustering as it permits the data points to be allocated to more than one cluster. Cluster assignment to the data points is carried out by utilizing cluster membership based on the spatial probabilities. Data points, in some circumstances, may exist in between the principal or central points or may be obscure in nature. In such circumstances, the application of fuzzy clustering is advantageous as it considers probability instead of distance. An example algorithm of this type of clustering is Fuzzy C-means (FCM). This algorithm is a fuzzy variant of K-means algorithm which allows allocation of data points to many clusters. It is helpful in managing overlapping data but requires 'c' number of clusters at the starting step which is a sensitive issue.

2.6 Constraint-Based or Semi-Supervised Clustering Method

Clustering is unsupervised machine learning and data analysis technique, and its learning approaches are entirely unguided. However, in some cases, the data analyst having prior knowledge about the domain can use some constraints to get the desired clustering results. The constraint-based clustering, also known as semi-supervised clustering, involves incorporation of domain-specific constraints into the clustering process. These constraints are typically derived from prior knowledge about the problem domain and influence the formation of clusters. This method is often used when traditional clustering techniques (e.g., K-means or hierarchical clustering) fail to account for specific requirements in the data [15].

2.7 Graph Theory Based Clustering Method

Graph theory studies the connections between objects using graphs. Schaeffer [71] specified that graphs are frameworks generated by a collection of vertices and edges that join pairs of vertices. These methods identify communities in the networks by utilizing the structural properties of graphs, such as node connectivity, edge weights, and topological relationships. These methods focus on identifying densely connected sub-graphs or communities within a network while minimizing inter-community connections.

Breadth First Search (BFS) [61] and Depth First Search (DFS) [74] are examples of graph theory-based algorithms which are based on graph traversal or search. BFS explores all the neighbors of a vertex before moving to the next level whereas DFS seeks all the vertices in one branch before retracing back to the source vertex and advancing to the next branch. Examples of algorithms based on the calculation of shortest path are 'Dijkstra's algorithm [14], finds the shortest path from a single source vertex to all the other vertices in a weighted graph, and Floyd-Warshall algorithm [28], that discovers shortest paths between all pairs of vertices in weighted graph.

Louvain algorithm [4] and Fast Greedy algorithm [21] are the graph theory-based algorithms which make clusters based on modularity. Louvain algorithm assigns nodes to the communities iteratively to maximize modularity, and Fast Greedy algorithm iteratively combines the communities that provide the largest increment in the modularity. Girvan-Newman algorithm, which recognizes communities by iteratively omitting edges with highest betweenness centrality, is another graph theory-based algorithm using betweenness centrality as a measure for community detection. Spectral clustering [76] is a graph-based clustering that utilizes eigen values and eigenvectors of matrices (e.g., Laplacian matrix of the graph) to partition the network. It relies on graph partitioning techniques like normalized cuts. Random Walk-based algorithms are another type of graph-based algorithms which are based on probability. These methods simulate random walks on the network and identify communities based on the likelihood of staying within a community.

The types of clustering methods and algorithms are depicted in Fig. 1.

3 Review of Application of Clustering in Different Areas of WDNs

Utilization of clustering in WDNs is to identify and search for solutions pertaining to the problems that arise in the water networks because of their intricate and enormous nature. Conception of clustering is realized extensively in diverse domains of WDNs comprising, district metered areas (DMAs) formation, leakage detection and localization, analysis of WDN, demand estimation, contamination detection and control, quality and pressure sensor placement, network calibration, operation & maintenance of WDNs, critical pipes identification and rehabilitation, decentralization of WDNs, in intermittent water supply systems, etc.

In this section, a comprehensive review of application of clustering in WDNs is provided. The readers will get insight into diverse application fields of WDNs, in which various types of clustering methods and algorithms are employed for accomplishing different objectives related to different problems that occur in WDNs, and the data set on which these algorithms are applied and tested is provided.

3.1 DMA Formation

The concept of DMAs is applied in the WDNs mainly for pressure management and leakage detection and localization. Creation of DMAs involves partitioning the network into sub-networks or clusters, and therefore, clustering methods are utilized vastly. After clustering, the network is partitioned physically by using boundary optimization in the subsequent step. Scarpa et al. [70] suggested graph theory-based algorithms such as BFS algorithm [61] and 'Dijkstra's algorithm [14] for clustering while forming DMAs. Clustering was applied based on the affinity matrix which contained all the relations between the network vertices and flow weighted adjacency matrix that included pipe flow data. The aim of network clustering was reduction in network complexity. The methodology was assessed on large WDN which is a part of real-life WDN containing multiple sources.

Louvain algorithm [4] was applied for community detection for the network simplification and partitioning in the study conducted by Zhang et al. [78]. Modularity [52] was used as the network partitioning index and the detection of communities in the network was accomplished based on multi resolution modularity utilizing Random Walk theory [35]. The battle of water sensor networks 2 (BWSN2) [54] was considered as case study for method testing. Nardo et al. [11], in their study, employed recursive spectral clustering algorithm by exploiting the properties of spectral algorithm [73, 76], to cluster the network for its simplification and to get topologically

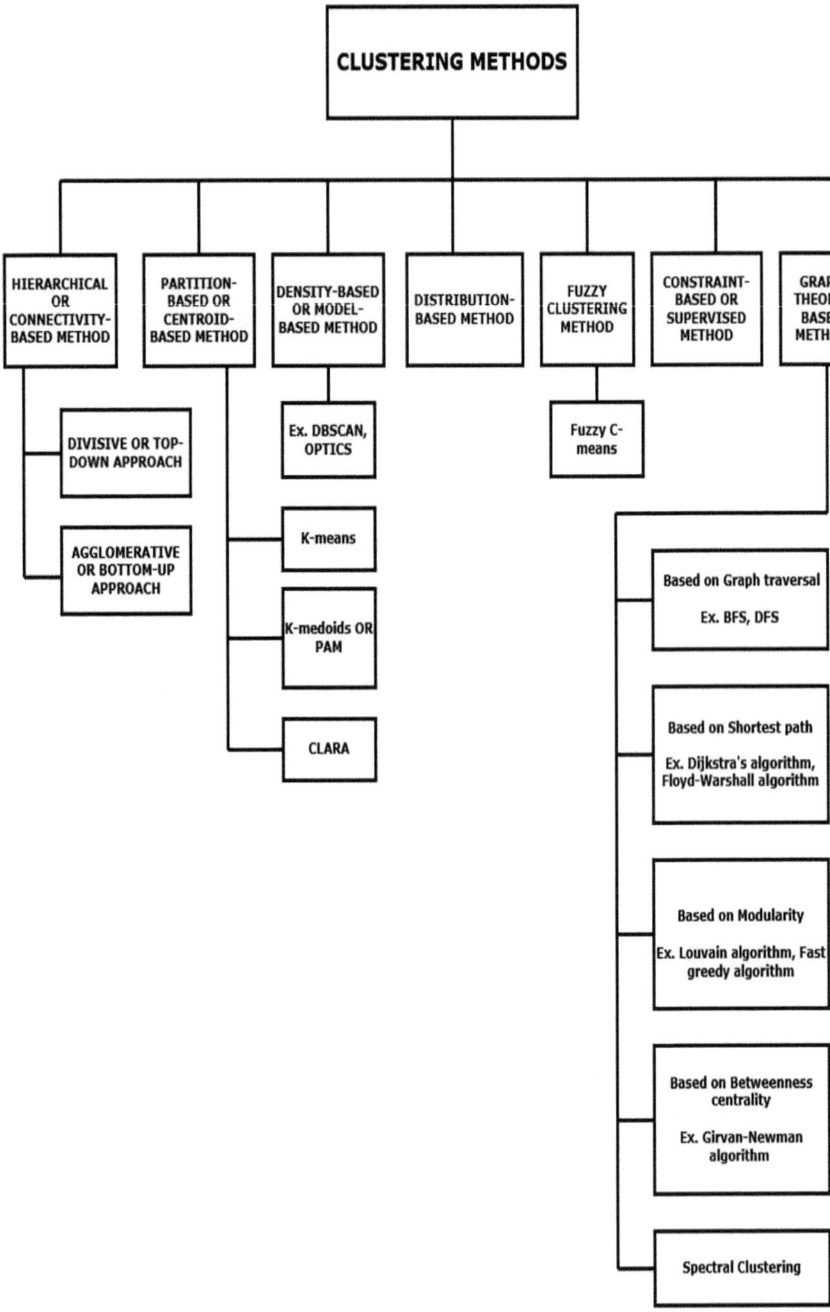

Fig. 1 Types of clustering methods and algorithms

optimal clusters. The clusters were formed based on the Laplacian matrix graph which is difference between degree of node and affinity matrix. Recursive spectral algorithm is a graph theory-based clustering algorithm that requires another clustering algorithm in one of its steps for node clustering and K-means algorithm was preferred because of its simplicity. A real world WDN of Parete city (Italy) was opted for testing the method.

The method of topology-distance-based clustering (TDBC) proposed by Du et al. [16], incorporates topology distance and network connectivity for the DMA creation of the WDN. The authors adopted Floyd and K-means clustering algorithms for discovering network clusters. The Floyd algorithm [28] is applied to formulate a matrix to compute the degree of node connectivity. The dominance of links comprising pressure reducing valves (PRVs) and pump stations is also integrated in the method by amplifying their topological distances. Grouping of the nodes is accomplished with the help of K-means algorithm by making use of the resulting topology-distance matrix. With the purpose of examining their distinctness in clustering mechanisms and check their effectiveness in systems comprising PRVs and pump stations, the authors compare three clustering algorithms that are geographical clustering-based partitioning (GCBP) method [25], the TDBC method and modularity-based fast greedy clustering (MFGC) method [13], and the TDBC method gave precise results for the network including the PRVs and pump stations. Four networks were considered for implementing the method which include artificial network which is Simple network [16], C-town network [55], Anytown network [77] and the Rural network [44].

Girvan-Newman, Fast Greedy and Short Random Walks algorithm were selected for detecting communities in the network to create sub-networks for DMAs with the objectives of network division and simplification, in the study performed by Shekofteh et al. [72]. Edge betweenness centrality is the measure based on which Girvan-Newman algorithm [50] detects communities within the network, Fast Greedy algorithm [21] maximizes modularity to cluster nodes of the network and create the sub-networks, and transition probability matrix is utilized by Short Random walks algorithm [36] for the network clustering. Girvan-Newman algorithm stood out amongst the three. Six complex benchmark systems were considered to test capability of the method, which are BIN WDN [66], Modified EXNET model [19], Modified Wolf-Cordera Ranch model [40], C-Town model [55], D-Town model [44] and Modena Network (MOD) model [5].

Fang et al. [18] employed weighted spectral clustering algorithm [10] for network division into clusters with the intention of minimizing boundary pipes and water requirement between areas, capturing nodes with common characteristics, improving hydraulic performance of the pipe network after division, and enhancing operation efficiency & accuracy of Genetic algorithm applied for boundary optimization in next step. K-medoids was applied as the clustering algorithm in the weighted spectral algorithm and similarity matrix, which is the integration of node affinity matrix and water demand affinity matrix, was adopted as basis of cluster creation. The networks opted for assessing the methodology were a real-world network medium sized city water system located in Modena, Italy [63] and a large KLmod network [3].

Girvan-Newman algorithm [50], which discovers communities in the network using modularity, was utilized in their study by Kakeshpour et al. [30]. Network complexity reduction and network partition are the targets for which clustering was applied. Modena WDN [5] and Wolf Cordera-Ranch WDN, established in Colorado and USA [40], were selected for the method implementation.

3.2 Analysis of WDN

WDNs are integral and crucial for the advancement of urban regions. They are underground infrastructures which consist of a massive number of components such as pipes, valves, reservoirs, junctions, pumps, etc. stretching over thousands of kilometers. Consequently, they are intricate in nature which makes their analysis difficult. Simplification of the WDNs is necessary for understanding the interaction between its components which helps in better analysis, operation & management, and monitoring of the networks.

To simplify the WDN, the concept of strongly connected components or clusters (SCCs) and weakly connected components or clusters (WCCs) can be used. A cluster is an SCC if it consists of at least one pair of vertices which has bi-directional flow in the links between them whereas a directed sub-graph is a WCC when its elemental undirected graph is connected. Therefore, for partitioning the network for its topological/connectivity analysis, Perelman and Ostfeld [58, 59], in their study, applied the graph traversal algorithms, DFS [74] and BFS [61] algorithms for community detection of the network to get stronger intra-connected clusters than inter-connected clusters. Perelman and Ostfeld [58] used the benchmark network of the Battle of the Water Sensors Network 1 (BWSN1) [54] for the proposed methodology demonstration. In their extension to the above study, Perelman and Ostfeld [53] selected modified version of EPANET Example1 and Dover Township network, New Jersey [45, 46] for applying the methodology. Lifshitz and Ostfeld [38] employed DFS [74] for clustering network with the intent of network aggregation and simplification based on the concept of SCCs discussed earlier. Here, they tested the method on KY5 [27] and BWSN1 [54].

3.3 Calibration

An important step in WDN modeling is calibration which is the process of adjusting the values of uncertain parameters in the network modelling to match values of modelled parameters with those observed in the field. Calibration problems are complex and difficult because of involvement of large number of components in the WDN. For reducing the number of variables and degree of freedom of the calibration problem, Freitas et al. [22], in their paper, suggested using K-means clustering algorithm for clustering network by considering pipe features such as age, material and diameter. The network used for the method testing was the benchmark Modena WDN [5].

Rajabi and Tabesh [64] employed multi-view K-means clustering algorithm for clustering vertices of network in which different aspects of node characteristics such as pressure sensitivity to roughness and leakage events were considered. Another community detection algorithm, i.e., the Louvain algorithm was used for clubbing nodes to reduce the dimensions of the Sobol sensitivity analysis problem, and each cluster of nodes was treated as a parameter. Decreasing the decision space and determining the number and arrangement of sensors were the purposes to be fulfilled by network clustering which was applied for identification of leakage. The case studies chosen here on which the method was implemented were Anytown network [53, 77], C-Town model and Modena network [63].

3.4 Pressure Sensor Placement

WDNs are designed to distribute water to consumers at adequate pressure with sufficient quantity. Fluctuations in pressure may indicate various problems occurring in the network such as leakage, pressure imbalances, pipe bursts, etc. Therefore, pressure management is necessary in WDN to prevent these issues from occurring. Sensors are placed at certain locations which help the water utilities to monitor pressure in the network and take preventions and actions accordingly. In their work, Cheng et al. [8] employed K-means algorithm for clustering the network by considering pressure sensitivity matrix of WDN that consists of the gradient knowledge related to nodal pressure to nodal outflow, which may be regarded as an order of measured vectors displaying the inter-relationships of nodal pressure alteration and utilized for the optimal employment of pressure sensor. The aims of clustering were to reduce network complexity and optimize placement of pressure sensors. A real-world network was selected for the implementation method.

Spectral clustering algorithm [76] for clustering WDN for burst detection, its localization and sensor employment were preferred by Presti et al. [62] in their study. With the goals of network simplification and minimizing the computational burden in the next optimization step involved, clustering was applied by utilizing graph Laplacian matrix (computed as the difference between degree diagonal matrix and similarity matrix) with usage of K-means algorithm [42] in the clustering step of spectral algorithm. WDN of Parete city (Italy) was selected for demonstrating the suggested approach.

3.5 Leakage Detection and Localization

Leakages in WDNs are a major concern worldwide which can cause substantial water losses and increase burden on water utilities that are accountable for transmitting potable water from source to the consumers at adequate pressure in sufficient quantity. These leakages can occur because of many reasons such as pipe breakages, pipe bursts, high pressures in the systems, etc. Urban WDNs are vast and complicated underground infrastructures which make the detection and localization of leakages within them a strenuous work.

Candelieri et al. [6] suggested using spectral clustering method [76], which divides network based on graph Laplacian matrix, with K-means as clustering algorithm required at one of its stages, for leakage localization in the network with objective of network simplification. The case study data set taken was real WDN (H2OLeak Italian project) for the demonstration of the method.

With the target of downsizing intricacy of the network for efficient leakage localization in the WDN, Li et al. [37] selected hierarchical clustering and K-means clustering algorithms [41, 42, 68] for cluster formation in the network by considering spatial distance between nodes as a measure for clustering.

3.6 Critical Pipes Identification and Rehabilitation

Ever rising population in urban regions is the reason for continuous expansion of WDNs. The constituents of network infrastructure require attention and regular service & maintenance as they are vulnerable to damage due to many reasons such as aging of the network infrastructure, pressure imbalances, etc. Recognition of vulnerable or critical pipes and their replacement or rehabilitation is required for proper functioning of the WDNs.

For obtaining clusters exhibiting stronger internal connections than external connections and network simplification, modularity-based Newman algorithm [9, 52] was preferred by Diao et al. [12] for identifying critical pipes in the network. The Newman algorithm discovers the communities in network by enhancing modularity. The benchmark network, New York Tunnels [69] and the Alpine WDN were opted for verifying the methodology.

Muhammed et al. [47] likewise selected modularity-based clustering technique [9, 51] that recognizes communities in the network based on modularity with the aim of decreasing complexity of rehabilitation problem and network structure simplification for analyzing closely interactions between the components of network. This method was implemented using Gephi software. The method was demonstrated on the EXNET network.

3.7 Demand Estimation

The fundamental purpose of urban WDN is to supply potable and safe water from source to the consumers existing in the city. A city can have different types of consumer demand such as residential, institutional, commercial, industrial, fire demand, etc. Consumer demand is an essential motivation in the modelling of drinking water system (DWS) networks. Rana et al. [65] employed Self-Organizing Map (SOM) algorithm [34], which is an artificial neural network (ANN) method, with aim to decrease the number of unknown demands to be evaluated. Clustering of the network was performed by utilizing the sensitivity of evaluations to changes in nodal demands. The Kentucky 2 (KY2) network [29] which is a synthetic WDN was selected for verifying the method.

3.8 Contamination Event Detection, Quality Analysis and Sensor Placement

Water security is one of the serious concerns regarding WDNs supplying water to the residents in an urban city. WDNs are susceptible to contamination which is the intrusion of pollutants or pathogenic microorganisms in the network. Contamination events occurring in the WDNs require prevention and mitigation measures such as contamination event detection strategies, water quality monitoring, etc. to minimize the contamination hazard exposed to the public health.

Chang et al. [7], in their study, selected few clustering algorithms such as Self-organizing map (SOM) [33], K-means clustering and Fuzzy C-mean (FCM) [2, 17] algorithms for dividing the network into sub-networks by considering different water quality indicators such as residual chlorine, water age, THMs (trihalomethanes), TAAs (trihaloacetic acids), TOC (total organic carbon) and BDOC (biodegradable dissolved organic carbon), and hydraulic indicators which are demand and pressure. The objectives of clustering were network simplification for water quality evaluation and analysis of the WDN. A large WDN was opted for the demonstration of the method.

Perelman and Ostfeld [60] utilized edge betweenness community algorithm [71] and K-means algorithm [26] for detection of communities using various parameters such as betweenness, eigenvector, and closeness centrality and page rank for achieving targets of network complexity minimization and optimizing number of sensors. The battle of the water sensor networks (BWSN1) [54] was used for applying the methodology.

Kirstein et al. [32] applied DFS algorithm [74] for water quality monitoring to discover communities in network by adopting the concept of topological clustering of network into SCCs and WCCs [58, 59] where the aim was to reduce the network intricacy. The investigated network model used was of Nørrebro district of Copenhagen, Denmark [31].

Lifshitz and Ostfeld [38] also adopted the concept of SCCs, in which the flow between at least one pair of nodes in the network is bidirectional, to address the problem of contamination spread control. DFS algorithm [74], with the target of network simplification, was preferred to detect the SCCs in the network. The networks selected for verifying the methodology were Scorpion network and BWSN1 water network [54].

Fasaee et al. [20], for contamination occurrence detection and sensor employment, suggested usage of Harmonic Search evolutionary optimization algorithm [24] with K-means algorithm [49] for creating network clusters. Grouping the contamination occurrences by utilizing spatial similarity in their propagation systems and sensor employment at the centroids of clusters identified were targets of network clustering. 'Jaccard's similarity index was utilized as the similarity measure for grouping the contamination events according to their propagation systems. The WDN of a virtual city, Mesopolis, was opted for applying the methodology.

By considering hamming distance and event detection matrix, Gautam et al. [23] for quality sensor placement, employed K-means algorithm for network clustering with goals of achieving reduction in intricacy of sensor employment problem for contamination detection and network simplification. The networks considered for verifying the study were BSWN Network1 and C-Town network.

3.9 Operation and Maintenance

Due to the enormous nature of the WDNs, their partition into sub-networks or clusters is essential for their better operation and maintenance as the smaller parts of network are much easier to manage. Aiming for minimizing intricacy of WDN and getting strong internally connected clusters than externally connected clusters, Perelman and Ostfeld [57] for WDN operation & maintenance, and security enhancement, suggested utilizing graph theory-based DFS and BFS algorithms for detecting communities within the network by adopting the concept of SCCs and WCCs. For application of clustering algorithm, EPANET example 1 and BWSN1 [54] were used.

3.10 Decentralization and Intermittent Water Supply

Urban utility systems around the world are designed in conventional fashion in which water is carried from source to consumers by using a well-designed infrastructure of pipes, junctions, reservoirs, pumps, etc. But this approach is proving to be less helpful in various regions of the world where water scarcity problems are prevalent. Thus, the idea of decentralization is being explored to tackle issues related to the centralized water utility systems. Decentralized WDNs are the systems where storage, treatment facilities, and other network components are scattered over multiple areas near the water source. These systems are more flexible and adaptable to changes than centralized ones.

Tsegaye et al. [75] suggested usage of K-means algorithm by using Euclidean distance to form clusters to decentralize the WDNs by aiming to minimize source demand distance and maximize intra-cluster homogeneity. The introduced cluster creation approach was implemented to an actual case study in Arua, Uganda.

WDNs are primarily designed for continuous water supply. But, due to scarcity of water in diverse areas of the world, these systems are also utilized to supply water to the consumers intermittently, which means that the water is delivered to the consumers for specific durations in a day rather than supplying water continuously.

The operation of intermittent water supply systems is more complex than continuous ones due to involvement of problems such as deterioration of the infrastructure, contamination threats, inequitable water supply and many more. To tackle this problem of operation of intermittent water supply systems, Ayyash et al. [1] applied a modularity-based community detection technique [9, 51] which detects communities in the network using modularity, which was implemented using Gephi software to get communities with good connectivity and simplify the network. A benchmark network of D-Town water distribution is used for verifying the approach.

4 Conclusions

Urban Water Distribution Networks (WDNs) are underground, intricate and massive infrastructures consisting of huge number of constituents stretching over extensive distances. Their principal role is to transmit sufficient water from source to the consumers at adequate pressure and of good quality. Simplification of WDNs is necessitated for their easier comprehension. Clustering is a well-known technique of machine learning and data mining which has wide application in solving problems regarding WDNs. In this chapter, introduction of WDNs and importance of clustering in the WDN are described. Then this study comprehensively reviews the application of different clustering algorithms in the diverse domains of WDNs for figuring out solutions to various issues that occur in WDNs. Conception of clustering is extensively realized in the different domains of WDNs, such as for understanding the network behaviour and analysis, calibration, contamination detection and control, water quality monitoring, quality and pressure sensor placement, demand estimation, district metered area (DMA) formation, leakage detection and localization, operation and maintenance, critical pipe identification and rehabilitation, intermittent water supply systems, decentralization of the networks.

References

1. Ayyash, F., Zhang, C., Javadi, A. A., & Farmani, R. (2024). Optimal operation of intermittent water supply systems under water scarcity. *Journal of Water Resources Planning and Management, 150*(3), 04024001.
2. Bezdek, J. C. (1981). *Pattern recognition with fuzzy objective function algorithms*. Plenum.

3. Bi, W., Dandy, G. C., & Maier, H. R. (2015). Improved genetic algorithm optimization of water distribution system design by incorporating domain knowledge. *Environmental Modelling and Software, 69*, 370–381.
4. Blondel, V. D., Guillaume, J. –L., Lambiotte, R., Lefebvre, E. (2008). Fast unfolding of communities in large networks. *J Stat Mech Theory Exp, 2008*(10), 10008.
5. Bragalli, C., D'Ambrosio, C., Lee, J., Lodi, A., & Toth, P. (2012). On the optimal design of water distribution networks: Practical MINLP approach. *Optimization and Engineering, 13*(2), 219–246.
6. Candelieri, A., Conti, D., & Archetti, F. (2014). Improving analytics in urban water management: A spectral clustering-based approach for leakage localization. *Procedia—Social and Behavioral Sciences, 108*, 235–248.
7. Chang, K., Gao, J. L., Wu, W. Y., & Yuan, Y. X. (2011). Water quality comprehensive evaluation method for large water distribution networks based on clustering analysis. *J Hydroinformatics, 13*(3), 390–400.
8. Cheng, L., Kun, D., Jia-peng, T., & Wei-xin, D. (2017). Optimal placement of pressure sensors in water distribution systems based on clustering analysis of pressure-sensitive matrix. *Procedia Eng, 186*, 405–411.
9. Clauset, A., Newman, M. E. J., & Moore, C. (2004). Finding community structure in very large networks. *Physical Review E, 70*, Article 066111.
10. Di Nardo, A., Di Natale, M., Giudicianni, C., Greco, R., & Santonastaso, G. F. (2017). Weighted spectral clustering for water distribution network partitioning. *Appl Netw Sci, 2*(1), 19.
11. Di Nardo, A., Di Natale, M., Giudicianni, C., Musmarra, D., & Santonastaso, G. F. (2018). An innovative approach based on recursive clustering to design optimal districts in water distribution networks. In *9th international congress on environmental modelling and software (iEMSs 2018), Fort Collins.*
12. Diao, K., Farmani, R., Fu, G., Astaraie-Imani, M., Ward, S., & Butler, D. (2014). Clustering analysis of water distribution systems: Identifying critical components and community impacts. *Water Science and Technology, 70*(11), 1764–1773.
13. Diao, K., Zhou, Y., & Rauch, W. (2013). Automated creation of district metered area boundaries in water distribution systems. *Journal of Water Resources Planning and Management, 139*(2), 184–190.
14. Dijkstra, E. W. (1959). A note on two problems in connection with graphs. *Numerische Mathematik, 1*, 269–271.
15. Dinler, D., & Tural, M. K. (2016). A survey of constrained clustering. *Unsupervised Learning Algorithms* (pp. 207–235). Springer.
16. Du, K., Li, J., Xu, W., Liu, Z., & Zheng, F. (2023). Topology-distance-based clustering method for water distribution network partitioning. *AQUA—Water Infrastruct Ecosyst Soc, 72*(11), 2186–2198.
17. Dunn, J. C. (1974). A fuzzy relative of the ISODATA process and its use in detecting compact well-separated clusters. *J Cybern, 3*(3), 32–57.
18. Fang, Q., Zhao, H., Xie, C., & Chen, T. (2023). A method for water supply network DMA partitioning planning based on improved spectral clustering. *Water Supply, 23*(8), 3432–3452.
19. Farmani, R., Savic, D. A., & Walters, G. A. (2004). Exnet benchmark problem for multi-objective optimization of large water systems. In *Modelling and control for participatory planning and managing water systems, IFAC Workshop, Venice, Italy.*
20. Fasaee, M. A. K., Monghasemi, S., Nikoo, M. R., Shafiee, M. E., Berglund, E. Z., & Bakhtiari, P. H. (2021). A K-sensor correlation-based evolutionary optimization algorithm to cluster contamination events and place sensors in water distribution systems. *Journal of Cleaner Production, 319*, Article 128763.
21. Fortunato, S. (2010). Community detection in graphs. *Physics Reports, 486*, 75–174.
22. Freitas, R., Brentan, B. M., & Meirelles, G. (2018). Dynamic clustering and a hybrid optimization for roughness calibration in water distribution model. In *WDSA/CCWI joint conference proceedings, Ontario, Canada.*

23. Gautam, D. K., Kotecha, P., & Subbiah, S. (2022). Efficient k-means clustering and greedy selection-based reduction of nodal search space for optimization of sensor placement in the water distribution networks. *Water Research, 220*, Article 118666.
24. Geem, Z. W., Kim, J. H., & Loganathan, G. V. (2001). A new heuristic optimization algorithm: Harmony search. *Simulation, 76*(2), 60–68.
25. Hajebi, S., Barrett, S., Clarke, A., & Clarke, S. (2013). Multi-agent simulation to support water distribution network partitioning. *European Simulation and Modelling Conference, 2013*, 163–168.
26. Hartigan, J. A., & Wong, M. A. (1979). Algorithm AS 136: A k-means clustering algorithm. *Applied Statistics, 29*, 100–108.
27. Hernandez, E., Hoagland, S., & Ormsbee, L. (2016). Water distribution database for research applications. In *World environmental and water resources congress 2016, FL, USA* (pp. 465–474).
28. Hougardy, S. (2010). The Floyd-Warshall algorithm on graphs with negative cycles. *Information Processing Letters, 110*(8–9), 279–281.
29. Jolly, M. D., Lothes, A. D., Bryson, L. S., Ormsbee, L. (2014). Research database of water distribution system models. *J Water Resour Plan Manag, 140*(4), 410–416.
30. Kakeshpour, M., Jalili Ghazizadeh, M., Hosseini, S. A., & Sharafati, A. (2024). Design of district metered areas for existing water distribution networks. *J Pipeline Syst Eng Pract, 15*(2), 04024006.
31. Kirstein, J. K., Albrechtsen, H.-J., & Rygaard, M. (2014). Simplification of water distribution network simulation by topological clustering—investigation of its potential use in Copenhagen's water supply monitoring and contamination contingency plans. *Procedia Eng, 89*, 1184–1191.
32. Kirstein, J. K., Albrechtsen, H. J., & Rygaard, M. (2015). Topological clustering as a tool for planning water quality monitoring in water distribution networks. *Water Sci Technol Water Supply, 15*(5), 1011–1018.
33. Kohonen, T. (1981). Automatic formation of topological maps of patterns in a self-organizing system. In *Proceedings of the 2nd Scandinavian conference on image analysis, Helsinki, Finland* (pp. 214–220).
34. Kohonen, T., Oja, E., Simula, O., Visa, A., & Kangas, J. (1996). Engineering applications of the self-organizing map. *Proceedings of the IEEE, 84*(10), 1358–1384.
35. Lambiotte, R., Delvenne, J.-C., & Barahona, M. (2014). Random walks, Markov processes and the multiscale modular organization of complex networks. *IEEE Trans Netw Sci Eng, 1*(2), 76–90.
36. Leskovec, J., Kleinberg, J., & Faloutsos, C. (2007). Graph evolution: Densification and shrinking diameters. *ACM Trans Knowl Discov Data (TKDD), 1*(1), 2.
37. Li, X., Chu, S., Zhang, T., Yu, T., & Shao, Y. (2022). Leakage localization using pressure sensors and spatial clustering in water distribution systems. *Water Supply, 22*(1), 1020–1034.
38. Lifshitz, R., & Ostfeld, A. (2017). A clustering model for contamination spreading control in water distribution systems. *World Environmental and Water Resources Congress, 2017*, 517–522.
40. Lippai, I. (2005). Colorado Springs utilities case study: water system calibration/optimization. In *Pipelines 2005: Optimizing pipeline design, operations, and maintenance in today's economy* (pp. 1047–1057).
41. MacKay, D. J. C. (2003). *Information theory, inference, and learning algorithms*. Cambridge University Press.
42. MacQueen, J. (1967). Some methods for classification and analysis of multivariate observations. *Proceedings of the Fifth Berkeley Symposium on Mathematical Statistics and Probability, University of California Press, Berkeley, 1*, 281–297.
43. Marchi, A., Dandy, G., Wilkins, A., & Rohrlach, H. (2014). Methodology for comparing evolutionary algorithms for optimization of water distribution systems. *Journal of Water Resources Planning and Management, 140*(1), 22–31.

44. Marchi, A., Salomons, E., Ostfeld, A., Kapelan, Z., Simpson, A. R., Zecchin, A. C., Maier, H. R., Wu, Z. Y., Elsayed, S. M., Song, Y., & Walski, T. (2014). Battle of the water networks II. *Journal of Water Resources Planning and Management, 140*(7), 04014009.
45. Maslia, M. L., Sautner, J. B., Aral, M. M., Gillig, R. E., Reyes, J. J., & Williams, R. C. (2001). *Historical reconstruction of the water distribution system serving the Dover Township area, New Jersey: January 1962–December 1996*. Rep. Agency for Toxic Substances and Disease Registry, Atlanta.
46. Maslia, M. L., Sautner, J. B., Aral, M. M., Reyes, J. J., Abraham, J. E., & Williams, R. C. (2000). Using water-distribution system modeling to assist epidemiologic investigations. *Journal of Water Resources Planning and Management, 126*(4), 180–198.
47. Muhammed, K., Farmani, R., Behzadian, K., Diao, K., & Butler, D. (2017). Optimal rehabilitation of water distribution systems using a cluster-based technique. *Journal of Water Resources Planning and Management, 143*(7), 04017022.
48. Nagpal, A., Jatain, A., & Gaur, D. (2013). Review based on data clustering algorithms. In *2013 IEEE conference on information & communication technologies, Thuckalay, India* (pp. 298–303). IEEE.
49. Naldi, M. C., Campello, R. J., Hruschka, E. R., & Carvalho, A. C. P. L. F. (2011). Efficiency issues of evolutionary k-means. *Applied Soft Computing, 11*(2), 1938–1952.
50. Newman, M. E. J. (2004). Fast algorithm for detecting community structure in networks. *Physical Review E, 69*, Article 066133.
51. Newman, M. E. J. (2006). Modularity and community structure in networks. *Proceedings of the National academy of Sciences of the United States of America, 103*(23), 8577–8582.
52. Newman, M. E. J., & Girvan, M. (2004). Finding and evaluating community structure in networks. *Physical Review E, 69*(2), Article 026113.
53. Ostfeld, A. (2012). Optimal reliable design and operation of water distribution systems through decomposition. *Water Resources Research, 48*(10), W10521.
54. Ostfeld, A., et al. (2008). The battle of the water sensor networks (BWSN): A design challenge for engineers and algorithms. *Journal of Water Resources Planning and Management, 134*(6), 556–568.
55. Ostfeld, A., Salomons, E., Ormsbee, L., Uber, J. G., & Bros, C. M. (2012). Battle of the water calibration networks. *Journal of Water Resources Planning and Management, 138*(5), 523–532.
56. Oyelade, J., Isewon, I., Oladipupo, O., Emebo, O., Omogbadegun, Z., Aromolaran, O., Uwoghiren, E., Olaniyan, D., & Olawole, O. (2019). Data clustering: algorithms and its applications. In *2019 19th International Conference on Computational Science and its Applications (ICCSA), St. Petersburg, Russia* (pp. 71–81). IEEE.
57. Perelman, L., & Ostfeld, A. (2010). Cluster analysis for water distribution systems security enhancement. In *World environmental and water resources congress 2010: Challenges of change* (pp. 4330–4338).
58. Perelman, L., & Ostfeld, A. (2011). Topological clustering for water distribution systems analysis. *Environmental Modelling and Software, 26*(7), 969–972.
59. Perelman, L., & Ostfeld, A. (2012). Water-distribution systems simplifications through clustering. *Journal of Water Resources Planning and Management, 138*(3), 218–229.
60. Perelman, L., & Ostfeld, A. (2013). Application of graph theory to sensor placement in water distribution systems. In *World environmental and water resources congress 2013: Showcasing the future* (pp. 617–625).
61. Pohl, I. S. (1969). *Bi-directional and heuristic search in path problems* (p. 169). Doctoral thesis. Stanford Linear Accelerator Center, Stanford University, Stanford, CA, USA.
62. Presti, J. L., Giudicianni, C., Toffanin, C., Creaco, E., Magni, L., & Galuppini, G. (2024). Combining clustering and regularised neural network for burst detection and localization and flow/pressure sensor placement in water distribution networks. *J Water Process Eng, 63*, Article 105473.
63. Quiñones-Grueiro, M., Ares Milián, M., Sánchez Rivero, M., Silva Neto, A. J., & Llanes-Santiago, O. (2021). Robust leak localization in water distribution networks using computational intelligence. *Neurocomputing, 438*, 195–208.

64. Rajabi, M., & Tabesh, M. (2024). Pressure sensor placement for leakage detection and calibration of water distribution networks based on multiview clustering and global sensitivity analysis. *Journal of Water Resources Planning and Management, 150*(5), 04024010.
65. Rana, S. M., Boccelli, D. L., Marchi, A., & Dandy, G. C. (2020). Drinking water distribution system network clustering using self-organizing map for real-time demand estimation. *Journal of Water Resources Planning and Management, 146*(12), 04020090.
66. Reca, J., & Martínez, J. (2006). Genetic algorithms for the design of looped irrigation water distribution networks. *Water Resources Research, 42*, 5.
67. Rokach, L., & Maimon, O. (2005). Clustering methods. In *Data mining and knowledge discovery handbook* (pp. 321–352)
68. Sarrate, R., Blesa, J., & Nejjari, F. (2014). Clustering techniques applied to sensor placement for leak detection and location in water distribution networks. In *2014 22nd Mediterranean conference on control and automation* (pp. 109–114). IEEE.
69. Savic, D. A. (2002). Single-objective vs. multiobjective optimisation for integrated decision support. In *1 st international congress on environmental modelling and software, Lugano, Switzerland.*
70. Scarpa, F., Lobba, A., & Becciu, G. (2016). Elementary DMA design of looped water distribution networks with multiple sources. *Journal of Water Resources Planning and Management, 142*(6), 04016011.
71. Schaeffer, S. E. (2007). Graph clustering. *Comput Sci Rev, 1*(1), 27–64.
72. Shekofteh, M. R., Yousefi-Khoshqalb, E., & Piratla, K. R. (2023). An efficient approach for partitioning water distribution networks using multi-objective optimization and graph theory. *Water Resour Manag, 37*(13), 5007–5022.
73. Shi, J., & Malik, J. (2000). Normalized cuts and image segmentation. *IEEE Transactions on Pattern Analysis and Machine Intelligence, 22*, 888–905.
74. Tarjan, R. (1972). Depth-first search and linear graph algorithms. *SIAM Journal on Computing, 1*(2), 146–160.
75. Tsegaye, S., Missimer, T. M., Kim, J. Y., & Hock, J. (2020). A clustered, decentralized approach to urban water management. *Water, 12*(1), 185.
76. Von Luxburg, U. (2007). A tutorial on spectral clustering. *Statistics and Computing, 17*, 395–416.
77. Walski, T. M., Brill Jr, E. D., Gessler, J., Goulter, I. C., Jeppson, R. M., Lansey, K., et al. (1987). Battle of networks models: Epilogue. *ASCE J Water Resour Plan Manag, 113*(2), 191–203.
78. Zhang, Q., Wu, Z. Y., Zhao, M., Qi, J., Huang, Y., & Zhao, H. (2017). Automatic partitioning of water distribution networks using multiscale community detection and multiobjective optimization. *J Water Resour Plan Manag, 143*(9), 04017057.

Role of Emerging Techniques in Data Mining

Manas Guleria, Lakshay Sharma, Uttam Malik, Naitik Singh, and Akhil Nigam

Abstract Data mining is a fundamental process in the field of data science and analytics, aimed at discovering useful patterns, trends, and insights from vast amounts of raw data. This multidisciplinary field leverages methodologies from statistics, machine learning, artificial intelligence, and database management to transform data into actionable knowledge. The significance of data mining lies in its ability to identify hidden relationships, make predictions, and aid in decision-making across various domains. Core techniques used in data mining include classification, clustering, regression, association rule mining, and anomaly detection. Classification is used to categorize data into predefined groups, while clustering identifies natural groupings within datasets. Regression analyzes relationships among variables, enabling predictive modeling, whereas association rule mining uncovers relationships between variables, often used in market basket analysis. Anomaly detection helps identify unusual patterns or outliers, which can be critical in fraud detection and cybersecurity. The applications of data mining span a wide array of fields. In business, it aids in customer segmentation, personalized marketing, sales forecasting, and supply chain optimization. In healthcare, it facilitates predictive diagnostics, patient risk assessment, and drug discovery. Financial institutions use data mining for credit scoring, fraud detection, and risk management. Scientific research benefits from it by uncovering complex patterns in genomic data, astronomical observations, and climate modeling. Additionally, data mining plays a significant role in social media analytics, e-commerce, and recommendation systems. Recent advancements in computational power and the emergence of big data technologies, such as distributed computing and cloud storage, have revolutionized data mining. These advancements have enabled the processing of structured, semi-structured, and unstructured data, including text, images, audio, and video, leading to more comprehensive insights. Techniques like deep learning have further enhanced data mining by enabling the analysis of high-dimensional and

M. Guleria · L. Sharma · U. Malik · N. Singh
CSE, Chandigarh University, Mohali, Punjab, India

A. Nigam (✉)
EE Chandigarh University, Mohali, Punjab, India
e-mail: akhilnigam03@gmail.com

M. A. Mellal et al. (eds.), *Computational Intelligence for a Greener Future: Innovations in Renewable Energy Systems*, Studies in Computational Intelligence 1240,
https://doi.org/10.1007/978-3-032-06732-6_23

complex datasets. Despite its transformative potential, data mining faces several challenges. Data quality remains a critical concern, as incomplete, noisy, or inconsistent data can significantly affect the accuracy of results. Scalability is another issue, as handling massive datasets requires efficient algorithms and computing infrastructure. Ethical concerns, such as data privacy, security, and algorithmic bias, have gained prominence, necessitating stringent measures to ensure responsible data usage. In conclusion, data mining is a powerful tool for extracting meaningful information from complex datasets, driving innovation, and enabling evidence-based decision-making. As technology evolves, the scope and impact of data mining will continue to expand, playing a vital role in shaping the future of industries and society. However, addressing its challenges will require ongoing interdisciplinary collaboration and ethical vigilance. This chapter focuses on the role of emerging techniques in data mining with its applications.

Keywords Data mining · Big data · Pattern recognition · Predictive analysis · Business intelligence

1 Introduction

The method to find valuable patterns and knowledge within big data collections is known as data mining. The data mining process means searching through extensive information for concealed valuable ideas that support wiser decisions. Through various statistical and machine learning and data-base system algorithms Data Mining extracts patterns along with trends and useful information from extensive datasets. The method resembles searching through vast amounts of information to obtain beneficial insights which guide business choices. Organizations leverage data mining tools to determine buying patterns from customers while seeking fraudulent financial transactions and predicting market trends. Data mining processes enormous data collections to extract beneficial knowledge from unorganized information. Through data mining companies extract hidden patterns from their information to create better decisions and forecast results as well as refine operational procedures.

We currently possess extensive data collections that include and surpass social media content and customer transaction and medical documentation. Through data mining organizations gain comprehension of vast databases to create valuable decisions from the collected data. The past buying patterns of customers enable retailers to forecast the future best-selling products. Purpose of This Chapter: The purpose of this chapter is to introduce basic data mining principles followed by an ex-planation of data mining functionality alongside its technical approaches and multiple real-world applications. Additionally, this chapter examines data mining obstacles as well as its projected path.

1.1 Data Driven Decision Making

The former method of company and organizational decision-making relied heavily on raw intuition together with restricted evidence and instinctive reactions. Such approaches create the risk of both substandard and slanted decision out-comes. Data mining methods allow organizations to produce decisions that rely on objective data analysis and statistical processing thereby increasing their accuracy level. Businesses operating in retail use past purchasing data through data mining techniques to predict future product demand instead of depending on guesswork for product predictions. The optimization of sales operates through effective inventory management which enables the organization to place merchandise in the proper quantity at proper moments.

1.2 Identifying Hidden Patterns & Trends

Data mining technology discovers relationships or patterns between multiple dataset components that traditional human examination would fail to identify. The analysis provides organizations with critical insights that identify fresh business prospects and discover conceal.

Data mining technologies identify patterns which show that customers who acquire laptops will probably buy laptop bags and headphones as matching accessories. The acquired business intelligence helps companies generate extra sales through cross-selling solutions which ultimately enlarges their revenue streams.

1.3 Predictive Analysis and Forecasting

Data mining achieves maximum power through its capability to forecast upcoming market behaviors. Organizations and businesses achieve accurate future predictions through their analysis of historical data records. These predictive models enable business organizations to take active instead of delayed decisions.

Data mining tools that analyze financial trends serve two purposes in the field of finance: stock market trend prediction and credit default risk assessment. Previous financial behavior analysis provides businesses insight into risks which helps them develop strategies for risk reduction.

1.4 Improved Customer Segmentation

Data mining enables organizations to divide their customer database into separate groups through behavioral and attribute analysis. Businesses use structured customer segments to provide specific marketing strategies matching personalized customer recommendations and customized services based on individual customer needs. Through data mining techniques the retailer creates three separate customer groups according to their shopping behavior patterns between frequent buyers and occasional shoppers and window shoppers. The tailored marketing information brings about increased customer commitment by speaking directly to each distinct customer segment. The sales figures show growth as a byproduct.

1.5 Fraud Detection and Risk Management

Data mining solutions in banking insurance and healthcare sectors identify non-typical behavioral patterns of customers to discover fraudulent activities while surpassing ordinary conduct. Early fraud detection allows businesses to protect themselves from major financial losses as well as prevent the destruction of their good reputation. Data mining algorithms help the analysis group at credit card companies identify suspicious activities through tracking customer transactions. During unexpected patterns of customer purchasing behavior such as big foreign transactions the data mining alert system detects suspicious behavior before it transforms into fraudulent activity.

1.6 Optimization of Operations

The utilization of data mining helps businesses to develop operational procedures to diminish waste fundamentals while optimizing operational results. Businesses attaining process optimization together with cost reduction benefit from operational bottleneck detection that exposes inefficiencies. The analysis of production data through data mining reveals manufacturers the optimal operational points which arise from perfect material-to-machine-worker arrangements. Organizations gain expanded operational productivity because these protocols create operational disruptions that remain shorter.

1.7 *Improved Healthcare Outcomes*

Healthcare operators use data mining methods to examine patient records to help medical professionals make better medical assessments and identify diseases before treatment selection. Healthcare providers apply data mining as an approach to deliver specialized patient healthcare solutions resulting in better patient outcomes and reduced healthcare expenditures. Medical records go through data mining processes that merge DNA profiles from patients with data about their personal behaviors together with healthcare history. Hospital doctors apply this technique to perform early disease prevention work for individualized medical care development of patient treatments.

1.8 *Enhancing Customer Experience*

Computer systems collect data to deliver superior customer satisfaction through personalized content recommendations that help companies forecast requirements and improve their service methods. Organizations gain better decision-making abilities regarding customer preferences which helps them build stronger relationships resulting in increased customer satisfaction. Film and TV program and music recommendations on streaming platforms Spotify and Netflix use the data mining application to analyze previously recorded user preferences. The platform maintains client engagement through personalized services which leads users to stay active on the platform.

1.9 *Handling Big Data*

Today's technology produces enormous quantities of daily information because the internet usage continues to expand. The current methods of data analysis fail to manage and process data amounts that fall under the "Big Data" category. Since it handles large-scale data effectively data mining provides essential management solutions for the technological areas of social media and telecommunications and e-commerce. Facebook and Twitter together with other social media platforms utilize data mining procedures to manage millions of user-created contents in order to identify sentiments and detect trends for targeted advertising purposes.

Authors have presented complete information about data preprocessing along with classification and clustering and association rule mining and pattern evaluation. They have delivered essential basic concepts together with highly developed techniques to provide students with foundational knowledge about extracting insights from big datasets [1]. Authors have done review of data mining techniques with its applications in educational data mining, finance, medical etc. [2]. Authors have discussed on data mining with different applications of trajectory data mining like location determination, movement behavior etc. [3]. By using different data mining techniques authors have improved educational filed with providing advanced skill towards developing new education scenario [4]. Sometimes there may be problem of fraud detection for saving data so enhanced version of data mining techniques has been approved so we can protect our data [5]. Authors have focused on diabetes prediction and its analysis using data mining techniques which help to predict future diabetes problem in humans [6]. Authors have focused on implementation of hybrid-based model which can perform on real time data mining system especially for tunnel boring machines [7]. Data mining techniques can become more efficient if any artificial intelligence technique has inserted so authors have inbuilt this for safety purpose [8]. Similarly for projects inspection data mining has played vital for determining the difference between types of defect and quality inspection grades [9]. In medical field authors have develop data mining techniques for breast cancer prediction and its analysis by preferring machine learning [10].

Aspect of data mining in renewable energy

Preferring data mining in the field of renewable energy is very important as user is not be able to collect data properly and to provide clear prediction of extracted electricity. There are many aspects such as follows:

A. **Load forecasting:** Estimating clearly future energy demand to optimize electricity production and grid functioning is very much important.
B. **Resource Allocation:** Data mining helps in proper allocation of suitable energy source at appropriate location so it can operate properly and can provide improved response.
C. **Predictive Maintenance:** Data mining helps to monitor each equipment for successful operation of renewable energy sources and can predict suitable maintenance cost.
D. **Land Utilize Optimization:** It is very important aspect from the point of renewable energy source because it is important to establish any renewable energy plant at suitable availability of land.

2 Background and History of Data Mining

2.1 Early Data Analysis

Data mining existed as a practice before humans exploited data through simple methods such as trend analysis and calculating averages with computers. The basic methods initially demonstrated success with smaller databases before data size increased so investigations became harder to conduct. Better computing power along with rising digital data during the 1990s created a need for improved advanced data analysis methods that led to the start of data mining. Data science progressed through the emergence of the data mining term that spurred researchers to create special techniques for producing valuable patterns from large datasets.

2.2 Development of Data Mining Techniques

The data mining method connects fundamental understanding between statistics and computer data learning (machine learning) while utilizing database system principles. A system capable of automatic data analysis must be built for handling large volumes of information while eliminating physical data handling processes.

2.3 How Technology Changed Data Mining

Data mining evolved through the technological advancements that occurred in the field. Through cloud computing and internet connectivity organizations and businesses gained the ability to process large data collections. The conversion of large information quantities into workable information relies on data mining as an essential operational component.

3 Key Concepts in Data Mining

3.1 Data Preprocessing (Cleaning and Pre-Paring Data)

Information analysis begins only after data preparation ensures its quality. Data professionals' clean data and remove errors before transforming the data through processes that allow for analysis. You should use trends and averages to estimate empty data points if you encounter missing information.

3.2 Data Cleaning

Cleaning data requires removing any mistakes that exist in the information. You should remove duplicated records of the same person because the confusion in analysis will diminish when you eliminate such duplicates. The process of read-ying your data for analysis resembles preparing information that is neat and precise.

3.3 Data Transformation

To simplify data analysis some information requires trans-formation into alternative data formats. The conversion of date data into standardized formats and the conversion of textual information into numeric values belongs to this process.

3.4 Data Visualization

The process of data visualization entails turning information into visual graphical elements such as charts or graphs that show patterns to the audience. A timeline depiction of sales growth allows viewers to identify patterns by simply looking at the presentation.

3.5 Association Rule Mining

The discovery of connections between various items of in-formation represents the core concept of this method. One typical case arises when customers who choose specific chips tend to purchase accompanying soda products. The identification of data relationships through data mining provides businesses with marketing and sales enhancement opportunities.

3.6 Classification of Clustering

The process of classification predicts which category or la-bel matches a certain item. The task involves determining whether an incoming email belongs to the legitimate mail category or functions as spam. Data classification utilizes boxes as containers for items through their distinctive features much like how email subject and sender information is utilized. When it comes to clustering systems generate groups that contain similar components. The application of customer groupings follows shopping behavior patterns that divide buyers between electronic products and apparel consumers.

4 Techniques and Algorithms in Data Mining

Data mining functions through multiple algorithms which implement step-by-step methods to study data.

4.1 Some Important Ones Include

Decision Trees operate through flowchart diagrams to per-form data-based decisions. The network structures work similarly to brain operation because they extract patterns from data to predict outcomes. Support Vector Machines (SVM) functions as a data classification technique that divides information across different groups. k-Nearest Neighbors (k-NN) recognizes new data points through their proximity to already classified items. Priory together with FP-Growth represent algorithms that discover relationships between market items. The predictive method of Regression Analysis lets users make future value predictions from historical data records. Based on different topologies of authors have proposed physical statical approach for data mining in wind turbine so that it can extract maximum power and they have done forecasting in some areas of Germany [11].

Similarly, authors have performed by preferring wind turbine for analyzing ramps in electric grid based on three different resolution level [12]. Authors have also suggested another technique Previento forecast approach for understanding different aspects of wind turbine like land, height etc. [13, 14].

5 Applications of Data Mining

Data mining serves numerous industries for their operations. Some examples include: Business organizations leverage data mining to learn about buyer preferences for enhancing their marketing campaigns. Medical practitioners depend on data mining to identify disease spread patterns while seeking appropriate treatments for patients. Through data mining social media platforms evaluate user behaviors as well as emotions (such as product sentiment analysis from user posts). Data mining allows banks together with insurance providers to detect fraud through identification of abnormal patterns in financial transactions. Research Scientists depend on data mining methods to analyze big datasets within scientific domains which include genomics and climate science among others and physics. The business sector alongside marketing operations uses data mining to analyze customer conduct together with market group formation and individualized marketing campaign development. Business organizations research customer buying habits to create product recommendations while improving their sales performance. Datamining techniques deliver financial and banking organizations capabilities for fraud prevention while providing credit risk evaluation and operational risk control. Financial institutions examine purchasing behavior to discover suspicious transactions and assess the lending risks of their business and customer prospects. The use of data mining in manufacturing enables process performance optimization then leads producers to reduce their operational expenses and boost manufacturing efficiency. Data mining supports scientific research through analyzing enormous datasets to discover patterns which results in making groundbreaking scientific discoveries in genomics and heat-related scientific fields and other areas of physics. Apart from this some authors have suggested privacy of data for renewable energy source as data security is very much important. For these operations some smart techniques have been developed like big data, machine learning, artificial intelligence etc. so we can save our data from outsiders. Authors have familiarized with Big Data as an emerging technique in terms of privacy preserving data mining model through which users can secure their data [15]. From the point of large-scale integration of renewable energy sources, authors have demonstrated about reinforcement model which performs on Deep Q-Network technique hence it is easy to secure data and can provide forecasting of renewable energy source [16].

6 Challenges in Data Mining

The method to find valuable patterns and knowledge within big data collections is known as data mining. The data mining process means searching through extensive information for concealed valuable ideas that support wiser decisions. Through various statistical and machine learning and data-base system algorithms Data Mining extracts patterns along with trends and useful information from extensive datasets. The method resembles searching through vast amounts of information to obtain beneficial insights which guide business choices. As a powerful information analytics tool data mining encounters multiple hurdles during its operation. The analysis becomes complicated by data quality issues because the information may be untidy or lacking essential pieces. Businesses must handle personal data with responsibility because privacy issues exist. Companies face questions regarding their methods of handling personal client information. The analysis of vast growing data presents significant scalability problems to researchers. Better algorithms need to run on advanced computer systems to achieve successful analysis. Acquiring clear understanding of data mining results becomes complicated because many methods produce outputs that are difficult to interpret. Business analysts require the ability to explain the methods used in making decisions. Data mining has the possibility to be misused for negative purposes such as advertising harmful content toward vulnerable populations. So, ethical guidelines are needed. For enhancing such challenges, the optimization is very important hence some authors have suggested smart optimization techniques. Regarding this, authors have developed cost effective based energy management model under MATLAB environment which can provide solution of data mining by framing polynomial function so it can provide smart grid solutions [17]. Similarly, authors have done review study of data mining techniques for the prediction of energy demand in smart grid by enhancing neural network, decision tree, vector support machine etc. that these can develop data mining model as well as can predict energy consumption data [18]. For better prediction of data mining authors have developed suitable framework based on different performance aspects like Principal Component Analysis Confirmatory Factor and cluster analysis so it can help through large loading on performing variables [19]. Hence predict and optimize framework can help only in data mining in any field. For renewable energy sources, some authors have developed stochastic optimization approaches that can help in forecast uncertainty [20].

7 Future Trends in Data Mining

The upcoming years in data mining show great promise. Some trends to look out for: Deep Learning technology along with AI will power up data mining by enabling computers to perform more advanced data learning since these tools will become smarter. The massive escalation of data volume will make possible through cloud computing to help businesses handle large-scale data processing and storage requirements. The field of data mining will transition toward performing instantaneous analysis for businesses to execute quick decisions on current information. The accuracy of data mining will increase via new algorithm development which also will boost operational speed thus making data mining more beneficial for users.

8 Conclusion

The valuable tool known as data mining transforms vast amounts of data to generate understandable insights. The technology holds unlimited prospects in various industrial domains and its value will surge due to growing data volumes. The complete achievement of data mining potential requires the solution of challenges that include data quality issues along with privacy and ethical concerns. The upcoming years hold optimistic prospects because better tools and techniques will continue to advance.

References

1. Jiwaie, H., Kamber, M., & Jian, P. (2012). Data mining: Concepts and techniques. The Morgan Kaufmann Series in Data Management Systems. 978–0–12–381479–1
2. H. A. M. (2017). Data mining techniques and applications—a decade review. In *2017 23rd International Conference on Automation and Computing (ICAC)* (pp. 1–7). IEEE
3. Feng, Z., Zhu, Y. (2016). A survey on trajectory data mining: Techniques and applications. *IEEE Access, 4*, 2056–2067.
4. Romero, C., & Ventura, S. (2010). Educational data mining: A review of the state of the art. *IEEE Trans Syst Man, Cybern Part C, 40*(6), 601–618.
5. Wang, S. (2010). A comprehensive survey of data mining-based accounting-fraud detection research. In *2010 Int Conf Intell Comput Technol Autom. ICICTA* (Vol. 2, No. 1, pp. 50–53)
6. Khan, F. A., Zeb, K., Al-Rakhami, M., Derhab, A., & Bukhari, S. A. C. (2021). Detection and prediction of diabetes using data mining: A comprehensive review. *IEEE Access, 9*, 43711–43735.

7. Leng, S., Lin, J. R., Hu, Z. Z., & Shen, X. (2020). A hybrid data mining method for tunnel engineering based on real-time monitoring data from tunnel boring machines. *IEEE Access, 8*, 90430–90449.
8. Ayhan, B. U., & Tokdemir, O. B. (2019). Safety assessment in megaprojects using artificial intelligence. *Safety Science, 118*, 273–287.
9. Lin, C.-L., & Fan, C.-L. (2018). Examining association between construction inspection grades and critical defects using data mining and fuzzy logic. *Journal of Civil Engineering and Management, 24*(4), 301–315.
10. Fatima, N., Li, L., Hong, S., & Ahmed, H. (2020) Prediction of breast cancer, comparative review of machine learning techniques, and their analysis. *IEEE Access, 8*, 150360–1503876
11. von Bremen, L., Saleck, N., & Heinemann, D. (2007). Enhanced regional forecasting considering single wind farm distribution for upscaling. *Journal of Physics: Conference Series, 75*(1), 1–9.
12. Drew, D. R., Cannon, D. J., Barlow, J. F., Coker, P. J., & Frame, T. H. (2017). The importance of forecasting regional wind power ramping: A case study for the UK. *Renewable Energy, 114*, 1201–1208.
13. Ulrich, F., Lange, M., & Waldl, H. P. (2001). Previento—A wind power prediction system with an innovative upscaling algorithm. In *Proc Eur Wind Energy Conf, Copenhagen, Denmark* (Vol. 276, pp. 1–4)
14. Matthias, L., Focken, U., Heinemann, D. (2002). Regional wind power prediction with risk control. In *Proc World Wind Energy Conf* (pp. 1–4)
15. Josphineleela, R., Kaliappan, S., Natrayan, L., & Garg, A. (2023). Big data security through Privacy—Preserving Data Mining (PPDM): A decentralization approach. In *2023 Second International Conference on Electronics and Renewable Systems (ICEARS), Tuticorin, India* (pp. 718–721)
16. Chen, T., Dai, Z., Chen, J., Hu, C., & Xue, Y. (2024). Intelligent identification method for device action frequency based on data mining. In *2024 6th International Conference on Frontier Technologies of Information and Computer (ICFTIC), Qingdao, China* (pp. 583–586).
17. Parvathareddy, S., Yahya, A., Amuhaya, L. L., Samikannu, R. (2025). Smart grid solutions for sustainable mining: A cost-effective energy management framework with optimization algorithms. In *2025 6th International Conference on Mobile Computing and Sustainable Informatics (ICMCSI), Goathgaun, Nepal* (pp. 420–426)
18. Soni, P., & Tandan, G. (2025). Data mining techniques for predicting energy demand in smart grids: Optimizing consumption and production. In *2025 International Conference on Automation and Computation (AUTOCOM), Dehradun, India* (pp. 1185–1189).
19. Nica, I., Oprea, S.-V., & Bâra, A. (2025). Measuring public opinions on renewable energy and climate change using structural equations modeling. *IEEE Access, 13*, 76981–77000.
20. Bergmeir, C., de Nijs, F., Genov, E., Sriramulu, A., Abolghasemi, M., & Bean, R. (2025). Centre for Energy Data Innovation, School of Information Technology and Electrical Engineering, The university of Queensland, Australia: Predict+Optimize Problem in Renewable Energy Scheduling. *IEEE Access, 13*, 60064–60087.

Development of Controllers for Dynamic Energy Systems

S. Swapna Kumar and K. Satyanarayan Reddy

Abstract Dynamic energy systems have trouble ensuring that security and performance are good because of uncertainty, nonlinear behavior, and changing operating conditions. Adding renewable energy sources requires advanced management strategies that can change as demand changes. This chapter examines how fuzzy logic techniques can be used to create strong and flexible controllers for dynamic energy systems. Fuzzy Logic provides a useful way to deal with uncertainty and imprecision, making it easier to make the right choice when things are changing. This chapter gives a full overview of how to design, build, and test fuzzy logic-based controllers. It also discusses how they can help improve power optimization, control, and reliability by using renewable energy sources. Fuzzy logic works well in real-life case studies and experiments that show how it can provide stable and efficient power control. The chapter also discusses how fuzzy logic could help improve energy systems by linking theory to practice.

Keywords Adaptive controllers · Dynamic energy systems · Fuzzy logic · Microgrids · Power optimization · Renewable energy

S. S. Kumar (✉)
Department of Electronics and Communication Engineering, Vidya Academy of Science and Technology, Thalakkottukara, Thrissur, Kerala, India
e-mail: drsswapnakumar@gmail.com; swapnakumar.s@vidyaacademy.ac.in

Department of Electronics and Communication Engineering, Srinivas University, Mukka, Surathkal, Mangalore, Karnataka, India

K. S. Reddy
Srinivas University, Mukka, Surathkal, Mangalore, Karnataka, India
e-mail: vicechancellor@srinivasuniversity.edu.in

M. A. Mellal et al. (eds.), *Computational Intelligence for a Greener Future: Innovations in Renewable Energy Systems*, Studies in Computational Intelligence 1240,
https://doi.org/10.1007/978-3-032-06732-6_24

1 Introduction

Dynamic energy systems require more complex controllers to maintain stability, effectiveness, and dependability. Nonlinear characteristics, ambiguity, and shifting operating conditions are common problems for traditional control techniques. System dynamics become more complex with the addition of renewable energy sources, necessitating intelligent and flexible control methods.

Fuzzy Logic (FL) has emerged as a key method for addressing these issues. It is a versatile and effective method for dealing with imprecision and uncertainty in control activities. Fuzzy Logic-based Controllers (FLCs) use heuristic knowledge and linguistic norms to produce fast control decisions based on exact mathematical models. This approach is especially useful for dynamic energy systems, where operational parameters are constantly changing.

Numerous research have shown that fuzzy logic is beneficial in energy system management. Altaş [1] demonstrated the flexibility and accuracy of fuzzy logic control by highlighting its application in a number of energy management scenarios. Additionally, research by Alsammak and Mohammed [2] found that FLCs play an important role in increasing power quality using Unified Power Flow Controllers (UPFC). Furthermore approaches for distributed power flow controllers using fuzzy logic concepts examined by Duvvuru et al. [3]. These studies collectively highlight the flexibility and effectiveness of fuzzy logic in dynamic energy systems.

Recent years have seen a further expansion of fuzzy logic in energy management. By creating a Dynamic Fuzzy Logic Energy Management System for a Multi-Energy Microgrid, Horrillo-Quintero et al. [4] illustrated the benefits of using FL to handle the complexities of contemporary energy systems. Similarly, Albasheri et al. [5] presented a supervisory energy management strategy for DC microgrids based on FL, highlighting its significance in distributing power among sources and extending the life of energy storage devices.

Figure 1 focuses on controlling load balancing, renewable energy integration, real-time decision-making, and system stability for dynamic energy systems. The controller uses intelligent and adaptive control algorithms while continually monitoring inputs such as energy demand, generation variability, and system restrictions to ensure efficient energy management, optimize performance, and increase dependability.

This chapter examines the design, implementation, and performance evaluation of fuzzy logic-based controllers for dynamic energy systems. By tying theoretical ideas to real-world applications, this chapter shows how fuzzy logic can be used to strengthen power management systems, increase their dependability, and aid in the integration of renewable energy sources. Fuzzy logic-based control mechanisms can greatly increase the stability and efficacy of energy systems, as demonstrated by case studies and statistical analyses.

The literature review highlighting significant research on dynamic energy systems is shown in Table 1:

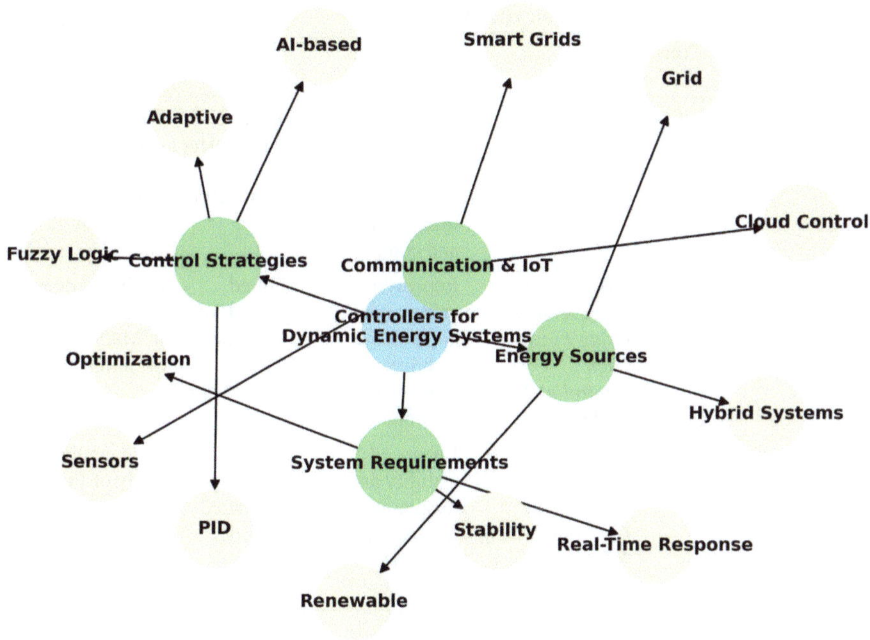

Fig. 1 Author-created representation of controller for dynamic energy systems

This table encapsulates the diverse applications and advancements in fuzzy logic-based controllers across multiple domains within dynamic energy systems.

Key Insights from the Literature

i. Fuzzy Logic domain: FL controllers are used in energy domain, such as smart grids, microgrids, renewable energy integration, and hybrid vehicles.
ii. Hybrid Approaches: Many studies have integrated FL with optimization algorithms (e.g., genetic algorithms, neural networks) for enhanced performance.
iii. Practical Implementations: Research demonstrates FL models in MATLAB/Simulink, real-time testing, and IoT-enabled systems.
iv. Performance Advantages: FL-based controllers outperform traditional control techniques in terms of power quality improvement, stability, efficiency, and adaptability.

The remainder of the chapter is structured as follows: Sect. 2 outlines the features of the dynamic energy systems. Section 3 addresses issues in the development of controllers, Sect. 4 offers insights into Fuzzy Logic: An Overview, Sect. 5 demonstrates the creation of Fuzzy Logic-based controllers, Sect. 6 emphasizes the uses in Dynamic Energy Systems, Sect. 7 presents the Case studies and Simulation outcomes, Sect. 8 details the Performance Evaluation, Lastly, Sect. 9 underscores emerging trends and future research paths in the field of energy system control. This succeeded by the conclusion of the chapter.

Table 1 Literature survey for dynamic energy systems

Author(s)	Year	Title	Key findings	Application area	Methodology used
İ. H. Altaş	2024	*Fuzzy Logic Control in Energy Systems*	Demonstrated the efficiency of fuzzy logic in energy management and control	Energy management & optimization	FL-based energy controllers, simulation & real-time validation
A. N. Alsammak et al.	2021	*Power Quality Improvement Using Fuzzy Logic Controller-Based UPFC*	Improved energy quality and voltage stability in transmission networks	Unified Power Flow Controller (UPFC)	FL-based control applied to UPFC in MATLAB/Simulink
R. N. Duvvuru et al.	2019	*Performance of Distributed Power Flow Controller Using Fuzzy Logic*	Enhanced stability and decreased transmission losses in distributed networks	Power flow control in distributed networks	Comparative study of FL vs. conventional controllers
Horrillo-Quintero et al.	2024	*Dynamic Fuzzy Logic Energy Management for Multi-Energy Microgrid*	Designed an adaptive FL-based EMS to regulate multi-energy microgrids amid variable demand	Microgrid energy management	Hybrid FL and optimization algorithms for energy dispatch
Albasheri et al.	2025	*Supervisory FL-Based Energy Management in DC Microgrids*	Proposed a FL-based system for balancing power among sources and extending battery lifespan	DC microgrids & energy storage	FL-based supervisory control, experimental validation
M. A. Hannan et al.	2019	*Energy Management for Hybrid Electric Vehicles Using Fuzzy Logic*	Fuzzy controllers for longer battery life and increased energy efficiency in HEVs	Hybrid Electric Vehicle (HEV) management	FL-based decision-making for energy storage system

(continued)

Table 1 (continued)

Author(s)	Year	Title	Key findings	Application area	Methodology used
J. He et al.	2015	*Islanding Microgrid Power Sharing with Fuzzy Logic*	Improved reactive power and unbalanced power sharing for islanded microgrids	Smart grid & microgrid integration	FL-based droop control strategy for microgrids
S. K. Kollimalla et al.	2014	*Adaptive P&O MPPT Algorithm with Fuzzy Logic*	Proposed a FL-enhanced MPPT algorithm for handling varying solar irradiance	Solar PV systems	Hybrid FL and perturb & observe (P&O) MPPT approach
R. Priyadarshini et al.	2022	*Fuzzy Logic Controller for Renewable Hybrid Microgrids*	Demonstrated FL-based controller optimizing energy distribution in hybrid systems	Renewable energy integration	FL model implemented in MATLAB with real-time scenarios
E. Kabir	2020	*Fuzzy Logic for Wind Energy Conversion Systems*	Boosted power extraction efficiency in wind energy systems using FL-based controllers	Wind energy optimization	Simulation-based analysis comparing FL with conventional PI controllers
F. Harun et al.	2021	*Smart Grid Fuzzy Logic Controller Based on IoT*	Merged IoT data with FL controllers for real-time energy optimization	IoT-enabled smart grids	IoT-FL hybrid approach with real-time cloud-based analysis

2 Characteristics of the Dynamic Energy Systems

Dynamic energy systems represent intricate networks that oversee the production, distribution, and use of energy. Here is a conceptual diagram in Fig. 2, depicting Characteristics of Dynamic Energy Systems. It highlights key elements that features like variability, uncertainty, nonlinearity, multi-source integration, adaptive control requirements, and real-time optimization, all of which are vital for effective energy management.

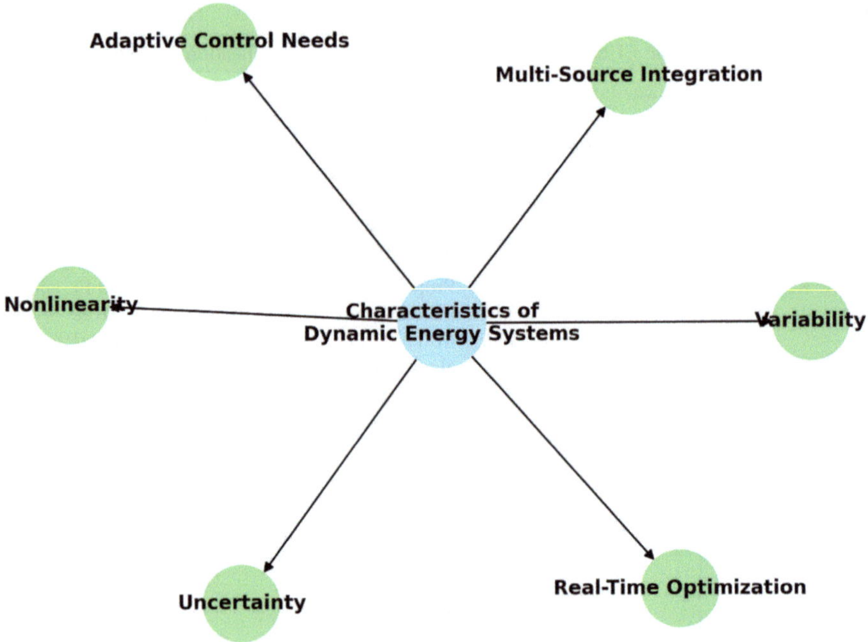

Fig. 2 Author-created diagram of characteristics of dynamic energy systems

Their characteristics offer particular difficulties as well as chances for management and improvement. According to the chapter's title, the following is a synopsis of the main traits of dynamic energy systems:

Key Characteristics of Dynamic Energy Systems

i. **Nonlinearity**

- In the dynamic energy systems nonlinear behaviors, particularly when integrating renewable energy sources like solar and wind during transient states.
- The output power of wind turbines, varies nonlinearly with wind speed Altaş [1].
- Impact: To handle complexity in system modeling and operation, sophisticated control strategies are needed.

ii. **Uncertainty**

- Unpredictability is introduced by variations in weather, demand trends, and equipment performance.
- As an illustration, wind and cloud cover variations affect the production of renewable energy Albasheri et al. [5].
- Impact: Requires adaptive controllers in order to guarantee system dependability and stability.

iii. **Time-Shift Conditions**

- Daily and seasonal cycles cause fluctuations in energy demand and resource availability over time.
- For instance, solar energy is available during the day and evenings will have the demand highest.
- Impact: Needs real-time control to dynamically balance supply and demand.

iv. **Integration of Renewable Energy Sources**

- The energy sources that is environment renewable like solar and wind add unpredictability.
- Problem: Renewable energy sources cannot be sent out when needed, in contrast to conventional power plants.
- Impact: To keep the grid stable, energy systems need to be adaptable and strong.

v. **High Dimensionality**

- Various interrelated parts with intricate interactions make up dynamic energy systems.
- Difficulty: Complex modeling and control methods are needed for large systems.
- Impact: For effective operation, sophisticated computational methods like AI-based control are crucial.

The main difficulties and specifications for creating controllers for dynamic energy systems are encapsulated in this synopsis. Additionally, Table 2 provides a statistical table that summarizes the characteristics of the dynamic energy systems along with pertinent descriptions and their effects.

Comprehending these attributes is essential for creating efficient controllers for dynamic energy systems. Fuzzy logic-based controllers and other advanced control strategies are well-suited to handle these difficulties because of their capacity to manage the complexity, nonlinearity, and uncertainty present in dynamic energy systems.

3 Challenges in Controller Development

The development of controllers for dynamic energy systems necessitates addressing critical issues such as robustness. This guarantees stable performance under uncertainties; scalability, that enables controllers to handle system complexity. Adaptability mechanisms, scalable systems, and advanced control approaches are required to get over these challenges and provide reliable and energy management.

i. **Robustness**: Despite unanticipated situations like variations in load, renewable energy, and component failures, the controllers must continue to operate steadily.

Table 2 Characteristics of dynamic energy systems

Characteristic	Description	Impact
Variability	Energy generation and demand fluctuate over time	Requires adaptive control to handle fluctuations efficiently
Uncertainty	Unpredictable factors like weather conditions affect energy supply	Necessitates robust forecasting and decision-making methods
Nonlinearity	System behavior is highly nonlinear due to complex interactions	Demands advanced control strategies like AI or fuzzy logic
Multi-source integration	Integration different energy sources (grid, solar, wind, etc.)	Coordination of management is required for the best possible power distribution
Adaptive control needs	System must dynamically adjust control strategies	Ensures real-time adjustments for stability and efficiency
Real-time optimization	Continuous monitoring and optimization of energy flows	Energy utilization rise, reduces waste, and enhances reliability

ii. **Scalability**: Effectively controllers must be able to effectively manage the growing complexity and data without experiencing performance deterioration as the energy systems grow.
iii. **Adaptability**: To maintain peak performance, controllers should dynamically adapt to shifting operating conditions, such as fluctuating energy generation and demand.
iv. **Efficiency**: To allow for real-time decision-making in complex systems, controllers must maximize performance while reducing computational overhead.

Table 3 lists the various difficulties in creating controllers for dynamic energy systems that must be overcome to guarantee dependable and efficient operation:

A comprehensive strategy combining sophisticated control theories, scalable architectures, adaptive mechanisms, and effective computational algorithms is needed to address these issues. Controllers can efficiently handle the intricacies in dynamic energy systems by overcoming these challenges, guaranteeing stability, dependability, and peak performance.

4 Fuzzy Logic: An Overview

A method of thinking called Fuzzy logic (FL) mimics human reasoning by using contradiction and ambiguity. Since FL operates in degrees of truth and "suspension of belief," it is distinct from classical binary logic, which only operates on true or false values. Therefore, it is perfect for intricate, unpredictable systems. FL framework provides a robust for dealing with the inherent uncertainty and variability

Table 3 Challenges in controller development for dynamic energy systems

Challenge	Description	Impact on the controller design	Proposed solutions
Robustness	Component failures, load variations, and fluctuations in renewable energy, controllers must operate reliably	It is crucial to guarantee performance and stability in various scenarios	Robust and adaptive control strategies to improve resilience [6]
Scalability	Controllers must effectively manage the growing complexity and data loads as energy systems expand	Large-scale systems are difficult for traditional centralized controllers to handle	It is enhanced by the use of decentralized and hierarchical control architectures [7]
Adaptability	Conditions in dynamic energy systems change over time, necessitating real-time controller adjustments	Fixed-parameter controllers might not continue to operate at their best	According to system feedback, adaptive control strategies modify the parameters [8]
Efficiency	Controllers must minimize computational overhead and maximize performance	Real-time decision-making may be delayed by the high computational complexity	Proper algorithms and control strategies achieve a balance between the processing speed and performance [9]

in dynamic energy systems, resulting in improving the control strategies and system performance.

4.1 *Key Components of the Fuzzy Logic*

i. **Fuzzification**: The procedure uses membership functions to convert the exact inputs into fuzzy sets. Fuzzification defines the level at which one input belongs to a given set, allowing the system to handle uncertainty and partial truths.
ii. **Inference Engine**: FL applies inference engine applies rules, typically expressed as IF–THEN, to fuzzy inputs. These rules by simulating, allow the system to infer suitable responses or actions from the input data.
iii. **Defuzzification**: This process turns the fuzzy outputs back into distinct values after the fuzzy inputs have been processed by the inference engine. This stage is essential for producing control signals or decisions that can be put into practice in practical applications.

4.2 *Advantages of Fuzzy Logic in Dynamic Energy Systems*

- **Effective Management of Imprecision and Uncertainty**: Demand and supply changes, particularly when renewable energy sources are used, frequently induced uncertainty in dynamic energy systems. By efficiently managing imprecise information, FL facilitates robust control strategies and increases resilience and adaptability [4].

- **Simplifying Nonlinear System Controller Design**: When energy systems exhibit nonlinear behaviors, conventional control techniques can become complicated. By using linguistic variables and rules, FL streamlines the design process and lessens the need for intricate mathematical models [10].
- **Decreased Reliance on Precise Mathematical Models**: FL controllers are more adaptable and simpler to apply in systems where modeling presents difficulties since they do not need exact mathematical representations of the system. This flexibility is especially useful in dynamic situations and involve changing system parameters over time [5].
- **Enhanced Adaptability and Robustness**: FL systems don't require constant recalibration to maintain performance despite shifting conditions and uncertainties. This ensure the control strategies remain effective even when the system's dynamics shift [11].

In conclusion, fuzzy logic offers a flexible and successful method for handling the intricacies of dynamic energy systems. It is a crucial tool in the creation of sophisticated control strategies for contemporary energy infrastructures because of its capacity to manage uncertainty, streamline controller design, and adjust to shifting circumstances.

5 Design of the Fuzzy Logic-Based Controllers

A methodical approach is required when designing a fuzzy logic-based controller (FLC) for dynamic energy systems in order to guarantee efficient control of system variables and accomplish desired control goals. Determining the controller architecture and adhering to defined development steps are both included in the design process. Below is a mathematical formulation for the fuzzification and defuzzification processes, which are central to FLC design.

Mathematical Statement

The fuzzy logic-based controller (FLC) for a dynamic energy system maps input variables $x = [x_1, x_2, \ldots .x_n]$ (load demand, generation capacity) to output control actions variables $y = [y_1, y, \ldots .y_n]$ (power dispatch, voltage regulation) through the following steps:

i. **Fuzzification**: For each input variable x_i, a membership function $\mu_{A_{ij}}(x_i)$ defines the degree of membership in a fuzzy set A_{ij} where j indexes the fuzzy sets (e.g., "low," "medium," "high"). The membership function is typically defined as:

$$\mu_{A_{ij}}(x_i) : \mathbb{R} \rightarrow [0, 1]$$

The $\mu_{A_{ij}}(x_i)$ is expressed as triangular membership function for x_i is:

$$\mu_{A_{ij}}(x_i) = \begin{cases} 0 & if\ x_i \leq a\ or\ x_i \geq c \\ \frac{x_i - a}{b-a} & if \quad a < x_i \leq b \\ \frac{c - x_i}{c-b} & if \quad a < x_i \leq b \end{cases}$$

where a, b, c are the lower, peak, and upper bounds of the triangular fuzzy set.

ii. **Rule Evaluation**: A rule base consists of $K\ IF - THEN$ rules of the form:

$$Rule\,k\ :\ IF\ x_i\ isA_{ij}\ AND\ x_2\ is\ A_{2j}\ AND\ \ldots\ldots THEN\ y\ is\ B_k$$

The firing strength of rule k is computed using a T-norm (minimum operator):

$$\omega_k = \min(\mu_{A_{ij}}(x_1), \mu_{A_{ij}}(x_2) \ldots\ldots \mu_{A_{ij}}(x_n))$$

iii. **Defuzzification**: The fuzzy $\mu_{B_k}(y)$ is aggregated across all rules, and a crisp output y^* is computed using the centroid method:

$$y^* = \frac{\int y * \mu_B(y)dy}{\int \mu_B(y)dy}$$

where $\mu_B(y) = max_k(\omega_k * \mu_B(y)dy)$ is an aggregated output membership function.

This formulation ensures the FLC can handle uncertainty and nonlinearity in dynamic energy systems by mapping imprecise inputs to precise control actions.

5.1 *Architecture of the Controller*

The following essential elements constitute a typical fuzzy logic controller (FLC) for dynamic energy systems, as shown in Fig. 3.

- **Input Variables**: These represent critical system parameters that influence the controller's decisions. Common input variables include:
  - Load demand
  - Generation capacity
  - Environmental conditions (e.g., temperature, irradiance)
- **Fuzzification Interface**: This module effectively manages imprecision and uncertainty in dynamic energy systems by converting precise input values into fuzzy sets using predefined membership functions.

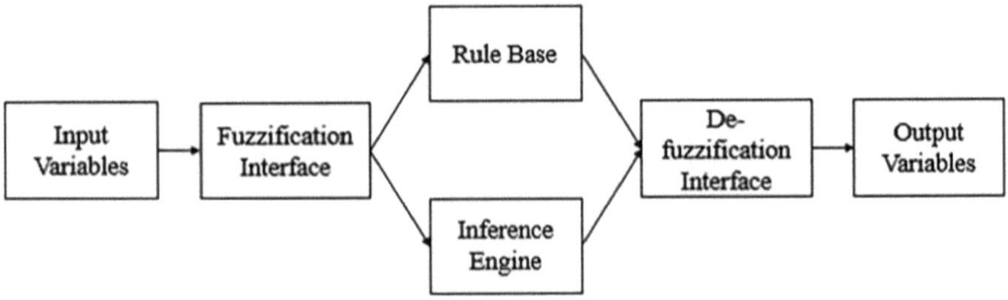

Fig. 3 Author-created block diagram of the fuzzy logic-based controller

- **Rule Base**: A systematic collection of IF–THEN rules developed using system analysis and expert insights that control how the system reacts to various circumstances.
- **Inference Engine**: Determines appropriate control decisions by analyzing the fuzzy input sets and applying the predefined rules to produce the fuzzy output sets.
- **Defuzzification Interface**: Converts the fuzzy output results back into accurate control signals so that the energy system can use them practically.
- **Output Variables**: These are the control actions executed to manage the energy system, such as:
  - Power dispatch
  - Voltage regulation
  - Load adjustment.

The FLC can interpret intricate system dynamics and make well-informed control decisions in real time thanks to this architecture Altas [1].

5.2 *Steps in Controller Development*

Developing an FLC involves several key steps:

i. **Critical System Variables**: Determine the critical input and output variables that significantly impact system performance.
ii. **Define Membership Functions**: Establish suitable membership functions for each input and output variable. These functions determine the extent to which a given input belongs to a fuzzy set, enabling the effective management of uncertainties.
iii. **Develop a Comprehensive Rule Base**: Develop a collection of IF–THEN rules that govern the system's behavior under different conditions. This rule base serves as a repository of expert knowledge and system dynamics.
iv. **Simulate the Controller in a Dynamic System Model**: To assess the FLC's performance, put it into practice in a simulation environment. For this, programs

like MATLAB/Simulink are frequently utilized, providing a platform for modeling and testing the controller in various scenarios Altas [1].

v. **Optimize Parameters for Performance**: To improve performance metrics like accuracy, stability, and response time, modify the rule base, membership functions, and other controller parameters. The controller may be adjusted using optimization techniques.

By taking these measures, engineers can create FLCs that guarantee reliable and flexible control while skillfully handling the intricacies of dynamic energy systems.

6 Applications of Dynamic Energy Systems

For dynamic energy systems to be more reliable and efficient, is a Fuzzy Logic Controllers (FLCs), are essential systems. Such systems applies to a variety of fields, such as demanding forecasting, load balancing, energy storage management, and the integration of renewable energy.

i. **Renewable Energy Integration**

The renewable energy sources variability makes it difficult to keep a stable grid system. FLCs manages the uncertainties associated with renewable energy output. Using FL dynamics in multi-energy microgrids enables the optimal use of renewable energy, thus ensuring a balanced energy supply and reducing dependence on traditional power grids [4].

ii. **Load Balancing**

The reliability of a system largely depends on the effective load distribution to prevent energy waste. Flexible Load Controllers (FLCs) enable load balancing in real-time by automatically adjusting power flows based on shifting system circumstances. This adaptability allows for efficient energy distribution management that accommodates fluctuations in demand and supply [1].

iii. **Energy Storage Management**

A thorough analysis of the complex mechanisms of energy storage systems like batteries is required to enhance the processes for recharging and discharging them. FLCs streamline these cycles by taking into account factors such as current load demands, the battery's charge status, and expected energy requirements. This method guarantees that energy storage systems operate within their most effective parameters, thereby improving both their performance and lifespan [4, 5].

iv. **Demand Forecasting**

The demand estimate are critical for both strategic energy planning and operational efficiency. Both strategic energy planning and operational effectiveness depend on precise demand forecasts. By processing imprecise and variable data, FLCs provide more reliable predictions, enabling better preparation for demand fluctuations [10].

In summary, FLCs offer versatile solutions to the challenges associated with dynamic energy systems. Their capability to handle uncertainty and respond to varying conditions makes them indispensable tools in modern energy management.

7 Case Studies and Simulation Results

The effectiveness of fuzzy logic-based controllers in dynamic energy systems is present in this section through simulated studies and real world experiences.

7.1 *Case Study: The Wind-Solar Hybrid System*

The stability of hybrid energy systems that include wind and solar power depends on controlling the erratic and intermittent nature of these renewable sources, is essential for ensuring system stability. A FLC for a standalone solar/wind/fuel cell hybrid system shown in Fig. 4.

The FLC was designed to manage the power distribution among the different energy sources and the battery bank, ensuring a balanced supply–demand scenario.

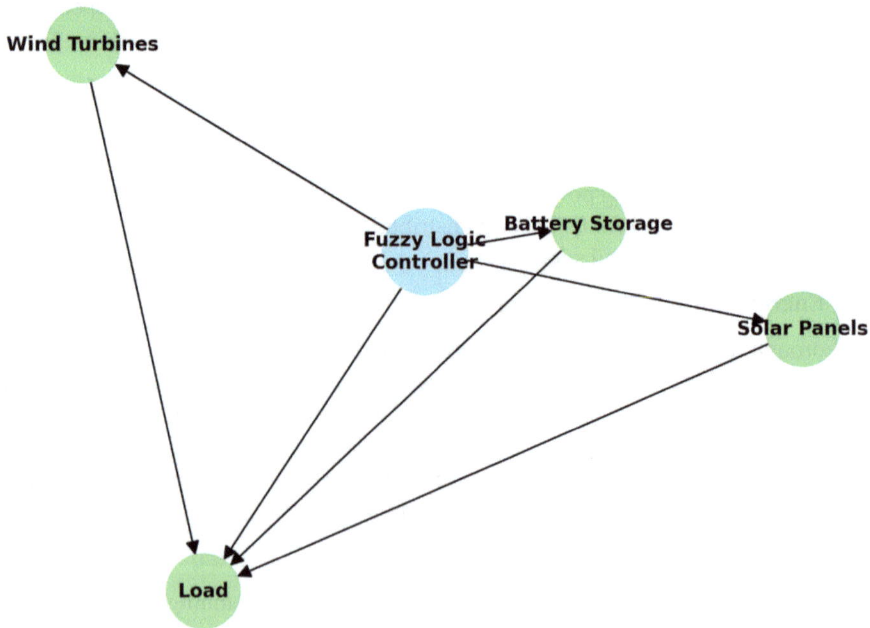

Fig. 4 Author-created schematic of a fuzzy logic-based controller for a wind-solar-fuel cell hybrid system

Simulation results demonstrated that the FLC effectively reduced variations in wind and solar power, thereby improving system reliability and maximizing battery efficiency [12].

In addition, another study focused on implementing a Fuzzy Logic-based Energy Management System (EMS) for both AC and DC microgrids. The EMS employed an FLC to control the power flow between renewable sources, energy storage units, and loads. Simulation results verified that the FLC successfully preserved voltage stability, maximized energy efficiency, and guaranteed a balanced power allocation throughout the system [13].

7.2 Simulation: Microgrid Energy Management

Advanced energy management strategies are required when integrating different renewable energy sources and storage systems into microgrids to ensure dependable and effective operation. Research has explored the use of a Fuzzy Logic-based Energy Management System (EMS) for both AC and DC microgrids. A fuzzy logic-based EMS for AC/DC microgrids, depicted in the author-created Fig. 5, controls power flow. Simulation results verified that such an FLC preserved voltage stability [13].

The case studies as shown in Fig. 5 and simulations demonstrate the effectiveness of Fuzzy Logic-based controllers in addressing the challenges of dynamic energy systems. Renewable energy source fluctuations can efficiently managed by FLCs, which in turn improves the stability, performance, and dependability of both hybrid energy systems and microgrids.

8 Performance Evaluation

Evaluating the performance of controllers based on Fuzzy Logic is essential in determining the effectiveness of such systems in dynamic energy management. The key performance metrics include.

Response Time: This measure on the controller adaptation to modification in the system's parameters. A shorter response time indicates a more agile system capable of promptly addressing fluctuations in energy demand or supply. Studies have shown that Fuzzy Logic controllers generally react faster than conventional PID controllers, leading to improved transient performance, even under time-delay conditions [14].

Stability: The controller's ability to maintain system equilibrium despite disturbances is described as stability. Fuzzy logic controllers have shown improved stability in managing the nonlinear and uncertain behaviors that are characteristic of dynamic energy systems. Fuzzy Logic controllers have surpassed conventional controllers in terms of stability when controlling dynamic systems [15].

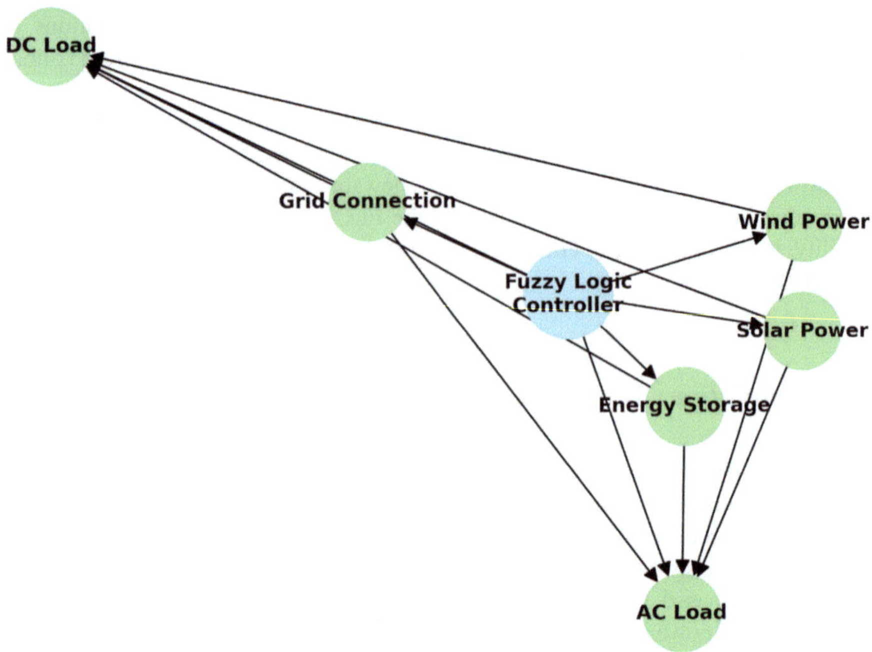

Fig. 5 Author-created schematic of a fuzzy logic-based energy management system for AC/DC microgrids

Efficiency: Efficiency measures the system's capability to minimize energy losses during operation. Fuzzy logic controllers efficiently handle imprecision and uncertainty, leading to optimized energy usage and consequently reducing waste [9].

Scalability: It assesses how the controller's performance when applied to larger or more complex systems. FLC can be designed to be adaptable and flexible, allowing them to manage complex systems without a decrease in performance. This adaptability is crucial for modern energy systems that are expanding in size and intricacy (Fig. 6).

In addition to these metrics, the Global Performance Assessment Index (GPI) employs Fuzzy Logic techniques to assess the overall performance of multiloop control systems, including factors such as energy usage and product quality, in line with Polito et al. [16].

Consistently evaluating these performance metrics allows stakeholders to gain a comprehensive understanding of the effectiveness of FL-based controllers in dynamic energy systems, thereby informing further optimization and implementation initiatives.

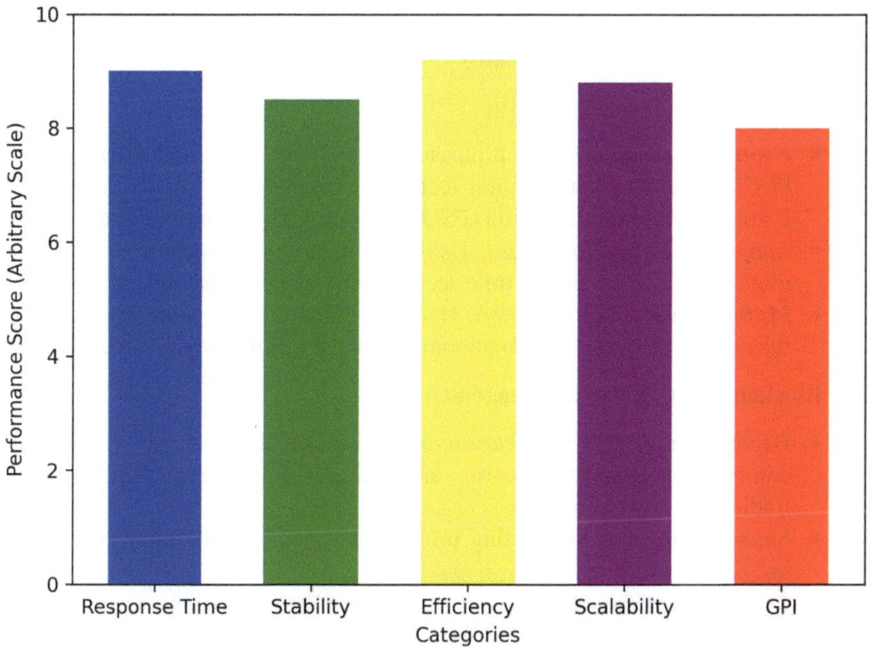

Fig. 6 Author-created diagram of performance metrics for fuzzy logic-based controllers

9 Future Trends and Research Opportunities

i. **Integration with Machine Learning**, [17]

- ***Hybrid Control Systems***: Energy systems' predictive and adaptive control can be improved by combining FL with machine learning (ML) approaches like reinforcement learning (RL) and artificial neural networks (ANNs).
- ***Self-Tuning Fuzzy Controllers***: In real-time applications, the performance of controllers can be enhanced by using machine learning algorithms to dynamically optimize fuzzy logic rules.
- ***Deep Learning Approaches***: To improve decision accuracy, fuzzy logic membership functions are refined using Deep Learning models.

ii. **IoT and Smart Grids**, [18]

- ***Real-Time Decision Making***: To improve power management, FL-based controllers can handle real-time data collecting enabled by the Internet of Things (IoT).
- ***Smart Grid Integration***: In order to achieve effective energy distribution and load balancing, fuzzy logic is essential to decentralized energy management in smart grids.

- ***Cybersecurity and Resilience***: Using fuzzy logic to identify irregularities and improve smart grid security.

iii. **Optimization Techniques**, [19]

- ***Evolutionary Algorithms***: It measures the effectiveness and adaptability of FLC by modern optimization techniques like Genetic Algorithms (GA), Particle Swarm Optimization (PSO), and Ant Colony Optimization (ACO).
- ***Adaptive Control Strategies***: Using optimization algorithms technique to modify fuzzy rules in real time according to system behaviour.
- ***Multi-Objective Optimization***: Handling trade-offs in controllers based on fuzzy logic between computational complexity, stability, and efficiency.

iv. **Blockchain for Energy Management**, [20]

- ***Decentralized Energy Transactions***: Blockchain and fuzzy logic are combined to improve security and transparency in peer-to-peer energy trading.
- ***Smart Contracts***: Automating pricing and energy distribution through the use of fuzzy logic in decision-making.

v. **Fuzzy Logic in Edge Computing**, [21]

- ***Low-Latency Control***: Using fuzzy logic in edge devices to manage energy in real time without depending on centralized servers.
- ***Energy Efficiency in IoT Devices***: Intelligent power management in IoT-based energy systems through the use of fuzzy logic.

The evolving trends underscore the increasing importance of Fuzzy Logic in developing intelligent energy management systems. The incorporation of AI, IoT, and optimization methods is anticipated to improve its adaptability and effectiveness in managing complex energy landscapes.

Figure 7 shows the visual representation showcases the key factors contributing to market growth, including technology integration, industry adoption, competitive differentiation, and regulatory compliance, in a comprehensive manner.

According to research, the need for intelligent automation, the need for sustainability, and technological advancements are the main factors driving the integration of FL-based controllers in energy systems. The high adoption of R&D investment (93%) and renewable energy adoption (91%) suggests that scopes place a high priority on ongoing innovation and green energy integration. There is a move towards data-driven, flexible, and effective energy solutions represented by the significance of AI & ML integration (83%) and smart grid optimization (80%).

The necessity for flexible applications is emphasized by the majority of industry-specific solutions (74%), while regulatory compliance (69%) demonstrates the laws and policies necessary for market acceptance.

The widespread adoption of fuzzy logic controllers in energy systems will depend on technological progress, AI-driven automation, and the incorporation of sustainable energy, necessitating a balance between customization and regulatory requirements.

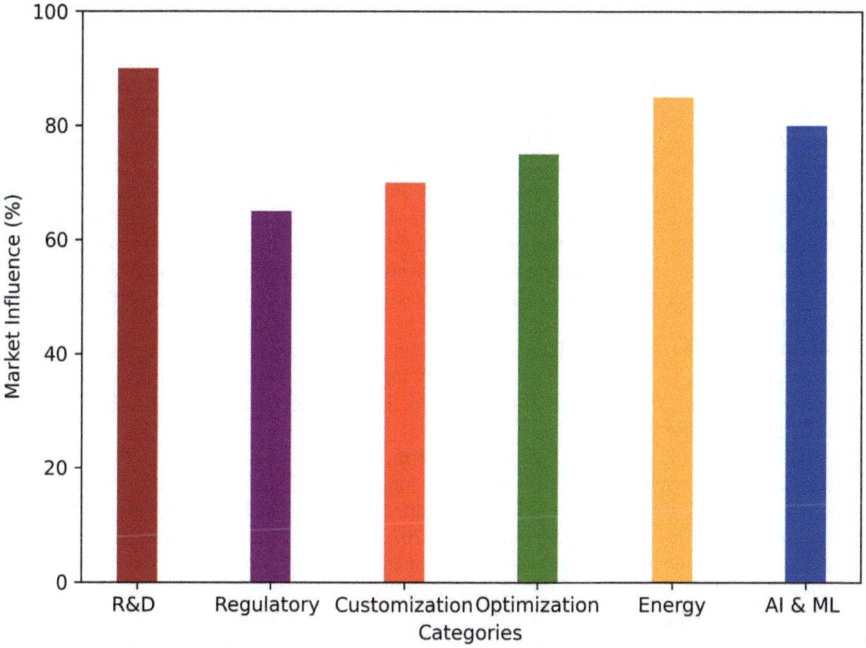

Fig. 7 Author-created diagram of market strategy factors for fuzzy logic-based controllers in energy systems

10 Conclusion

Developing controllers for dynamic energy systems necessitates addressing issues such as uncertainty, nonlinearity, and changing operating conditions. Fuzzy logic-based controllers have demonstrated the ability to provide accurate decision-making capabilities in the presence of complex system interactions and the uncertainty of renewable energy sources. In contrast to traditional control methods, fuzzy logic offers an advanced and adaptable framework that improves energy reliability, load management, and power optimization.

In this chapter, the performance of fuzzy logic-based controllers has been assessed in terms of their ability to optimize energy distribution, stabilize power systems, and enhance operational efficiency. Simulation results of real-world case studies provide validation of their outstanding performance in integrating renewable energy sources, and safe control mechanism in contemporary energy systems.

Advances in ML, the IoT, and evolutionary algorithms will enlarge the potential of these controllers, ultimately closing the gap between theoretical breakthroughs and practical implementations. Logic-based control systems will remain essential for establishing a more robust, effective, and sustainable energy infrastructure, as long as innovation persists.

References

1. Altas, I. H. (2024). *Fuzzy logic control in energy systems: With design examples in MATLAB®/ Simulink®*. The Instituion of Engineering and Technology. ISBN-13: 978–1–83953–882–7. https://doi.org/10.1049/PBPO254E
2. Alsammak, A. N., & Mohammed, H. A. (2021). Power quality improvement using fuzzy logic controller based unified power flow controller. *Indonesian Journal of Electrical Engineering and Computer Science*, *21*(1), 1–9. ISSN: 2502–4752. https://doi.org/10.11591/ijeecs.v21.i1
3. Duvvuru, R., Rajeswaran, N., Sanjeeva Rao, T. (2019). Performance of distributed power flow controller in transmission system based on fuzzy logic controller. *International Journal of Recent Technology and Engineering (IJRTE)*, *8*(3), 2039–2043. https://doi.org/10.35940/ijrte.C4530.098319
4. Horrillo-Quintero, P., García-Triviño, P., Hosseini, E., García-Vázquez, C. A., Sánchez-Sainz, H., Ugalde-Loo, C. E. (2024). Dynamic fuzzy logic energy management system for a multi-energy microgrid. In *IEEE access* (Vol. 12, pp. 93221–93234). https://doi.org/10.1109/ACCESS.2024.3422009
5. Albasheri, M. A., Bouchhida, O., Soufi, Y., Cherifi, A., & Mujammal, M. A. H. (2025). Supervisory energy management strategy based-fuzzy logic for a DC microgrid. *International Journal of Dynamics and Control, Springer*, *13*, 31. https://doi.org/10.1007/s40435-024-01510-8
6. Pinthurat, W., Kongsuk, P., & Marungsri, B. (2023). Robust-adaptive controllers designed for grid-forming converters ensuring various low-inertia microgrid conditions. *Smart Cities, 6*(5), 2944–2959. https://doi.org/10.3390/smartcities6050132
7. Bakare, M. S., Abdulkarim, A., Shuaibu, A. N. (2024). Energy management controllers: strategies, coordination, and applications. *Energy Inform, 7. SpringerOpen,* 57. https://doi.org/10.1186/s42162-024-00357-9
8. Sadamoto, T., Chakrabortty, A., Ishizaki, T., & Imura, J. I. (2019). Dynamic modeling, stability, and control of power systems with distributed energy resources: Handling faults using two control methods in Tandem. In *IEEE control systems magazine* (Vol. 39, No. 2, pp. 34–65). https://doi.org/10.1109/MCS.2018.2888680
9. Zambrana, P., Fernandez-Quijano, J., Mayorga Rubio, P. M., Fernandez-Lozano, J. J., & Garcia-Cerezo, A. (2023). Development and evaluation of fuzzy logic controllers for improving performance of wind turbines on semi-submersible platforms under different wind scenarios. *MDPI, Appl Sci*, *13*(4), 2422. https://doi.org/10.3390/app13042422
10. Zhang, Z. (2024). Energy system optimization based on fuzzy decision support system and unstructured data. *Energy Inform 7, SpringerOpen, 82.* https://doi.org/10.1186/s42162-024-00396-2
11. Mathews, M. A., & Rajeev, T. (2020). Fuzzy based management of hybrid energy storage system for improved dynamic response of DC microgrid. In *2020 IEEE International Conference on Power Electronics, Smart Grid and Renewable Energy (PESGRE2020), Cochin, India* (pp. 1–6). https://doi.org/10.1109/PESGRE45664.2020.9070645
12. Saravanan, S., & Thangavel, S. (2013). Fuzzy logic controller based power management for a standalone solar/wind/fuel cell fed hybrid system. *J Renewable Sustainable Energy, 5*, Article 053147. https://doi.org/10.1063/1.4827315
13. Prathyush, M., & Jasmin, E. A. (2019). Fuzzy logic based energy management system design for AC and DC microgrid. *International Journal of Engineering Research & Technology (IJERT), 8*(10). ISSN: 2278–0181. http://www.ijert.org
14. Omar, F. A. (2021). Performance comparison of PID controller and fuzzy logic controller for water level control with applying time delay. *Konya Journal of Engineering Sciences, 9*(4), 858–871. https://doi.org/10.36306/konjes.976918
15. Iqbal, S., & Ayyub, M. (2018). Improved performance of fuzzy logic controller to control dynamical systems: A comparative study. In *2018 International Conference on Computational and Characterization Techniques in Engineering & Sciences (CCTES), Lucknow, India* (pp. 122–126). https://doi.org/10.1109/CCTES.2018.8674137

16. Polito, C. M., Jota, F. G., Braga, A. R., & Pena, R. T. (1996). Fuzzy logic techniques to evaluate global dynamic performance of control systems. In *Proceedings of the 39th Midwest Symposium on Circuits and Systems, Ames, IA, USA* (Vol. 2, pp. 587–590). IEEE Access, Proceedings of IEEE. https://doi.org/10.1109/MWSCAS.1996.587795
17. Li, F., Gao, L., Zhang, Y., & Liu, Y. (2025). Integrated energy management for hybrid electric vehicles: A Bellman neural network approach. *Engineering Applications of Artificial Intelligence, 145*. ISSN 0952-1976. https://doi.org/10.1016/j.engappai.2025.110166
18. Al-Ali, A. R., & Aburukba, R. (2015). Role of internet of things in the smart grid technology. *Journal of Computer and Communications, 3*, 229–233. https://doi.org/10.4236/jcc.2015.35029
19. Castillo, O., & Melin, P. (2013). Bio-inspired optimization of interval Type-2 fuzzy controllers. In *Soft computing: State of the art theory and novel applications* (pp. 241–254). Springer. https://doi.org/10.1007/978-3-642-34922-5_17
20. Cantillo-Luna, R., Moreno-Chuquen, H. R., Chamorro, V. K., Sood, S., & Badsha, C. K. (2022). Blockchain for distributed energy resources management and integration. *IEEE Access, 10*, 68598–68617. https://doi.org/10.1109/ACCESS.2022.3184704
21. Kolomvatsos, K. (2020). An intelligent scheme for uncertainty management of data synopses management in pervasive computing applications. https://doi.org/10.48550/arXiv.2007.12648

IoT Framework for Data Management Using Computing Technologies

Sujata Bhavikatti and Maya B. Upadhye

Abstract The Internet of Things (IoT) generates large amount of data from networked devices. The IoT framework integrates IoT devices and gateways while leveraging modern computing technologies. It addresses the challenge of interoperability among sensors, edge computing processors, and network devices while ensuring seamless cloud connectivity. The use of standardized communication protocols for secure interactions with the cloud, end-user devices, and authorized personnel is detailed. The framework is illustrated through an air Cleanliness Monitoring Systems, smart city case study, highlighting the role of edge and cloud computing in data management and processing. Additionally, it emphasizes the need for a cost-effective IoT framework that supports environmental sustainability, including applications in renewable energy systems.

Keywords Computing technologies · IoT framework · Data management · Sustainability

1 Introduction

The applications—such as smart traffic management, smart public transportation, and smart management of energy, waste, water, environmental monitoring, smart governance/e-governance, and smartness applied to security and surveillance, buildings and infrastructure, smart healthcare, smart parking, smart education, and disaster management systems—are developed and deployed as integral components of the smart city ecosystem. Among these, certain applications like smart buildings, hospitals, industrial systems, water management, and smart parking typically do not require wide-area information exchange [1–3], as their data flow is confined to localized environments. The smart air monitoring system can be deployed city-wide without geographical limitations, and its data flow occupies a large geographical area.

S. Bhavikatti (✉) · M. B. Upadhye
Department of Electronics and Communication Engineering, KLS GIT, Belagavi, India
e-mail: snbhavikatti@git.edu

M. A. Mellal et al. (eds.), *Computational Intelligence for a Greener Future: Innovations in Renewable Energy Systems*, Studies in Computational Intelligence 1240,
https://doi.org/10.1007/978-3-032-06732-6_25

The clean air monitoring system is built using an IoT network. The IoT network consists of sensors, gateways, network devices, and storage devices. During the process, a large amount of data is generated by the sensors. This is required at local storage and also needs to be stored at the cloud platform for later use. In such cases, the data from the IoT network [4–6] is transferred to cloud storage. The volume of data must be optimized to reduce storage overhead and minimize energy consumption during the data flow over the network. Data filtering using machine learning algorithms is leveraged to optimize the volume of data for later processing. Edge computing algorithms are provided at the application end, while machine learning algorithms [2, 7] operate at the cloud service to provide a centralized solution. It is important to determine which necessities must be addressed at the IoT network level and which services should be provided centrally.

An IoT framework is a structured architecture that enables efficient data collection, processing, and management in IoT ecosystems. It integrates sensors, [8, 9] gateways, edge computing, and cloud platforms to ensure real-time data analysis, secure communication, and scalable storage. These key components work together to enable seamless data flow and intelligent decision-making. Edge computing plays a vital role in reducing latency and supporting localized analytics for smart city applications [10].

Key Design Considerations for IoT Framework:

- Real-time data sensing and analysis of air quality metrics is needed to support timely responses and decision-making.
- Framework is capable of scaling to support a large number of sensors deployed across the entire city without performance degradation.
- The system must support a diverse array of IoT devices, communication protocols, and data formats, while ensuring appropriate placement and location services for seamless operation.
- Robust measures like encryption and device authentication are adapted to protect sensitive environmental data to ensure trust in the system.
- Energy-efficient operation and low resource consumption are key to maintaining long-term sustainability.
- It should allow for either centralized analytics or edge computing for faster, localized responses depending on deployment scale.
- The network should support uninterrupted data transmission from multiple nodes across the city, regardless of location.

1.1 Layered Structure of IoT Framework

The IoT framework is composed of several interconnected components working together to ensure efficient, secure, and scalable data handling [8, 11]. This multi-layered approach optimizes resource usage, reduces network congestion, and enhances real-time responses in IoT applications. A structured IoT framework is vital for enabling reliable, secure, and scalable IoT ecosystems [3]. Table 1 gives the layered structure that integrates edge, fog, and cloud computing with IoT network devices to efficiently manage data. By distributing data processing across these layers, the system enhances performance, scalability, and responsiveness.

The table outlines how each layer contributes to this architecture, with specific roles in sensing, processing, and decision-making. Further, the chapter flow includes: Chapter 2 introduces the key concepts necessary to understand the framework, focusing on network architecture. Chapter 3 discusses the various network devices used at different layers—IoT nodes, edge devices, and the connections between edge and cloud—all in the context of industry standards. Chapter 4 covers the relevant network protocols. Finally, the chapter concludes by addressing strategies across all these components aimed at reducing power consumption and promoting green computing and a sustainable environment.

Further, the chapter flow includes Chap. 2 introduces the key concepts necessary to understand the framework, focusing on network architecture. Chapter 3 discusses the various network devices used at different layers: IoT nodes, edge devices, and the connections between edge and cloud, all in the context of industry

Table 1 IoT framework layers and functions

Layer	Description	Functions and responsibilities
Perception	Comprises IoT devices (sensors, actuators, RFID tags, and cameras) that collect real-time data	– Data sensing and collection – Actuating and controlling IoT devices – Initial data transmission
Edge computing	Processes raw data locally on edge devices or gateways to reduce latency and filter unnecessary data	– Preprocessing and data filtering – Reducing network congestion – Ensuring faster responses
Fog computing	Bridge the edge devices and the cloud, performing data aggregation and analytics closer to the data source	– Localized data storage and processing – Aggregating and refining edge-processed data – Reducing the load on cloud infrastructure
Cloud computing layer	Provides scalable storage, advanced analytics, machine learning, and decision-making capabilities	– Large-scale data storage and analysis – Machine learning and AI-driven insights – Centralized decision-making and automation
Application layer	Interfaces with end-users through dashboards, mobile applications, and APIs, *enabling* interaction with IoT data	– Real-time monitoring and visualization – Automation and remote device control – Decision support and predictive analytics

tandards. Chapter 4 covers the relevant network protocols. Finally, the chapter concludes by addressing strategies across all these components aimed at reducing power consumption and promoting green computing and a sustainable environment.

2 IoT Framework Architecture

The IoT framework [1, 5] architecture is defined in four key layers shown in Fig. 1. Each layer plays an important role in ensuring seamless data collection, transmission, processing, and storage, enabling efficient smart city applications. This structured approach enhances security, scalability, and cost efficiency while promoting environmental sustainability. Fog computing bridges the edge and cloud by offering localized processing and storage, which has been successfully applied in healthcare frameworks as well [12].

The **Perception Layer** consists of IoT sensors and actuators that gather real-time environmental and operational data. These sensors monitor various parameters

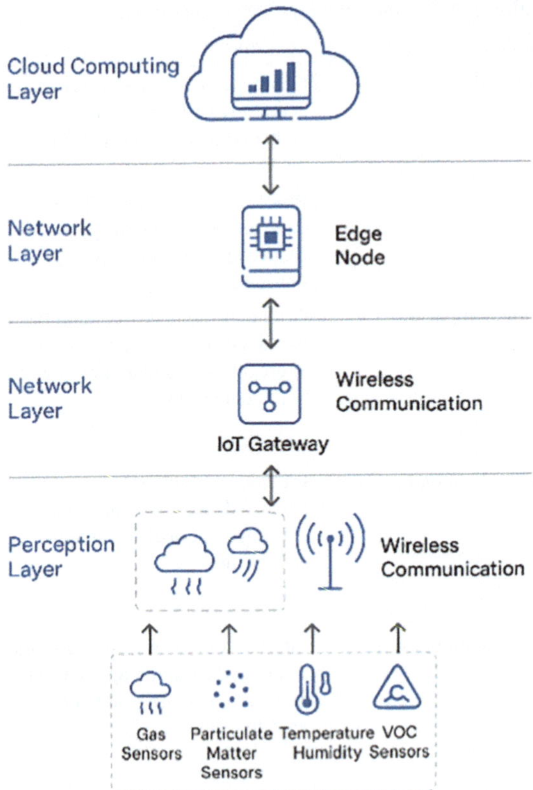

Fig. 1 IoT framework architecture for air monitoring system

such as air quality, temperature, traffic flow, and energy consumption. Actuators respond to the processed data by triggering specific actions, such as adjusting street lighting based on pedestrian movement or optimizing water supply in smart irrigation systems.

The **Network Layer** acts as a communication bridge between IoT devices and processing units. IoT gateways function as intermediaries, aggregating sensor data and using wireless protocols—such as Wi-Fi, LoRaWAN, 5G, and MQTT—to securely transmit information to edge and cloud computing infrastructures.

The **Edge Computing Layer** processes data closer to the source, reducing latency and bandwidth usage. Edge nodes, such as embedded AI systems or industrial IoT controllers, perform preliminary data analysis before forwarding relevant insights to the cloud.

The **Cloud Computing Layer** handles large scale data storage, advanced analytics, and machine learning-driven decision making often enhanced by block chain based mechanisms for scalability and trust [13]. It provides dashboards for visualization, predictive modeling, and optimization of urban infrastructure. Cloud services process the aggregated data, derive meaningful insights, and enable remote monitoring and control of smart city applications.

Interaction Between Layers

The interaction between these four layers follows a structured flow that ensures smooth data collection, processing, and action execution. The Perception Layer captures real-time data, which is transmitted through the Network Layer using IoT gateways. Once the data reaches the Edge Computing Layer, local AI models analyze and preprocess it, filtering out redundant or low-priority information. Only relevant insights are sent to the Cloud Computing Layer, where advanced analytics and machine learning algorithms are applied to detect patterns and predict trends.

Decisions made at the cloud level—such as adjusting energy distribution or optimizing traffic signals—are communicated back through the Edge Computing and Network Layers to actuators in the Perception Layer. This structured data flow ensures real-time responsiveness, minimizes network congestion, and enhances system efficiency [2, 14].

Technologies Involved in Each Layer

Each IoT framework layer is powered by specific technologies tailored to its function:

Perception Layer: This layer relies on IoT sensors (temperature, motion, pollution, humidity) and actuators (motors, automated lighting, smart grids). Technologies such as RFID, LiDAR, and Bluetooth Low Energy (BLE) enhance sensing capabilities.

Network Layer: The transmission of data relies on wireless and wired communication protocols, including 5G, LoRaWAN, Zigbee, MQTT, and CoAP. **IoT gateways** equipped with edge AI chips and security protocols (TLS/SSL encryption) ensure secure and efficient communication.

Edge Computing Layer: Key technologies include NVIDIA Jetson, Intel Movidius, and Google Edge TPU for AI inference at the edge. Fog computing platforms, Docker containers, and machine learning models optimize local data processing.

Cloud Computing Layer: Cloud platforms such as Google Cloud IoT Core, AWS IoT, and Microsoft Azure IoT Hub provide large-scale data storage, advanced analytics, and AI-powered decision-making. BigQuery, TensorFlow, and cloud-based databases enable predictive modeling and smart city automation.

Architecture Plan for Cost-Effective and Environmentally Sustainable IoT Framework

The IoT framework architecture is designed to be cost-effective and environmentally sustainable by optimizing resource utilization, reducing redundant data transmission, and leveraging green computing technologies.

Cost Efficiency: The use of edge computing minimizes cloud dependency, reducing data transmission costs. AI-powered predictive maintenance helps reduce unnecessary repairs and operational expenses. Open-source technologies like Kubernetes and MQTT brokers further cut infrastructure costs.

Energy Efficiency: Smart energy management techniques, such as adaptive power allocation in smart grids and solar-powered IoT sensors, reduce overall power consumption. Low Power Wide Area Networks (LPWAN) such as LoRaWAN extend battery life for IoT devices.

Environmental Sustainability: Deploying biodegradable sensors, implementing AI-driven waste management, and optimizing smart water distribution [2] contribute to sustainable urban development. By reducing energy wastage and improving city-wide resource allocation, the IoT framework supports eco-friendly initiatives [15]. Additionally, IoT plays a pivotal role in renewable energy systems by enabling real-time monitoring and control of solar panels, wind turbines, and energy storage units. Smart meters and IoT-enabled grids help balance supply and demand dynamically, enhancing the efficiency and reliability of green energy integration into urban infrastructure. This layered architecture ensures that smart city applications operate efficiently, securely, and sustainably, leveraging the latest advancements in IoT, edge computing, and cloud technologies [7].

3 Network Device Architecture in IoT-Based Air Cleanliness Monitoring Systems

An air cleanliness monitoring system built on the IoT leverages a multi-layered network architecture that spans from low-power sensors at the edge to high-performance cloud analytics platforms. This section explores the network devices deployed at different layers of such a system—IoT nodes, edge devices, and cloud connectivity mechanisms for data integrity, communication, and system interoperability.

At the foundational layer are the IoT nodes, which consist of environmental sensors embedded in or connected to microcontroller-based devices. These nodes collects real-time data on parameters such as PM2.5/PM10 concentrations, CO_2 levels, NOx, O_3, temperature, and humidity. Devices like the SDS011, Bosch BME680, and Sensirion SPS30 are widely used due to their precision and low power consumption. These sensors are integrated with microcontrollers such as ESP32 or STM32, which support wireless communication protocols like Bluetooth Low Energy (BLE), Zigbee, or Wi-Fi, depending on range and power needs. The IEEE 802.15.4 standard under pins many of these short-range wireless protocols and ensures consistent performance and reliability in dense sensor networks [16].

Data from multiple nodes is aggregated and processed by edge devices, forming the system's intermediary layer. These devices may be single board computers (e.g., Raspberry Pi, NVIDIA Jetson Nano) or industrial-grade gateways that offer more robust performance and reliability. Edge devices often perform preliminary data filtering, noise reduction, and threshold-based alerting, reducing the volume of data sent to the cloud. Importantly, these devices also handle **protocol** translation, bridging low-power sensor data to standard internet communication formats. Protocols like **Open Platform Communications Unified Architecture** are essential at this layer, enabling secure, vendor-agnostic interoperability. Additionally, edge devices support local data storage and remote device management—functions that are crucial in remote or high-risk environments where cloud connectivity might be intermittent [7].

The final layer in the architecture involves connectivity between edge devices and cloud platforms. Data is transmitted using standard IP-based protocols such as MQTT (Message Queuing Telemetry Transport), HTTP/HTTPS, or CoAP (Constrained Application Protocol). MQTT, standardized under ISO/IEC 20,922, is often the preferred choice due to its lightweight, publish-subscribe model, which conserves bandwidth and power—critical considerations in IoT deployments. These communication pathways are secured using TLS (Transport Layer Security), and authentication mechanisms such as X.509 certificates are implemented to prevent unauthorized access and ensure data confidentiality.

From a standards perspective, frameworks such as IEC 62,443 guide the secure design and deployment of industrial automation and control systems, including air quality monitoring. Meanwhile, environmental data formats may adhere to standards from Ogc SensorThings API or EPA's Air Quality System (AQS), ensuring data interoperability across platforms and regions.

The architecture of an air cleanliness monitoring IoT system relies on a well-orchestrated network of devices and protocols across multiple layers. IoT nodes gather precise environmental data, edge devices handle real-time processing and integration, and cloud connections enable broad analytics and decision-making. Industry standards across all layers ensure interoperability, scalability, and security, forming a resilient foundation for smarter environmental monitoring.

3.1 Network Protocols Across IoT Layers in Air Cleanliness Monitoring Systems

In an IoT-based air cleanliness monitoring system, seamless and reliable communication between devices across various layers is essential for effective data collection, processing, and analysis. Each layer of system architecture employs specific network protocols suited to its role, considering factors like data volume, latency, power efficiency, and security. This section discusses the relevant protocols at each of these layers within the context of air quality monitoring. A variety of communication protocols such as Zigbee, LoRa, and MQTT enable connectivity across IoT layers, each chosen based on range, power, and latency requirements [17].

a. **Perception Layer: Short-Range Communication Protocols**

The perception layer consists of the actual sensing devices—IoT nodes that collect air quality data such as particulate matter (PM2.5, PM10), carbon monoxide (CO), nitrogen dioxide (NO_2), ozone (O_3), temperature, and humidity. These sensors are often embedded in compact, battery-powered devices that prioritize energy efficiency and low data rates.

To enable communication between these devices and nearby aggregators (e.g., gateways or edge devices), short-range wireless protocols are used. These protocols support basic data encryption and device authentication to ensure secure operation at the edge of the network. The most common among these include:

- Zigbee (IEEE 802.15.4): A low-power, mesh-based protocol that supports multiple sensors over a network and is ideal for fixed indoor or urban deployments.
- Bluetooth Low Energy (BLE): Suitable for low-data, short-distance transmissions. BLE is used in portable or mobile air monitoring units.
- Wi-Fi (IEEE 802.11): While more power-intensive, Wi-Fi is used in fixed, powered installations where higher bandwidth is needed.
- LoRa/LoRaWAN: A long-range, low-power protocol used in outdoor and wide-area environmental monitoring, although it is also suitable for perception-to-edge communication in low-density urban scenarios.

b. **Edge Layer: Data Aggregation and Intermediate Processing**

The edge layer acts as a bridge between the perception layer and the cloud. Devices at this layer, such as industrial gateways or embedded edge computers, aggregate data from multiple sensors and often perform real-time analytics, filtering, or threshold-based alerts.

Protocols at this layer serve both upstream communication to the cloud and downstream control or configuration. Key protocols include:

- MQTT (Message Queuing Telemetry Transport): A lightweight, publish-subscribe protocol ideal for transmitting small packets of sensor data with minimal overhead.

- CoAP (Constrained Application Protocol): Designed for constrained devices, it operates over UDP and supports RESTful interactions, useful for sensor configuration and status queries.
- OPC UA (Open Platform Communications Unified Architecture): A platform-independent, secure protocol that provides structured data exchange in industrial environments.

These protocols often run on top of Transport Layer Security (TLS) to ensure secure transmission between edge and cloud or edge and local nodes.

c. **Cloud Computing Layer: Data Ingestion and Analytics**

The cloud layer hosts centralized services for data storage, analytics, machine learning and visualization. To support reliable ingestion and processing of high volumes of environmental data, cloud platforms utilize following protocols:

- Hypertext Transfer Protocol Secure (HTTPS): Widely used for RESTful APIs and web-based dashboards.
- AMQP (Advanced Message Queuing Protocol): Suitable for large-scale enterprise environments requiring message brokering, queuing, and reliability.
- MQTT (Cloud Broker Integration): Also widely supported by platforms like AWS IoT, Azure IoT Hub, and Google Cloud IoT Core for secure, scalable IoT communication.

An air cleanliness monitoring system relies on a carefully selected stack of communication protocols across each layer. The perception layer focuses on low-power, short-range protocols; the edge layer balances real-time data handling with upstream communication; and the cloud layer supports high-volume, secure data management. Together, these protocols form the backbone of a reliable and scalable IoT solution for environmental monitoring.

4 Design Considerations for Machine Learning Algorithms in Air Cleanliness Monitoring

At the **edge computing layer**, resource constraints like limited processing power, memory, and energy availability must be considered. A commonly implemented ML approach at the edge is anomaly detection to identify and filter out redundant or irrelevant data before it's sent to the cloud. Isolation Forest is suitable for identifying anomalies (e.g., sudden spikes in PM2.5) in sensor readings. It works by isolating anomalies rather than profiling normal data.

An IoT-based air cleanliness monitoring system involves a seamless flow of data across multiple layers—perception, edge, and cloud. To ensure efficient and sustainable operation, addressing mechanisms, connectivity strategies, and intelligent data processing techniques must be carefully designed. Additionally, leveraging machine learning (ML) algorithms plays a vital role in minimizing power consumption and promoting green computing, contributing to a more sustainable environmental monitoring ecosystem.

At the heart of any IoT system lies the need for unique device identification. In air cleanliness monitoring, where hundreds or thousands of sensors may be deployed across a city or industrial area, each device must be uniquely addressable. Common strategies include:

- IPv6 addressing: Offers a practically unlimited address space, allowing each sensor and edge device to have a unique IP address. This is particularly beneficial for large-scale deployments.
- MAC address-based identifiers: Used at the data link layer for localized addressing within short-range communication networks like Zigbee or BLE.
- Logical addressing: In MQTT or CoAP, devices are often identified by topic paths or URIs, which can include metadata such as location, type of pollutant monitored, or sensor health status.

These addressing strategies ensure that devices can communicate effectively across layers and that data can be traced back to its source for validation and analysis.

Effective connectivity between the perception layer, edge layer, and cloud layer is essential for real-time air quality monitoring. Data flows from low-power sensors to edge gateways using protocols like Zigbee, BLE, or LoRaWAN. The edge layer aggregates and processes this data before transmitting it to the cloud via MQTT or HTTPS. To optimize connectivity:

- Duty cycling and data throttling are implemented to control how often sensors transmit data, thereby saving energy.
- Edge caching and filtering, reduces the amount of data sent to the cloud, minimizing bandwidth and energy usage.
- Event-driven communication, where data is sent only when thresholds are crossed, helps reduce unnecessary transmissions.

Machine learning enhances system efficiency by enabling adaptive sensing and intelligent data management. Key applications include:

- Anomaly detection models at the edge can flag unusual air quality patterns without needing to transmit raw data continuously.
- Predictive sampling algorithms adjust sensor operation frequency based on environmental patterns (e.g., increasing sampling during rush hours or industrial activity).
- Energy-aware ML models optimize sensor activity by learning when high-resolution data is necessary, allowing sensors to enter sleep modes during low-variance periods.

Moreover, Tiny ML—machine learning optimized for microcontrollers—allows inference directly on low-power devices, reducing the need to send data to the cloud and thereby cutting down on energy usage and latency.

Beyond power savings, intelligent data processing and reduced cloud dependency help in minimizing the carbon footprint of the system. Efficient use of computing resources through lightweight protocols, on-device processing, and targeted cloud analytics supports the principles of green computing. Air cleanliness data itself contributes to sustainability efforts by enabling policy makers, urban planners, and communities to take informed action against pollution sources.

4.1 Scalable Machine Learning Approaches for Energy-Efficient Air Monitoring

In an IoT-enabled air cleanliness monitoring system, machine learning plays a critical role in optimizing data management and promoting energy-efficient, sustainable operations. At the **edge computing layer**, the primary objective is to manage and filter sensor data locally to minimize bandwidth usage and power consumption. One effective machine learning algorithm for this purpose is a lightweight version of the **Isolation Forest**, an unsupervised anomaly detection model. This algorithm identifies unusual patterns in air quality data—such as sudden spikes in PM2.5 or unexpected chemical concentrations—by isolating anomalies rather than modeling normal behavior. Since edge devices are resource-constrained, the model is trained offline using historical data, then deployed in a minimal form with a reduced number of shallow decision trees to ensure it can run in real time. The edge device continuously applies the model to a sliding window of recent data, transmitting only flagged anomalies or summary statistics to the cloud. This selective data transmission strategy significantly reduces network traffic and extends device battery life while ensuring that critical environmental events are captured.

At the cloud computing layer, the focus shifts toward efficient data storage and long-term scalability. Given the potentially massive influx of sensor data from distributed locations, a deep learning-based autoencoder is implemented to compress data before storage.

Auto encoders, learn to encode input data into a compact latent representation and the reconstruct as closely as possible to the original. In the context of air quality monitoring, time-series sensor readings—such as levels of PM2.5, NO_2, and CO_2—are compressed using the autoencoder, which is trained to preserve essential features of the data while significantly reducing its size. This approach allows the system to maintain long-term historical records with a fraction of the storage space typically required. The model's reconstruction loss is monitored to ensure data integrity, and only encodings with acceptable loss levels are stored. To stay relevant, the autoencoder is periodically retrained using updated datasets that reflect seasonal or behavioral shifts in air pollution patterns. Compressed data is also tagged with relevant metadata, including timestamps, geolocation, and anomaly flags from the edge, facilitating fast and meaningful retrieval.

Isolation Forest Algorithm at the Edge Layer for Anomaly Detection

In an air cleanliness monitoring IoT framework, real-time anomaly detection is crucial for identifying sudden and potentially harmful changes in environmental conditions. The **Isolation Forest** algorithm is an efficient and lightweight unsupervised machine learning technique that is well-suited for deployment at the **edge layer**, where computational and power resources are limited.

Impact on Data Collection Rate Optimization

By incorporating Isolation Forest, ML algorithms at the edge layer, the system can intelligently identify periods of stability in air quality data and reduce the frequency of data collection accordingly. For example, during times when pollutant levels are consistently low—such as at night or in rural zones—the sensor sampling rate can be reduced from once every 30 s to once every 2 min, results in a 75% reduction in number of data points collected. This adaptive approach not only conserves device energy and extends sensor life but also minimizes the volume of unnecessary data being processed and transmitted. When combined with edge-based anomaly detection and cloud-level compression using auto encoders, the overall data volume can be reduced by over 95% from its original size without sacrificing meaningful insights. These strategies significantly enhance the efficiency of the air cleanliness monitoring system, lowering bandwidth requirements and storage costs while promoting sustainable operation. In conclusion, optimizing the data collection rate through intelligent edge analytics and machine learning not only ensures timely detection of critical environmental changes but also aligns the system with the principles of green computing and environmental sustainability.

4.2 Leveraging IoT and Machine Learning for Renewable Energy Optimization

In addition to air cleanliness monitoring, IoT frameworks play a pivotal role in optimizing renewable energy systems, a critical dimension of environmental sustainability. By integrating smart sensors with solar panels, wind turbines, and energy storage units, IoT enables real-time monitoring of generation, distribution, and consumption patterns. At the edge layer, local devices can predict short-term solar irradiance or wind fluctuations using lightweight machine learning models, adjusting panel orientation or turbine operation accordingly. These actions reduce energy losses and ensure efficient usage of natural resources. The localized data processing also minimizes transmission overhead and improves system responsiveness, supporting energy-efficient operations even in remote or off-grid installations.

At the cloud layer, IoT-collected data from distributed renewable energy assets is aggregated and analyzed to optimize grid balancing and energy trading. Deep learning models, such as LSTM networks or autoencoders, can forecast energy demand and production, detect anomalies in power output, and compress historical datasets for long-term analysis. This predictive capability allows utilities to shift loads, manage demand response programs, and integrate intermittent sources like solar and wind more effectively. As with air monitoring, selective data transmission from the edge and cloud-based learning pipelines ensure the solution remains both scalable and sustainable. Together, these IoT-driven strategies significantly enhance the resilience and efficiency of renewable energy infrastructure, contributing directly to global climate goals.

5 Conclusion

Integrating edge computing and machine learning algorithms improves the efficiency of air cleanliness monitoring. The system uses Isolation Forest at the edge to detect anomalies in real time. It also uses auto encoders in the cloud to compress data before storage. Together, these techniques enable fast, accurate, and efficient data management within the IoT framework. By detecting anomalies locally and compressing data before cloud storage, the system significantly reduces data volume, transmission costs, and energy consumption.

This layered, intelligent architecture not only ensures accurate environmental monitoring but also aligns with green computing principles, contributing to a scalable, sustainable, and responsive air quality monitoring solution. Moreover, the modular nature of the framework allows for easy adaptation across various domains, including renewable energy systems. For instance, IoT-enabled solar or wind farms can leverage similar edge-cloud architectures for real-time monitoring, performance prediction, and energy optimization. Such adaptability underscores the framework's potential in supporting broader environmental sustainability goals, advancing the development of smart, resource-efficient infrastructures across smart cities and beyond.

References

1. Asano, M., Miyoshi, T., & Yamazaki, T. (2024). Internet-of-Things traffic analysis and device identification based on two-stage clustering in smart home environments. *Future Internet, 16*(1), 17. https://doi.org/10.3390/fi16010017MDPI
2. Khan, L. U., Yaqoob, I., Tran, N. H., Kazmi, S. A., Dang, T. N., & Hong, C. S. (2020). Edge-computing-enabled smart cities: A comprehensive survey. *IEEE Internet of Things Journal, 7*(10), 10200–10232. https://doi.org/10.1109/JIOT.2020.2993605
3. Kumar, N., Chilamkurti, N., & Zeadally, S. (2018). Industrial internet of things: Performance evaluation and open issues. *Computers & Electrical Engineering, 64*, 111–117. https://doi.org/10.1016/j.compeleceng.2017.06.016
4. Andriulo, F. C., Fiore, M., Mongiello, M., Traversa, E., & Zizzo, V. (2024). *Edge computing.* Abdelhamid, M. M., Sliman, L., & Ben Djemaa, R. (2024). AI-enhanced blockchain for scalable IoT-based supply chain. *Logistics, 8*(4), 109. https://doi.org/10.3390/logistics8040109MDPI

5. Elhanashi, A., Dini, P., Saponara, S., & Zheng, Q. (2023). Integration of deep learning into the IoT: A survey of techniques and challenges for real-world applications. *Electronics, 12*(24), 4925. https://doi.org/10.3390/electronics12244925MDPI
6. Elkhodr, M., Khan, S., & Gide, E. (2024). A novel semantic IoT middleware for secure data management: Blockchain and AI-driven context awareness. *Future Internet, 16*(1), 22. https://doi.org/10.3390/fi16010022MDPI
7. Ficili, I., Giacobbe, M., Tricomi, G., & Puliafito, A. (2025). From sensors to data intelligence: Leveraging IoT, cloud, and edge computing with AI. *Sensors, 25*(6), 1763. https://doi.org/10.3390/s25061763MDPI
8. Al-Fuqaha, A., Guizani, M., Mohammadi, M., Aledhari, M., & Ayyash, M. (2015). Internet of Things: A survey on enabling technologies, protocols, and applications. *IEEE Communications Surveys & Tutorials, 17*(4), 2347–2376. https://doi.org/10.1109/COMST.2015.2444095
9. Cloud Computing for Internet of Things: A Review. *Informatics, 11*(4), 71. https://doi.org/10.3390/informatics11040071MDPI
10. Leng, J., Chen, Z., Huang, Z., Zhu, X., Su, H., Lin, Z., & Zhang, D. (2022). Secure blockchain middleware for decentralized IIoT towards industry 5.0: A review of architecture, enablers, challenges, and directions. *Machines, 10*(10), 858. https://doi.org/10.3390/machines10100858MDPI
11. Zhang, C., Xu, L., & Xu, Y. (2023). A survey on secure data management in the internet of things. *Journal of Network and Computer Applications, 202*, Article 103377. https://doi.org/10.1016/j.jnca.2022.103377
12. Sood, S. K., & Mahajan, I. (2021). A fog-based healthcare framework for Chikungunya. *IEEE Internet of Things Journal, 8*(2), 1216–1228. https://doi.org/10.1109/JIOT.2020.3015688
13. García, Á., & Berrocal, J. (2023). A survey on IoT-edge-cloud orchestration approaches. *Sensors, 23*(5), 2521. https://doi.org/10.3390/s23052521
14. Misirli, J., & Casalicchio, E. (2024). An analysis of methods and metrics for task scheduling in fog computing. *Future Internet, 16*(1), 16. https://doi.org/10.3390/fi16010016MDPI
15. Gkonis, P., Giannopoulos, A., Trakadas, P., Masip-Bruin, X., & D'Andria, F. (2023). A survey on IoT-edge-cloud continuum systems: status, challenges, use cases, and open issues. *Future Internet, 15*(12), 383. https://doi.org/10.3390/fi15120383MDPI
16. Li, S., Da Xu, L., & Zhao, S. (2018). 5G internet of things: A survey. *Journal of Industrial Information Integration, 10*, 1–9. https://doi.org/10.1016/j.jii.2018.01.005
17. Pliatsios, A., Lymperis, D., & Goumopoulos, C. (2023). S2NetM: A semantic social network of things middleware for developing smart and collaborative IoT-based solutions. *Future Internet, 15*(6), 207. https://doi.org/10.3390/fi15060207MDPI

The Future of AI and Earth's Future

Prachi Pandey, Rashmika, Parnika Shukla, Daksh Sikri, and Akhil Nigam

Abstract Artificial intelligence (AI) has become a part of our lives, changing the way we work, connect, and care for the planet. It holds incredible promise, offering solutions to some of the world's toughest challenges, like fighting climate change, making education and healthcare accessible, and helping us live more sustainably. Imagine AI helping farmers grow food more efficiently, designing cities that waste less energy, or enabling scientists to predict natural disasters with greater accuracy. It is already transforming industries, making breakthroughs in medicine, and improving our understanding of complex ecosystems. At the same time, AI's rapid growth raises important questions that we must address. How do we address the loss of conventional jobs? What if technology gets misused for harmful considerations? And how can we ensure AI techniques are fair and trustworthy? While the possibilities are arising, the risks are equally significant. Misaligned with AI technique developments could widen social divides, exacerbate inequalities, or even hazardous threats to humans. This chapter gives insights into how AI technology is shaping into our environment, society, & collective future. It monitors both forthcoming possibilities and serious concerns, reminding us that we are choosing AI techniques which may become a source of good or harm. By prioritizing on responsible innovation, ethical frameworks, and global collaboration, we can ensure that AI technique not only advances technology but can also help the people and can support towards thriving universe.

Keywords Introduction · Environmental sustainability · Climate change · Food security · Energy optimization · Biodiversity protection · Waste management · Challenges · Future prospects

---

P. Pandey · Rashmika · P. Shukla · D. Sikri
CSE Cyber Security, Chandigarh University, Gharuan, India

A. Nigam (✉)
EE, Chandigarh University, Gharuan, India
e-mail: akhilnigam03@gmail.com

M. A. Mellal et al. (eds.), *Computational Intelligence for a Greener Future: Innovations in Renewable Energy Systems*, Studies in Computational Intelligence 1240,
https://doi.org/10.1007/978-3-032-06732-6_26

1 Introduction

Artificial Intelligence (AI) is the frontier of technology, transforming industries and redefining how human beings engage with the world. From monitoring the environment to resource management, AI plays a central role in addressing global issues such as climate change, pollution, deforestation, and loss of biodiversity. However, AI has its drawbacks, including high energy usage, ethical issues, and socio-economic disruption.

This chapter examines the role of artificial intelligence techniques in making the future of worldwide through concerning its role, challenges, and ethical considerations in environmental sustainability. Through the analysis of real-world applications, this study highlights the potential of AI to protect & preserve the environment and to emphasize responsible and ethical use.

2 AI's Role in Environmental Sustainability

AI is being rapidly utilizing to tackle with climatic challenges, positioning it as an important tool for earth's future. In the direction of providing optimal solutions, AI technology enables the analysis of large data sets and can enhance resource management. By preferring these capabilities, it may support through confronting these issues and can enhance proper waste management system.

2.1 Climate Change Mitigation and Adaptation

Artificial intelligence is leading the way in understanding and acting on climate change because it can facilitate efficient prediction, reduce carbon emissions, and enhance climate resilience.

- Climate Modeling and Prediction

  Artificial intelligence computer programs analyze past climate data to build credible models that predict future climate patterns and extreme weather events, such as hurricanes, floods, and droughts. Google's DeepMind uses machine learning to improve weather forecasting systems, hence increasing preparedness for natural disasters.

- Carbon Emissions Monitoring and Reduction

  Artificial intelligence systems monitor carbon emissions caused by industrial processes and offer suggestions on reducing such emissions.

Carbon capture technologies using artificial intelligence optimize the efficiency of industrial source carbon dioxide capture and storage.

Carbon Tracker, a computer platform, conducts satellite data analysis to identify methane leaks and monitor global carbon emissions.

- Optimization of Renewable Energy

  Artificial intelligence enhances the performance of renewable energy systems such as solar farms and wind farms through energy generation forecasting and load management of grids.

  Intelligent grids with AI control the demand and supply of energy in balance, minimizing wastage and ensuring use of clean energy. Siemens employs artificial intelligence to forecast the performance of wind turbines and modify their configuration to optimize their performance.

2.2 *Precision Farming and Sustainable Agriculture*

Artificial intelligence is revolutionizing agriculture by facilitating data-based decisions, which boost productivity while reducing environment devastation.

- Smart Irrigation and Water Management

  AI takes into account the current weather, soil moisture, and what the plants need to maximize irrigation, conserving water. Artificial intelligence irrigation systems, like NetBeat by Netafim, track and control how water is delivered to crops. This discourages excess water use and saves water.

- Pest and Disease Management

  Artificial intelligence systems identify and forecast pest issues and crop disease by image recognition and pattern recognition. Through the early detection of pests, farmers can employ focused solutions, hence decreasing the need for chemical pesticides.

  Example: Plantix, a computer vision app, detects plant disease from farmer photographs and recommends treatment.

- Crop Yield Prediction and Optimization

  AI considers weather, soil condition, and environmental conditions to forecast crop yields and enhance planting schedules.

The farmer can make harvesting and resource allocation decisions based on data, thus reducing wastage of food. OneSoil utilizes artificial intelligence to provide farmers with suggestions and real-time information for better crop care.

2.3 *Monitoring of Ecosystems and Conservation of Biodiversity*

Because AI can analyze environmental data and identify changes in real time, it is essential for monitoring ecosystems and preserving biodiversity.

- Monitoring Wildlife and Combating Poaching

  Drones with AI and camera traps track the movements of animals and identify instances of poaching. AI systems examine information from sound recorders and motion sensors to find possible dangers to wildlife.

  As an illustration, the PAWS (Protection Assistant for Wildlife Security) system employs artificial intelligence (AI) to forecast areas where poaching is likely to occur and help rangers stop it.

- Evaluation of Ecosystem Health

  AI examines ecological data and satellite imagery to evaluate the condition of ecosystems like wetlands, forests, and oceans.

  Early detection of changes in water bodies, vegetation cover, and soil quality allows for prompt interventions.

  Global Forest Watch, for instance uses AI to monitor deforestation and notify authorities of illegal logging activities.

- Marine Life Monitoring and Ocean Health AI-powered underwater drones analyze ocean data to track marine biodiversity and monitor coral reef health.

  AI aids in determining the sources of pollution and evaluating how climate change is affecting marine ecosystems.

2.4 *Smart Cities and Urban Sustainability*

AI is playing vital role in developing smart cities which can focus on environmental sustainability by achieving best smart energy management, transportation systems, & waste management.

- Energy-Efficient Infrastructure

  AI emerging techniques provide suitable energy usage in buildings by cooling, heating and lighting system based on availability and climate patterns. For an example: DeepMind of Google has reduced power consumption in data centers by preferring 40% AI technology.

- Waste and Resource Management in Urban Areas

  AI-driven waste management systems may keep track record of waste generation and can predict when regarding this data is collected which can further keep track for disposal. Smart bins enabled with smart sensors and AI techniques can optimize data collection schedule and can overcome environmental impact.

2.5 *Environmental Education and Awareness*

AI framework is able to enhance environmental education and to provide awareness by developing complex environmental data which is accessible and engaging to the public.

- AI-Powered Environmental Analytics Platforms

  As AI is able to transform complex environmental data into correctly visual insights which may further help to the scientist and people that they can understand the clarity towards climatic changes.

  Example: Earth.ai provides real-time environmental data and insights through interactive dashboards.

- Citizen Science and Community Engagement

  AI platforms encourage public participation in environmental monitoring by analysing data collected by citizen scientists.

  Example: iNaturalist uses AI to help users identify plant and animal species from photographs, contributing to global biodiversity data.

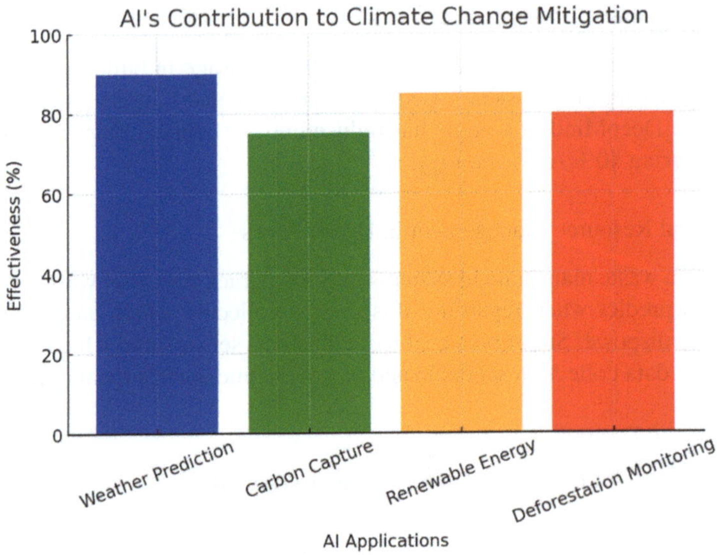

There are many authors who have proposed different models for the study of earth's future in numerous terminologies. These different models or approaches help to predict and analyse critical issues and by utilizing intelligent techniques we have achieved improved response. From the point of economic savings and financial cost authors have proposed Deep learning-based NLP model which is able to reduce cost and enhance quality. By using this model, It helps to determine carbon gas emission into an environment and electricity cost over different countries [1]. By proposing conventional neural network technique, authors have estimated sea surface temperature with its height and salinity. This model has run for 12 months and they have analyzed and improved the accuracy of subsurface temperature [2]. Similarly for the study of subsurface temperature from remote sensing, researchers have developed a new approach-based pre-cluster neural network. This approach has been tested depth upto 1900 meter and determined coefficient of (R2) of approximate 90 percent [3]. Authors have also estimated single-guide RNAs by using conventional neural network technique. They have implemented such a model that predict sgRNA behaviour in Escherichia coli and have improved accuracy [4]. Authors have analysed data for achieving sustainable goal through the technological enhancement. For reaching such goal authors have utilized artificial intelligence indifferent aspects through the world and by preparing estimated data they have achieved sustainable goal [5]. In terms of agriculture, authors have analysed by using emerging technique like Internet of Things (IoT) and Machine Learning. Through these techniques, they have established wireless sensor network where each farm region will be covered by numerous sensor modules [6]. To identify earthquake location is a very important aspect of earth because in previous times, there was very difficult to predict and to save human and environment. Authors have given overview of earthquake location approaches with computational complexities [7].

For the study of wildfire sciences, authors have proposed machine learning techniques in which they have analyzed different aspects like characterization of fuels, detection of fire, climate change and fire weather etc. [8]. In terms of financial, authors have Deep-Learning based techniques for verification of taxonomy aspects. They have utilized data for investigation of conventional neural network model with its operation [9]. Authors have also used geospatial datasets for their analysis by using artificial intelligence methods. They have proposed Geoweaver to enhance the efficiency of full-stack AI workflow management [10]. From the progressively usage of AI authors have developed fusion based conventional neural network method for the analysis of climatic conditions [11]. Further authors have discussed controller design approaches for sampled-data non-linear system with uncertain parameters and disturbances [12]. Authors have also proposed backpropagation neural network technique for determining subsurface temperature of the North Pacific Ocean (NPO) by selecting suitable parameters from satellite system. They have identified different parameters of subsurface like sea surface temperature, height, salinity, wind etc. and have improved the accuracy [13].

Further authors have analyzed data from e NASA Earth Science Data Systems for providing comprehensive overview of artificial intelligence in Earth sciences [14]. Similarly, authors have proposed a new model like conservation area prioritization through the use of artificial intelligence trade-off between cost & advantages of area and protection of biodiversity [15]. Authors have studied about European Destination Earth (DestinE) initiative which aims to implement a digital twin of the Earth system. This initiative helps in developing opportunities in various sectors among renewable resources, water, energy food and public health [16]. Authors have also aimed at earth's ocean by developing You Only Look Once (YOLO) framework. This model helps in identifying location of small-scale ocean using satellite images and through this it has given direction towards for handling earth sciences [17]. Rare earth materials play an important role for next gen memory applications so authors have studied about these materials and with integration of rare earth materials they have enhanced response speed, efficiency and retention. Their review study has established material science and neuromorphic framework for supporting future research directions [18]. Authors have studied about AI powered tool for developing improved efficiency in agriculture land areas by preferring earth observation data. This study has been done under various machine learning techniques and analyzed suitable result [19]. In addition, with this, authors have done comprehensive review of rare earth materials for improving microstructure properties. They have identified that rare earth elements can encourage metallurgical reaction and restrains the defects [20].

3 Challenges and Ethical Considerations

While Artificial Intelligence (AI) holds incredible promise for tackling environmental issues and fostering sustainability, it also presents a host of challenges and ethical dilemmas that we need to confront to make sure it has a positive effect. From its hefty energy demands to the potential for algorithmic biases and socio-economic upheavals, using AI in the realm of environmental sustainability calls for a careful balance between pushing technological boundaries and upholding ethical standards. Despite its advantages, AI presents several challenges and ethical dilemmas:

{%} of your text is likely AI-generated

3.1 *Energy Consumption of AI Models*

- Energy-Intensive AI Models

  Training advanced AI models, especially deep learning ones like GPT-4 and various neural networks, requires a huge amount of energy. This high energy use often leads to a considerable carbon footprint. Research shows that training just one large AI model can produce as much carbon dioxide as five cars do over their entire lifespan. As more people adopt AI, the overall environmental impact of the data centers that host these models is becoming a serious issue.

- Possible Solutions

  Encouraging Green AI may involve enhancing energy efficient models and optimizing techniques to overcome computational demand. Transitioning towards renewable energy source, it may improve energy efficiency through safety techniques.

3.2 *Algorithmic Bias and Fairness*

- Bias in Environmental Data and Models

  AI frameworks built on incomplete datasets may destroy climatic decision making. For instance, environmental algorithms based on specific region data may operate poorly in terms of different ecological variables.

- Possible Solutions

  Ensuring AI frameworks are trained towards diverse dataset to avoid existing bias. Engaging local communities and integrating indigenous information into AI powered environmental decision-making procedure.

3.3 *Socio-Economic Disruptions and Job Displacement*

- Automation in Agriculture and Waste Management

  The growing utilization of AI-powered automation in different sectors like agriculture, waste management & resource optimization turns into significant employment losses, especially in developing countries where numerous people rely on conventional jobs for fulfilling their daily needs. Farmers, factory workers, and waste management staff might find themselves out of work as AI takes over essential tasks like monitoring crops, sorting materials, and recycling.

- Widening Digital Divide

  As AI becomes more integrated into efforts for environmental sustainability, there's a risk that the digital divide will grow even wider. Developing nations often struggle with the infrastructure and expertise necessary to harness AI effectively. Without access to these technologies, they may fall behind in reaching their sustainable development goals.

- Possible Solutions

  Launch reskilling and upskilling initiatives to help workers gain the skills needed for jobs in AI-driven industries. Advocate for inclusive AI policies that ensure everyone, especially in developing areas, can benefit from the advancements AI brings.

3.4 *Data Privacy and Security Risks*

- Sensitive Environmental and Geographic Data

  AI systems that monitor the environment gather a huge amount of information from satellites, sensors, and citizen science initiatives. If this data is mishandled or not properly secured, it could lead to serious privacy issues and even national security threats. Unauthorized access to geospatial information about natural resources, biodiversity hotspots, or deforestation trends could allow private companies to exploit these valuable insights.

- Citizen Science and Privacy Concerns

  When AI platforms involve the public in collecting environmental data, there's a risk that user privacy might be compromised if the data isn't managed carefully. Personal information from mobile apps used for environmental monitoring or species identification can be at risk of being misused.

- Possible Solutions

  Create robust data governance frameworks to safeguard environmental data from being misused. Adopt privacy-by-design principles in AI systems that are used for environmental monitoring.

3.5 Lack of Transparency and Explainability

- Opaque Decision-Making in Environmental Management

  A lot of AI models, particularly those using deep learning, tend to function like "black boxes." This makes it tough for stakeholders to grasp how decisions are actually made.

- Possible Solutions

  Developing explainable AI frameworks which offer transparent and clear insights to support making decisions. It must be ensured about human oversight in AI decision-making processes to preserve accountability and foster trust.

3.6 Ethical Implications and Responsibility

- Accountability and Human Oversight

  AI frameworks in environmental management must be accountable to human decision-makers. Relying on AI for significant decisions may results into unforeseen affects.

- Inclusive and Ethical AI Design

  AI frameworks should be configured significantly with variables from diverse stakeholders. Inclusive AI design ensures that solutions consider the needs of vulnerable communities and promote environmental justice.

- Possible Solutions

  Establish ethical AI guidelines for environmental applications that prioritize inclusivity and accountability. Incorporate human-in-the-loop systems to ensure human oversight in AI-powered decision-making.

4 Future Prospects and Innovations

As Artificial Intelligence (AI) continues to evolve, its potential to address environmental challenges and promote sustainable practices is expanding rapidly. Emerging innovations in AI are paving the way for smarter climate models, enhanced biodiversity protection, circular economies, and sustainable urban development. These future prospects hold the promise of mitigating climate change, protecting natural ecosystems, and optimizing resource management on a global scale.

Below are the key future prospects and innovations in AI for environmental sustainability:

4.1 Advanced Climate Modelling and Disaster Prediction

AI is set to change the game in climate modelling and disaster management by offering more accurate, real-time insights into weather patterns and natural disasters.

- Hyper-Accurate Climate Forecasting

In the future, AI models will pull together massive amounts of climate data from satellites, sensors, and IoT devices to create incredibly precise climate predictions.

- What's Coming

    1. AI-driven Earth system models that can predict regional climate changes with much greater accuracy.
    2. A better grasp of extreme weather patterns, which will help develop early warning systems to safeguard at-risk communities.

4.2 AI-Powered Disaster Prediction and Management

- AI will predict the occurrence and severity of natural disasters, such as hurricanes, floods, wildfires, and droughts, with improved accuracy.
- Real-time data from satellites and drones combined with AI's predictive capabilities will enhance emergency response strategies.
- What's Coming

    1. Autonomous AI systems that analyze environmental parameters and automatically issue disaster warnings.
    2. AI-enabled disaster simulation models that help governments develop more effective disaster management plans

4.3 AI-Driven Precision Agriculture and Food Security

As the global population grows, the demand for sustainable food production increases. AI is set to revolutionize agriculture through precision techniques that optimize resource use and enhance food security.

- AI in Soil Health Monitoring and Regenerative Agriculture

In the future, AI systems will be able to analyze soil health data in real time, helping farmers adopt regenerative agricultural practices that boost soil fertility and capture carbon. AI-powered predictive models will assist farmers in embracing climate-smart farming techniques, ultimately lessening the environmental impact of agriculture.

- What's on the Horizon

  1. AI models designed to monitor and enhance soil biodiversity, supporting long-term agricultural sustainability.
  2. Autonomous drones and robotic systems will keep an eye on soil conditions and apply organic fertilizers or biochar to enhance soil health.

4.4 Smart Farming with AI and IoT Integration

- By merging AI with IoT devices, we can keep a close eye on crop health, weather patterns, and soil moisture levels in real-time. Automated systems for irrigation and fertilization will help us use resources more efficiently, cutting down on water waste and lessening our impact on the environment.
- What's on the Horizon

We can expect AI-driven autonomous farming systems that minimize the need for human involvement while promoting sustainable crop production. Plus, AI frameworks will analyse pest threats and can recommend eco-friendly pest control techniques.

4.5 Ocean Health Monitoring and Marine Conservation

AI is set to revolutionize marine conservation by analysing their data and is able to provide safe and reliable marine ecosystem.

- AI for Marine Life Conservation and Monitoring

  1. Imagine AI-driven underwater drones and machinery can explore ocean mapping ecosystem and it may help to monitor marine life in real time system. It can be equipped with advanced intelligent tools that can help in tracking and helps to identify specific solutions.

2. AI based systems will be able to identify illegal fishing activities and can alert to higher authorities immediately. In addition with this, AI framework can be able to analyse ocean data and forecasting the impact of environmental changes.

- AI-Powered Plastic Waste Detection and Cleanup

1. With the help of AI based satellites and drones, it also helps to identify plastic pollution in oceans and rivers. Such activities enable people to take participation in cleaning up activities.
2. What's Coming: Integration of AI with systems with capability or handling such cleaning up activities. These also help in extracting microplastic materials from rivers and oceans.

4.6 AI for Biodiversity Protection and Ecosystem Restoration

The ability of AI is to process environmental database towards the protection of biodiversity and restoring enhanced biosystem.

- AI-Enabled Habitat Mapping and Restoration

  AI approaches will examine imagery of satellite and ecological information for the identification of degraded ecosystem and needs to give priority restoration areas.

  With the help of AI and drones, autonomous reforestation systems will plant native species and keep an eye on their growth over time.

- AI in Wildlife Protection and Anti-Poaching

  AI-powered monitoring systems will keep tabs on the movements of endangered species and spot potential threats as they happen. Sophisticated AI models will analyze audio and visual data to catch poaching activities and notify the authorities.

4.7 Waste Management and Circular Economy Innovations

AI is expected to play a pivotal role in promoting a circular economy by optimizing waste management processes and reducing environmental impact.

- AI-Powered Smart Waste Sorting Systems

Advanced AI-driven waste sorting systems will enhance recycling efficiency by automatically identifying and separating materials. AI-powered robotic systems will sort waste with higher accuracy, reducing contamination in recycling streams.

- AI in Circular Supply Chains and Resource Optimization

AI will optimize supply chains by predicting demand, reducing overproduction, and minimizing waste. Future AI systems will identify opportunities to repurpose waste materials, closing the loop on resource consumption.

- AI-powered platforms that assess the lifecycle of products and recommend sustainable alternatives.
- Integration of AI with blockchain technology to ensure transparency and traceability in supply chains.

4.8 Smart Cities and Sustainable Urban Development

AI is set to play a crucial role in shaping smart cities that focus on sustainability, making the most of resources, and cutting down on carbon emissions.

- AI-Enabled Energy Management in Smart Cities

AI will enhance energy efficiency in smart buildings by studying occupancy trends and fine-tuning heating, cooling, and lighting systems accordingly. Smart grids, driven by AI, will help balance energy supply and demand while incorporating renewable energy sources to lower carbon footprints.

AI models designed to forecast and manage energy needs in urban settings, decreasing our dependence on fossil fuels. AI-driven microgrids that support decentralized and resilient energy solutions.

- AI for Sustainable Urban Mobility

AI will streamline public transportation systems, easing congestion and cutting down on air pollution. Traffic management systems powered by AI will analyze real-time data to enhance traffic flow and minimize carbon emissions.

- Autonomous electric vehicles, guided by AI, that adjust to shifting traffic conditions and encourage sustainable transportation.
- AI models that help predict and mitigate urban heat islands by optimizing green infrastructure.

4.9 AI-Powered Environmental Education and Awareness

AI is set to revolutionize environmental education and boost public involvement in sustainability initiatives.

- AI-Driven Environmental Analytics and Visualization

With AI-powered platforms, we can visualize intricate environmental data, making it much more accessible for scientists, policymakers, and everyday people to grasp climate trends and ecological shifts. Interactive simulations will help aware everyone about how human actions impact on environment. What's Coming:

1. AI-powered digital twins of ecosystems which may be able to stimulate environmental changes and to provide practical insights.
2. Virtual reality experienced by AI-powered which transport users into endangered ecosystems, which raise an awareness towards environmental concerns.

- Citizen Science and Public Participation

AI framework will actively engage public in environmental observing by analysing and monitoring data collected by citizen scientists. Real-time feedback from AI framework will empower individuals to perform a role in the protection of environment.

What's Coming

1. AI-driven apps which enable users or citizens to observe their carbon footprints & adoption of sustainable lifestyle.
2. AI-driven framework which support community-led reforestation & conservation models.

4.10 AI in Space Exploration for Earth Sustainability

AI is poised to play a crucial role in space exploration with enabling the data collection on environmental changes, availability of resources and climatic patterns.

- AI for Earth Observation and Climate Monitoring

With the help of artificial intelligence techniques, it provides availability of space data from satellite and helps us to monitor atmosphere of earth.

AI-driven framework will monitor how environmental changes are affecting polar ice caps, increasing sea levels, and numerous ecosystems. What's Coming:

1. An AI-powered platforms which collaborate satellite data with ground based environmental monitoring data for a holistic view of climate concerns.
2. In addition with this, AI will be able to enhance early detection and warning threats by using data collected by space satellites.

5 Conclusion

Artificial Intelligence (AI) is an emerging technology which plays vital role as transformative force with providing optimal and reliable solutions to environmental challenges like climate variation, resource management and loss of biodiversity. From enhancing climate models and forecasting natural tragedy to an advance precise agriculture, wastage management, water management and employment of smart cities. AI techniques have sufficient potential to drive substantial environmental development. However, such advancements have accompanied by remarkable concerns, including higher power consumption, data security issue and socio-economic concerns. So, these are essential to address such challenges with implement of new policies. These AI approaches continue to evolve desired responsibilities and applications which can enhance communities and to strengthen towards reliable planet. Hence overall future of sustainability and reliability depends on how wisely users integrate of AI into existing models.

References

1. Strubell, E., Ganesh, A., & McCallum, A. (2019). *Energy and policy considerations for deep learning in NLP* (pp. 1–6). arXiv preprint.
2. Han, M., Feng, Y., Zhao, X., Sun, C., Hong, F., & Liu, C. (2019). A convolutional neural network using surface data to predict subsurface temperatures in the Pacific Ocean. *IEEE Access, 7*, 172816–172829.
3. Lu, W., Su, H., Yang, X., & Yan, X. H. (2019). Subsurface temperature estimation from remote sensing data using a clustering-neural network method. *Remote Sensing of Environment, 229*, 213–222.
4. Wang, L., & Zhang, J. (2019). Prediction of sgRNA on-target activity in bacteria by deep learning. *BMC Bioinformatics, 20*(517), 1–14.
5. Vinuesa, R., Azizpour, H., Leite, I., Balaam, M., Dignum, V., Domisch, S., Fellander, A., Langhans, S. D., Tagmark, M., & Nerini, F. F. (2020). The role of artificial intelligence in achieving the sustainable development goals. *Nature Communications, 11*(1), 1–10.
6. Vij, A., Vijendra, S., Jain, A., Bajaj, S., Bassi, A., & Sharma, A. (2020). IoT and machine learning approaches for automation of farm irrigation system. *Procedia Computer Science, 167*, 1250–1257.
7. Karasözen, E., & Karasözen, B. (2020). Earthquake location methods. *International Journal on Geomathematics, 11*(13).
8. Jain, P., Coogan, S. C., Subramanian, S. G., Crowley, M., Taylor, S., & Flannigan, M. D. (2020). A review of machine learning applications in wildfire science and management. *Environmental Reviews, 28*(4), 478–505.
9. Kloster, M., Langenkämper, D., Zurowietz, M., Beszteri, B., & Nattkemper, T. W. (2020). Deep learning-based diatom taxonomy on virtual slides. *Scientific Reports, 10*, 1–13.
10. Sun, Z., Di, L., Burgess, A., Tullis, J. A., & Magill, A. B. (2020). Geoweaver: Advanced cyberinfrastructure for managing hybrid geoscientific AI workflows. *ISPRS International Journal of Geo-Information, 9*(2), 1–20.
11. Qiu, C., Tong, X., Schmitt, M., Bechtel, B., & Zhu, X. X. (2020). Multilevel feature fusion-based cnn for local climate zone classification from sentinel-2 images: Benchmark results on the So2Sat LCZ42 dataset. *IEEE Journal of Selected Topics in Applied Earth Observations and Remote Sensing, 13*, 2793–2806
12. Lee, J., Moon, J. H., & Lee, H. J. (2021). Continuous-time synthesizing robust sampled-data dynamic output-feedback controllers for uncertain nonlinear systems in Takagi–Sugeno form: A descriptor representation approach. *Information Sciences, 565*, 456–468.
13. Cheng, H., Sun, L., & Li, J. (2021). Neural network approach to retrieving ocean subsurface temperatures from surface parameters observed by satellites. *Water, 13*(3), 1–20.

14. Sun, Z., et al. (2022). A review of Earth Artificial Intelligence. *Computers & Geosciences, 159*, 1–16.
15. Silvestro, D., Goria, S., Sterner, T., & Antonelli, A. (2022). Improving biodiversity protection through artificial intelligence. *Nature Sustainability, 5*, 415–424.
16. Hoffmann, J., Bauer, P., Sandu, I., Wedi, N., Geenen, T., & Thiemert, D. (2023). Destination Earth—a digital twin in support of climate services. *Climate Services, 30*, 1–3.
17. Al Mahmud Mostafa, S., Wang, J., Holt, B., & Wang, J. (2024). YOLO based Ocean Eddy Localization with AWS Sage maker. In *2024 IEEE International Conference on Big Data (BigData), Washington, DC, USA* (pp. 3720–3728)
18. Ghazanfar, H., Kim, H., Rabeel, M., Ahmad, M., Nisar, S., Zulfiqar, M.W., Rehman, A., Dastgeer, G., & Kim, D. (2025). A review of rare earth materials for emerging memory devices for neuromorphic computing. *Materials Today Physics, 55*
19. Kaur, A., Singh, G., & Singh, S. (2025). AI-driven tools and technologies for agriculture land use & land cover classification using earth observation data analytics. *Google Earth Engine and Artificial Intelligence for Earth Observation,* 527–541.
20. Zhang, B., Liu, K., Li, J., Chen, B., Huang, C. J., & Soboleva, N. (2025). A comprehensive review on the rare earth elements improving microstructure and properties of laser cladded coatings. *Journal of Alloys and Compounds* 1036, (2025)

Statistical Study of Regula Falsi Method for Cubic Equations with Random Coefficients with Application in System Optimization

Bhargavi Patel and Soubhik Chakraborty

Abstract Regula Falsi Method, is an iterative numerical technique used to approximate solutions to nonlinear equations (algebraic or transcendental) by progressively narrowing down the intervals where the root of the equation is likely to be found. The primary goal of this work is to investigate how the probability distribution's parameter(s) that influence a cubic polynomial's coefficients can affect the Regula Falsi method's convergence. This chapter examines both continuous and discrete distributions, specifically on continuous uniform distributions, discrete uniform distributions, and normal distributions. Key findings indicate that for a given parameter r, which represents the $[-r, r]$ interval, the average number of iterations can be predicted by a second-degree polynomial equation for the uniform distributions. The coefficients of this polynomial are almost the same for both continuous and discrete uniform distributions, showing that the average iteration count depends on the distribution range rather than its type. For normal distributions, the average iteration count is influenced by the mean for fixed standard deviation, indicating the mean's critical role. In contrast, varying the standard deviation while fixing the mean shows no significant impact on the average iterations. Overall, the study concludes that for uniform distributions, the iteration count depends on the range of the distribution, while for normal distribution, it depends on the mean. Finally, a new avenue for future research is suggested, where we propose combining the Regula Falsi method with other techniques to enhance the rate of convergence and application of Regula Fasi method for system optimization is also discussed.

Keywords Regula Falsi method · Cubic equation · Applied regression analysis · Simulation of variates · System optimization

B. Patel · S. Chakraborty (✉)
Department of Mathematics, Birla Institute of Technology, Mesra, Ranchi, Jharkhand, India
e-mail: soubhikc@yahoo.co.in

M. A. Mellal et al. (eds.), *Computational Intelligence for a Greener Future: Innovations in Renewable Energy Systems*, Studies in Computational Intelligence 1240,
https://doi.org/10.1007/978-3-032-06732-6_27

1 Introduction

1.1 Introduction to Regula Falsi Method

Root finding is a crucial part of solving math problems in many areas, from engineering to finance. Sometimes, we can solve equations with algebra, but often real-world problems are too complicated for that.

For example:-

$$e^x - x = 0 \tag{1}$$

where

- e^x is Euler's number (approximately 2.71828), the base of the natural logarithm
- x is the unknown variable to be solved.

This equation involves the exponential function and the variable x. Finding an exact algebraic solution for x in terms of elementary functions is not possible, making it a type of equation known as a transcendental equation. This is where numerical methods come into play.

While solving complicated equations sometimes they can't be solved with algebraic methods. That's where numerical methods like Regula Falsi help. It's a pretty straight forward method. We pick two points and keep adjusting them until you land close to the actual answer. Even for slightly tricky equations, it usually works quite well. It is used a lot in fields like engineering and science, especially when we need to find root is difficult or if it takes too much time. Here Regula Falsi helps by giving decent approximation. It's also useful in optimisation tasks. For example, [1] came up with a modified version to improve power tracking in solar systems is just one example of how it's applied outside textbooks.

1.2 Regula Falsi Method

The method of Regula Falsi is a simple numerical method that is used to solve nonlinear equations. The first step of the Regula Falsi method involves guessing two initial values, assuming that the root lies between them. Then this interval is gradually narrowed down at each step with the help of Linear interpolation. This method is balanced between simplicity and precision. Although it may be a little slower than other techniques such as Newton–Raphson, but it is still good enough. Therefore, with the application of both interpolation and bracketing, this method is very useful when the precise root is not readily available.

1.2.1 Steps for the Regula Falsi Method

1. Choose initial points i and j such that evaluating the function f at points i and j results in having $f(i)$ and $f(j)$ opposite signs.

   If $f(i)$ and $f(j)$ is having opposite signs, it indicates that the root lies between i and j, allowing us to proceed with the Regula Falsi method to narrow down the interval and approximate the root.
2. Compute the next approximation k using the formula:

$$c = \frac{i \cdot f(j) - j \cdot f(i)}{f(j) - f(i)}$$

3. Evaluate the function f at the point k to obtain $f(k)$.
4. Determine the sign of $f(c)$:

   a. If $f(k)$ have same sign as of $f(i)$, change i with k.
   b. If $f(k)$ have same sign as of $f(j)$, change j with k.

5. Repeat steps 2–4 until the difference between i and j becomes sufficiently small or until the intended level of accuracy is attained. Then, return to step 2 and recalculate k using updated i or j.

1.3 Literature Review

The Regula Falsi method, is a step-by-step way of finding the solutions to equations within a certain range. In recent years, the Regula Falsi method is applied to different-different fields, including to clear road blockage. In A Study on Regula Falsi Graphs to Clear Road Blockage by Ambrish Pandey and Chitra Singh [2], it is mentioned that when the roads of a city get blocked due to snow, salt has to be spread to clear them; proper planning is needed, cannot be done randomly. For this, the authors suggest using Euler paths or circuits to plan the route in such a way that all roads are covered efficiently without repeating the same paths again and again.

The earliest studies on the Regula Falsi method were done on deterministic setups, where the coefficients in the polynomial equations are already known. In such cases, researchers mainly focused on how well the method converges, how efficient it is, and how much error it produces. These studies helped to build a basic understanding of how Regula Falsi works under stable conditions and gave useful insights into its theory and behaviour [3].

In recent years, researchers have also attemptedto observe what Regula Falsi method does in case coefficients are random rather than constant. They mainly focused on its convergence and stability. In their finding, they discovered that method's behavior relies on cofficients' statical behavior. Because of this, more complex techniques are used to improve the results [4]. To handle this, adaptive technique are used. These adaptive technique adjust certain values as the method is in progress, depending on how the coefficients are evolving, improving performance and becomes more stable. Making it very useful in real-life problems where the values keep changing from one case to another [5].

Some researchers haven't just focused on improving Regula Falsi using adaptive techniques. They've also compared it with other methods like bisection and secant to see how it works in different scenarios. Especially when the coefficients are random, these comparisons have revealed that no one method is best all the time. Depending on the situation, one might perform better than the others. And that's the point. Insights like these really help when you're trying to build hybrid methods that pick the best parts of each approach. It makes the overall method work faster and more reliably [4].

Regula Falsi has been pretty useful in solving real world problems, especially when there's some randomness involved. In areas like engineering, finance and science and most of the time, it manages equations fairly well. Even in some cases, it beats other methods. That's why there's still room to improve it so it can deal better with the uncertainty and messy conditions that usually come up outside textbook scenarios.

1.4 Motivation and Problem Statement

In mathematics, people often encounter problems that can be solved easily and accurately, such as quadratic and linear equations which typically yield precise solutions. However, some equations are far more challenging to solve exactly, and finding an accurate solution can be exceedingly difficult. In school, we mostly deal with simple, clean equations like linear or quadratic ones, and they usually give exact answers. But once you're outside the classroom, math isn't always that tidy. Real world problems are often messier, and getting an exact answer isn't always possible or even needed. Most of the time, we just need something close enough to work with. That's where numerical methods come in. They help when equations don't have perfect solutions. And honestly, a rough but reliable answer is often better than spending too much time trying to be exact.

In this chapter, I worked with the Regula Falsi method a numerical way to find roots of equations. What I wanted to check was how the shape of a probability distribution (which determines the coefficients of a cubic equation) changes how fast the method converges. The thinking is: if we already know something about that distribution, we might be able to predict how the method behaves. Maybe even estimate how many steps it'll take, on average, to get close to the root.

Looking at Regula Falsi from a statistical lens, rather than just a numerical one, might actually make the whole process quicker especially when you're dealing with tougher problems that don't have clear solutions.

2 Study on Cubic Equation for Discrete Uniform Coefficients

2.1 Overview

In statistics, a "uniform distribution" is one of the type of probability distribution where every possible outcome within a specified range has an equal chance of occurring. This means that each potential outcome is just as likely as any other. In simpler terms, in a uniform distribution, the probability is evenly spread across all possible values, ensuring that each outcome has the same likelihood of being observed.

For instance, imagine a lottery where each ticket has an identical chance of winning. If there are 100 tickets sold, each ticket has a probability of 1/100 of being the winning ticket. This situation exemplifies a uniform distribution because every ticket, regardless of its number, has the same chance of being drawn. However, if the lottery organizer decided to favour certain tickets based on arbitrary criteria, such as the color of the ticket or the number printed on it, those favoured tickets would have a higher chance of winning compared to others. This favouritism would violate the principle of uniform probability, where every ticket should have an equal chance of winning, without any biases.

Uniform distribution can be classified into two main types depending on the nature of the possible outcomes: Discrete distribution and Continuous distribution. Following section will delve into detail of the discrete uniform distribution.

2.2 Discrete Uniform Distribution

In the discrete uniform distribution, every potential outcomes are equally likely to occur within a finite set of possibilities. For instance, imagine a situation where you are selecting a marble from a bag containing five marbles, each of a distinct color. In this scenario, every marble has an equal likelihood of being chosen, showcasing a discrete uniform distribution where each possible outcome (selecting a specific color marble) has an identical probability of happening.

To clarify, consider A as a discrete random variable that takes r values within the interval [i, j]. Let A has a discrete uniform distribution if its probability mass function can be stated as:

$$f(a) = \frac{1}{r}, a = 1, 2, 3, \ldots r \tag{2}$$

2.2.1 Expected Value and Variance

The expected value and the variance are two frequently computed statistical measures. For a discrete uniform random variable, the expected value of A is computed by:

$$\varepsilon(A) = \sum_{a=1}^{r} a.P(A = a) \tag{3}$$

which for discrete uniform variate A is

$$\varepsilon(A) = \frac{r+1}{2} \tag{4}$$

The expression for variance is given by:

$$\vartheta(A) = \varepsilon\left(A^2\right) - [\varepsilon(A)]^2 \tag{5}$$

where $\varepsilon\left(A^2\right)$ is given by:

$$\varepsilon\left(A^2\right) = \sum_{a=1}^{r} a^2 \cdot P(A = a) \tag{6}$$

which, in our case, gives

$$\sigma^2 = \frac{r^2 - 1}{12} \tag{7}$$

Here:-

- σ is the standard deviation,
- $\varepsilon(A)$ is the expected value of A,
- $\vartheta(A)$ is the variance of A,
- $\varepsilon\left(A^2\right)$ is the second raw moment of A,
- r is the maximum value of A, assuming $A \in \{1, 2, \ldots, r\}$.

2.3 Approach

This section explains the steps we followed to reach our goal. To generate the coefficients for the cubic equation, we used a function called disunirand(). It's built using MATLAB's rand and rng functions to create random values for the coefficients.

2.3.1 Disunirand()

U = disunirand(r) is designed to produce random numbers distributed according to a discrete uniform distribution. This specific distribution is characterized by its range, which extends symmetrically from –r to r.

To achieve this, disunirand() includes:

rand

A = rand() generates a single random number, specifically a scalar, within the interval (0,1). This number is selected randomly from a uniform distribution, ensuring that any value within the range has an equal probability of being chosen.

Ex:

rand ()

0.9058

A = rand(n) produces an n-by-n matrix filled with randomly generated integers that follow a uniform distribution.

Ex:

rand (2)

$0.1270 \cdots 0.6324$

$0.9134 \cdots 0.0975$

A = rand(n_1; $n_2 \cdots$; n_n) produces an array of random numbers of dimensions n_1–by– n_2–by–...–by–n_n, Here n_1; n_2; $\cdots n_n$, specify the size of each dimension.

Rng

In MATLAB, the rng function plays a crucial role in random number generation by allowing users to modify the seed of the random number generator. This helps prevent the repetition of the same sequence of random numbers. By default, rng uses the current time to generate a new seed, ensuring randomness with each invocation. When we execute the "Shuffle" command, rng is automatically reseeded with a fresh seed, further enhancing randomness.

Example:

rng shuffle

r = rand()

r = 0.9808

Refer to Appendix A for the MATLAB code.

2.3.2 Steps for Matlab Code

1. First, we assume a tolerance value of 0.001 and specify the range r for generating random number.
2. Then, coefficients of the cubic equation are randomly generated from a discrete uniform distribution using custom function disunirand().

   Ex:

   a0 = disunirand(10)

   a0 = −7

   In the same way, the other coefficients a1, a2, and a3 are generated.
3. Now, we check two conditions using a while loop:

   I. If a3 = 0, we generate a new value until it's not zero.
   II. If $a3 < 0$, we make all the coefficients negative so the leading term becomes positive.

4. After that, we choose initial guess values i and j such that $f(i) \cdot f(j) < 0$.
5. Compute the value of $k = \frac{i \cdot f(j) - j \cdot f(i)}{f(j) - f(i)}$ and then, we calculate $f(k)$.
6. If-else conditional statement is now utilized to evaluate a condition. If $f(k) < 0$, set $i = k$; otherwise, set $j =$ k.
7. Increment the iteration count h by 1.
8. Continuously execute steps 5 to 7 within a while loop until the value of $|f(k)|$ becomes smaller than the given tolerance value.
9. Record the final iteration count h in an array named arr.
10. Display the values of i, j, k, and $f(k)$ along with the root.
11. Repeat steps 2–10 for a total of 100 iterations using a for loop.
12. Display the array of iterations (ar).
13. The ar's mean and standard deviation have been computed and presented.

2.4 Results

Based on the instructions stated in Appendix A, we conducted 100 trials in MATLAB to determine the mean and standard deviation of iterations (h) across a specified parameter range [–r, r]. The outcomes of these trials are presented in Table 1.

2.5 Statistical Analysis

Graph in Fig. 1 shows how mean iterations in the Regula Falsi method changes with respect to the parameters r of discrete uniform U(−r, r) distribution characterizing the coefficients of cubic equations.

Table 1 Data of the mean and standard deviation for iterations (h) derived from a discrete uniform distribution

S. num	Parameter range (r)	Mean iteration (mean h)	Standard deviation of iteration
1	1000	12.55	5.4853
2	2000	12.64	7.5539
3	3000	13.44	7.9358
4	4000	13.67	7.4793
5	5000	13.71	9.1831
6	6000	14.3	7.1499
7	7000	14.47	7.6863
8	8000	14.59	8.2842
9	9000	14.97	7.5591
10	10,000	15.56	7.9128

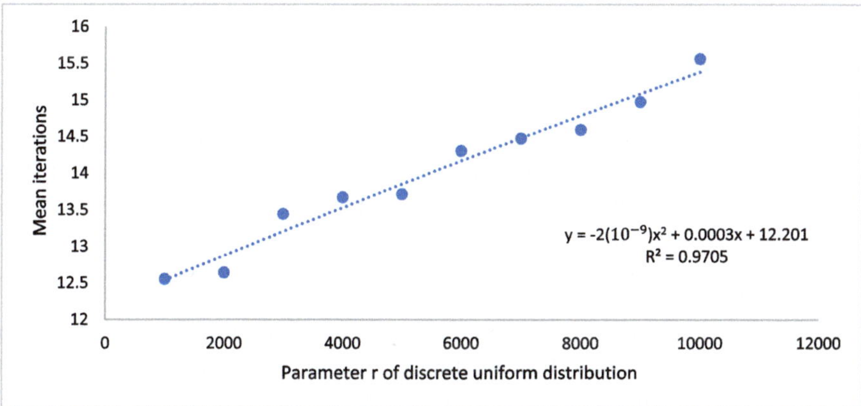

Fig. 1 Graph illustrating the mean iterations for the Regula Falsi method against the parameter r of the discrete uniform U(−r, r) distribution for the coefficients of a cubic equation

Figure 1 clearly shows that for a given parameter, it's average number of iterations can be predicted accurately by a second-degree polynomial equation.

3 Study on Cubic Equation for Continuous Uniform Coefficients

3.1 Overview

In addition to discrete uniform distributions, continuous uniform distributions also exist and are more commonly used. In a continuous uniform distribution, every possible outcome within a specified range has an equal probability of occurring, similar to the discrete uniform distribution.

However, the key difference is that the continuous uniform distribution encompasses an infinite number of possible outcomes within the range.

While the previous section thoroughly examined the discrete uniform distribution, this section will look deeper into the continuous uniform distribution, exploring its characteristics and applications in greater detail.

3.2 Continuous Uniform Distribution

In a continuous uniform distribution, the random variable A can assume any value within a specified range, determined by the lower limit α and the upper limit β. The two limits α and β set the boundaries of the range. The values are spread evenly between these two points, and nothing falls outside them. Every value inside that range has the same chance of showing up, which is why the distribution is flat. Because of that flat shape, it's sometimes called a rectangular distribution.

Unlike the discrete uniform distribution, a continuous uniform distribution doesn't have a fixed set of outcomes. Instead, it covers an endless number of values within a set range. Think of it like measuring rainfall during a day say it can fall anywhere between 0 and 10 mm. Any number in that range is just as likely as any other. Even though there are infinite possible values, each one has the same chance of showing up.

More specifically, let A be a continuous random variable within the interval $[\alpha, \beta]$, denoted as $A \sim \mathrm{U}(\alpha, \beta)$. The probability distribution function (p.d.f) of A can be expressed as follows:

$$f_A(a) = \begin{cases} \frac{1}{\beta-\alpha}, & \alpha < a < \beta \\ 0, & otherwise \end{cases} \tag{8}$$

3.2.1 Expected Value and Variance

The expected value of a continuous uniform random variable can be determined as follows.

$$\varepsilon(A) = \int_{-\infty}^{\infty} a \cdot f_A(a)da \tag{9}$$

Hence, the expected value is

$$\varepsilon(A) = \frac{\alpha + \beta}{2} \tag{10}$$

The variance of a continuous uniform distribution can be expressed as follows:

$$\vartheta(A) = \varepsilon\left(A^2\right) - [\varepsilon(A)]^2 \tag{11}$$

where $\varepsilon\left(A^2\right)$ is given by:

$$\varepsilon\left(A^2\right) = \int_{\alpha}^{\beta} a^2 \cdot f_A(a)da \tag{12}$$

Hence, the variance is

$$\sigma^2 = \frac{(\alpha - \beta)^2}{12} \tag{13}$$

where:-

- σ is standard deviation.
- A is the continuous random variable following a uniform distribution.
- α, β are the lower and upper bounds of the distribution interval.
- $f_A(a)$ is the probability density function (p.d.f.) of A.
- $\varepsilon(A)$ is the expected value (mean) of A.
- $\vartheta(A) =$ is the variance of A.

3.3 Approach

This section explains the steps we followed to reach our goal. To generate the coefficients for the cubic equation, we used a function called conunirand(). It's built using MATLAB's rand and rng functions to create random values for the coefficients.

3.3.1 Conunirand()

U = conunirand(r) is designed to produce random numbers distributed according to a continuous uniform distribution. This specific distribution is characterized by its range, which extends symmetrically from -r to r.

To achieve this, conunirand() includes:

Rand

A = rand() generates a single random number, specifically a scalar, within the interval (0,1). This number is selected randomly from a uniform distribution, ensuring that any value within the range has an equal probability of being chosen.

Ex:

rand ()

0.9058

A = rand(n) produces an n-by-n matrix filled with randomly generated integers that follow a uniform distribution.

Ex:

rand (2)

$0.1270 \cdots 0.6324$

$0.9134 \cdots 0.0975$

A = rand(n_1; $n_2 \cdots$; n_n), produces an array of random numbers of dimensions n_1–by– n_2–by–...–by–n_n, Here n_1; n_2; $\cdots n_n$ specify the size of each dimension.

Rng

In MATLAB, the `rng` function plays a crucial role in random number generation by allowing users to modify the seed of the random number generator. This helps prevent the repetition of the same sequence of random numbers. By default, `rng` uses the current time to generate a new seed, ensuring randomness with each invocation. When we execute the "Shuffle" command, `rng` is automatically reseeded with a fresh seed, further enhancing randomness.

Example:

rng shuffle

r = rand()

r = 0.9808

Refer to Appendix B for the MATLAB code.

3.3.2 Steps to Follow in Matlab Code

1. Assume a tolerance value of 0.001 and specify the range r for the random number.
2. The coefficients of the cubic equation are randomly generated from a continuous uniform distribution using conunirand().
   Example: a0= conunirand(10)
   $a0 = 5.3115$
   In the same way, the other coefficients a1, a2, and a3 are generated at random.
3. Implement the following checks using a while loop:
   I. If a3 = 0, generate a new value for $a3$ using conunirand() until this condition is no longer true.
   II. If $a3 < 0$, negate all coefficients $a0$, $a1$, $a2$, $a3$ until this condition is no longer true.
4. Calculate the initial guess values i and j such that $f(i)\cdot f(j) < 0$.
5. Compute the value of $k = \frac{i\cdot f(j)-j\cdot f(i)}{f(j)-f(i)}$ and $f(k)$.
6. If-else conditional statement is now utilized to evaluate a condition. If $f(k) < 0$, set $i = k$; otherwise, set $j =$ k.
7. Execute step 5 once more.
8. Update h by adding 1, tracking the number of iterations necessary to reach the root.
9. Continuously execute steps 6–8 within a while loop until the value of $|f(k)|$ surpasses the specified tolerance.
10. Record the final iteration count h in an array named ar.
11. Display the values of i, j, k, and $f(k)$ along with the root.
12. Repeat steps 2–11 for a total of 100 iterations using a for loop.
13. Display the array of iterations (ar).
14. The ar's mean and standard deviation have been computed and presented.

3.4 *Results*

Based on the instructions stated in Appendix B, we conducted 100 trials in MATLAB to determine the mean and standard deviation of iterations (h) across a specified parameter range [–r, r]. The outcomes of these trials are presented in Table 2.

3.5 *Statistical Analysis and Discussion*

Graph in Fig. 2 illustrates the relationship between the mean iterations in the Regula Falsi method with respect to the parameter r of the continuous uniform U(−r, r) distribution, which characterizes cubic equation's coefficients.

Table 2 Data of the mean and standard deviation for iterations (h) derived from a continuous uniform distribution

S. no	Range of parameters (r)	Mean iteration (mean h)	Standard deviation of iteration
1	1000	12.09	5.6284
2	2000	12.49	4.9676
3	3000	13.09	6.2508
4	4000	14.39	7.5731
5	5000	13.81	6.8471
6	6000	14.9	8.8859
7	7000	14.79	8.0945
8	8000	15.04	6.8045
9	9000	15.57	7.1467
10	10,000	16.42	8.7677

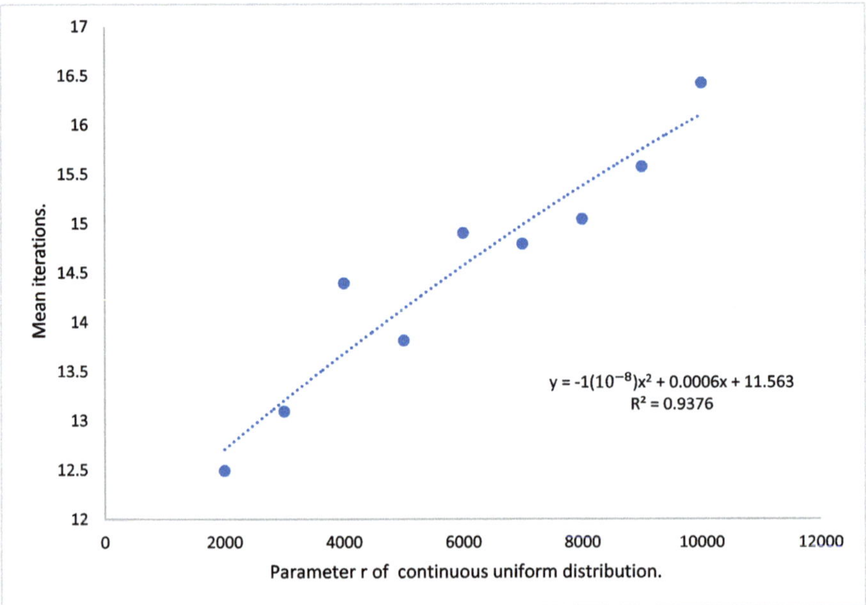

Fig. 2 Graph illustrating the mean iterations for the Regula Falsi method against the parameter r of the continuous uniform U(−r, r) distribution for the coefficients of a cubic equation

Figure 2 clearly shows that for a given parameter, it's average number of iterations can be predicted accurately by a second-degree polynomial equation.

3.6 Comparison Between Discrete and Continuous Uniform Distribution Results

Both continuous and discrete uniform distributions can be effectively analyzed using second-degree polynomial equations to estimate the average number of iterations required for a given parameter value. Remarkably, the second-degree polynomial equation consistently proves to be the best model for predicting average iterations in both types of distributions.

Furthermore, a comparison reveals that for both continuous and discrete distributions the coefficients of the second-degree polynomial equations are almost identical. This indicates that the average number of iterations is not significantly affected by the type of distribution. Instead, it largely depends on the distribution's range of values, which is a key parameter.

In summary, whether the distribution is discrete or continuous, the range of values within the distribution plays a crucial role in determining the average number of iterations needed for root-finding using methods like the Regula Falsi method.

4 Study on Cubic Equation for Coefficients Normally Distributed

4.1 Overview

The Gaussian distribution, commonly known as the normal distribution, plays a crucial role in statistics, particularly for independent random variables. Its distinctive bell-shaped curve is widely recognized and often encountered in statistical analyses.

In a normal distribution, the majority of data points cluster around the central peak, creating a bell-shaped curve. This distribution is continuous and symmetrically centered around its mean. As you move farther from the mean in either direction, the probability of observing those values decreases uniformly. Extreme values at the tails of the distribution are rare occurrences. It is important to note that although a normal distribution is symmetric, not every symmetric distribution follows the normal distribution pattern.

Normal distribution characterizes how values of a variable are spread out, similar to other probability distributions. It is regarded as the most significant probability distribution in statistics because it effectively represents the distribution of values for numerous natural phenomena. Normal distributions are frequently used to represent traits that result from multiple independent factors. For instance, the normal distribution often describes characteristics such as exam scores, reaction times, daily temperatures, and income levels.

4.2 Parameters of the Normal Distribution

The normal distribution, like other distributions, is defined by its parameters. The mean and standard deviation are the primary factors that determine the shape and position of the normal distribution. It is crucial to understand that there isn't a single, fixed normal distribution; rather, its appearance varies depending on the specific values of the mean and standard deviation.

Mean

The mean of a normal distribution indicates its central tendency, identifying the peak of the bell curve. Most data points are concentrated around this mean. A change in the mean results in the whole curve shifting left or right along the X-axis, as shown in Fig. 3 [6] → as shown in [6].

Standard Deviation

The standard deviation quantifies the variability of a normal distribution, defining its spread. It measures the typical deviation of data from the mean, representing the average distance between individual observations and the mean. A larger standard deviation suggests that the data points are more spread out, while a smaller standard deviation indicates that the data points are more tightly clustered around the mean. This metric is crucial for understanding the dispersion of data in relation to the mean.

As illustrated in Fig. 4 [7] → as shown in [6], altering the standard deviation impacts the distribution's width along the X-axis. Higher standard deviations result in wider distributions, while lower standard deviations cause the distribution to tighten around the mean.

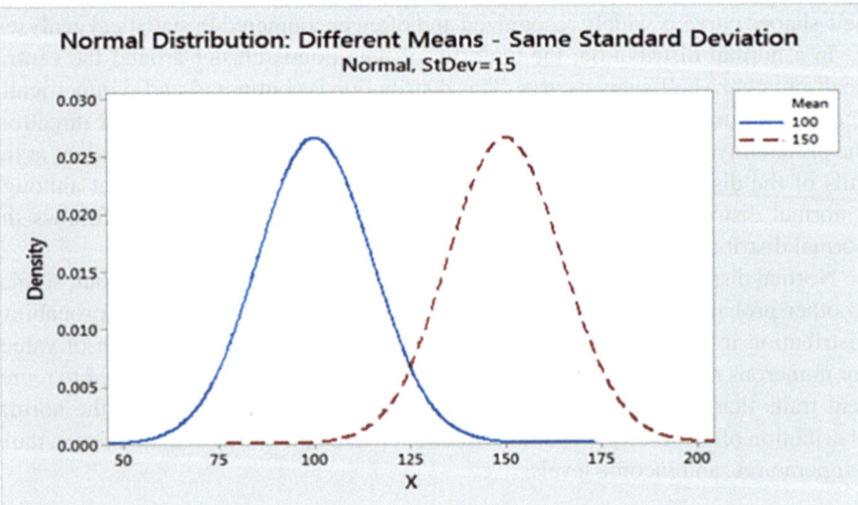

Fig. 3 The graph illustrates the shift of the normal distribution curve as the mean varies

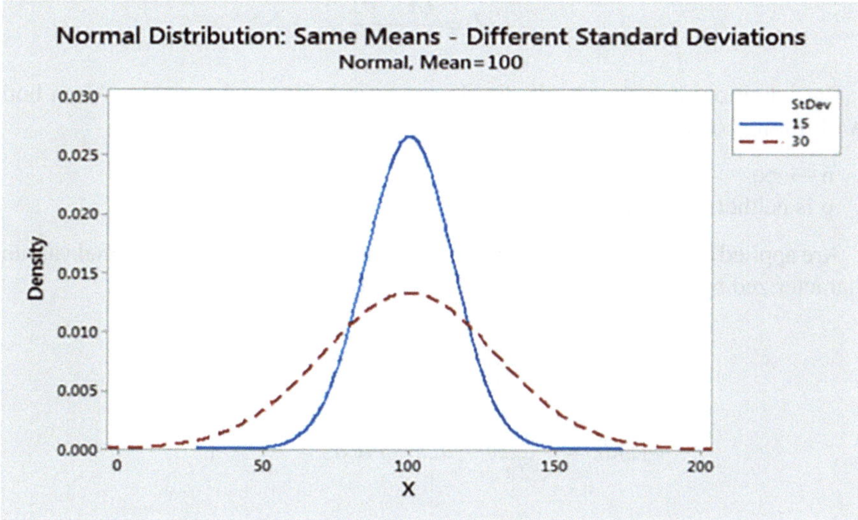

Fig. 4 The graphs depict the shape of the normal distribution curve as the standard deviation varies

Figure 4 depicts the shape of the normal distribution curve as the standard deviation varies.

When distributions are narrow, most of the values are clustered near the mean, making it likely that any randomly selected value will be close to the average. Conversely, when the distribution widens, there is a broader range of values that can deviate further from the mean. Thus, as the normal distribution spreads out, the probability of encountering values significantly different from the average increases.

4.3 Normal Distribution Probability Density Function

Normal distribution can be derived as a limiting case of binomial distribution. Let *A* be a binomial variate with parameters n and p. Let

$$\mathrm{Z} = \frac{(A - np)}{\sqrt{npq}}$$

where:-

- $np = \theta$ (the mean of the binomial distribution),
- $\sqrt{npq} = \sigma$ (standard deviation.),
- $\mathrm{q} = 1\text{–}\mathrm{p}$,

Which means

$$Z = \frac{(A - \theta)}{\sigma}$$

By definition, Z is the standardized version of a binomial variable. When both two limiting conditions are satisfied:

a. $n \to \infty$
b. p is neither small nor large.

Are applied on Z, then Z can be proven to transform into a standard normal variate, characterized by the following probability density function (pdf):

$$\varphi(z) = \frac{1}{\sqrt{2\pi}} e^{\frac{-z^2}{2}}, \quad -\infty < z < \infty$$

$$\varphi(z)dz = \frac{1}{\sqrt{2\pi}} e^{\frac{-z^2}{2}} dz, \textit{ where } dz = \frac{da}{\sigma} - +$$

$\varphi(z)$ is the probability density function of Z.

$\varphi(z)dz$ is called the probability differential of Z.

A becomes a normal variate with probability density function given by:

$$f(x) = \frac{1}{\sigma\sqrt{2\pi}} e^{\frac{-(x-\theta)^2}{2\sigma^2}}, \textit{ where } -\infty < x < \infty, -\infty < \theta < \infty, \sigma > 0$$

$$f(x)dx = \frac{1}{\sigma\sqrt{2\pi}} e^{\frac{-(x-\theta)^2}{2\sigma^2}dx}$$

where:-

- *A* is the binomial variate
- $f(x)$ is the probability density function of *A*;
- $f(x)dx$ is called the probability differential of *A*.
- *A* is a normal variate with mean θ and variance σ^2. In symbols, $A \sim N(\theta, \sigma^2)$.
- Z is a standard normal variate. In symbols, $Z \sim N(0, 1)$.

The term "normal distribution" originated from efforts to establish it as the standard for all continuous probability distributions. Although this attempt was unsuccessful, as many continuous distributions significantly deviate from the normal distribution, the name "normal" persisted.

4.4 Approach

This section explains the steps we followed to reach our goal. To generate the coefficients for the cubic equation, we used a function called disnormal(). It's built using MATLAB's rand and rng functions to create random values for the coefficients.

4.4.1 Disnormal()

The u = disnormal (mean, sigma) function generates random values from a normal distribution with the given mean and standard deviation. This normal() function utilizes several built-in functions, including:

rand

A = rand() generates a single random number, specifically a scalar, within the interval (0, 1). This number is selected randomly from a uniform distribution, ensuring that any value within the range has an equal probability of being chosen.

Ex:

rand ()

0.9058

A = rand(n) produces an n-by-n matrix filled with randomly generated integers that follow a uniform distribution.

Ex:

rand (2)

$0.1270 \cdots 0.6324$

$0.9134 \cdots 0.0975$

A = rand(n_1; $n_2 \cdots$; n_n) produces an array of random numbers of dimensions n_1–by– n_2–by–...–by–n_n, Here n_1; n_2; $\cdots n_n$ specify the size of each dimension.

Rng

In MATLAB, the `rng` function plays a crucial role in random number generation by allowing users to modify the seed of the random number generator. This helps prevent the repetition of the same sequence of random numbers. By default, `rng` uses the current time to generate a new seed, ensuring randomness with each invocation. When we execute the "Shuffle" command, `rng` is automatically reseeded with a fresh seed, further enhancing randomness.

Example:

rng shuffle

r = rand()

r = 0.9808.

Refer to Appendix C for the MATLAB code.

4.4.2 Steps to Follow in Matlab Code

1. Provide the mean and the standard deviation as sigma, with the tolerance value set to 0.001.
2. The coefficients of the cubic equation are randomly generated from a normal distribution using disnormal().

   Example: a0 =normal (100, 10)

   $a0 = 101.7212$
   In the same way, the other coefficients a1, a2, and a3 are generated at random.
3. Implement the following checks using a while loop:
   If a3 = 0, generate a new value for $a3$ using disnormal() until this condition is no longer true.
4. Calculate the initial guess values i and j such that $f(i)\cdot f(j) < 0$.
5. Compute the value of $k = \frac{i\cdot f(j)-j\cdot f(i)}{f(j)-f(i)}$ and $f(k)$.
6. If-else conditional statement is now utilized to evaluate a condition. If $f(k) < 0$, set $i = k$; otherwise, set j = k.
7. Execute step 5 once more.
8. Update h by adding 1, tracking the number of iterations necessary to reach the root.
9. Continuously execute steps 6–8 within a while loop until the value of $|f(k)|$ surpasses the specified tolerance.
10. Record the final iteration count h in an array named ar.
11. Display the values of i, j, k, and $f(k)$ along with the root.
12. Repeat steps 2–11 for a total of 100 iterations using a for loop.
13. Display the array of iterations (ar).
14. The ar's mean and standard deviation have been computed and presented.

4.5 Results

Following the MATLAB code detailed in Appendix C, the mean and standard deviation of iterations (h) were obtained over 100 trials for two scenarios: One scenario features a fixed mean with a changing standard deviation (sigma) of the normal distribution, detailed in Tables 3 and 4, while the other scenario maintains a fixed standard deviation (sigma) with a varying mean, as outlined in Tables 5 and 6.

Table 3 Mean and standard deviation data for iterations (h) in a normal distribution, with a fixed mean = 100

S. no	Standard deviation (sigma) of normal distribution	Mean iteration (mean h) (over 100 trials)	Standard deviation of iteration (over 100 trials)
1	2	10.7	6.0244
2	4	10.92	6.4097
3	6	11.45	6.1896
4	8	11.18	6.3012
5	10	11.11	6.1495
6	12	11.14	5.9187
7	14	11.48	6.2125
8	16	11.78	5.6525
9	18	11.75	5.4428
10	20	11.95	5.2056

Table 4 Mean and standard deviation data for iterations (h) in a normal distribution, with a fixed mean = 30,000

S. no	Standard deviation (sigma) of normal distribution	Mean iteration (mean h) (over 100 trials)	Standard deviation of iteration (over 100 trials)
1	1000	17.37	10.8764
2	2000	17.25	10.6395
3	3000	17.6	9.9494
4	4000	17.59	9.993
5	5000	17.17	9.4677
6	6000	17.8	8.8031
7	7000	17.44	7.8087
8	8000	17.58	8.47
9	9000	17.3	7.6838
10	10,000	17.68	8.0112

4.6 Statistical Analysis and Discussion

In Table 3, the derived equation is y = 10.69, which shows an almost constant value for the given inputs. The mean is held constant at 100, while the standard deviation varies slightly.

In Table 4, the derived equation is y = 17.37, indicating an almost constant value for the given inputs. The mean is fixed at 30,000, while the standard deviation varies over large intervals.

In Table 5 above, the resulting equation is $y = -3(10^{-6})x^2 + 0.005x + 7.8352$ with $R^2 = 0.9819$. This indicates that a second-degree polynomial equation can be

Table 5 Mean and standard deviation data for iterations (h) in a normal distribution, with a fixed sigma = 2

S. no	Mean of normal distribution	Mean iteration (mean h) (over 100 trials)	Standard deviation of iteration (over 100 trials)
1	100	8.3	6.0126
2	200	8.85	6.2626
3	300	9.04	6.2683
4	400	9.56	6.4028
5	500	9.94	6.5224
6	600	9.95	6.3808
7	700	10.02	6.5056
8	800	10.21	6.4874
9	900	10.37	6.4818
10	1000	10.31	6.491

Table 6 Mean and standard deviation data for iterations (h) in a normal distribution, with a fixed sigma = 1000

S. no	Mean of normal distribution	Mean iteration (mean $\boldsymbol{h}$) (over 100 trials)	Standard deviation of iteration (over 100 trials)
1	10,000	15.33	9.5273
2	20,000	15.77	10.4976
3	30,000	16.03	11.0201
4	40,000	16.2	11.207
5	50,000	16.24	11.2574
6	60,000	16.34	11.3887
7	70,000	16.28	11.5396
8	80,000	16.48	11.6346
9	90,000	16.56	11.7862
10	100,000	16.61	11.8624

used to predict the average iterations for a given parameter. In this case, the standard deviation is fixed at 2, while the mean varies over small intervals, as depicted in Fig. 5.

Table 6 presents the equation $y = -2(10^{-10})x^2 + 3(10^{-5})x + 15.524$ with $R^2 = 0.9391878917$. This suggests that a second-degree polynomial equation can predict the average iterations for the given parameter. In this scenario, the standard deviation is fixed at 1000, and the mean changes across large intervals, as illustrated in Fig. 6.

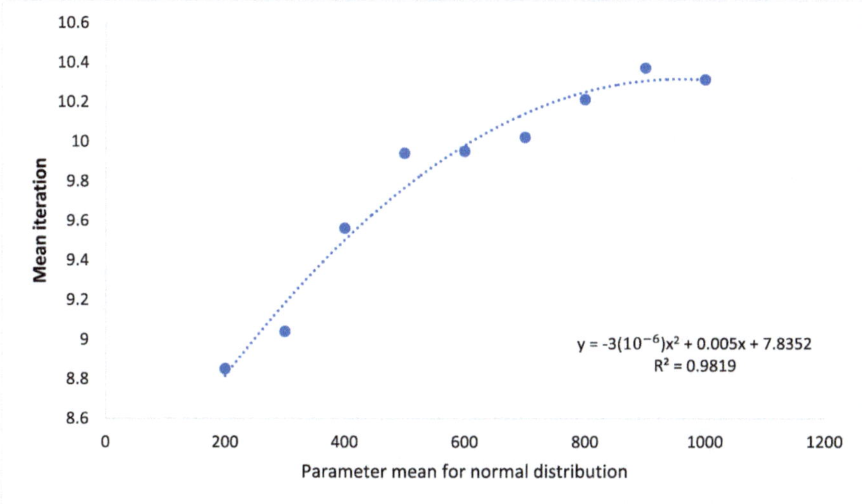

Fig. 5 Graph showing mean iterations for Regula Falsi method versus parameter mean, while keeping sigma fixed at 2, for a cubic equation with coefficients distributed normally

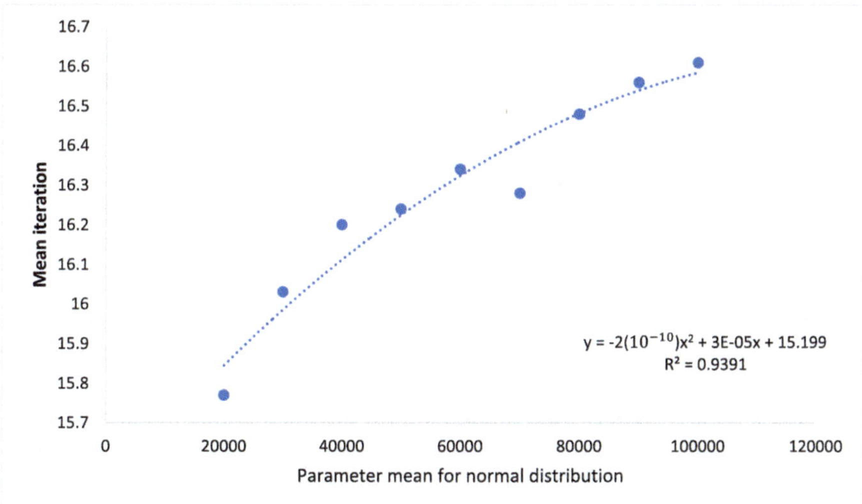

Fig. 6 Graph showing mean iterations for Regula Falsi method versus parameter mean, while keeping sigma fixed at 1000, for a cubic equation with coefficients distributed normally

5 Conclusion and Future Scope of Work

5.1 Conclusion

For researchers to make thoughtful decisions about how to collect data, use the right methods to study it, and clearly understand and share the results, having a solid grasp of statistics is important. Having knowledge of statistics really helps in making data-driven decisions, predicting future outcomes, and supporting scientific research. That's why statistical tools are really helpful. They allow us to find useful patterns in data and make decisions based on facts, not guesses.

In our study, we have used statistical analysis to see how the Regula Falsi method works while solving cubic equations, where the coefficients are randomly taken from different probability distributions. As we all know, the convergence of such methods mostly depends on the values of the coefficients, so it's quite natural to think that the parameters of the chosen distributions may have an effect on how quickly or accurately the method gives the result. To study this, we worked with both uniform and non-uniform types of distributions. These include discrete uniform, continuous uniform, and normal distributions. To generate random coefficients from the distributions, we developed custom MATLAB functions such as disunirand(), conunirand(), and disnormal(), using MATLAB's built-in rand and rng functions. This helped us run the analysis in a controlled and repeatable way, so we could properly study how the distribution parameters affect the convergence of the method.

After going through the results, we noticed that a quadratic polynomial does a pretty good job of predicting the average number of iterations based on the given parameter no matter which of the three distributions we looked at. What stood out was that the coefficients of this polynomialwere nearly the same for both discrete and continuous uniform distributions, but not for the normal one. Based on that, we arrived at the following conclusions:

(i) When we used uniform distributions whether discrete or continuous the average number of iterations didn't really change. Whatactually made a difference was the range of the distribution. That turned out to be the key factor.
(ii) When we used inputs from a non-uniform distribution like the normal distribution, we noticed that the mean had an effect on the average number of iterations, but the standard deviation didn't. So it looks like the number of iterations depends more on where the distribution is centered than how spread out it is.

These results show that different things influence how fast the Regula Falsi method converges, depending on the type of distribution used for the coefficients.

5.2 *Future Scope of Work*

You could even try running Regula Falsi with other distributions like Bernoulli or exponential to check if the results change much. It might not behave the same way, and that alone could be interesting to look into. While this method is guaranteed to converge, it often does so more slowly than some of the other available techniques. Mixing it with a faster method like Newton–Raphson could be a smart move. A hybrid like that might speed things up and make the overall process more efficient.

A good next step could be to actually test out these hybrid methods and see how they stack up against the original Regula Falsi. Both in how quickly they converge and how well they perform overall. Combining methods isn't without challenges for example the Newton–Raphson method fails when the derivative is zero. This means future work will need to consider dataset selection more carefully to ensure the conditions required by such hybrid approaches are actually met.

The bigger goal here is to make root-finding methods more useful and reliable in real-world situations. One clear example is renewable energy. In solar PV systems, there's a well-known problem called maximum power point tracking (MPPT), which is basically about finding the point where the derivative of power with respect to voltage is zero. A classic root-finding scenario. Chun and Kwasinski (2011) demonstrated that a Modified Regula Falsi Method (MRFM) can be applied for digital maximum power point tracking (MPPT) in photovoltaic systems. Building on that, the statistical analysis in this study such as using a quadratic model to estimate the number of iterations could support MPPT controllers in adjusting more effectively to changes in sunlight or temperature. This might help the method work faster and use energy more efficiently especially useful in systems like wind power, where behavior is harder to predict and the equations don't stay simple.

Remarks

1. This work is motivated by a previous paper by Kumari & Chakraborty, (2024). → as shown in [8].
2. For further literature on Regula Falsi and other numerical methods, the reader may consult [7, 9–12].
3. For the computational part, [13] may be consulted.

Ethical Declaration The authors hereby declare that this research did not receive any funding. They further declare that they do not have any conflict of interest.

Appendix A

```
% clearing screen
clc
% input
r = input('Enter range of r: ');
e = 0.001;
l = 1;
ar = zeros (1, 100);

initializing coefficients of polynomial
for n = 1:1:100
    a0 = disunirand(r); a1 = disunirand(r);
    a2 = disunirand (r); a3 = disunirand(r);
    while a3 == 0
    a3 = disunirand(r);
end
while a3 < 0
    a0 = -a0; a1 = -a1; a2 = -a2; a3 = - a3;
end

y = @(x)(a3*x^3 + a2*x^2 + a1*x + a0) ;


% calculating guess value
m = -1000;
fi = y(m);
fj = y(m+1);
while fi * fi > 0
    m = m + 1:
    fi = y(m);
    fj = y(m+1);
end

i = m;
j = m+1;


% implementing Regula Falsi Method

h = 1:
fprintf('\n\na\t\tb\t\tc\t\tf(c)\n');
while (j - i) >
    k = (i * y(j) - j * y(i)) / (y(j) - y(i));
    fk = y(k);
    fprintf('%f\t%f\t%f\t%f\n', i, j, k, fk);
    if abs(fk) < e
        break;
    end
    if fk * fi < 0
        j = k;
    else
        i = k;
    end
    h = h + 1;
end
ar(l, 1) = h;
l = l + 1;
fprintf('%f\t%f\t%f\t%f\n', i, j, k, fk);
fprintf('\nRoot is: %f\n', k);
end
fprintf('\niteration is:\n');
fprintt('%d\t', ar);
```

```
% finding mean and standard deviation of ar

Fprint('nMean or iterations: \n');
disp(mean(ar));
fprintf('\nStandard deviation of iterations: \n');
disp(std(ar));

% generating R.N

function u = disunirand(r)
rng shuffle


     ul = rand();
     u = (-r) + fix(ul * (2 * r + 1));
     end
```

Appendix B

```
% clearing screen
 clc
% input
r = input( 'Enter range of r: ");
e = 0.001;
l = 1;
ar = zeros(1, 100);

%initializing coefficients of polynomial

for n = 1:1:100
    a0 = conunirand(r);
    a1 = conunirand(r);
    a2 = conunirand(r);
    a3 = conunirand(r):
    while a3 == 0
        a3 = conunirand (r);
    end
    while a3 < 0
        a0 = -a0; a1 = -a1; a2 = -a2; a3 = -a3;
    end
    y = @(x) (a3*x^3 + a2*x^2 + a1*x + a0);
% calculating quess value
   m = -1000;
   fi = y(m);
   fj = y(m+1);
   while fi * fj > 0
       m = m + 1;
       fi = y (m);
       fj = y (m+1);

   end

   i = m;

    j = m+1;
```

```
% implementing Regula Falsi Method

h = 1;
fprintf('\n\na\t\tb\t\tc\t\tf(c)\n');
while (j - i) > e
    k = (i * y(j) - j * y(i)) / (y(j) - y(i));
    fk = y(k);
    fprintf('%f\t%f\t%f\t%f\n', i, j, k, fk);
    if abs(fk) < e
        break;
    end
    if fk * fi < 0
        j = K;
    else
            i = k;
       end
       h = h + 1;
   end
   ar(l, 1) = h;
   l = l + 1:
   fprintf('%f\t%f\t%f\t%f\n', i, j, k, fk);
   fprintf('\nRoot is: %f\n', k);

end
fprintf('\niteration is:\n');
fprintf(' %d\t', ar);
fprintf('\nMean of iterations: \n');
disp(mean (ar));
fprintf'\nStandard deviation of iterations: \n');
disp(std(ar));

% generating R.N

function u = conunirand(r)
rng shuffle
u = (-1)*r +(r-(-1)*r)*rand();

end
```

Appendix C

```
%initializing coefficients of polynomial
for n = 1:1:100
    ar2 = disnormal (p, sigma);
    ar3 = disnormal(p, sigma);
    a0 = ar2(1, 1);
    al = ar2(1, 2);
    a2 = ar3(1, 1);
    a3 = ar3(1, 2);
    while a3 == 0
        ar3 = disnormal (p, sigma);
        a2 = ar3(1, 1);
        a3 = ar3(1, 2);
    end
    y = @(x) (a3*x^3 + a2*x^2 + al*x + a0);

    % calculating guess value
     m = - 1000;
     fi = y(m);
     fj = y (m + 1) ;
     while fi * fi > 0
         m = m + 1;
         fi = y(m);
         fj = y(m + 1);
     end
     i = m;
     j = m + 1;

     % implementing Regula Falsi Method

     h = 1;
     fprintf('\n\na\t\tb\t\tc\t\tf(c)\n');
     while (j - i) > e
         k = (i * y(j) - j * y(i)) / (y(j) - y(i));
         fk = y (k);
         fprintf('8f\t®f\t&f\t&f\n', i, j, k, fk);
         if abs(fk) < e
             break;
         end
```

```
Clc      % to clear screen

% taking input

p = input('Enter the value of mean: ');
sigma = input('Enter the value of standard deviation: ');
e = 0.001;
l = 1;
ar = zeros (1, 100);
```

```
        if fk * fi < 0
            j = k;
        else
            i = k;
        end
        h = h + 1;
    end
    ar(l, 1) = h;
    l = l + 1;
    fprintf('%f\t%f\t%f\t%f\n', i, j, k, fk);
    fprintf('\nRoot is: %f\n', k);
end
fprintf('\niteration is: \n');
fprintf('%d\t' , ar);
fprintf('\nMean of iterations: \n');
disp (mean(ar));
fprintf('\nStandard deviation of iterations: \n');
disp(std(ar));

% generating R.N

function ar1 = disnormal(p , sigma)
    rng shuffle
    ar1 = zeros (1, 2);
    u1 = rand ():
    u2 = rand();
    z1 = sqrt(-2 * log (u1)) * cos (2 * pi * u2) ;
    z2 = sqrt(-2 * log(u1)) * sin(2 * pi * u2);
    X1 = p + z1 * sigma;
    X2 = D + Z2 * sigma:
    ar1(1, 1) = x1;
    ar1(1, 2) = ×2;
end
```

References

1. Vrahatis MN (2020) Generalizations of the intermediate value theorem for approximating fixed points and zeros of continuous functions. In *NUMTA 2019: Numerical computations: Theory and algorithms* (pp. 223–238). https://doi.org/10.1007/978-3-030-40616-5_17
2. Pandey, A., & Singh, C. (2023). A study on regula falsi graphs to clear road blockage. *Quest Journal of Research in Applied Mathematics, 9*(4), 10–15
3. Atkinson, K. E. (1989). *An introduction to numerical analysis*. Wiley.
4. Williams, M., & Thompson, R. (2017). Applications of numerical methods in real-world problems. *Statistical Science, 32*(3), 250–265.
5. Lee, S., & Kim, H. (2019). Adaptive numerical techniques for random environments. *Computational Statistics, 37*(1), 45–60.
6. Frost, J. (2023a). *Normal distribution: Different means-same standard deviations*. https://statisticsbyjim.com/basics/normal-distribution/
7. Frost, J. (2023b). *Normal distribution: Same means-different standard deviations*. https://statisticsbyjim.com/basics/normal-distribution
8. Kumari, S., Chakraborty, S. (2024). Statistical study of bisection method for cubic equations with random coefficients. *Archives of Advanced Engineering Science, 2*(1), 37–52, https://doi.org/10.47852/bonviewAAES32021322
9. Chapra, S. C., & Canale, R. P. (1985). *Numerical methods for engineers: With personal computer applications*. McGraw Hill.
10. Epperson, J. F. (2007). *An introduction to numerical methods and analysis*. John Wiley & Sons.
11. Kreyszig, E. (2006). *Advanced engineering mathematics*. John Wiley & Sons.

12. Sastry, S. S. (2003). *Introductory methods of numerical analysis*. PHI Learning Ptd.
13. Burden, R. L., & Faires, J. D. (2010). *Numerical analysis*. Cengage Learning.
14. Chun, S., Kwasinski, A.: Modified Regula Falsi optimization method approach to digital maximum power point tracking for photovoltaic application. In *2011 Twenty-Sixth Annual IEEE Applied Power Electronics Conference and Exposition (APEC), Fort Worth, TX, USA* (pp. 280–286). https://doi.org/10.1109/APEC.2011.5744609

MIX
Papier aus verantwortungsvollen Quellen
Paper from responsible sources
FSC® C105338

If you have any concerns about our products,
you can contact us on
ProductSafety@springernature.com

In case Publisher is established outside the EU,
the EU authorized representative is:
Springer Nature Customer Service Center GmbH
Europaplatz 3, 69115 Heidelberg, Germany

Printed by Libri Plureos GmbH
in Hamburg, Germany